Fundamentals of Noise and Vibration

Fundamentals of Noise and Vibration

Edited by F.J. Fahy and J.G. Walker

E & FN SPON
An imprint of Routledge
London and New York

This edition published 1998 by E & FN Spon,
an imprint of Routledge
11 New Fetter Lane, London EC4P 4EE

Simultaneously published in the USA and Canada by Routledge
29 West 35th Street, New York, NY 10001

Typeset in 10/12pt Times by SPS, India
Printed and bound in Great Britain by
TJ International Ltd, Padstow, Cornwall

British Library Cataloguing in Publication Data
A catalogue record for this book is available from the British Library

Library of Congress Cataloguing in Publication Data
A catalog record for this book has been requested

ISBN 0 419 27700 8 (Pbk)
ISBN 0 419 241809 (Hbk)

Contents

Preface

This book is based on material presented during the first semester of the postgraduate Masters Course in Sound and Vibration Studies at the Institute of Sound and Vibration Research in the University of Southampton, UK. The main objective of the course is to provide students with the skills and knowledge required to practise in the field of noise and vibration control technology and its related environmental and human effects.

The first semester component of the course is designed as a conversion course for students who have graduated with first degrees in engineering, physics or mathematics, but who have received little, or no, prior specialized instruction in acoustics. This is reflected in the title, technical level and focus of the book. Although the pedagogical function is paramount, most chapters use examples of models and forms of analysis to illustrate the principles which they introduce.

In contrast to many introductory academic text books, extensive reference is made to recent technical publications for the benefit of those who wish to extend their knowledge and understanding of acoustic and vibration technology for professional purposes.

A companion volume dealing with advanced topics in noise and vibration will follow; it comprises material presented in the second semester of the Masters Course, together with chapters covering selected topics from the ISVR Advanced Course in Noise and Vibration.

A multi-author book inevitably lacks the uniformity and style of a single author volume. However, we have attempted to maintain coherence and consistency of presentation by cross-referencing between chapters and by complementarity of the material. The matter of mathematical notation has proved to be problematic, particularly in respect of complex mathematical quantities. Complete uniformity has proved to be impracticable on account of the diversity of conventions employed in the various specializations represented in the book. In order to assist the reader, a separate list of symbols is presented in an appendix. It is felt to be unnecessary to indicate that quantities are complex by means of explicit notation, but rather to make such indication at the point of introduction in the text. Complex exponential representation of harmonic-time variation uses the convention $e^{j\omega t}$, rather than $e^{-i\omega t}$ which is the convention normally used by physicists and mathematicians. Readers are asked to note this usage carefully, particularly with regard to impedances, mobilities and associated ratios of complex harmonic amplitudes.

The pedagogical nature of the book is emphasized by the inclusion of examples at the end of most of the chapters; an associated Solutions Manual is obtainable from the publishers, which may be helpful to both students and academic colleagues alike.

Acknowledgements

The editors and authors wish to acknowledge the very considerable effort and time spent in the preparation of this volume by our academic colleagues and the vital contributions made by the members of the secretarial staff who typed numerous drafts and revisions of most of the chapters. We wish in particular to thank Sally Abrams, Sue Brindle, Sue Greenwood, Hilary Smith and Maureen Strickland, and Marilyn Hicks for preparing many of the figures.

We also wish to place on record our appreciation of the professional assistance from the staff of the publishers, E & FN Spon, including Tim Robinson and Michael Doggwiler, who provided so much helpful advice and encouragement, and who exercised their editorial skills in transforming the raw material into a fine publication.

List of Contributors

Frank J. Fahy

Frank J. Fahy, Emeritus Professor of Engineering Acoustics, began his career in acoustics in 1963 when he joined ISVR after having spent five years at the Aircraft Research Association. His training as an aeronautical engineer motivated his early research in aircraft interior noise which later expanded into vibroacoustic studies of aerospace structures such as the payload fairing of the European Rocket Launcher *Ariane*. He completed his PhD on Statistical Energy Analysis in 1969 and subsequently spent thirteen years leading a small research team on vibroacoustic problems in nuclear reactors. He began developing sound intensity measurement systems in 1973 which culminated in the book *Sound Intensity*, first published in 1991 and in a second edition published by E & FN Spon in 1995. He was chairman of the ISO Committee which developed two standards for the determination of sound power using intensity measurement. He has also written a widely used advanced text book on vibroacoustics entitled *Sound and Structural Vibration* (1987). He has taught courses on many aspects of acoustics and currently teaches Noise and Vibration Control, Statistical Energy Analysis, Advanced Measurement Techniques and Room Acoustics and Music. His current research concerns the development of new techniques in SEA, development of new surface vibration transducers, applications of vibroacoustic reciprocity, violin acoustics, and the improvement of sound packages for passenger cars. Frank Fahy was awarded the Tyndall Silver Medal of the UK Institute of Acoustics in 1982, the IOA Rayleigh Gold Medal in 1990, the Medaille Etrangère by the Société Française d'Acoustique in 1995 and was elected to Honorary Fellowship of the IOA in 1998. He has acted as Acoustic Consultant to the architects of over 20 theatres and opera houses in the UK.

Ian H. Flindell

Ian Flindell, lecturer in ISVR, graduated from Reading University and worked as an Electronics Technician at the Applied Acoustics Laboratory at Chelsea College, University of London while taking his MSc there on a part-time basis. He then spent a year obtaining a Postgraduate Certificate in Education at Garnett College, University of London before taking up a post as a Research Fellow (studying environmental noise effects) in ISVR at the University of Southampton where he obtained his PhD. He won a United States National Research Council Fellowship to carry out postdoctoral research at the NASA Langley Research Centre in Virginia from 1982 to 1983, and then returned to ISVR as a research consultant in the Auditory Communications and Hearing Conservation Unit until 1986 when he joined the academic staff at ISVR as a Continuing Education Lecturer with particular responsibility for running short courses for industry and government departments. He is currently working part-time at ISVR while acting as a consultant to BAA plc in connection with the Heathrow Terminal 5 public inquiry. He has served on numerous standards committees, conference programme committees etc and has managed ISVR involvement in numerous medium to large scale research contracts for the EPSRC, the European Commission, the Departments of the Environment and Transport, the Ministry of Defence, the United States Air Force, the Canadian Department of Defense, and a number of industrial and commerical research sponsors. He is currently working on the noise assessment aspects of the EC COMMUTE programme of research into methods for the Strategic Environmental Assessment of the Trans-European Transport Network proposals; for the Department of Environment, Transport and the Regions on two projects about environmental noise assessment; and for the Ministry of Defence on Noise and Health issues. He has extensive acoustic consultancy interests in environmental noise assessment, audio public address system design, etc and recently carried out all the acoustic design work for the British Airways Compass Centre at Heathrow which won a Civic Trust Award on completion. He has written over 40 papers and articles in conference proceedings.

Michael J. Griffin

Michael J. Griffin, Professor of Human Factors in ISVR, has directed ISVR research concerned with human factors for the past 30 years. He is an Associate Director of ISVR and Chairman of the Human Sciences Group. Professor Griffin obtained a PhD in 1972. His subsequent research has been conducted in the laboratory, in industry and in transport. He has supervised research for Governmental and Industrial organisations in Britain, Europe, the USA and Japan. He has conducted research in many aspects of human responses (including human performance, comfort, health and motion sickness). His research has resulted in the publication of more than 250 scientific papers and the *Handbook of Human Vibration*, published in 1990. Professor Griffin is the Chairman of the British Standards Institution Sub-Committee concerned with human response to mechanical vibration and shock and is also a member of relevant committees of the International Organization for Standardization and the European Committee for Standardization. He is a member of the Medical Research Council/Royal Navy Ship Motion Working Party, a member of the Royal College of Physicians, Faculty of Occupational Medicine Working Party on hand-transmitted vibration and a member of the International Advisory Committee of the International Conference on Hand-Arm Vibration. He is a member of the Human Factors Society, the Aerospace Medical Association and the Ergonomics Society. He was awarded the Bartlett Medal of the Ergonomics Society for contributions to research in 1984.

Professor J.K. Hammond

J.K. Hammond is the Director of the ISVR and Professor of Signal Analysis. His research interests are in the areas of digital signal processing and the theory and application of time series analysis. Topics of current study include the modelling and analysis of non-stationary phenomena using time-frequency methods, inverse problems, nonlinear and non-Gaussian processes. He has published 40 journal papers and over 100 conference papers.

Nicholas Lalor

Nicholas Lalor, Senior Lecturer in ISVR, was trained as an aeronautical engineer, having spent five years as a student apprentice at the Royal Aircraft Establishment, Farnborough. During this period he also obtained an external BSc (Eng) degree from London University. After working for a further five years with British Aerospace and the National Gas Turbine Establishment, he joined ISVR in 1965 as a Research Assistant to carry out research into diesel engine noise. He was awarded a MSc degree in 1966 and obtained his PhD in 1970, at which point he was promoted to the post of lecturer. Since then Dr Lalor has worked on the development of models for predicting the dynamic response of a variety of structures mainly in the automotive field. He has pioneered the use of Finite Element Analysis for minimising the noise of engines by first developing predictive models and then applying optimisation algorithms to them. Dr Lalor has also made a considerable contribution towards the application of Statistical Energy Analysis to the diagnosis of noise and vibration problems in cars and engines. The experimental techniques developed during several UK government, EC BRITE and private industry sponsored projects have been almost universally adopted by other practitioners in the field as well as by an increasing number of vehicle manufacturers. These techniques also form the basis of a recently released commercial software package. The optimisation algorithms used with the FEM models have also been successfully applied to the SEA models. Dr Lalor is the author of over 70 conference and journal papers and has given many invited lecturers to industry, academic institutions and conferences throughout the world.

Timothy G. Leighton

Tim Leighton, is Reader in Underwater Acoustics in the ISVR, University of Southampton, UK. He graduated in 1985 from the University of Cambridge with a Double First Class degree in Physics and Theoretical Physics. He obtained a PhD at the Cavendish Laboratory, Cambridge, in 1988. After Senior and Advanced Research Fellowships at Magdalene College (Cambridge University) and the Engineering Physics and Science Research Council (EPSCR), and three periods at the Institut de Machines Hydrauliques et de Mechanique des Fluides (Ecole Polytechnic Federale de Lausanne, Switzerland), he was appointed to a lectureship at ISVR in 1992. He is author of over 100 publications including, in 1994, *The Acoustic Bubble*. His team currently comprises 2 Research Fellows and 7 Research Assistants. Formal post-contract assessment of both the science and management of his research contracts with EPSRC and NERC (the Natural Environment Research Council) scored highly. He is a member of Work Group 22 of Accredited Standards Committee S1 of the Acoustical Society of America (ASA) on '*Cavitation Detection and Monitoring*' and is a member of the team which, through meetings in Japan and Germany, is producing the World Federation for Ultrasound in Medicine and Biology report on '*The Safety of Ultrasound in Medicine: Emphasis on Non-Thermal Mechanisms*'. In 1996 he was commissioned by the National Measurement System Policy Units of the UK Department of Trade and Industry, in partnership with members of the National Physical Labartory, to produce the '*Study report to review progress, identify measurement methods, and assess implementation for remote and local sensing methods for the measurement of high power/cavitating ultrasonic fields*'. He is a Fellow of the Acoustical Society of America, and was awarded the 1994 A B Wood Medal for Underwater Acoustics by the Institute of Acoustics.

Philip A. Nelson

Philip Nelson, Professor of Acoustics in the ISVR, graduated from the University of Southampton in 1974 with first class honours in Mechanical Engineering. He went on to study the aerodynamic production of sound for the degree of PhD in the Institute of Sound and Vibration Research at the University of Southampton. From 1978 he worked for four years in industry on a wide range of practical problems in noise and vibration control. He was appointed lecturer in ISVR in 1982, being promoted to Senior Lecturer in 1988, and Professor in 1994. His main research interests are in the fields of acoustics, vibrations, fluid dynamics and signal processing and he is author or co-author of 2 books, over 60 papers in refereed journals, 12 patents, and 180 other technical publications. He is particularly interested in the active control of sound and vibration, the automatic control of boundary layer flow, inverse problems in underwater acoustics and multi-channel signal processing. His current administrative duties include the Chairmanship of the Fluid Dynamics and Acoustics Group of ISVR and the Associate Directorship of ISVR. He currently teaches Dynamics of Fluids, coordinates the MEng course in Acoustical Engineering Design, and teaches an introductory course in acoustics to students on the MSc courses in Sound and Vibration Studies and in Automotive Dynamics, Noise and Vibration. He also teaches specialist options in Analytical and Numerical Acoustics (with Professor M. Petyt) and in Active Control of Sound and Vibration (with Professor S. J. Elliott). Professor Nelson is a Chartered Engineer, a Fellow of the Institution of Mechanical Engineers, a Fellow of the Institute of Acoustics and a Member of the Institute of Electrical and Electronics Engineers. He serves on the Council of the Institute of Acoustics and was awarded (jointly with S. J. Elliott) the 1992 Tyndall Medal of the Institute of Acoustics.

Roger J. Pinnington

Roger Pinnington, Senior Lecturer in the ISVR, entered the Mechanical Engineering Department at the University of Southampton in 1971, completing his degree with an upper second in 1974. He then took a Masters degree in Sound and Vibration Studies at the ISVR, writing a thesis on *Acoustic Finite Elements* to graduate in 1975. Following this he was employed for two years at the Plessey Marine Research Unit in Templecombe where he worked on the control of sound radiation from submarines. He returned to the ISVR in 1977 to do a PhD under the guidance of Professor R.G. White, the topic of which was vibrational power flow from a machine through isolators to a receiving structure. This has proved to be a central theme in his research work ever since, where attempts have been made to predict, measure and control the power in each path between a source and receiver. As a result, work has been carried out on vibration isolators, pipes, acoustic enclosures, sound radiation, damping, approximate modelling of complex structures, rotating systems, and shock modelling. He is the author of 100 research publications.

1

An introduction to acoustics*

P.A. NELSON

1.1 PROPAGATION OF ACOUSTIC DISTURBANCES

There are many books on acoustics that serve as an introduction to the subject. Particularly useful basic texts written primarily for undergraduate students are those by Hall (1987), Kinsler *et al.* (1982) and Dowling and Ffowcs-Williams (1983). More advanced treatments of the subject are presented, for example, by Pierce (1981), Morse and Ingard (1968) and Skudrzyk (1971). The material presented in this chapter is intended as a brief summary of the basic principles of acoustics that form the foundation for much of the work described in later chapters. The treatment given is by no means comprehensive and is relatively superficial compared to many of these other texts. Here we will discuss briefly the physical processes involved in sound propagation and introduce some elementary ideas that will be useful in later chapters.

The physical nature of the process of propagation of acoustic waves is illustrated in Fig. 1.1. This shows a fluid (a gas or a liquid) enclosed in a semi-infinite tube with a sliding piston at one end. We will describe exactly what happens to the fluid when the piston begins to move. Assume the piston moves from rest with a constant velocity u. (Since a finite velocity cannot be achieved instantaneously we will assume that the piston 'very rapidly' reaches velocity u). Since the fluid inside the tube is compressible then immediately in front of the piston the fluid becomes squeezed and its pressure is increased. The compressibility of the fluid plays a central role in the propagation of acoustic disturbances. If, for example, the fluid in the tube were incompressible and did not deform at all, then all of the fluid in

---

* Large parts of this chapter are reprinted from *Active Control of Sound* by P.A. Nelson and S.J. Elliott (1992).The kind permission of Academic Press to reprint this work is gratefully acknowledged.

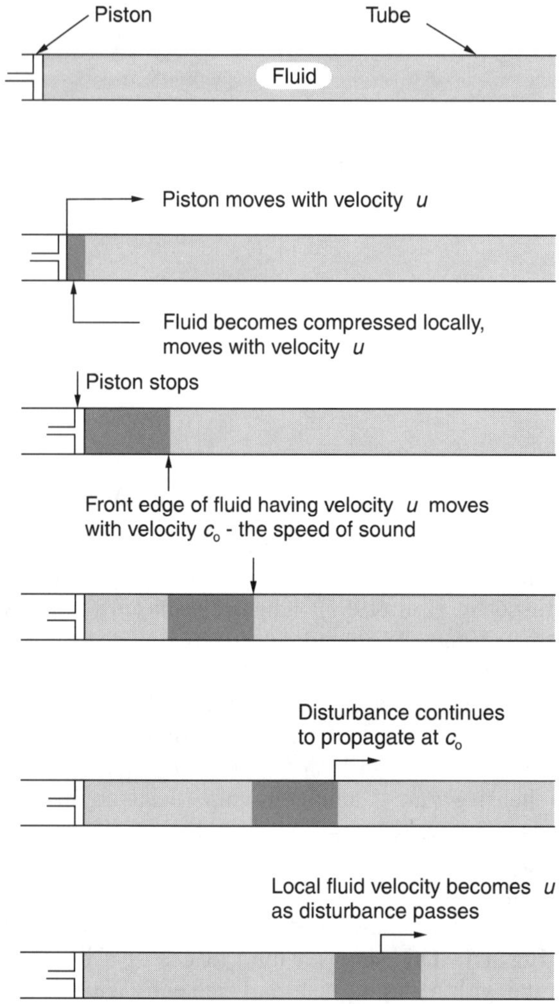

Figure 1.1 An illustration of the propagation of a sound pulse in a semi-infinite tube. Note that the shaded area denotes a region of increased pressure in the tube.

the tube would move at the same instant as the piston. However, if the fluid is compressible and has inertia, it takes a finite time for the motion of the piston to be transmitted to the fluid at the right-hand end of the tube. The speed at which this motion is transmitted is the *speed of sound*. In air at a temperature of 20 °C, the sound speed is around 343 m s^{-1}. Thus, if the tube were 100 m long the motion of the fluid due to the piston would be transmitted from one end to the other within about one-third of a second. The speed of sound in water is even faster. Although the sound speed again depends on the ambient temperature of the fluid, and indeed the salinity in the case of sea water (see, for example, Pierce, 1981, Ch.1 and Ch. 7 of this text), it is typically about 1500 m s^{-1}.

The physical process involved is illustrated in Fig. 1.1, where the propagation of an acoustic disturbance has been 'slowed down' to show what happens to the fluid in the tube after the piston is set in motion. First the fluid immediately in front of the piston becomes compressed. The increase in pressure is uniform across the cross-section of the tube. Acoustic disturbances for which the pressure is uniform on planes normal to the direction of propagation are known as *plane waves*. The compression is then transmitted down the tube. The leading edge of the disturbance, having an increased pressure over the ambient pressure, moves at the speed c_0, the speed of sound. As the compression moves down the tube more and more of the fluid reaches the velocity u at which the piston is moving. If the piston is now suddenly brought to rest then the fluid immediately in front of the piston will also come to rest; nevertheless, both the leading and trailing edges of the disturbance continue to propagate at the velocity c_0, and as the compression passes down the tube it imparts a velocity u to the fluid as it passes. This illustrates the important distinction between the velocity c_0 at which the compression propagates and the velocity u which the fluid reaches as the compression passes through a given local region of fluid. The latter is known as the *particle velocity* which results from the passage of an acoustic wave. As we shall see later, the particle velocity generally has a magnitude which is very much less than the sound speed.

It is usually the motion of solid surfaces which is responsible for radiating sound, although turbulent flow can also generate sound through a process which to this day is still not perfectly understood. In practice, many surfaces which give rise to sound radiation move rapidly to and fro, usually vibrating about an equilibrium position. One particular form of surface vibration which is useful in the study of acoustic waves is the sinusoidal displacement of the piston depicted in Fig. 1.2. One cycle of the piston motion is illustrated which takes place during the period of time T. At time $t = 0$, the piston is at its equilibrium position but is moving with its maximum forward velocity. Since, as we shall see later, in this special case of plane wave propagation in one direction, the acoustic pressure increase in the fluid is directly proportional to the velocity of the piston, this forward velocity produces an equivalent pressure rise. As the piston continues to move forward it slows down and the pressure increase produced in front of it becomes proportionately smaller. Meanwhile however, the initial increase in pressure has propagated through the fluid in the tube at the speed of sound. Thus once the piston reaches its maximum forward displacement (after a time $t = T/4$) there is a continuous distribution of pressure produced through the fluid in front of the piston. As the piston moves backwards it gathers speed and progressively *rarefies* the fluid and decreases the pressure

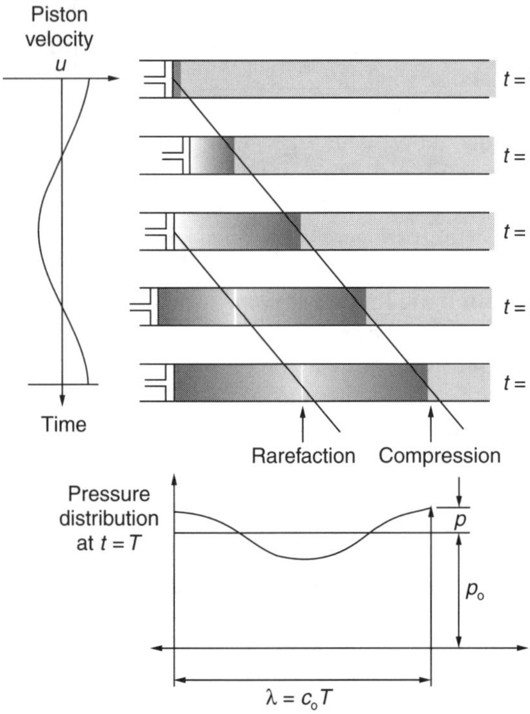

Figure 1.2 The distribution of pressure in the tube during one cycle of harmonic piston motion. Note that at $t = T$, the spatial dependence of the acoustic pressure is of the form cos kx where $k = 2\pi\lambda$.

of the fluid immediately in front of it. After a time $t = T/2$ (after half a cycle of the motion) the piston has returned to its equilibrium position and is moving with its maximum backwards velocity; it is at this time that the piston produces the maximum rarefaction of the fluid in front of it. All the time during this process, the disturbances produced are propagated away from the piston surface at the speed of sound. After one complete cycle of the motion (at time $t = T$) a continuous spatial distribution of pressure has been produced in the tube which replicates the pattern of piston velocity produced in time. The distance travelled by the sound during one cycle of motion is given by c_0T; this distance is one *wavelength* λ. Since the frequency of the motion $f = 1/T$ Hz (cycles per second), then frequency and wavelength are related by $c_0 = f\lambda$. The frequencies in which we are interested are usually those of sounds which can be detected by the human ear. This range of (audio) frequencies lies between 20 Hz and 20 kHz. This corresponds to a range of wavelengths of approximately 17 m to 17 mm in air (or 75 m to 75 mm in water).

1.2 THE DECIBEL SCALE FOR THE MEASUREMENT OF SOUND PRESSURE

Figure 1.2 also shows the pressure distribution in the tube after one cycle of the piston motion. The pressure is shown as a variation either above or below the ambient (atmospheric) pressure in the tube. We generally denote the ambient pressure as p_0. The increase or decrease in pressure either above or below this value that is produced by the passage of an acoustic wave we shall denote as p, the acoustic pressure fluctuation. Thus, at a fixed point in the tube, as disturbances propagate past, the pressure will fluctuate with time. For the case illustrated in Fig. 1.2, this fluctuation will show a sinusoidal form of time dependence. If, however, the piston in the tube were made to vibrate in an irregular manner, for example with a random variation in its displacement from its equilibrium position, the pressure fluctuation at a given position in the tube would also exhibit the same random variation with time but at a little time later. A suitable measure of the magnitude of these fluctuations is given by the mean square value of the fluctuation defined by the time average

$$\overline{p^2(t)} = \lim_{T\to\infty} \frac{1}{T} \int_{-T/2}^{T/2} p^2(t)\mathrm{d}t \tag{1.1}$$

where the averaging time T is in practice chosen to be very large compared to the period of the fluctuation, and in principle should extend to infinity for random fluctuations whose statistical properties remain stationary with time. Since this measure of magnitude has dimensions of pressure squared, we often work with the 'root mean square' (rms) value associated with the fluctuation, which we denote by p_{rms} and is given by the square root of the mean square pressure.

The range of rms pressure fluctuation which we encounter in practice is vast; its values vary from as small as 10^{-5} Pa to as large as 10^3 Pa and it is useful to compress this range of numbers by adopting a logarithmic measurement scale. In addition, the human perception of loudness exhibits an approximately logarithmic behaviour, with human beings judging the relative loudness of two sounds by approximately the ratio of their mean square pressures. Thus, the quantity we use for the measurement of the sound pressure is the *sound pressure level* which is defined by

$$L_{\mathrm{p}} = 10\log_{10}\frac{\overline{p^2(t)}}{p_{\mathrm{ref}}^2} = 20\log_{10}\frac{p_{\mathrm{rms}}}{p_{\mathrm{ref}}}, \tag{1.2}$$

where p_{ref} is a reference pressure of 20 μPa. This has been selected as the reference pressure used for sound propagation in air since it corresponds to the rms pressure of a pure tone at 1 kHz which is just audible to the ear of the average young human being. This 'threshold of hearing' at 1 kHz thus

Table 1.1 Typical rms pressure fluctuations and their sound pressure levels. The values shown are only rough approximations to the actual levels produced.

	p_{rms} (Pa)	L_p (dB re 2×10^{-5} Pa)
3 m from jet engine	200	140
Rock concert	20	120
2 m from pneumatic hammer	2	100
Vacuum cleaner	0.2	80
Conversational speech	0.02	60
Residential area at night	0.002	40
Rustling of leaves	0.0002	20
Threshold of hearing	0.00002	0

corresponds to a sound pressure level of 0 dB. Some typical values of the approximate sound pressure level produced by various sources are shown in Table 1.1.

1.3 PROPERTIES OF ACOUSTIC DISTURBANCES

We will now return to the situation depicted in Fig. 1.1 and describe briefly the behaviour of the fluid immediately in front of the piston as the piston is set in motion with a constant velocity u. As illustrated in Fig. 1.4, after a small time t, the piston will have moved a distance ut and the leading edge of the disturbance will be a distance c_0t from the starting position of the piston. Assume that the fluid initially has uniform pressure and density denoted by p_0 and ρ_0. The compressed fluid will be increased in pressure and density by p and ρ, respectively, where ρ is the acoustic density fluctuation. For the most part, acoustic compressions do not involve appreciable heat transfer from the element of fluid that becomes compressed. Under these circumstances, where the compression can be considered isentropic (see, for example, Pierce, 1981, Ch.1 for a full discussion) we can assume that the density change in the fluid is determined solely by the pressure change in the fluid and is not dependent on the temperature change, even though the temperature of the fluid will be increased slightly during the course of the compression. The form of the relationship between the density and pressure will depend on the properties of the fluid. If we express the total instantaneous pressure in the fluid as $p_{tot} = p_0 + p$ and similarly the total instantaneous density as $\rho_{tot} = \rho_0 + \rho$, we can express the relationship between them as

$$p_{tot} = f(\rho_{tot}). \tag{1.3}$$

where $f(\ )$ is used to denote 'a function of'.

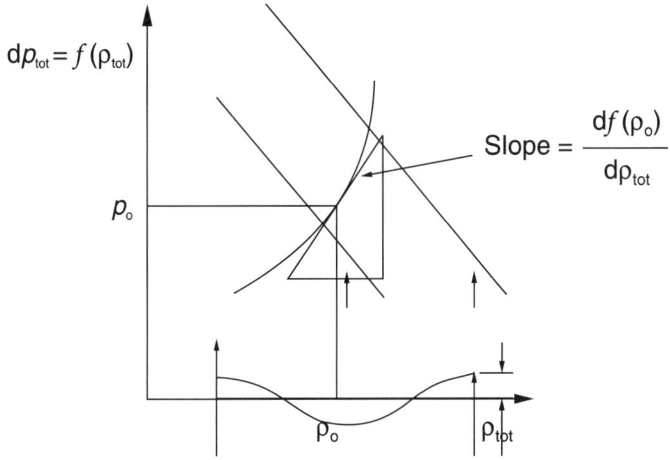

Figure 1.3 The general form of relationship between total instantaneous pressure and density.

The second assumption that we make is that the increases p and ρ in the pressure and density are sufficiently small to be considered to be *linearly* related. That p and ρ are generally very small compared to their ambient values p_0 and ρ_0 may be demonstrated using a simple example. The maximum acoustic pressure fluctuation p measured close to the vibrating surface of a typical automobile engine running at full speed is unlikely to exceed around 10 Pa, even though it is one of the louder sounds we have to deal with (see Table 1.1). This compares with a value of atmospheric pressure p_0 of around 10^5 Pa. Similarly, the density fluctuation ρ will be of the order of 10^{-4} kg m^{-3} compared to an ambient value of around 1.2 kg m^{-3}. A graph showing the typical relationship between p_{tot} and ρ_{tot} is sketched in Fig. 1.3. For small changes p and ρ above and below the ambient values p_0 and ρ_0, we assume that p and ρ are directly proportional, the constant of proportionality being given by the slope of the curve $f(\rho_{tot})$ at the ambient values p_0 and ρ_0. This can be demonstrated formally using a Taylor series expansion of the function $f(\rho_{tot})$. Thus we can write

$$f(\rho_{tot}) = f(\rho_0) + \rho\frac{df(\rho_0)}{d\rho_{tot}} + \frac{\rho^2}{2!}\frac{d^2f(\rho_0)}{d\rho_{tot}^2} + \cdots\cdots, \tag{1.4}$$

where $df(\rho_0)/d\rho_{tot}$ is the gradient of the function evaluated at the density ρ_0. For small increases ρ in the density (and for a function whose rate of change of gradient $d^2f(\rho_0)/d\rho_{tot}^2$ is not too large), the terms proportional to ρ^2 and higher powers of ρ may be neglected in the series expansion. Since $f(\rho_{tot}) = p_0 + p$ and $f(\rho_0) = p_0$, then with only the term proportional to ρ retained, equation (1.4) reduces to

$$p = \rho \frac{\mathrm{d}f(\rho_0)}{\mathrm{d}\rho_{\mathrm{tot}}}. \tag{1.5}$$

This expresses the linear relationship between the changes in pressure and density, the constant of proportionality being given by the rate of change of pressure with density evaluated with the fluid in the ambient (uncompressed) state.

The pressure increase p is also linearly related to the piston velocity and resulting particle velocity u. This can be demonstrated following Morse and Ingard (1968, see Ch. 6) by applying the principle of momentum conservation to the region of compressed fluid immediately in front of the piston. The net force applied to the element by the piston is given by pS where S is the cross-sectional area of the element. This can be equated to the rate of change of momentum of the fluid which shows that

$$pS = \frac{\mathrm{d}}{\mathrm{d}t}(\rho_0 c_0 t S u), \tag{1.6}$$

where $(\rho_0 c_0 t S)$ quantifies the total mass of the fluid in the uncompressed element (see Fig. 1.4). It follows from equation (1.6) that

$$p = \rho_0 c_0 u. \tag{1.7}$$

The constant of proportionality in this linear relationship is $\rho_0 c_0$. This quantity is known as the *characteristic acoustic impedance* of the medium. In

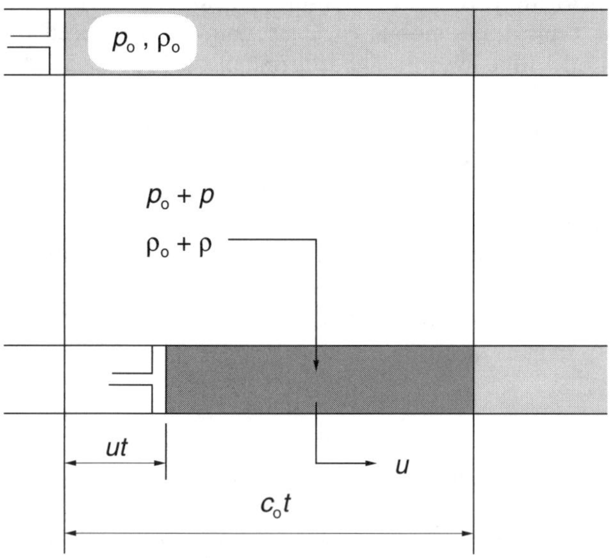

Figure 1.4 The compression of a volume of fluid in front of the piston after the piston has been moving with a velocity u for a time t.

air at atmospheric pressure and a temperature of 20 °C, this takes a value of 415 kg m^{-2} s^{-1} while in fresh water at 20 °C, it is equal to 1.48×10^6 kg m^{-2} s^{-1}. Thus a given surface velocity will produce a much larger pressure in water than it will in air.

Finally we can establish a further linear relationship between the small density increase ρ and the particle velocity u. Again following Morse and Ingard (1968, see Ch. 6) we can apply the principle of mass conservation to the fluid element considered and equate the mass of the fluid in the element prior to its compression to its value after the time t. This gives

$$\rho_0 c_0 t S = (\rho_0 + \rho)(c_0 t - ut)S. \tag{1.8}$$

As we have noted above, the acoustic density increase ρ is very small compared to ρ_0. Also, in general, the particle velocity is very much smaller than c_0 (the typical velocity of the vibrating surface of the automobile engine referred to above is less than 0.1 m s^{-1}). Thus in expanding the bracketed terms on the right-hand side of this equation, the term proportional to ρu can be neglected, since it involves the product of two very small quantities and will be very much smaller than the other terms on the right-hand side of the equation. The neglect of this term leads directly to the relationship

$$u = \rho c_0/\rho_0. \tag{1.9}$$

Equations (1.5), (1.7) and (1.9) can be combined to show that

$$p = c_0^2 \rho, \qquad c_0^2 = \frac{\mathrm{d}f(\rho_0)}{\mathrm{d}\rho_{\mathrm{tot}}}. \tag{1.10a,b}$$

The sound speed is therefore determined solely by the ambient properties of the fluid and is derived from the relationship between the pressure and the density defined by equation (1.1). Recall that a relationship of this type holds for compressions in which no heat escapes from the compressed fluid. In the particular case of a perfect gas (to which air is a good approximation) undergoing such an adiabatic compression, the relationship between pressure and density is given by

$$p_{\mathrm{tot}} = f(\rho_{\mathrm{tot}}) = p_0 \left(\frac{\rho_{\mathrm{tot}}}{\rho_0}\right)^{\gamma}, \tag{1.11}$$

where γ is the ratio of the specific heats of the gas. (See, for example, Kinsler *et al.*, 1982, Appendix 9, for a detailed account of the thermodynamics of such compressions.) Differentiation of this expression and evaluation of the derivative at density ρ_0 shows that

$$c_0^2 = \frac{\mathrm{d}f(\rho_0)}{\mathrm{d}\rho_{\mathrm{tot}}} = \frac{\gamma p_0}{\rho_0}. \tag{1.12}$$

The sound speed in a perfect gas is therefore very simply related to the ambient fluid pressure and density. It also follows from the ideal gas equation $p_{tot} = \rho_{tot} R T_a$, where R is the gas constant and T_a is the absolute temperature, that $c_0^2 = \gamma R T_a$. This relationship thus defines the dependence of the sound speed on the temperature of the gas. More thorough discussions of the dependence of the sound speed on ambient fluid properties are given by, for example, Pierce (1981, Ch. 1) and Lighthill (1978, Ch. 1). The speed of sound in liquids is dealt with in Chapter 7.

1.4 THE ONE-DIMENSIONAL WAVE EQUATION

We will now proceed to derive in a more formal way, the relationships between the acoustic variables p, ρ and u when they vary continuously in space and time as, for example, in the case of harmonic wave motion illustrated in Fig. 1.2. First, consider an arbitrary fixed volume of the tube through which acoustic disturbances pass. The volume has a cross-section S and a length dx in the x direction. This is illustrated in Fig. 1.5. We will again denote the total instantaneous density of the fluid in the element by $\rho_{tot} = \rho_0 + \rho$ where ρ is the increase in density produced by the passage of an acoustic wave. Consider the application of the principle of conservation of mass to the element shown. The rate of mass inflow into the control volume is given by

$$\rho_{tot} u S - \left[\rho_{tot} u + \frac{\partial(\rho_{tot} u)}{\partial x} \mathrm{d}x\right] S = -\frac{\partial(\rho_{tot} u)}{\partial x} \mathrm{d}x S. \tag{1.13}$$

The principle of mass conservation requires that this net inflow must be balanced by an increase in mass in the element. The rate of increase of mass inside the control volume is given by $(\partial \rho_{tot}/\partial t)\mathrm{d}xS$. If we now equate the rate of mass inflow and the rate of increase of mass then it follows that

$$\frac{\partial(\rho_{tot})}{\partial t} + \frac{\partial(\rho_{tot} u)}{\partial x} = 0. \tag{1.14}$$

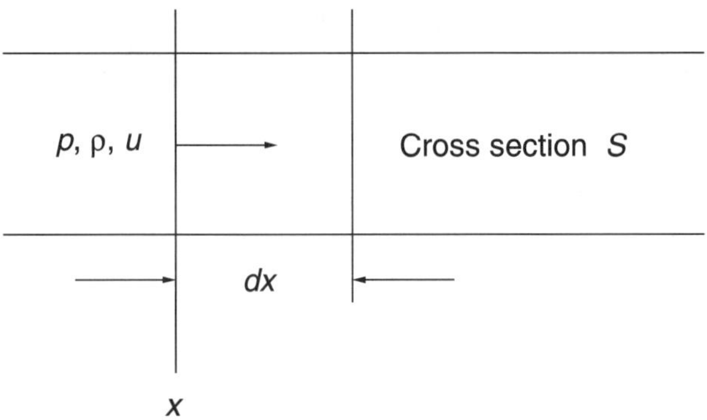

Figure 1.5 A control volume in the tube where the acoustic variables are given by p, ρ, u at the position x.

First consider the derivative $\partial(\rho_{tot}u)/\partial x$. The product $(\rho_{tot}u)$ is equal to $(\rho_0 + \rho)u$ and we again make the assumption that since the term ρu is the product of two small quantities, it is sufficiently small to be neglected. We therefore retain only the term $\rho_0 u$ and since ρ_0 is not a function of x, the derivative of this product with respect to x reduces to $\rho_0(\partial u/\partial x)$. Since ρ_0 is also not a function of t, the derivative $\partial(\rho_{tot})/\partial t$ becomes $\partial\rho/\partial t$ and equation (1.14) reduces to

$$\frac{\partial \rho}{\partial t} + \rho_0 \frac{\partial u}{\partial x} = 0. \tag{1.15}$$

This relationship is known as the *linearized* equation of mass conservation, a title which originates from the fact that the terms in the equation are only linear functions of the fluctuating quantities ρ and u.

We can also apply the principle of momentum conservation to a fluid element. In this case we will apply the principle to an element which moves with the fluid and again we will assume that the net instantaneous pressure in the fluid is given by $p_{tot} = p_0 + p$ where p is the acoustic pressure. Thus the net force on the fluid element in the positive x direction is given by

$$p_{tot}S - \left(p_{tot} + \frac{\partial p_{tot}}{\partial x}\mathrm{d}x\right)S = -\frac{\partial p_{tot}}{\partial x}\mathrm{d}xS. \tag{1.16}$$

The momentum conservation principle states that this net applied force must be balanced by the mass times the acceleration of the fluid element. The acceleration of the fluid element is given by $\mathrm{d}u/\mathrm{d}t = u\partial u/\partial x + \partial u/\partial t$ which accounts for the fact that the velocity of the fluid is also a function of x (see, for example, Kinsler *et al.*, 1982, Ch.5, for a fuller discussion). However, the term $u\partial u/\mathrm{d}x$ turns out to be much smaller (by a factor of u/c_0) than $\partial u/\partial t$ and thus we can approximate the mass times the acceleration of the fluid element by $\rho_0 \mathrm{d}xS(\partial u/\partial t)$. Equating this rate of change of momentum to the net force on the fluid element therefore results in

$$\rho_0 \frac{\partial u}{\partial t} + \frac{\partial p}{\partial x} = 0. \tag{1.17}$$

This relationship is known as the linearized equation of momentum conservation, again because the terms in the equation are only linearly dependent on the small fluctuating quantities.

If we now differentiate equation (1.15) with respect to t and differentiate equation (1.17) with respect to x then a combination of the two resulting equations shows that

$$\frac{\partial^2 p}{\partial x^2} - \frac{\partial^2 \rho}{\partial t^2} = 0. \tag{1.18}$$

We again assume that the acoustic compressions can be considered isentropic such that the increase in pressure and increase in density are linearly related and that the square of the sound speed is the constant of proportionality such that $p = c_0^2\rho$. Substitution of this relationship into equation (1.18) yields the one-dimensional wave equation for the propagation of acoustic pressure fluctuations. This is given by

$$\frac{\partial^2 p}{\partial x^2} - \frac{1}{c_0^2}\frac{\partial^2 p}{\partial t^2} = 0. \tag{1.19}$$

This equation relates the way in which the acoustic pressure fluctuations behave with respect to space (as a function of the coordinate distance x) and with respect to time t. In the derivation of this equation, the process of linearization was a central assumption. We will now proceed to find solutions that satisfy the wave equation and demonstrate that the solutions bear out the description of the propagation of acoustic disturbances presented in section 1.1.

1.5 SOLUTIONS OF THE ONE-DIMENSIONAL WAVE EQUATION

It is easy to show that pressure fluctuations which have a dependence on space and time described by functions f and g such that

$$p(x,t) = f(t - x/c_0) \quad \text{and} \quad p(x,t) = g(t + x/c_0), \tag{1.20}$$

are solutions of the one-dimensional wave equation. This can be demonstrated by differentiating the functions f and g twice with respect to x and twice with respect to t and then checking that $\partial^2 p/\partial x^2 = (1/c_0^2)\partial^2 p/\partial t^2$. These two functions also have a well defined physical interpretation. The first term $f(t - x/c_0)$ describes a pressure fluctuation which is travelling in the positive x direction, i.e. from left to right in Fig. 1.1. Conversely the function $g(t + x/c_0)$ describes fluctuations which travel in the negative x direction (i.e. from right to left in Fig. 1.1) For waves travelling in the positive x direction then equation (1.20) implies that the fluctuation $f(t)$ generated at $x = 0$ at the face of the piston in Fig. 1.1 is reproduced at a position at a distance x along the tube at a time x/c_0 later. The form of the fluctuation is preserved exactly as it travels at constant speed through the tube. This is illustrated in Fig. 1.6. This also applies to fluctuations $g(t)$ travelling in the negative x direction.

A particular form of function which describes harmonic waves travelling in the positive x direction is given by the complex exponential representation of harmonic motion described by

$$f(t - x/c_0) = \mathrm{Re}\left\{Ae^{j\omega(t - x/c_0)}\right\} = \mathrm{Re}\left\{Ae^{j(\omega t - kx)}\right\}, \tag{1.21}$$

where $\omega = 2\pi/T$ is the angular frequency of the harmonic fluctuation having period T and $k = \omega/c_0 = 2\pi/\lambda$ is the spatial equivalent of ω and is known as the *wavenumber*. The wavenumber is thus a measure of 'spatial frequency'.

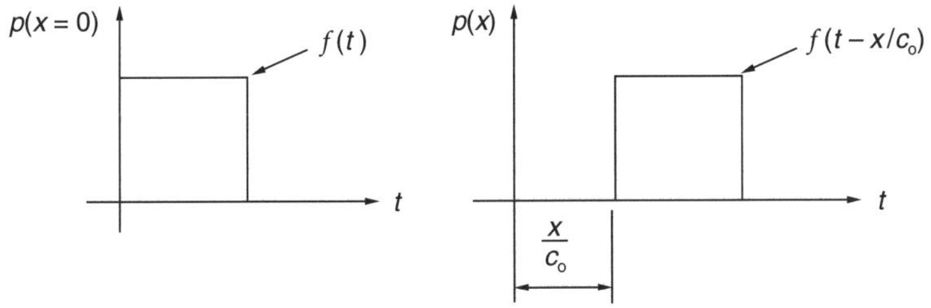

Figure 1.6 The pressure fluctuation generated as a function of time by the propagation of the acoustic disturbance depicted in Fig. 1.1. The fluctuation generated at $x = 0$ (the piston face) is reproduced at the position a distance x along the tube at a time x/c_0 later.

The term $e^{j(\omega t-kx)}$ is a complex number which, by using the identity $e^{j\theta} = \cos\theta + j\sin\theta$, can be written as

$$e^{j(\omega t-kx)} = \cos(\omega t - kx) + j\,\sin(\omega t - kx). \tag{1.22}$$

Thus $\cos(\omega t - kx)$ and $\sin(\omega t - kx)$ define the real and imaginary parts of the complex number. The term A is also assumed to be a complex number with real and imaginary parts $A_R = \mathrm{Re}\{A\}$ and $A_I = \mathrm{Im}\{A\}$ such that

$$A = A_R + jA_I = |A|e^{j\phi_A}, \tag{1.23}$$

where $|A|$ is the modulus of the complex number A and ϕ_A is its phase. These are in turn related to A_R and A_I by

$$|A| = \sqrt{(A_R^2 + A_I^2)}, \qquad \phi_A = \tan^{-1}\frac{A_I}{A_R}. \tag{1.24a,b}$$

Note that A_R and A_I are both *real* numbers. In dealing with a representation of the form given by equation (1.21) we always recover the actual pressure fluctuation by taking the real part of the complex representation. Thus the physical pressure fluctuation is given by

$$p(x,t) = \mathrm{Re}\left\{Ae^{j(\omega t-kx)}\right\}, \tag{1.25}$$

and using $A = |A|e^{j\phi_A}$ together with $e^{j\theta} = \cos\theta + j\sin\theta$ shows that

$$p(x,t) = |A|\ \cos(\omega t - kx + \phi_A). \tag{1.26}$$

Thus the modulus $|A|$ of the complex number A defines the amplitude of the harmonic pressure fluctuation and ϕ_A is an arbitrary phase angle. Differentiation of this function twice with respect to both x and t will demonstrate that it is a solution to the wave equation.

Similarly, we can use complex exponential representation to describe harmonic waves travelling in the negative x direction. Thus we can write

$$g(t + x/c_0) = \mathrm{Re}\left\{Be^{j\omega(t+x/c_0)}\right\} = \mathrm{Re}\left\{Be^{j(\omega t+kx)}\right\}, \tag{1.27}$$

where B is again a complex number such that

$$B = B_R + jB_I = |B|e^{j\phi_B}. \tag{1.28}$$

The pressure fluctuation associated with a negative-going harmonic wave can thus be written as

$$p(x,t) = \text{Re}\left\{Be^{j(\omega t+kx)}\right\} = |B|\ \cos(\omega t + kx + \phi_B). \tag{1.29}$$

It is also useful in dealing with acoustical problems at a single frequency to define a spatially dependent *complex pressure* which in this work we will denote simply by $p(x)$. This is a complex number which is a function of position. The real pressure fluctuation can be recovered by using

$$p(x,t) = \text{Re}\{p(x)e^{j\omega t}\}. \tag{1.30}$$

In the particular case of positive-going harmonic plane waves discussed above, then the complex pressure $p(x) = Ae^{-jkx}$, and for negative-going waves $p(x) = Be^{jkx}$. Substitution of equation (1.30) into the wave equation shows that

$$\left(\frac{\partial^2}{\partial x^2} - \frac{1}{c_0^2}\frac{\partial^2}{\partial t^2}\right)\text{Re}\{p(x)e^{j\omega t}\} = 0. \tag{1.31}$$

Since the operation of taking the real part commutes with the differential operators, we can write this equation as

$$\text{Re}\left\{\left[\frac{d^2p(x)}{dx^2} + \frac{\omega^2}{c_0^2}p(x)\right]e^{j\omega t}\right\} = 0, \tag{1.32}$$

where the double differentiation with respect to time has been undertaken. This equation must be satisfied for all values of time t and in particular when $e^{j\omega t} = 1$ or $e^{j\omega t} = j$. Both these conditions can be satisfied only if the term in square brackets is zero. That is

$$\frac{d^2p(x)}{dx^2} + k^2p(x) = 0, \tag{1.33}$$

where $k^2 = \omega^2/c_0^2$. This equation is the one-dimensional *Helmholtz equation* which must be satisfied by the complex pressure $p(x)$. It is easy to verify that $p(x) = Ae^{-jkx}$ and $p(x) = Be^{jkx}$ are solutions of this equation. A fuller discussion of the use of complex spatially dependent field variables is presented by, for example, Pierce (1981, see Ch. 1).

1.6 LINEARITY AND THE SUPERPOSITION PRINCIPLE

In our discussion of acoustic wave motion we have emphasized repeatedly that the acoustic variables pressure p, density ρ and particle velocity u are *small* perturbations. That is to say, p and ρ are very small compared to their ambient values p_0 and ρ_0 and u is very small compared to the speed of sound c_0. We have seen in section 1.3 that, as a direct consequence of this, the variables p, ρ and u are, to a very good approximation, all *linearly* related to one another. We have also seen in section 1.4 that we can neglect the products of small fluctuating quantities in deriving a wave equation which governs the form of acoustic pressure fluctuations in one dimension. This equation is therefore linear in the pressure fluctuation $p(x,t)$ and does not contain any nonlinear terms like $p^2(x,t)$. In addition, the (partial) differential operators in the wave equation are also 'linear' operators. That is to say, given a pressure fluctuation $p_1(x,t)$ and another pressure fluctuation $p_2(x,t)$, then

$$\frac{\partial^2}{\partial x^2}[p_1(x,t)+p_2(x,t)] = \frac{\partial^2}{\partial x^2}p_1(x,t)+\frac{\partial^2}{\partial x^2}p_2(x,t), \tag{1.34}$$

$$\frac{\partial^2}{\partial t^2}[p_1(x,t)+p_2(x,t)] = \frac{\partial^2}{\partial t^2}p_1(x,t)+\frac{\partial^2}{\partial t^2}p_2(x,t). \tag{1.35}$$

As a direct consequence of the wave equation both being linear in the fluctuation $p(x,t)$ and containing only linear differential operators, we can show that two sound-pressure fluctuations, each of which satisfies the wave equation, can also be simply added together to produce another pressure fluctuation which also satisfies the wave equation. Thus, if we have

$$\left(\frac{\partial^2}{\partial x^2}-\frac{1}{c_0^2}\frac{\partial^2}{\partial t^2}\right)p_1(x,t)=0, \qquad \left(\frac{\partial^2}{\partial x^2}-\frac{1}{c_0^2}\frac{\partial^2}{\partial t^2}\right)p_2(x,t)=0, \tag{1.36a,b}$$

then it also follows that

$$\left(\frac{\partial^2}{\partial x^2}-\frac{1}{c_0^2}\frac{\partial^2}{\partial t^2}\right)[p_1(x,t)+p_2(x,t)]=0. \tag{1.37}$$

This is known as the *principle of superposition*. That is, two sound fields which have pressure fluctuations that vary both as a function of the spatial coordinate x and as a function of time t, can simply be added at each position x and time t to yield the net pressure fluctuation produced. Thus, the resulting pressure fluctuation will be given by

$$p(x,t)=p_1(x,t)+p_2(x,t). \tag{1.38}$$

In practical terms, this means that one pressure fluctuation will not become distorted by the presence of another; having the radio on will not distort the sound of another person's voice in the same room; the two sound pressure fluctuations simply add up at each instant. There are, of course, circumstances where the neglect of the nonlinear terms in the basic equations used to derive the wave equation will become significant and sound propagation is no longer governed by the wave equation derived here. This occurs in particular when dealing with pressure fluctuations of very large amplitude and with propagation over very large distances. Typically, nonlinear effects will produce significant distortion of an acoustic waveform in air having a sound pressure level of 120 dB when the wave propagates over a distance of a few kilometres (see the elementary discussion presented by Hall, 1987, Ch. 9 or the more detailed accounts presented by Pierce, 1981, Ch. 11 or Morse and Ingard, 1968, Ch. 14). However, for most circumstances of practical interest, sound propagation can safely be assumed to be linear and the principle of superposition can be applied.

1.7 SPECIFIC ACOUSTIC IMPEDANCE

In addition to defining a complex acoustic pressure we can also define a spatially dependent complex acoustic particle velocity associated with the passage of a plane wave. Thus the real particle velocity fluctuation is expressed as

$$u(x,t) = \mathrm{Re}\{u(x)\mathrm{e}^{j\omega t}\}, \tag{1.39}$$

where $u(x)$ is a complex number. The linearized equation of momentum conservation defined by equation (1.17) can therefore be written as

$$\rho_0 \frac{\partial}{\partial t}\mathrm{Re}\{u(x)\mathrm{e}^{j\omega t}\} + \frac{\partial}{\partial x}\mathrm{Re}\{p(x)\mathrm{e}^{j\omega t}\} = 0, \tag{1.40}$$

which, on using the same arguments used in section 1.5, can also be written in the form

$$\mathrm{Re}\left\{\left(j\omega\rho_0 u(x) + \frac{\mathrm{d}p(x)}{\mathrm{d}x}\right)\mathrm{e}^{j\omega t}\right\} = 0. \tag{1.41}$$

Again this equation can only be satisfied for all time provided

$$j\omega\rho_0 u(x) + \frac{\mathrm{d}p(x)}{\mathrm{d}x} = 0. \tag{1.42}$$

The complex particle velocity can thus be derived directly from the gradient of the complex pressure with respect to x. Thus in the case of the plane travelling wave where $p(x) = A\mathrm{e}^{-jkx}$,

$$u(x) = \frac{-1}{j\omega\rho_0}\frac{\mathrm{d}p(x)}{\mathrm{d}x} = \frac{1}{\rho_0 c_0}p(x), \tag{1.43}$$

since $k = \omega/c_0$. It also follows therefore that the actual particle velocity fluctuation is given by

$$u(x,t) = \mathrm{Re}\left\{\frac{p(x)}{\rho_0 c_0}\right\} = \frac{|A|}{\rho_0 c_0}\cos(\omega t - kx + \phi_A). \tag{1.44}$$

Therefore, in this particular case, the particle velocity fluctuation is of a form identical to the acoustic pressure fluctuation and its magnitude is related to that of the pressure by a factor of $1/\rho_0 c_0$ where $\rho_0 c_0$ is the characteristic acoustic impedance of the medium. Thus in the harmonic travelling plane wave the acoustic pressure and particle velocity are always in phase. We shall see later that this is not always necessarily the case in acoustic propagation and more generally there is a complicated relationship between acoustic pressure and acoustic particle velocity in terms of the ratio of their amplitudes and their relative phases. One way in which we quantify this relationship is by defining a more general acoustic impedance which is a ratio of the complex pressure to the complex particle velocity. The *specific acoustic impedance* is then defined for harmonic fluctuations as

$$z(x) = \frac{p(x)}{u(x)}. \tag{1.45}$$

This ratio of complex numbers therefore directly quantifies the relative amplitude and phase of the pressure and particle velocity. This can be seen more clearly if we define $p(x) = |p(x)|e^{j\phi_p(x)}$ and $u(x) = |u(x)|e^{j\phi_u(x)}$. Then the expression for the impedance reduces to

$$z(x) = \frac{|p(x)|}{|u(x)|}e^{j[\phi_p(x)-\phi_u(x)]}. \tag{1.46}$$

Thus the modulus of the complex impedance is given by the ratio of the moduli of the complex pressure and particle velocity and its phase quantifies the difference in phase between these variables.

1.8 ACOUSTIC ENERGY DENSITY AND INTENSITY

To return to our basic description of acoustic wave propagation, it is evident that as compressions pass through the fluid then there is a transfer of energy from the source (the vibrating piston) to other positions in the tube. Fluid that was previously at rest gains kinetic energy, since there is a particle velocity associated with the local motion of the fluid, and also it gains potential energy as there are pressure fluctuations induced in the fluid. It is often useful to quantify these fluctuations in kinetic and potential energy and this can be done by returning to the simple acoustic compression illustrated in Fig. 1.4.

First it is straightforward to write down an expression for the kinetic energy associated with the element of fluid which is squeezed by the action of the piston. If the original volume of the uncompressed fluid is given by V_0, the instantaneous kinetic energy of the element can be written as

$$E_K(t) = \tfrac{1}{2}\rho_0 V_0 u^2(t), \tag{1.47}$$

which is simply one half times the mass of the element times its velocity squared. An expression for the potential energy is a little more difficult to derive. The potential energy produced by compressing the element of fluid from a volume V_0 to a volume V is given by

$$E_p(t) = -\int_{V_0}^{V} p(t)\mathrm{d}V, \tag{1.48}$$

where the minus sign is included since work is done on the element in decreasing its volume from V_0 to V. As usual it is assumed that the change in volume and density of the fluid is relatively small such that by mass conservation $(\rho_0 + \rho)V = \rho_0 V_0$ which, when products of small quantities are neglected, can be written as $\rho_0 V + \rho V_0 = \rho_0 V_0$. Thus differentiating this expression we can write $\mathrm{d}V/\mathrm{d}\rho = -V_0/\rho_0$ and since the compression is adiabatic, using equation (1.10a) we have $\mathrm{d}V = (V_0/\rho_0 c_0^2)\mathrm{d}p$. Thus we can substitute for $\mathrm{d}V$ in equation (1.48) and integrate with respect to the acoustic pressure p. The integral is given by

$$E_p(t) = \frac{V_0}{\rho_0 c_0^2}\int_0^p p(t)\mathrm{d}p = \frac{V_0}{2\rho_0 c_0^2}p^2(t). \tag{1.49}$$

This is the expression which defines the potential energy of the element resulting from the increase in acoustic pressure p. The total instantaneous energy *density* of the fluid (i.e. the energy per unit volume) is given by the sum of $e_k(t) = E_k(t)/V_0$ and $e_p(t) = E_p(t)/V_0$ and can be written as

$$e(t) = e_k(t) + e_p(t) = \frac{1}{2}\rho_0\left[u^2(t) + \frac{p^2(t)}{(\rho_0 c_0)^2}\right]. \tag{1.50}$$

This expression applies at any instant in time and since both u and p are time variable, then this energy quantity also fluctuates as the pressure and velocity rise and fall due to the passage of acoustic disturbances. Note that this expression for the acoustic energy density holds for all linear acoustic fields (see, for example, Pierce, 1981, Ch. 1). Note also that for *plane progressive wave* disturbances then $p(t)$ and $u(t)$ are simply related by $u(t) = p(t)/\rho_0 c_0$ and therefore the instantaneous energy density associated with the passage of a plane wave is given simply by $e(t) = [p^2(t)/\rho_0 c_0^2]$. It is also useful to assess the time average of these fluctuations in energy density. In the case of a plane wave whose pressure can be written as $\mathrm{Re}\{p(x)\mathrm{e}^{j\omega t}\}$, if $p(x)$ is the complex number $p_R + jp_I$ we can write

$$\mathrm{Re}\{p(x)\mathrm{e}^{j\omega t}\} = \mathrm{Re}\{(p_R + jp_I)(\cos\omega t + j\sin\omega t)\} = p_R\cos\omega t - p_I\sin\omega t. \tag{1.51}$$

Thus the time averaged energy density can be written as

$$e = \frac{1}{T}\int_{-T/2}^{T/2}\frac{p^2(t)}{\rho_0 c_0^2}\mathrm{d}t = \frac{1}{\rho_0 c_0^2 T}\int_{-T/2}^{T/2}(p_R\cos\omega t - p_I\sin\omega t)^2\mathrm{d}t, \tag{1.52}$$

where T is one period of the harmonic motion. The product containing the term $\cos \omega t \sin \omega t$ integrates to zero over one period while the terms in $\cos^2 \omega t$ and $\sin^2 \omega t$ integrate to a value of $(T/2)$. This gives

$$e = \frac{1}{2\rho_0 c_0^2}(p_R^2 + p_I^2) = \frac{|p(x)|^2}{2\rho_0 c_0^2} \qquad \text{(for plane waves).} \tag{1.53}$$

The time averaged energy density due to the passage of the plane wave is therefore related simply to the amplitude of the pressure fluctuations. It is also evident that as the waves propagate through the fluid then there is a transmission of energy from one part of the fluid to another. The time averaged rate at which this energy is transferred is the *acoustic intensity*. This is defined as

$$I = \frac{1}{T}\int_{-T/2}^{T/2} p(t)u(t)\mathrm{d}t, \tag{1.54}$$

where T is again the period of one cycle of harmonic motion. Note that the acoustic intensity therefore has dimensions of force per unit area times velocity which is equivalent to the rate at which work is done per unit area by the acoustic pressure. Acoustic intensity is thus measured in Wm^{-2}. In the case of plane progressive waves then $u(t)$ and $p(t)$ are simply related by $u(t) = p(t)/\rho_0 c_0$ and undertaking the time average in the same way as in the case of energy density produces an expression for the intensity. This is related to the amplitude of the pressure fluctuations by

$$I = \frac{|p(x)|^2}{2\rho_0 c_0} \qquad \text{(for plane waves).} \tag{1.55}$$

Thus for plane progressive waves the time averaged intensity and energy density are related by $I = c_0 e$. In general, however, the pressure and velocity are not always simply related as in the case of a plane progressive wave. If we define a complex pressure and particle velocity at a given frequency it is useful to derive an expression for the time averaged intensity in terms of the complex numbers $p(x)$ and $u(x)$. When undertaking time averages we again have to be careful to choose only the real part of these two quantities. We can write the real part of the pressure fluctuation as in equation (1.51) and similarly the real part of the velocity fluctuation is given by

$$\mathrm{Re}\{u(x)\mathrm{e}^{j\omega t}\} = u_R \cos \omega t - u_I \sin \omega t. \tag{1.56}$$

Therefore the time averaged acoustic intensity for general harmonic fluctuation in pressure and velocity is given by

$$\frac{1}{T}\int_{-T/2}^{T/2} p(t)u(t)\mathrm{d}t = \frac{1}{T}\int_{-T/2}^{T/2} (p_R \cos \omega t - p_I \sin \omega t)(u_R \cos \omega t - u_I \sin \omega t)\mathrm{d}t. \tag{1.57}$$

Note that only the products involving $\cos^2 \omega t$ and $\sin^2 \omega t$ will result in a non-zero value of the integral over the period T. The result of these integrations is therefore given by

$$\frac{1}{T}\int_{-T/2}^{T/2} p(t)u(t)\mathrm{d}t = \tfrac{1}{2}(p_\mathrm{R}u_\mathrm{R} + p_\mathrm{I}u_\mathrm{I}). \tag{1.58}$$

The right-hand side of this equation is simply equal to $(1/2)\mathrm{Re}[p^*(x)u(x)] = (1/2)\mathrm{Re}\{p(x)u^*(x)\}$, where $p^*(x)$ and $u^*(x)$ are the complex conjugates of $p(x)$ given by $p^*(x) = p_\mathrm{R} - jp_\mathrm{I}$ and $u^*(x) = u_\mathrm{R} - ju_\mathrm{I}$. Thus in general we can use the expression

$$I = \tfrac{1}{2}\mathrm{Re}\{p^*(x)u(x)\} = \tfrac{1}{2}\mathrm{Re}\{p(x)u^*(x)\}, \tag{1.59}$$

to evaluate the time averaged acoustic intensity associated with harmonic fluctuations in pressure and velocity.

1.9 STANDING WAVES

Hitherto we have considered only plane propagating acoustic disturbances of the type depicted in Fig. 1.1, where there is no possibility of any reflection of the disturbances from the right-hand end of the tube. Let us now deal with the case where we have a fluid in a tube which is excited by a piston at the left-hand end but which is terminated by a rigid boundary at the right-hand end. This is illustrated in Fig. 1.7. For harmonic excitation, one solution which will satisfy the wave equation is given by the superposition of the complex pressures associated with a positive-going and a negative-going plane wave. Thus we can write

$$p(x) = A\mathrm{e}^{-jkx} + B\mathrm{e}^{jkx}, \tag{1.60}$$

where A and B are arbitrary complex numbers which represent the amplitude and relative phases of waves travelling in the positive x and negative x directions respectively. Use of the linearized momentum equation (1.42) enables the complex particle velocities associated with each of these component travelling plane waves to be deduced. This is undertaken by evaluation of the pressure gradient associated with each of the waves as described in section 1.7. This yields an expression for the complex particle velocity which is given by

$$u(x) = \frac{A\mathrm{e}^{-jkx}}{\rho_0 c_0} - \frac{B\mathrm{e}^{jkx}}{\rho_0 c_0}. \tag{1.61}$$

We can now determine the values of the complex constants A and B by application of the boundary conditions. Thus since we know that there must be zero particle velocity at the right-hand end of the tube where $x = L$, then the condition $u(L) = 0$ in equation (1.61) results in the relationship

$$Ae^{-jkL} = Be^{jkL}. \tag{1.62}$$

Similarly, we know that the particle velocity fluctuation at the left-hand end of the tube is such that it must exactly match the fluctuating velocity of the piston. Thus we can use the boundary condition $u(0) = U$ where U is the complex velocity fluctuation associated with the piston. Application of this boundary condition to equation (1.61) results in

$$\rho_0 c_0 U = A - B. \tag{1.63}$$

Equations (1.62) and (1.63) can now be combined to give expressions for the complex amplitudes of the plane waves. These are given by

$$A = \frac{\rho_0 c_0 U e^{jkL}}{2j \sin kL}, \quad B = \frac{\rho_0 c_0 U e^{-jkL}}{2j \sin kL}. \tag{1.64a,b}$$

Note that in deriving these relationships use has been made of the expression $e^{j\theta} - e^{-j\theta} = 2j \sin \theta$. If we now substitute these expressions into equation (1.60) we can derive a relationship between the amplitudes of the pressure fluctuation in the tube and the amplitude of the harmonic fluctuation of the piston. Using the relationship $e^{j\theta} + e^{-j\theta} = 2 \cos \theta$, together with some algebra, results in

$$p(x) = \frac{-j\rho_0 c_0 U \cos k(L - x)}{\sin kL}. \tag{1.65}$$

Similarly substitution of equations (1.64a) and (1.64b) into equation (1.61) yields the complex particle velocity given by

$$u(x) = \frac{U \sin k(L - x)}{\sin kL}. \tag{1.66}$$

The two plane travelling waves therefore interfere to produce a *standing wave* whose pressure and particle velocity are a function of distance x along the tube.

First note that for harmonic fluctuations in the piston such that $\sin kL = 0$ then both the complex pressure and particle velocity become infinite. This infinite response of the acoustic pressure fluctuations in the tube occurs at frequencies where $kL = n\pi$ where n is an integer. This corresponds to the relationship $(2\pi/\lambda)L = n\lambda/2$ or $L = n\lambda/2$, i.e. at frequencies where an integer number of half wavelengths of sound can fit exactly into the length of the

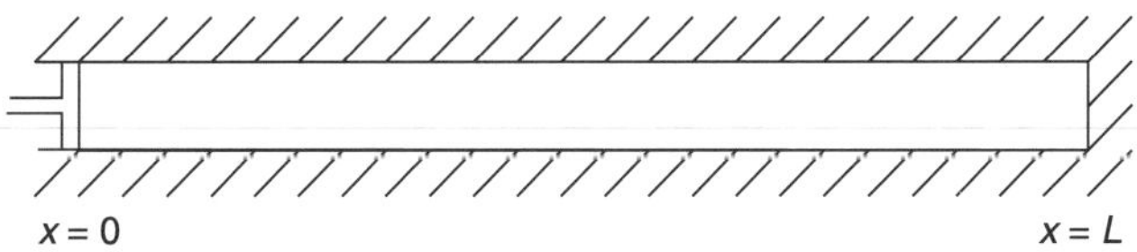

Figure 1.7 A tube that is rigidly terminated at $x = L$ and in which the fluid is driven by a vibrating piston at $x = 0$.

tube. Frequencies where this occurs are known as the *natural frequencies* or *resonance frequencies* of the fluid in the tube. Of course, in reality the response of the fluid in the tube will not be infinite at these frequencies since the presence of some dissipation (such as the presence of some absorbing material) will prevent the response from becoming infinite. It is also interesting to note the form of the pressure fluctuation at the natural frequencies as a function of the distance x along the tube. Thus when $kL = n\pi$ the amplitude of the pressure fluctuation as a function of x has the form $\cos n\pi(L - x)/L$ which is equivalent to $\pm\cos(n\pi x/L)$. These particular cosinusoidal functions of the pressure amplitude along the tube represent *mode shapes* corresponding to the natural frequencies of the fluid in the tube, and each natural frequency specified by a value of the integer n is associated with a different mode shape.

Also note that the specific acoustic impedance varies along the length of the tube. This is given by

$$z(x) = \frac{p(x)}{u(x)} = -j\rho_0 c_0 \cot k(L - x). \qquad (1.67)$$

Note that it is a purely imaginary number, i.e. the impedance has only a reactive part and no resistive part. This is in direct contrast to the case where only propagating waves in one direction are present. A purely imaginary impedance is associated with the lack of any net flow of energy since the particle velocity is in quadrature with the pressure. Thus if we evaluate the acoustic intensity at any frequency of excitation of the piston, at $x = 0$ (the piston face) then this has a value given by

$$I = (1/2)\mathrm{Re}\{p^*(0)U\} = (1/2)|U|^2\mathrm{Re}\{z(0)\} = 0, \qquad (1.68)$$

and therefore, under steady-state conditions, at no frequency is there any net flow of energy from the piston surface. Of course, this only applies under steady-state harmonic conditions and there will be an initial flow of energy from the piston as the motion is started which has not been described by these relationships. Again this zero intensity is in contrast to the steady flow of energy associated with the propagation of travelling plane waves.

1.10 THE THREE-DIMENSIONAL WAVE EQUATION

In order to derive the relationship between the spatial and temporal variation of the acoustic pressure fluctuations in cases where wave propagation is not just constrained to the one-dimensional situations described so far, we again use the principles of mass and momentum conservation in order to relate the acoustic variables pressure, density and velocity in the fluid. For the three-dimensional case depicted in Fig. 1.8, any position in space can be defined by the vector $\boldsymbol{x}$ which has components $x_1\boldsymbol{i}$, $x_2\boldsymbol{j}$ and $x_3\boldsymbol{k}$, where $\boldsymbol{i}$, $\boldsymbol{j}$ and $\boldsymbol{k}$ are the unit vectors in the x_1, x_2 and x_3

directions, respectively. The velocity of the fluid at any given position can be denoted by the vector $\boldsymbol{u}$, with components $u_1\boldsymbol{i}$, $u_2\boldsymbol{j}$ and $u_3\boldsymbol{k}$ in the three coordinate directions specified by x_1, x_2 and x_3. Consideration of the balance between the net inflow of mass into a small three-dimensional element and the net increase of mass of the element leads to a similar mass conservation equation to that derived in section 1.4. Again to the accuracy of linear approximation, the equation that results for the mass balance in a small element is given by

$$\frac{\partial \rho}{\partial t} + \rho_0\left(\frac{\partial u_1}{\partial x_1} + \frac{\partial u_2}{\partial x_2} + \frac{\partial u_3}{\partial x_3}\right) = 0. \tag{1.69}$$

This can alternatively be written in vector notation as

$$\frac{\partial \rho}{\partial t} + \rho_0 \nabla.\boldsymbol{u} = 0, \tag{1.70}$$

where $\nabla.$ is the divergence operator defined by $[\boldsymbol{i}(\partial/\partial x_1) + \boldsymbol{j}(\partial/\partial x_2) + \boldsymbol{k}(\partial/\partial x_3)]$ where $\boldsymbol{i}$, $\boldsymbol{j}$, $\boldsymbol{k}$ are again the unit vectors in the x_1, x_2 and x_3 directions respectively.

This equation expresses essentially the same relationship between the rate of increase of density and the rate of inflow of mass that was derived for the one-dimensional case considered in more detail above. Similarly, one can relate the acceleration of a given element of fluid in three coordinate directions to the pressure gradients in those three directions by applying the principle of momentum conservation to a fluid element. In this case there are three independent momentum balances associated with each coordinate direction, which again to a linear approximation can be written as

$$\rho_0\frac{\partial u_1}{\partial t} + \frac{\partial p}{\partial x_1} = 0, \quad \rho_0\frac{\partial u_2}{\partial t} + \frac{\partial p}{\partial x_2} = 0, \quad \rho_0\frac{\partial u_3}{\partial t} + \frac{\partial p}{\partial x_3} = 0. \tag{1.71a,b,c}$$

These three equations can be written as one vector equation defined by

$$\rho_0\frac{\partial \boldsymbol{u}}{\partial t} + \nabla p = 0, \tag{1.72}$$

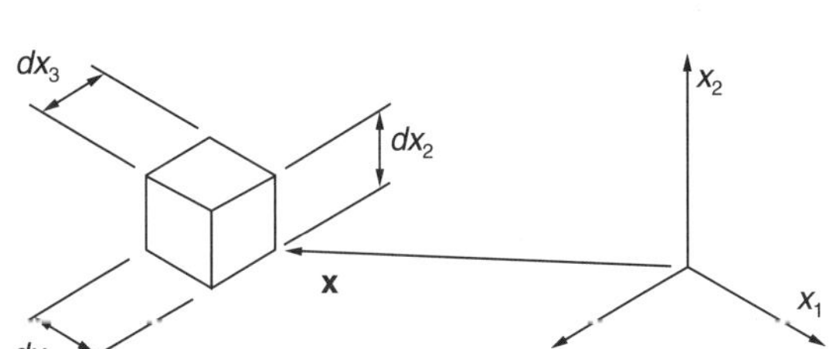

Figure 1.8 A small control volume used to derive the three-dimensional linearized continuity equation.

where ∇_p has the three components $\boldsymbol{i}(\partial p/\partial x_1)$, $\boldsymbol{j}(\partial p/\partial x_2)$, $\boldsymbol{k}(\partial p/\partial x_3)$. This is the three-dimensional linearized equation of conservation of momentum. If one now takes the divergence of this equation (i.e. operates on the equation with divergence operator $\nabla.$) and also if we differentiate equation (1.10.2) with respect to time we can eliminate $\nabla.\boldsymbol{u}$ from the two resulting equations to give

$$\nabla^2 p - \frac{\partial^2 \rho}{\partial t^2} = 0. \tag{1.73}$$

Since the pressure and density are linearly related by the equation $p = c_0^2 \rho$ then we have

$$\nabla^2 p - \frac{1}{c_0^2}\frac{\partial^2 p}{\partial t^2} = 0, \tag{1.74}$$

which is the wave equation which relates the spatial and temporal dependence of the pressure in a three-dimensional sound field. Note that in rectangular Cartesian coordinates $\nabla^2 p$ is given by the summation

$$\partial^2 p/\partial x_1^2 + \partial^2 p/\partial x_2^2 + \partial^2 p/\partial x_3^2.$$

This wave equation must be satisfied by acoustic disturbances propagated in the medium. However, in order to find solutions of this equation we must specify in more detail the form that the sound field takes. In particular, as we have demonstrated in the one-dimensional case, we have to apply boundary conditions in order to find the form of the sound field.

Finally, by using an argument that is identical to that used in section 1.5, we can define a spatially dependent complex pressure $p(\boldsymbol{x})$ that, for harmonic sound, must satisfy

$$(\nabla^2 + k^2)p(\boldsymbol{x}) = 0, \tag{1.75}$$

which is the three-dimensional Helmholtz equation. We should also note that the principle of linear superposition also applies to solutions of both the wave equation (1.74) and the Helmholtz equation (1.75). This is easily justified by following the argument presented in section 1.6.

1.11 SOLUTIONS OF THE THREE-DIMENSIONAL WAVE EQUATION

The simplest kind of three-dimensional waves can be found if we assume that the propagation taking place is spherically symmetric. This is the type of wave motion that would be generated by, for example, a sphere whose surface pulsates uniformly and generates sound waves which travel radially outward from the sphere into free space (i.e. a 'free field'). If the radiation is assumed to be spherically symmetric then the spatial operator on the pressure in the three-dimensional wave equation can be written as

$$\nabla^2 p = \frac{1}{r^2}\frac{\partial}{\partial r}\left(r^2 \frac{\partial p}{\partial r}\right), \tag{1.76}$$

and the wave equation for spherically symmetric wave propagation is given by

$$\frac{1}{r^2}\frac{\partial}{\partial r}\left(r^2 \frac{\partial p}{\partial r}\right) - \frac{1}{c_0^2}\frac{\partial^2 p}{\partial t^2} = 0. \tag{1.77}$$

By using the laws of the derivatives of the product of two functions it can be shown that this equation is identical to

$$\frac{\partial^2}{\partial r^2}(rp) - \frac{1}{c_0^2}\frac{\partial^2 (rp)}{\partial t^2} = 0. \tag{1.78}$$

This is a very useful relationship since it has a form which is equivalent to that of the one-dimensional wave equation discussed above. Note however that in this case it is not just the pressure p which is the variable considered, but the product of r and p. However, we know that the solution of this equation can be written as

$$rp(r,t) = f(t - r/c_0) + g(t + r/c_0), \tag{1.79}$$

where again f and g are arbitrary functions. It therefore follows that the solution for the pressure can be written in the form

$$p(r,t) = \frac{f(t - r/c_0)}{r} + \frac{g(t + r/c_0)}{r}. \tag{1.80}$$

By analogy with the one-dimensional case, the first term in this equation represents pressure fluctuations which travel outwards from the origin of the spherical coordinates, while the second term represents pressure fluctuations which travel inwards to the origin. The second type of wave is unlikely to be encountered in practice and we are generally concerned with solutions in the three-dimensional case which consist of only the first term of equation (1.80). Again it is useful to use a harmonic function of $(t - r/c_0)$ of the type described by

$$p(r,t) = \mathrm{Re}\left\{\frac{A\mathrm{e}^{j(\omega t - kr)}}{r}\right\}, \tag{1.81}$$

where A is again an arbitrary complex number which specifies both the amplitude and phase of the pressure fluctuation. We can again recover the particle velocity associated with this pressure fluctuation by applying the momentum conservation equation. For the spherically symmetric case where u_r denotes the radial particle velocity component, the momentum conservation equation reduces to

$$\rho_0 \frac{\partial u_\mathrm{r}}{\partial t} + \frac{\partial p}{\partial r} = 0. \tag{1.82}$$

First note that we can define a radially dependent complex pressure $p(r)$ as

$$p(r) = \frac{Ae^{-jkr}}{r}, \tag{1.83}$$

and again we recover the actual pressure fluctuation from $p(r,t) = \mathrm{Re}\{p(r)e^{j\omega t}\}$. We can define similarly a complex particle velocity which has a radial component $u_r(r)$ where the actual radial particle velocity fluctuation is given by $u_r(r,t) = \mathrm{Re}\{u_r(r)e^{j\omega t}\}$. In the same way that we used the one-dimensional equation of conservation of momentum in section 1.5, it can be shown that equation (1.82) can be used to relate the complex pressure and particle velocity. In this case the equation reduces to

$$j\omega\rho_0 u_r(r) + \frac{\partial p(r)}{\partial r} = 0. \tag{1.84}$$

Differentiation of equation (1.83) with respect to r and substitution into equation (1.84) results in the expression for the complex radial particle velocity which is

$$u_r(r) = \frac{A}{j\omega\rho_0}\left(\frac{jk}{r} + \frac{1}{r^2}\right)e^{-jkr}. \tag{1.85}$$

Note that this has both real and imaginary parts and therefore there are components of the particle velocity which are both in quadrature and in phase with the acoustic pressure. As the radial distance r becomes increasingly large then the second term in the brackets of equation (1.85) tends to zero and the relationship between the pressure and particle velocity reduces to that between the pressure and particle velocity in a plane one-dimensional progressive wave ($u_r(r) = p(r)/\rho_0 c_0$). This makes physical sense since at a very long distance from the source of the waves then locally the wavefronts will appear plane. However when r becomes very small, then the velocity will tend to become 90° out of phase with the pressure. This can be demonstrated more clearly by writing an expression for the specific acoustic impedance as a function of the radial distance r. This is given by

$$z(r) = \frac{p(r)}{u_r(r)} = \rho_0 c_0\left(\frac{jkr}{1 + jkr}\right). \tag{1.86}$$

Thus as kr becomes large compared with unity then the acoustic impedance tends to the value $\rho_0 c_0$ whereas when kr is very small then the acoustic impedance becomes increasingly imaginary. Note that it is the product kr which is the important parameter. This is equivalent to the ratio $2\pi r/\lambda$ and therefore it is the radial distance relative to the acoustic wavelength which is important in determining the specific acoustic impedance in spherical wave propagation.

1.12 POINT SOURCES OF SPHERICAL RADIATION

It is also useful to relate the pressure in the spherically symmetric outwardly propagating wave to the fluctuations in the velocity of the surface of the sphere which we assume to be the source of these waves. If we now apply boundary conditions such that the particle velocity described by equation (1.85) must exactly match the surface velocity of the pulsating sphere, and further assume that this pulsating sphere has a radius a and a complex velocity given by U_a, then the expression for the complex pressure at a can be written as

$$p(a) = \rho_0 c_0 \left(\frac{jka}{1 + jka} \right) U_a = \frac{A e^{-jka}}{a}. \tag{1.87}$$

From this we can deduce the relationship between A and U_a and it follows that the complex pressure as a function of the radial distance r from the centre of the sphere can be written as

$$p(r) = \rho_0 c_0 a \left[\frac{jka}{1 + jka} \right] \frac{U_a e^{-jk(r-a)}}{r}, \tag{1.88}$$

A very useful concept in acoustics is that of the point source. We can think of a point source by considering the sphere to be reduced steadily in radius but the surface velocity being increased such that the product of the surface velocity and the surface area of the sphere remains constant. We then define an effective complex *source strength*, such that, as $a \to 0$,

$$q = U_a 4\pi a^2 = \text{const.} \tag{1.89}$$

As the radius of the sphere becomes vanishingly small we can relate the pressure at a radial distance r from the origin of the waves to the source strength of this point source. Thus

$$p(r) = \rho_0 c_0 \left(\frac{jk}{1 + jka} \right) \frac{q e^{-jk(r-a)}}{4\pi r}, \tag{1.90}$$

which in the limit of ka tending to zero reduces to

$$p(r) = \frac{j\omega \rho_0 q e^{-jkr}}{4\pi r}, \quad ka \ll 1. \tag{1.91}$$

Note that the dimensions of complex source strength q are those of volume velocity, i.e. the product of velocity multiplied by surface area. It is sometimes useful to introduce the concept of an *acoustic transfer impedance* $Z(r)$ which is the complex number relating the complex pressure to the complex strength of a point source. Thus we write

$$p(r) = qZ(r), \tag{1.92}$$

where it is clear from equation (1.91) that for free field propagation $Z(r) = j\omega\rho_0 e^{-jkr}/4\pi r$. Note that this acoustic impedance has dimensions which differ from those of *specific* acoustic impedance since $Z(r)$ is a ratio of pressure to volume velocity. Using the identity $e^{j\theta} = \cos\theta + j\sin\theta$ also enables the expression for this transfer impedance to be written as

$$Z(r) = Z_0\left(\frac{\sin kr}{kr} + j\frac{\cos kr}{kr}\right), \tag{1.93}$$

where $Z_0 = \omega^2\rho_0/4\pi c_0$ is the real part of this transfer impedance in the limit $kr \to 0$.

1.13 ACOUSTIC POWER OUTPUT

Now consider the time averaged acoustic intensity associated with spherical wave propagation. In three-dimensional harmonic wave propagation of period T, the intensity is given by

$$\boldsymbol{I} = \frac{1}{T}\int_{-T/2}^{T/2} p(t)\boldsymbol{u}(t)\mathrm{d}t, \tag{1.94}$$

which is a vector quantity which depends both on the magnitude and direction of the vector $\boldsymbol{u}$. For the spherically symmetric propagation described above we can evaluate the intensity vector in the direction of the radial coordinate r. This can be expressed in terms of the complex pressure and radial particle velocity given by

$$I_\mathrm{r} = \tfrac{1}{2}\mathrm{Re}\{p^*(r)u_\mathrm{r}(r)\}. \tag{1.95}$$

Using the relationship between the complex pressure and radial particle velocity given by equation (1.86) shows that

$$I_\mathrm{r} = \frac{|p(r)|^2}{2\rho_0 c_0}\mathrm{Re}\left\{\frac{1 + jkr}{jkr}\right\} = \frac{|p(r)|^2}{2\rho_0 c_0}. \tag{1.96}$$

Recall that the intensity represents the average rate of energy flow per unit area of cross-section normal to the direction of propagation. This therefore shows that the intensity depends only on the modulus squared of the complex pressure but, of course, this value decreases as the square of the radial distance from the source as is evident from equation (1.83). The total power output of a source is given by integrating the intensity travelling across a surface which totally encloses the source. The sound power output is defined by

$$W = \int_S \boldsymbol{I}.\boldsymbol{n}\,\mathrm{d}S, \tag{1.97}$$

where $\boldsymbol{n}$ is the unit normal vector pointing outwards from the closed surface S. In the case of the spherically symmetric propagation described here, the expression of the power output can be written as

$$W = I_\mathrm{r}(4\pi r^2), \tag{1.98}$$

and is therefore related to the amplitude of the pressure wave by

$$W = \frac{|A|^2}{2\rho_0 c_0 r^2} 4\pi r^2 = \frac{2\pi |A|^2}{\rho_0 c_0}. \tag{1.99}$$

This is clearly a constant and, as one would expect under conditions of steady harmonic motion, there is a constant time-averaged rate of power flowing from the spherically radiating source. We can also evaluate the total power flowing from the source from the time average product of the pressure at the surface of the pulsating sphere and the velocity of the surface of the sphere. The power flow is given by

$$W = 4\pi a^2 \tfrac{1}{2} \mathrm{Re}\{p^*(a) U_\mathrm{a}\} \tag{1.100}$$

which can be written as

$$W = 4\pi a^2 \rho_0 c_0 |U_\mathrm{a}|^2 \frac{1}{2} \mathrm{Re}\left\{ \frac{-jka}{1 - jka} \right\}. \tag{1.101}$$

Evaluating the real part of the term in curly brackets then shows that

$$W = 2\pi a^2 \rho_0 c_0 |U_\mathrm{a}|^2 \frac{(ka)^2}{1 + (ka)^2}. \tag{1.102}$$

This describes the efficiency with which a given surface velocity can radiate acoustic power. As the product ka becomes very large then the expression for the power output tends to the expression for a power output of a plane vibrating surface of area $4\pi a^2$ radiating plane acoustic waves. However, as the product ka becomes very small then the expression for the power output can be approximated by

$$W = 2\pi a^2 \rho_0 c_0 |U_\mathrm{a}|^2 (ka)^2, \quad ka \ll 1. \tag{1.103}$$

The product ka is simply the ratio $2\pi a/\lambda$ and therefore the radiation efficiency of the pulsating sphere depends on the ratio of the circumference of the sphere to the wavelength at the frequency of excitation. In general, for bodies which are large compared to the wavelength, sound can be very effectively radiated by the vibrations of the surface but bodies which are very small compared to the wavelength are very inefficient radiators of acoustic power. In the limit of $ka \to 0$, the acoustic power output can be related to the strength of the equivalent point source and the expression for the power output reduces to

$$W = \frac{\omega^2 \rho_0}{8\pi c_0} |q|^2 = \frac{1}{2} Z_0 |q|^2, \tag{1.104}$$

where $Z_0 = \omega^2 \rho_0 / 4\pi c_0$ as defined above. Thus for a single point source in the free field the sound power output depends on the amplitude of the source strength, the frequency and the properties of the medium.

As in the case of the mean square sound pressures encountered in practice, the range of acoustic power outputs is similarly vast. For example, the power output of a whispered voice is as small as 10^{-9} W whereas the power output of a jet engine can be as large as 10^4 W. A useful means of compressing this range is to define a logarithmic measurement scale as was the case with sound pressure. We therefore define the sound power level by

$$L_W = 10 \log_{10} \frac{W}{W_{ref}}, \tag{1.105}$$

where the reference sound power level W_{ref} is given by 10^{-12} W. Some typical sound power levels are shown in Table 1.2.

1.14 ENCLOSED SOUND FIELDS AT LOW FREQUENCIES

Now consider the solution to the three-dimensional wave equation when the sound field is in a fluid enclosed by a rigid-walled rectangular box. In this case, for harmonic pressure fluctuations of a particular frequency in the enclosure, the complex pressure must satisfy the Helmholtz equation

$$\left(\frac{\partial^2}{\partial x_1^2} + \frac{\partial^2}{\partial x_2^2} + \frac{\partial^2}{\partial x_3^2}\right) p(\mathbf{x}) + k^2 p(\mathbf{x}) = 0. \tag{1.106}$$

In order to solve this equation we assume that the pressure fluctuation can be expressed as three separate functions of x_1, x_2 and x_3, such that

$$p(\mathbf{x}) = F_1(x_1) F_2(x_2) F_3(x_3). \tag{1.107}$$

Substitution of equation (1.107) into equation (1.106) yields

$$\frac{1}{F_1(x_1)} \frac{\partial^2 F_1(x_1)}{\partial x_1^2} + \frac{1}{F_2(x_2)} \frac{\partial^2 F_2(x_2)}{\partial x_2^2} + \frac{1}{F_3(x_3)} \frac{\partial^2 F_3(x_3)}{\partial x_3^2} + k^2 = 0. \tag{1.108}$$

This sum of k^2 plus three functions which respectively depend only on one of the three independent variables can vanish in general only if k^2 is the sum of three independent constants,

$$k^2 = k_1^2 + k_1^2 + k_3^2, \tag{1.109}$$

where the constant k_1, for example, is related to the function F_1 through the ordinary differential equation given by

$$\frac{1}{F_1(x_1)} \frac{d^2 F_1(x_1)}{dx_1^2} + k_1^2 = 0. \tag{1.110}$$

Table 1.2 Typical sound power outputs.

	(W)	(dB re 10^{-12} W)
Jet aircraft	10000	160
Chain saw	1	120
Normal voice	0.00001	70
Whisper	0.000000001	30

Two other ordinary differential equations also follow for the dependence of the pressure in the x_2 and x_3 directions. The function $F_1(x_1)$ in particular must also satisfy the rigid wall boundary conditions of zero particle velocity at $x_1 = 0$ and $x_1 = L_1$. Since the complex particle velocity is given by $u(\boldsymbol{x}) = (-1/j\omega\rho_0)\partial p(\boldsymbol{x})/\partial x_1$, a zero particle velocity requires a zero pressure gradient $\partial p(\boldsymbol{x})/\partial x_1$. It is relatively easy to show that a set of solutions of equation (1.110) each of which satisfies the boundary conditions is $F_1(x_1) = \cos k_1 x_1$ where $k_1 = n_1\pi/L_1$ and n_1 is an integer. Similar solutions of the other two differential equations for $F_2(x_2)$ and $F_3(x_3)$ also follow. This shows that we can write a general solution of the wave equation (1.107) which also satisfies the boundary conditions as

$$p(\boldsymbol{x}) = \sum_{n=0}^{\infty} A_n \cos\frac{n_1\pi x_1}{L_1}\cos\frac{n_2\pi x_2}{L_2}\cos\frac{n_3\pi x_3}{L_3}, \tag{1.111}$$

where the integer n denotes the trio of integers $(n_1,\ n_2,\ n_3)$ and A_n is an arbitrary complex constant. Note the similarity of this solution to that of the problem considered in section 1.9, where a fluid in a one-dimensional closed tube was excited by a piston. In that case we saw that at certain frequencies the pressure amplitude would become theoretically infinite and that at these frequencies an integer number of half wavelengths would define a given mode shape associated with the pressure fluctuations in the tube. In this three-dimensional case, exactly similar pressure fluctuations are generated but now for a given natural frequency there is not only a spatial dependence on the x_1 coordinate but also on the x_2 and x_3 coordinates. In fact the natural frequency associated with each of these mode shapes is found by noting that for a given combination of the integers n_1, n_2 and n_3 equation (1.109) can be written in the form

$$\omega_n = c_0\sqrt{\left[\left(\frac{n_1\pi}{L_1}\right)^2 + \left(\frac{n_2\pi}{L_2}\right)^2 + \left(\frac{n_3\pi}{L_3}\right)^2\right]}. \tag{1.112}$$

This defines the *natural frequencies* of the enclosure associated with each of the mode shapes defined by each combination of integers. In fact the solution defined by equation (1.111) is applicable *only* to unforced motion at these particular natural frequencies. For the perfectly rigid walled enclosure considered here with no source forcing the sound field, a sound field can exist only at the particular values of ω_n specified by equation (1.112). These particular values of wavenumber $k_n = \omega_n/c_0$ are known as *eigenvalues* of the Helmholtz equation. Each of these eigenvalues is associated with an *eigenfunction* which defines the particular spatial pattern associated with the given value of n and which is specified by equation (1.111). These eigenfunctions associated with rigid wall boundary conditions can be used to construct a solution for the pressure field in an enclosure with walls having a high but finite impedance (see for example Nelson and Elliott,

1992, Ch. 10). The practical usefulness of the above solution is further limited by the fact that in many enclosures at the frequencies of interest there are potentially so many modal contributions. For example, Fig. 1.9 shows, for a typical enclosure, the number of modes having natural frequencies in a given bandwidth. Also, when some sound-absorbing material is introduced onto the walls of the enclosure, the individual modes become damped. In addition, each mode can be excited over a range of frequencies other than that corresponding to its natural frequency. Thus at high-excitation frequencies, a great many enclosure modes may be excited. Computation of the sound field by using this modal model then becomes extremely laborious and we adopt the approach described in the next section where the sound field is assumed to consist of an infinitely large number of plane wave contributions. This is not inconsistent with the modal model presented here since each mode may be considered to consist of a number of plane wave contributions. Just as we saw in section 1.9 that a one-dimensional standing wave can be considered to consist of two plane propagating waves, it can also be shown that the three-dimensional standing wave pattern associated with a given mode can be considered to consist of a sum of eight travelling plane wave contributions (see, for example, Kinsler *et al.*, 1982, Ch. 9 for a full discussion).

1.15 ENCLOSED SOUND FIELDS AT HIGH FREQUENCIES

As we discussed in the last section, in order to deal with enclosed sound fields at high frequencies (or strictly speaking, enclosures in which many modes have natural frequencies within the bandwidth of any one mode) we can adopt a model in which we assume that the sound field at a point consists of a superposition of the contributions from a number of plane waves. The plane waves are next assumed to be arriving at the point considered from *all possible propagation directions*. One can visualize this by assuming that a point in the sound field is at the centre of a sphere whose surface is divided into a very large number of segments of *equal* area. The line passing through the centre of one of these segments and the centre of the sphere defines the propagation direction associated with a particular plane wave. The field is then assumed to consist of an *infinite* number of plane waves, each associated with a *different* propagation direction. These propagation directions are defined by assuming that all the segments of equal area on the surface of the sphere become infinitesimally small. Thus the sound field at a single frequency has a complex pressure at a point (defined by the position vector $\boldsymbol{x}$) that is given by

$$p(\boldsymbol{x}) = \lim_{N\to\infty} \frac{1}{\sqrt{N}} \sum_{n=1}^{N} p_n(\boldsymbol{x}), \tag{1.113}$$

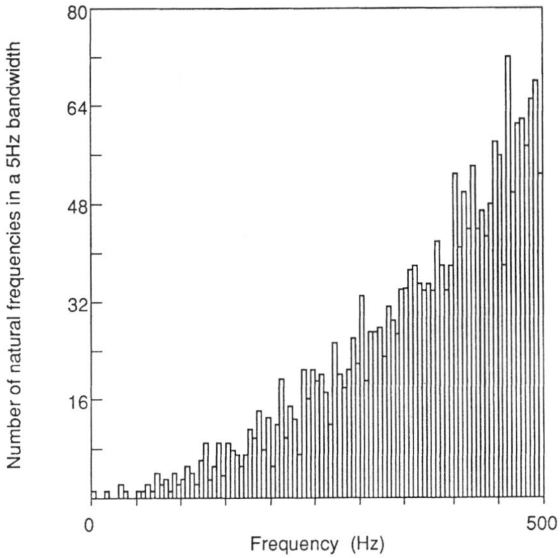

Figure 1.9 The number of natural frequencies in each 5 Hz bandwidth up to 500 Hz for a rectangular enclosure of dimensions 10 m × 5 m × 3 m.

where $p_n(\boldsymbol{x}) = |p_n| e^{j\phi_n(\boldsymbol{x})}$ is the complex pressure associated with the n-th plane wave and where we have normalized the amplitudes of the constituent plane waves by the square root of the total number N of contributing waves. Thus the constituent plane waves have moduli $|p_n|$ and phases $\phi_n(\boldsymbol{x})$, the latter changing as a function of position $\boldsymbol{x}$ as a result of the propagation of the waves through space. It is also assumed that at a given position, the phases of the individual plane waves are in a random relationship to one another. We can compute the modulus squared of the complex pressure at the position $\boldsymbol{x}$ by taking the product of $p(\boldsymbol{x})$ and its complex conjugate. Thus

$$|p(\boldsymbol{x})|^2 = \lim_{N\to\infty} \frac{1}{N} \left(\sum_{n=1}^{N} p_n(\boldsymbol{x}) \right) \left(\sum_{m=1}^{N} p_m(\boldsymbol{x}) \right)^*, \tag{1.114}$$

where the additional index m has been introduced to denote the m-th plane wave. Now note that the product of terms for which $n \neq m$ takes the form

$$p_n(\boldsymbol{x}) p_m^*(\boldsymbol{x}) = |p_n||p_m| e^{j[\phi_n(\boldsymbol{x}) - \phi_m(\boldsymbol{x})]}, \tag{1.115}$$

which can also be written as

$$p_n(\boldsymbol{x}) p_m^*(\boldsymbol{x}) = |p_n||p_m| \{\cos[\phi_n(\boldsymbol{x}) - \phi_m(\boldsymbol{x})) + j \sin(\phi_n(\boldsymbol{x}) - \phi_m(\boldsymbol{x})]\}. \tag{1.116}$$

If we now move through a volume of space which has dimensions that are large compared to the acoustic wavelength at the frequency considered (but still much smaller than the dimensions of the enclosure) we will see that the terms in the curly brackets of equation (1.116) will have an average value of zero, since both positive or negative values of the sine and cosine terms are equally likely to occur. This is a direct result of the assumption that ϕ_n and ϕ_m are unrelated and that the two wave propagation directions are different. Thus, if we define a *local spatial average* of the modulus squared pressure, where we compute the average value of $|p(\boldsymbol{x})|^2$ by taking samples of the squared pressure field over a volume of space with dimensions much larger than an acoustic wavelength, then all the terms for which $n \neq m$ will average to zero and only terms for which $n = m$ contribute. The space averaged modulus squared pressure thus becomes

$$\langle |p(\boldsymbol{x})|^2 \rangle = \lim_{N\to\infty} \frac{1}{N} \sum_{n=1}^{N} |p_n|^2, \tag{1.117}$$

where we have used $\langle\ \rangle$ to denote the operation of space averaging. We can use very similar arguments to this in deducing expressions for the space averaged energy density and intensity in the sound field. First, we can use the expression derived in section 1.8 for the energy density of a three-dimensional sound field provided we interpret $u(t)$ as the instantaneous modulus of the particle velocity vector at a position in the sound field. For harmonic excitation we can use an argument identical to that presented in section 1.8 in order to compute an expression for the time averaged energy density. This is given by

$$e(\boldsymbol{x}) = \frac{1}{4}\rho_0 \left[|u(\boldsymbol{x})|^2 + \frac{|p(\boldsymbol{x})|^2}{(\rho_0 c_0)^2} \right], \tag{1.118}$$

where $p(\boldsymbol{x})$ and $u(\boldsymbol{x})$ are the complex pressure and complex particle velocity modulus respectively. If we now compute a local spatial average of this time averaged energy density, we will again find that the contribution of terms arising from dissimilar propagation directions will average to zero. It follows that

$$\langle e(\boldsymbol{x}) \rangle = \lim_{N\to\infty} \frac{1}{N} \sum_{n=1}^{N} \frac{1}{4}\rho_0 \left[|u_n|^2 + \frac{|p_n|^2}{(\rho_0 c_0)^2} \right]. \tag{1.119}$$

Since we know that for the n-th plane wave, $|u_n| = |p_n|/\rho_0 c_0$, this shows that

$$\langle e(\boldsymbol{x}) \rangle = \lim_{N\to\infty} \frac{1}{N} \sum_{n=1}^{N} \frac{|p_n|^2}{2\rho_0 c_0^2} = \frac{\langle |p(\boldsymbol{x})|^2 \rangle}{2\rho_0 c_0^2}. \tag{1.120}$$

We now make a further assumption regarding the form of the sound field model. Equation (1.120) shows that the local space-averaged, time-average energy density can be considered to be the sum of contributions from an infinite number of plane waves. We assume that all these directions of propagation contribute *equally* to the space averaged energy density. A sound field model comprising plane waves of random phases and coming from all possible directions which also satisfies this criterion is said to be *diffuse*. (This is the definition adopted, for example, by Pierce, 1981, see Ch. 6). Thus, if we consider the contribution to the summation in equation (1.120) from a number of plane waves propagating in a range of directions (again by specifying this range of directions by dividing up the surface area of a sphere into segments of equal area) then each contribution associated with a given segment will remain the same as the areas of the segments are allowed to become infinitesimal.

1.16 THE ENERGY BALANCE EQUATION FOR AN ENCLOSURE

We can use the diffuse field assumption to derive an expression for the local space average of the time-averaged intensity in the diffuse sound field model. By this we mean the intensity produced by the waves whose directions are specified by a *hemisphere* centred on a point in space, i.e. the intensity due to the waves falling on *one side* of a hypothetical surface in the sound field, since the total intensity found by adding the contributions from both sides of the surface will be zero. The time averaged intensity in a direction normal to this surface can be written as

$$I(\boldsymbol{x}) = \tfrac{1}{2}\mathrm{Re}[p^*(\boldsymbol{x})u(\boldsymbol{x})], \tag{1.121}$$

where $u(\boldsymbol{x})$ is the component of the complex particle velocity in the direction normal to the surface. This velocity component can be written as

$$u(\boldsymbol{x}) = \lim_{N\to\infty} \frac{1}{\sqrt{N/2}} \sum_{n=1}^{N/2} \frac{|p_n|}{\rho_0 c_0} \cos\theta_n, \tag{1.122}$$

where θ_n defines the angle of propagation direction of the n-th plane wave using the coordinate system specified in Fig. 1.10. If we now evaluate the local space average value of $I(\boldsymbol{x})$, again assuming that the terms associated with different propagation directions average to zero, the expression that results is given by

$$\langle I(\boldsymbol{x})\rangle = \lim_{N\to\infty} \frac{1}{(N/2)} \sum_{n=1}^{N/2} \frac{|p_n|^2}{2\rho_0 c_0} \cos\theta_n. \tag{1.123}$$

We now have to make the important assumption that the contributions to the summation from all the values of $|p_n|^2$ are equal for all the directions of propagation considered. A detailed and rigorous discussion of this point is presented by Pierce (1981, Ch. 6, pp. 255–8). This amounts to assuming that, as discussed at the end of section 1.15, the contributions to the average energy density are the same for all directions of propagation. Under these conditions, equation (1.123) reduces to

$$\langle I(\boldsymbol{x})\rangle = \frac{\langle |p_i(\boldsymbol{x})|^2\rangle}{2\rho_0 c_0} \lim_{N\to\infty} \frac{1}{(N/2)} \sum_{n=1}^{N/2} \cos\theta_n, \qquad (1.124)$$

where we have used $p_i(\boldsymbol{x})$ to denote the net complex pressure due only to waves incident on one side of the hypothetical surface in the field. Again, since we assume that all the contributions to the total space averaged mean square pressure are equal for all propagation directions, it follows that $\langle |p(x)|^2\rangle = 2\langle |p_i(\boldsymbol{x})|^2\rangle$ and therefore that

$$\langle I(\boldsymbol{x})\rangle = \frac{\langle |p(\boldsymbol{x})|^2\rangle}{4\rho_0 c_0} \lim_{N\to\infty} \frac{1}{(N/2)} \sum_{n=1}^{N/2} \cos\theta_n, \qquad (1.125)$$

We can evaluate the summation over $\cos\theta_n$ by reducing the summation to an integral. Figure 1.10 shows an element of surface area of a sphere surrounding a point in space. The elemental area is defined by the angles $\Delta\theta$ and $\Delta\psi$. Since the number of plane waves passing through a unit area of the sphere is given by $N/4\pi r^2$, where r is the radius of the sphere, the number of plane waves passing through the element of surface area ΔS is given by

$$\frac{N}{4\pi r^2}\Delta S = \left(\frac{N}{4\pi r^2}\right) r^2 \sin\theta\ \Delta\theta\ \Delta\psi. \qquad (1.126)$$

We therefore write the summation over all directions of propagation as a summation over all the elements of area. Thus

$$\lim_{N\to\infty} \frac{1}{(N/2)} \sum_{n=1}^{N/2} \cos\theta_n = \lim_{\Delta\theta,\ \Delta\psi\to 0} \frac{1}{(N/2)} \sum_{\Delta\psi}\sum_{\Delta\theta} \frac{N}{4\pi} \cos\theta \sin\theta\ \Delta\theta\ \Delta\psi. \qquad (1.127)$$

The summations over $\Delta\psi$ and $\Delta\theta$ can thus be reduced to integrals and

$$\lim_{N\to\infty} \frac{1}{(N/2)} \sum_{n=1}^{N/2} \cos\ \theta_n = \frac{1}{2\pi} \int_0^{2\pi}\int_0^{\pi/2} \cos\theta \sin\theta\ \mathrm{d}\theta\, \mathrm{d}\psi. \qquad (1.128)$$

Performing the integration shows that the result is simply (1/2) and therefore the expression for the ‘one-sided’ space averaged, time average intensity is given by

$$\langle I(\boldsymbol{x})\rangle = \frac{\langle |p(\boldsymbol{x})|^2\rangle}{8\rho_0 c_0} = \frac{c_0}{4}\langle e(\boldsymbol{x})\rangle, \qquad (1.129)$$

where we have used equation (1.120) in order to relate the intensity to the space averaged, time average energy density.

We can use this result to write an energy balance equation for the whole enclosure (of volume V, say). The central assumption that we make is that the difference in the rate of energy input (power input) to the enclosure and the rate of energy lost by the absorption of sound at the enclosure walls

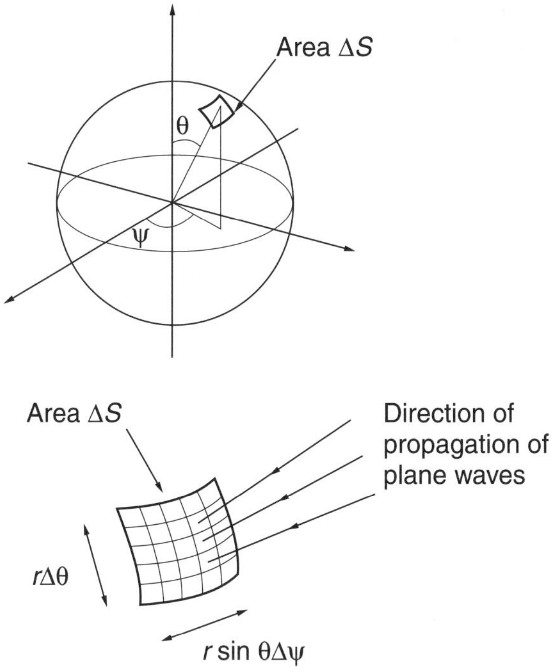

Figure 1.10 The plane wave construction of a diffuse sound field model. The surface of the sphere of radius r surrounding a point is divided into N tiny segments of equal area. Each area element has an associated plane wave propagating in a direction defined by the normal to the surface area element. Thus the segment of area ΔS contains $(N/4\pi r^2)$ tiny elements.

must be balanced by the rate of change of total room energy. The rate of energy lost as a result of the incidence of the diffuse field on a given area S_M of the wall of the enclosure can be written as $\langle I(\boldsymbol{x})\rangle S_M \alpha_M$. The term α_M is an *absorption coefficient* which defines the ratio of the intensity absorbed by the surface to the intensity incident on the surface. Note that here we are assuming that the intensity incident on an enclosure surface is equal to the 'one-sided' intensity in the diffuse field model. The total power absorbed by all the surfaces of the enclosure can thus be written as

$$\langle I(\boldsymbol{x})\rangle [S_1\alpha_1 + S_2\alpha_2 \ldots S_M\alpha_M], \tag{1.130}$$

where we add all the powers absorbed by the M different surfaces comprising the walls of the enclosure. We often define an *average absorption coefficient* $\bar{\alpha}$ given by

$$\bar{\alpha} = \frac{S_1\alpha_1 + S_2\alpha_2 \ldots S_M\alpha_M}{S}, \tag{1.131}$$

where S is the total surface area of the enclosure walls. The energy balance equation for the enclosure can then be written as

$$V\frac{\mathrm{d}}{\mathrm{d}t}\langle e(\boldsymbol{x})\rangle = W - \langle I(\boldsymbol{x})\rangle S\,\overline{\alpha}, \tag{1.132}$$

where W is the power input to the diffuse field and V is the volume of the enclosure. Under steady-state conditions, when the power input to the diffuse field is balanced by the power lost, this leads directly to a relationship between the space averaged squared pressure and power input to the diffuse field. This is given by

$$\langle |p(\boldsymbol{x})|^2\rangle = \frac{8\rho_0 c_0}{S\,\overline{\alpha}}W. \tag{1.133}$$

The transient solution to the energy balance equation is also of significance. If the power input to the enclosure is suddenly switched off, use of equation (1.129) and the energy balance equation (1.132) shows that the space averaged squared pressure will satisfy

$$\frac{4V}{S\,\overline{\alpha}c_0}\frac{\mathrm{d}}{\mathrm{d}t}\langle |p(\boldsymbol{x})|^2\rangle = -\langle |p(\boldsymbol{x})|^2\rangle. \tag{1.134}$$

The solution of this equation is an exponential decay of the form

$$\langle |p(\boldsymbol{x})|^2\rangle = \langle |p(\boldsymbol{x})|^2\rangle_0 \mathrm{e}^{-(S\,\overline{\alpha}c_0 t/4V)}, \tag{1.135}$$

where $\langle |p(\boldsymbol{x})|^2\rangle_0$ is the initial value of the space averaged squared pressure. If we evaluate the time taken for the squared pressure to decay to 10^{-6} times its original value (i.e. the sound pressure level decays by 60 dB) then it follows from equation (1.125) that this time is given by

$$T_{60} = \frac{0.161V}{S\,\overline{\alpha}}, \tag{1.136}$$

where the constant $0.161 = 4\ln 10/c_0$ holds for air at 20 °C. This is known as the *reverberation time* of the enclosure and equation (1.136) is known as the *Sabine equation* after W.C. Sabine, who first established this relationship empirically (see Sabine, 1898). This equation is useful in characterizing the acoustical properties of enclosed sound fields. In fact direct measurements of the reverberation time of an enclosure can be made in order to quantify its total room absorption $S\overline{\alpha}$. This is one technique for measuring the effective (or equivalent) diffuse field absorption coefficient. Thus the reverberation time of an empty reverberant room is first measured and its absorption calculated. A sample of material (typically ten square metres in area) is then laid in the room and the reverberation time measured again. The difference in the calculated room absorption then enables the calculation to be made of the diffuse field absorption coefficient of the material. A good discussion of the limitations of the Sabine equation is presented by Pierce (1981) who also describes the contributions made by later workers such as Eyring who proposed a refinement of the Sabine model which results in the replacement of $\bar{\alpha}$ in the denominator of equation (1.136) by $-\ln(1-\bar{\alpha})$.

1.17 THE POINT DIPOLE SOURCE

Before introducing a more sophisticated treatment of the sound fields radiated by complex source distributions, it is useful to describe the nature of the point dipole field which results from two monopole sources of equal and opposite strength placed a distance apart which is vanishingly small compared to the acoustic wavelength. The point dipole results when the product of the strength of the component sources and their separation distance is held constant as the separation distance is allowed to vanish. With reference to Fig. 1.11 the net complex pressure produced by this arrangement of two equal and opposite sources can be written as

$$p = j\omega\rho_0 q\left[\frac{\mathrm{e}^{-jkr_1}}{4\pi r_1} - \frac{\mathrm{e}^{-jkr_2}}{4\pi r_2}\right]. \tag{1.137}$$

If we assume that the vector $\boldsymbol{x}$ (corresponding to coordinates x_1, x_1, x_3) specifies the position at which we wish to calculate the sound field and that the vector $\boldsymbol{y}$ (corresponding to coordinates y_1, y_2, y_3) specifies the position of the source, then for the particular arrangement illustrated in Fig. 1.11, we can write the radial distances r_1 and r_2 from the two component monopole sources as

$$r_1 = \sqrt{[(x_1 - (y_1 + \varepsilon_1))^2 + x_2^2 + x_3^2]} \tag{1.138}$$

$$r_2 = \sqrt{[(x_1 - (y_1 - \varepsilon_1))^2 + x_2^2 + x_3^2]} \tag{1.139}$$

where ε_1 is a distance which is small compared to the acoustic wavelength. The terms in the square brackets of equation (1.137) can now be regarded as functions of the type $f(y_1 \pm \varepsilon_1)$, such that we can use a Taylor series expansion of the form given by

$$f(y_1 + \varepsilon_1) = f(y_1) + \varepsilon_1\frac{\partial}{\partial y_1}f(y_1) + \frac{1}{2!}\varepsilon_1^2\frac{\partial^2}{\partial y_1^2}f(y_1)\ldots \tag{1.140}$$

where ε_1 is the small expansion parameter. Thus we can write

$$\frac{\mathrm{e}^{-jkr_1}}{4\pi r_1} \approx \frac{\mathrm{e}^{-jkr}}{4\pi r} + \varepsilon_1\frac{\partial}{\partial y_1}\left[\frac{\mathrm{e}^{-jkr}}{4\pi r}\right] + \frac{1}{2!}\varepsilon_1^2\frac{\partial^2}{\partial y_1^2}\left[\frac{\mathrm{e}^{-jkr}}{4\pi r}\right]\cdots \tag{1.141}$$

$$\frac{\mathrm{e}^{-jkr_2}}{4\pi r_2} \approx \frac{\mathrm{e}^{-jkr}}{4\pi r} - \varepsilon_1\frac{\partial}{\partial y_1}\left[\frac{\mathrm{e}^{-jkr}}{4\pi r}\right] + \frac{1}{2!}\varepsilon_1^2\frac{\partial^2}{\partial y_1^2}\left[\frac{\mathrm{e}^{-jkr}}{4\pi r}\right]\cdots \tag{1.142}$$

In the limit of $\varepsilon_1 \to 0$, then we need only consider the leading order terms in the series, and the expression for the radiated pressure reduces to

$$p = j\omega\rho_0 q 2\varepsilon_1 \frac{\partial}{\partial y_1}\left[\frac{e^{-jkr}}{4\pi r}\right]. \tag{1.143}$$

Denoting $2\varepsilon_1$ as the distance 'd' between the two sources we can also write this expression as

$$p = j\omega\rho_0 q d \frac{\partial r}{\partial y_1} \cdot \frac{\partial}{\partial r}\left[\frac{e^{-jkr}}{4\pi r}\right]. \tag{1.144}$$

Since $r = \sqrt{[(x_1 - y_1)^2 + x_2^2 + x_3^2]}$, then $\partial r/\partial y_1 = (y_1 - x_1)/r = -\cos\theta$. In addition, we have

$$\frac{\partial}{\partial r}\left[\frac{e^{-jkr}}{4\pi r}\right] = -\frac{jke^{-jkr}}{4\pi r}\left[1 - \frac{j}{kr}\right] \tag{1.145}$$

and therefore the net radiated field can be written as

$$p = -\frac{\omega\rho_0 q e^{-jkr}}{4\pi r}(kd\cos\theta)\left[1 + \frac{1}{jkr}\right]. \tag{1.146}$$

This is the sound field produced by a *point dipole* source. It is the sound field radiated in the limit when the distance d tends to zero but the product (qd) is held constant. Thus the strengths of the component monopoles are increased as the distance between them is reduced. The product (qd) is known as the 'strength' of the dipole. Firstly note that in the far field, when

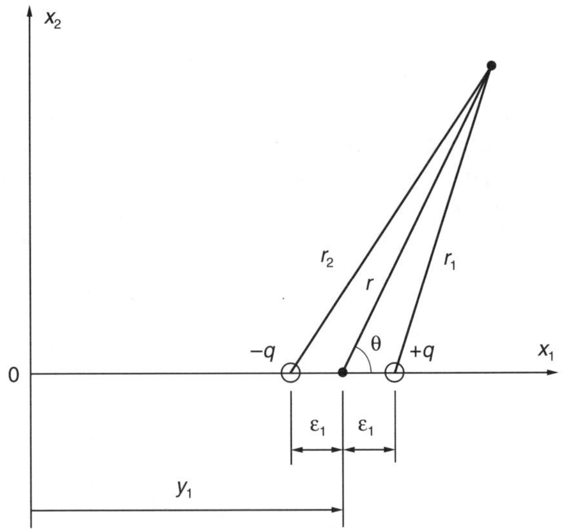

Figure 1.11 Geometrical arrangement of two sources of equal and opposite strength separated by a distance $d = 2\varepsilon_1$.

one is many wavelengths from the source, such that $kr \to \infty$, then this expression reduces to $p = p_M(jkd \cos\theta)$, where p_M represents the far field pressure radiated by a point monopole source of strength q. Thus, the far field pressure radiated differs from that of a monopole in two respects. First, it is reduced in amplitude by a factor kd and secondly it has a directional dependence specified by the term $\cos\theta$. In addition, close to the source, then the term $1/jkr$ becomes important and characterizes the more intense near field of the dipole source. This near field has large velocity fluctuations associated with it as liquid oscillates to and fro between the two component sources of the dipole. We can calculate the form of the velocity fluctuation by evaluating the radial component of particle velocity u_r. We know from the momentum conservation equation that $\rho_0 \partial u_r/\partial t = -\partial p/\partial r$, and therefore for harmonic fluctuations $u_r = (j/\omega\rho_0)\partial p/\partial r$. It follows from differentiation of equation (1.146) that the radial component of the particle velocity due to the dipole can be written as

$$u_r = -\frac{qe^{-jkr}}{4\pi r}k^2 d\cos\theta\left[1 - \frac{2j}{kr} - \frac{2}{(kr)^2}\right]. \tag{1.147}$$

Similarly we can calculate the tangential component of particle velocity u_θ which is given by (see Morse and Ingard, 1968).

$$u_\theta = -\frac{qe^{-jkr}}{4\pi r}k^2 d\sin\theta\left[\frac{j}{kr} + \frac{1}{(kr)^2}\right]. \tag{1.148}$$

We shall show later that an identical sound field is produced by the application of a point force to the medium. The origin of this force is associated with the rate of change of momentum of the fluid flowing between the two component monopoles (see, for example, Nelson and Elliott, 1992, Ch. 5).

It is also important to note that in the three-dimensional case the sound field produced is a function of the orientation of the axis on which the two component monopole sources lie. This can be explained with reference to the case depicted in Fig. 1.12, where the radial distance from the two component monopole sources to the field point are given by $r_1 = |\boldsymbol{x} - (\boldsymbol{y} + \boldsymbol{\varepsilon})|$ and $r_2 = |\boldsymbol{x} - (\boldsymbol{y} - \boldsymbol{\varepsilon})|$, where the vector ε has components $(\varepsilon_1\boldsymbol{i}, \varepsilon_2\boldsymbol{j}, \varepsilon_3\boldsymbol{k})$ and $\boldsymbol{i},\boldsymbol{j},\boldsymbol{k}$ specify the unit vectors in the three Cartesian coordinate directions. In this case the terms in the square brackets of equation (1.147) can be regarded as a function of the three variables y_1, y_2, y_3 and can be expanded in the multi-dimensional Taylor series given by

$$f(y_1+\varepsilon_1, y_2+\varepsilon_2, y_3+\varepsilon_3) = f(y_1, y_2, y_3) + \left(\varepsilon_1\frac{\partial}{\partial y_1} + \varepsilon_2\frac{\partial}{\partial y_2} + \varepsilon_3\frac{\partial}{\partial y_3}\right) f(y_1, y_2, y_3) + \frac{1}{2!}\left(\varepsilon_1\frac{\partial}{\partial y_1} + \varepsilon_2\frac{\partial}{\partial y_2} + \varepsilon_3\frac{\partial}{\partial y_3}\right)^2 f(y_1, y_2, y_3) + \cdots \qquad (1.149)$$

See, for example, Spencer *et al.* (1977) for a description of this form of series expansion. Note that we can also write this series expansion more compactly as

$$f(\boldsymbol{y}+\boldsymbol{\varepsilon}) = f(\boldsymbol{y}) + \boldsymbol{\varepsilon}.\nabla_y f(\boldsymbol{y}) + \frac{1}{2!}(\boldsymbol{\varepsilon}.\nabla_y)^2 f(\boldsymbol{y}) \cdots \qquad (1.150)$$

where $\nabla_y f = \partial f/\partial y_1 \boldsymbol{i} + \partial f/\partial y_2 \boldsymbol{j} + \partial f/\partial y_3 \boldsymbol{k}$. Thus to leading order in the small parameter vector $\boldsymbol{\varepsilon}$ we can write the terms in the square brackets of (1.137) as

$$\frac{\mathrm{e}^{-jkr_1}}{4\pi r_1} \approx \frac{\mathrm{e}^{-jkr}}{4\pi r} + \boldsymbol{\varepsilon}.\nabla_y\left(\frac{\mathrm{e}^{-jkr}}{4\pi r}\right) \qquad (1.151)$$

$$\frac{\mathrm{e}^{-jkr_2}}{4\pi r_2} \approx \frac{\mathrm{e}^{-jkr}}{4\pi r} - \boldsymbol{\varepsilon}.\nabla_y\left(\frac{\mathrm{e}^{-jkr}}{4\pi r}\right). \qquad (1.152)$$

Since $r = |\boldsymbol{x} - \boldsymbol{y}|$ then the expression for the pressure field can be written in the form

$$p = j\omega\rho_0 q\boldsymbol{d}.\nabla_y\left(\frac{\mathrm{e}^{-jkr}}{4\pi r}\right). \qquad (1.153)$$

The product $q\boldsymbol{d}$ is now the vector dipole strength which, as we shall see later, can be directly related to the force vector applied to the fluid. Finally, it is sometimes useful to note that

$$\nabla_y\left(\frac{\mathrm{e}^{-jkr}}{4\pi r}\right) = \nabla_y r \frac{\partial}{\partial r}\left(\frac{\mathrm{e}^{-jkr}}{4\pi r}\right) \qquad (1.154)$$

and also that

$$\nabla_y r = -\nabla_x r = -\frac{1}{r}(\boldsymbol{x} - \boldsymbol{y}). \qquad (1.155)$$

Since $\boldsymbol{n}_\mathrm{r} = (\boldsymbol{x} - \boldsymbol{y})/r$ is the unit vector pointing in the direction of the dipole axis, then the expression for the pressure field can be written as

$$p = -j\omega\rho_0\left[\boldsymbol{d}.\boldsymbol{n}_\mathrm{r}\frac{\partial}{\partial r}\left(\frac{\mathrm{e}^{-jkr}}{4\pi r}\right)\right]. \qquad (1.156)$$

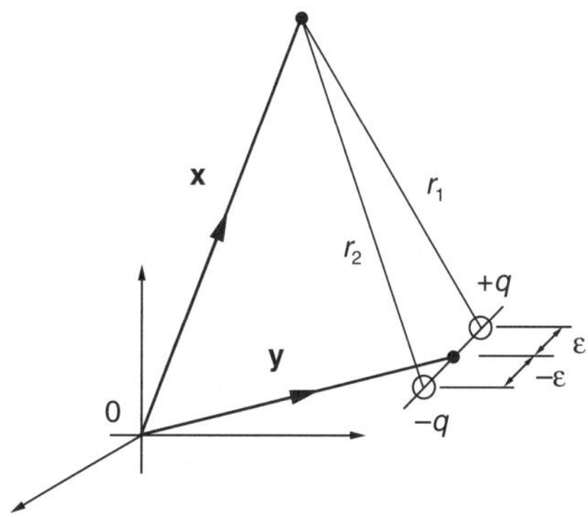

Figure 1.12 Co-ordinate system for the analysis of the dipole field when the component monopole sources have arbitrary orientation.

1.18 POINT QUADRUPOLE SOURCES

A further sound field of particular interest is that produced by the source arrangement depicted in Fig. 1.13(a), where we have a source of strength $-2q$ at the position $(y_1, 0, 0)$ and further sources of strength $+q$ at positions $(y_1 + \varepsilon_1, 0, 0)$ and $(y_1 - \varepsilon_1, 0, 0)$. The pressure field radiated is

$$p = j\omega\rho_0 q\left[\frac{e^{-jkr_1}}{4\pi r_1} + \frac{e^{-jkr_2}}{4\pi r_2} - \frac{2e^{-jkr}}{4\pi r}\right]. \tag{1.157}$$

If we use the series expansions given by equation (1.141) and (1.142) then the terms $O(\varepsilon_1)$ add to zero and only the terms $O(\varepsilon_1^2)$ remain. The expression for the pressure field that results is thus given by

$$p = j\omega\rho_0 qd^2\frac{\partial^2}{\partial y_1^2}\left[\frac{e^{-jkr}}{4\pi r}\right] \tag{1.158}$$

where we have assumed that the distance between the component sources is again given by $d = \varepsilon_1$. This sound field is that produced by a *longitudinal quadrupole* source. The product qd^2 is known as the strength of the quadrupole. An explicit expression for the sound field radiated can be derived by evaluating the double derivative with respect to y_1 in equation (1.158). After some algebraic effort, the expression that results is given by

$$p = -\frac{j\omega\rho_0 qe^{-jkr}}{4\pi r}(kd)^2\left[\cos^2\theta\left\{1 + \frac{3}{jkr} + \frac{3}{(jkr)^2}\right\} - \frac{1}{jkr} - \frac{1}{(jkr)^2}\right]. \tag{1.159}$$

Note the form of this expression in the far field. Thus, when $kr \to \infty$, the expression takes the form $p = -p_M(kd\cos\theta)^2$, where p_M is again the pressure radiated by a point monopole source of strength q. Thus the amplitude of the pressure is now a factor $(kd)^2$ less than that of a monopole, and we now have a directivity which is dependent on $\cos^2\theta$. In addition, as kr becomes small, then terms $O(1/kr)^2$ dominate the near field of the quadrupole. The directivity of the longitudinal quadrupole source is sketched in Fig. 1.14.

Another form of quadrupole field is that produced by the source arrangement depicted in Fig. 1.13(b). This type of arrangement is known as a *lateral quadrupole* and has the pressure field given by

$$p = j\omega\rho_0 q\left[\frac{e^{-jkr_1}}{4\pi r_1} - \frac{e^{-jkr_2}}{4\pi r_2} - \frac{e^{-jkr_3}}{4\pi r_3} + \frac{e^{-jkr_4}}{4\pi r_4}\right]. \tag{1.160}$$

If, as illustrated in Fig. 1.13(b), we regard the centre of the quadrupole source array to have coordinates $(y_1, y_2, 0)$, then the expression for the radial distance r_1, for example, can be written as

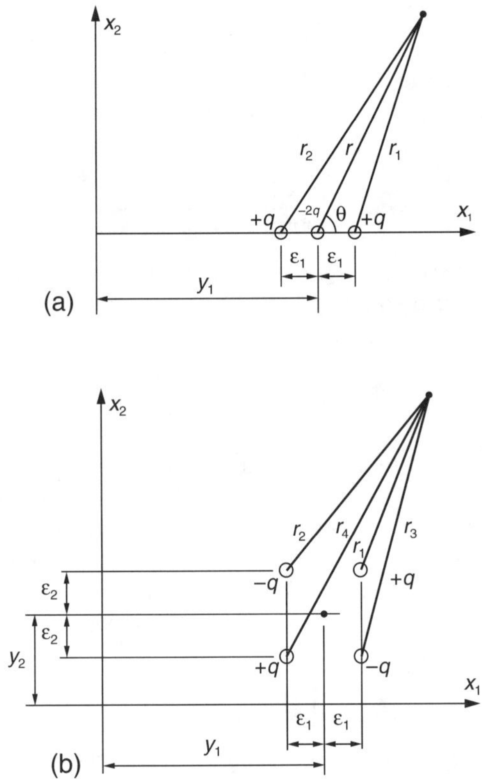

Figure 1.13 Co-ordinate systems used for the analysis of (a) a longitudinal quadrupole and (b) a lateral quadrupole.

$$r_1 = \sqrt{[(x_1 - (y_1 + \varepsilon_1))^2 + (x_2 - (y_2 + \varepsilon_2))^2 + x_3^2]}. \tag{1.161}$$

Similar expressions can be written for the other distances r_2, r_3 and r_4. Use of the Taylor series expansion specified in equation (1.149) then shows that we can write

$$\frac{e^{-jkr_1}}{4\pi r_1} \approx \frac{e^{-jkr}}{4\pi r} + \left(\varepsilon_1 \frac{\partial}{\partial y_1} + \varepsilon_2 \frac{\partial}{\partial y_2}\right) \frac{e^{-jkr}}{4\pi r} + \frac{1}{2!}\left(\varepsilon_1 \frac{\partial}{\partial y_1} + \varepsilon_2 \frac{\partial}{\partial y_2}\right)^2 \frac{e^{-jkr}}{4\pi r}. \tag{1.162}$$

Similar series expansions can be derived for the other terms in the square brackets of equation (1.160). The summation of these series expansions again shows that terms $O(\varepsilon)$ vanish and the expression for the pressure field reduces to

$$p = j\omega\rho_0 q d^2 \frac{\partial^2}{\partial y_1 \partial y_2}\left[\frac{e^{-jkr}}{4\pi r}\right] \tag{1.163}$$

where the distances between the sources in this case has been assumed to be such that $d = 2\varepsilon_1 = 2\varepsilon_2$. Evaluation of the partial derivatives with respect to y_1 and y_2 shows, again after some effort, that the pressure field can be written as

$$p = -\frac{j\omega\rho_0 q e^{-jkr}}{4\pi r}(kd)^2\left[\frac{(x_1 - y_1)(x_2 - y_2)}{r^2}\left(1 + \frac{3}{jkr} + \frac{3}{(jkr)^2}\right)\right]. \tag{1.164}$$

Note again that the far field pressure due to this source array is a factor $(kd)^2$ less than that due to a monopole source of strength q. The directivity of the far field radiation from a lateral quadrupole of this type is shown in Fig. 1.15.

In addition to the particular quadrupole arrangements considered here, one can also conceive of longitudinal quadrupoles which are oriented with their axes parallel with the y_2 and y_3 axes. Thus, there are three types of longitudinal quadrupole, each of which have the pressure field specified by equation (1.158) but containing one of the differential operators $\partial^2/\partial y_1^2$, $\partial^2/\partial y_2^2$, $\partial^2/\partial y_3^2$. Similarly in addition to the particular lateral

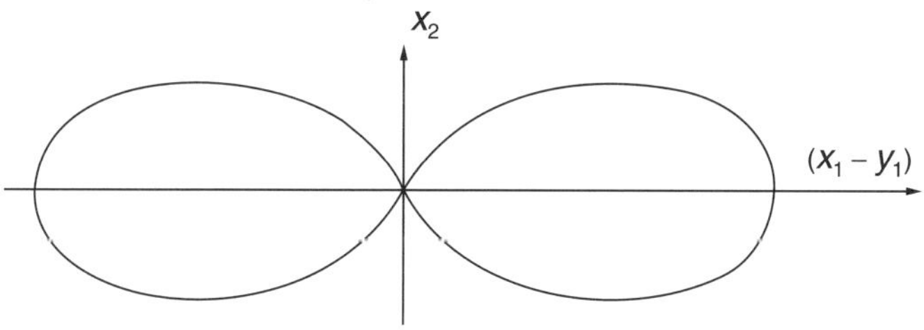

Figure 1.14 Directivity of a longitudinal quadrupole showing the angular variation of the far field pressure amplitude.

quadrupole arrangement considered above that consists of a distribution of monopole sources in the y_1, y_2 plane, one can equally conceive of identical source arrangements but in the y_1, y_3 plane, and the y_2, y_3 plane. Thus, there are in all three types of lateral quadrupoles corresponding to the differential operators, given by $\partial^2/\partial y_1 \partial y_2$, $\partial^2/\partial y_1 \partial y_3$, and $\partial^2/\partial y_2 \partial y_3$. (Note that these three operators are also equivalent to the operators $\partial^2/\partial y_2 \partial y_1$, $\partial^2/\partial y_3 \partial y_1$ and $\partial^2/\partial y_3 \partial y_2$.)

Finally it is interesting to note that the combination of three longitudinal quadrupoles having the same source strength produces a monopole-like radiation field. Thus, if we combine the three fields due to the three component longitudinal quadrupoles, the expression for the radiated pressure can be written as

$$p_Q = d^2 \left(\frac{\partial^2}{\partial y_1^2} + \frac{\partial^2}{\partial y_2^2} + \frac{\partial^2}{\partial y_3^2} \right) \frac{j\omega\rho_0 q e^{-jkr}}{4\pi r}. \tag{1.165}$$

This can be written more compactly as

$$p_Q = d^2 \nabla_y^2 p_M \tag{1.166}$$

where p_M is again the pressure field radiated by a single point monopole. We know, however, from the wave equation that $(\nabla_y^2 + k^2)p_M = 0$ and therefore it follows that

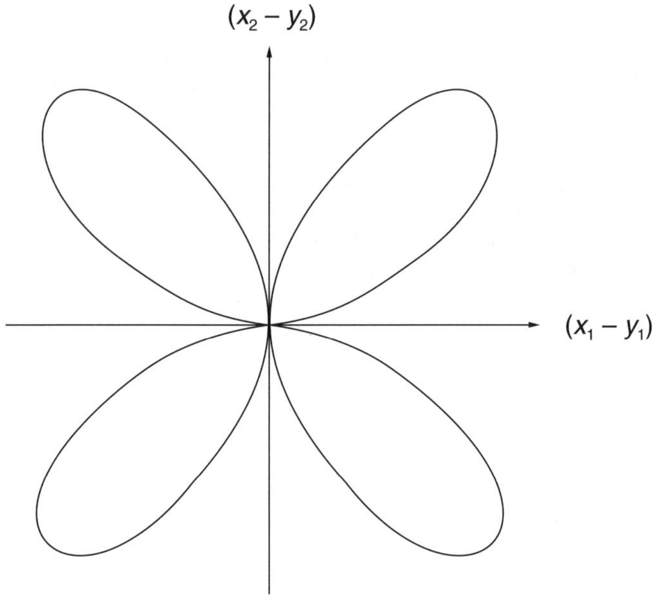

Figure 1.15 Directivity of a lateral quadrupole showing the angular variation of the far field pressure amplitude.

$$p_Q = -(kd)^2 p_M. \tag{1.167}$$

Thus the field of the three longitudinal quadrupoles is identical to that of a single point monopole source but reduced in amplitude by the factor $(kd)^2$.

1.19 MULTIPOLE ANALYSIS

If we have a collection of point monopole sources clustered around a given position specified by the vector $\boldsymbol{y}$ and the region occupied by the sources is small compared to the acoustic wavelength, we can again use a series expansion technique to express the net field of the source distribution in terms of an equivalent series of multipoles located at the position $\boldsymbol{y}$. The net field due to the combination of monopoles each having strength q_n is given by

$$p(r) = \sum_{n=1}^{N} \frac{j\omega\rho_0 q_n \mathrm{e}^{-jkr_n}}{4\pi r_n}. \tag{1.168}$$

The pressure field due to the n-th monopole can be expanded as

$$\frac{\mathrm{e}^{-jkr_n}}{4\pi r_n} = \frac{\mathrm{e}^{-jkr}}{4\pi r} + \boldsymbol{d}_n.\nabla_y\left(\frac{\mathrm{e}^{-jkr}}{4\pi r}\right) + \frac{1}{2!}(\boldsymbol{d}_n.\nabla_y)^2\left(\frac{\mathrm{e}^{-jkr}}{4\pi r}\right) + \cdots \tag{1.169}$$

where $\boldsymbol{d}_n$ is the position vector of the n-th monopole relative to the point specified by $\boldsymbol{y}$. This is illustrated in Fig. 1.16. Thus the net field can be expressed as

$$p(\bar{r}) = \left[\sum_{n=1}^{N} q_n + \left(\sum_{n=1}^{N} q_n \boldsymbol{d}_n\right)\nabla_y + \frac{1}{2}\sum_{n=1}^{N} q_n(\boldsymbol{d}_n.\nabla_y)^2 \cdots\right]\frac{j\omega\rho_0 \mathrm{e}^{-jkr}}{4\pi r}. \tag{1.170}$$

Following the nomenclature adopted by Pierce (1981), we can now define the monopole and dipole strengths respectively of the source distribution as

$$Q = \sum_{n=1}^{N} q_n \quad \boldsymbol{D} = \sum_{n=1}^{N} q_n \boldsymbol{d}_n. \tag{1.171}$$

Note that the third term in the series results from the operator

$$(\boldsymbol{d}_n.\nabla_y)^2 = \left(d_{n1}\frac{\partial}{\partial y_1} + d_{n2}\frac{\partial}{\partial y_2} + d_{n3}\frac{\partial}{\partial y_3}\right)\left(d_{n1}\frac{\partial}{\partial y_1} + d_{n2}\frac{\partial}{\partial y_2} + d_{n3}\frac{\partial}{\partial y_3}\right) \tag{1.172}$$

and that there are nine possible combinations of differential operator which result from this product. Six of these terms express the fields of longitudinal quadrupoles (i.e. the operators $\partial^2/\partial y_1\partial y_2 = \partial^2/\partial y_2\partial y_1$, $\partial^2/\partial y_1\partial y_3 = \partial^2/\partial y_3\partial y_1$ and $\partial^2/\partial y_2\partial y_3 = \partial^2/\partial y_3\partial y_2$) while the other three terms are associated with longitudinal quadrupoles (i.e. the operators

$\partial^2/\partial y_1^2$, $\partial^2/\partial y_2^2$, $\partial^2/\partial y_3^2$). Thus, if we define the quadrupole strengths of the source distribution as

$$Q_{\mu\nu} = \frac{1}{2}\sum_{n=1}^{N} d_{n\mu} d_{n\nu} q_n \tag{1.173}$$

where we have again adopted the notation of Pierce (1981), the net field can be written in the form

$$p(r) = \left[Q + \boldsymbol{D}.\nabla_y + Q_{\mu\nu} \sum_{\mu=1,\nu=1}^{\mu=3,\nu=3} \frac{\partial^2}{\partial y_\mu \partial y_\nu} \cdots \right] \frac{j\omega\rho_0 \mathrm{e}^{-jkr}}{4\pi r}. \tag{1.174}$$

It is also useful to note that when applied to a function $f(r)$ then $\nabla(f(r)) = -\nabla_y(f(r))$ and the form of the radiated field can be expressed as

$$p(r) = \left[Q - \boldsymbol{D}.\nabla + Q_{\mu\nu} \sum_{\mu=1,\ \nu=1}^{\mu=3,\ \nu=3} \frac{\partial^2}{\partial x_\mu \partial x_\nu} \cdots \right] \frac{j\omega\rho_0 \mathrm{e}^{-jkr}}{4\pi r}. \tag{1.175}$$

Finally, we should point out that the technique of multipole analysis can be extended to deal with more complex compact regions having an arbitrary distribution of source strength, and the formulation presented here can be made more general (see, for example, the work of Doak, 1973).

1.20 THE INHOMOGENEOUS WAVE EQUATION

Thus far in our treatment of free field acoustic radiation we have dealt only with point monopole sources and the particular combinations of point

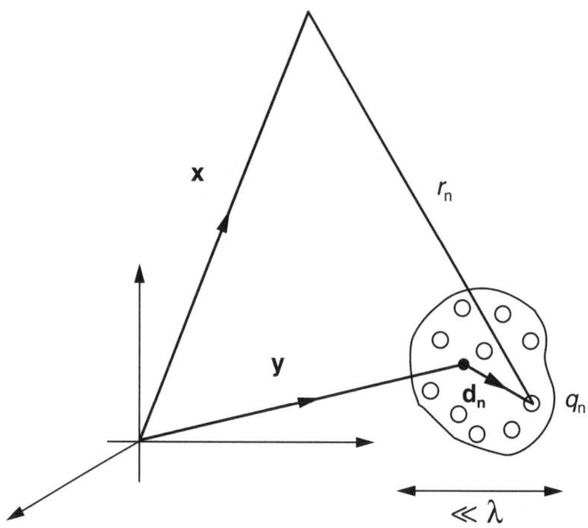

Figure 1.16 Co-ordinate system used to express the field at $\boldsymbol{x}$ due to a number of multipoles at $\boldsymbol{y}$.

monopole sources which reduce to higher-order multipole sources. Now we will discuss continuous distributions of source strength. A source distribution of monopole type can be visualized by dividing a region of the medium into a number of tiny three-dimensional elements and assuming that each element contains an inhomogeneity (such as a sphere or a bubble) which pulsates and thereby introduces a fluctuating volume flow into the element. The magnitude of the additional volume flow will vary from element to element. This fluctuating volume flow *per unit volume* is specified by $q_{\text{vol}}(\boldsymbol{x}, t)$ and in the limit (when the elements in the medium are infinitesimally small) this fluctuating rate of addition of volume flow is a continuous function of position, $\boldsymbol{x}$, and time, t. Under these circumstances the mass conservation equation dealt with in section 1.4 will be modified by the addition of a further term which quantifies this local rate of addition of volume. The mass conservation equation can be written to first order in the fluctuating quantities as

$$\frac{\partial \rho(\boldsymbol{x}, t)}{\partial t} + \rho_0 \nabla . \boldsymbol{u}(\boldsymbol{x}, t) = \rho_0 q_{\text{vol}}(\boldsymbol{x}, t). \tag{1.176}$$

In a similar way we can conceive of a distribution of fluctuating force per unit volume acting on the medium, specified by the vector $\boldsymbol{f}_{\text{vol}}(\boldsymbol{x}, t)$. This can be visualized, for example, by assuming that the fluid elements referred to above contain tiny spheres which oscillate to and fro and thereby apply a net fluctuating force to each element. The linearized momentum conservation equation under these circumstances reduces to

$$\rho_0 \frac{\partial \boldsymbol{u}(\boldsymbol{x}, t)}{\partial t} + \nabla_{\text{p}}(\boldsymbol{x}, t) = \boldsymbol{f}_{\text{vol}}(\boldsymbol{x}, t). \tag{1.177}$$

If we use the adiabatic relationship between the pressure and density fluctuations given by $p(\boldsymbol{x}, t) = c_0^2 \rho(\boldsymbol{x}, t)$ and we operate on the momentum conservation equation with the divergence operator ($\nabla.$) then subtract the time derivative of the mass conservation equation, then we can derive an inhomogeneous wave equation given by

$$\left(\nabla^2 - \frac{1}{c_0^2} \frac{\partial}{\partial t^2} \right) p(\boldsymbol{x}, t) = \nabla . \boldsymbol{f}_{\text{vol}}(\boldsymbol{x}, t) - \frac{\rho_0 \partial q_{\text{vol}}(\boldsymbol{x}, t)}{\partial t}. \tag{1.178}$$

The terms on the right-hand side of this equation can be interpreted as sources of sound (see, for example, the discussion by Dowling and Ffowcs Williams, 1983). In the particular case of harmonic fluctuations in the field quantities, then this equation reduces to the inhomogeneous Helmholtz equation given by

$$(\nabla^2 + k^2) p(\boldsymbol{x}) = \nabla . \boldsymbol{f}_{\text{vol}}(\boldsymbol{x}) - j\omega \rho_0 q_{\text{vol}}(\boldsymbol{x}). \tag{1.179}$$

Now $f_{\text{vol}}(\boldsymbol{x})$ and $q_{\text{vol}}(\boldsymbol{x})$ are interpreted as the distribution of complex force per unit volume and complex volume velocity per unit volume applied to the medium. We shall demonstrate that these source terms may be respectively identified as dipole and monopole source distributions. First, however, we will discuss briefly the well-established techniques for solving the inhomogeneous scalar Helmholtz equation.

1.21 THE GREEN FUNCTION AND THE PRINCIPLE OF RECIPROCITY

The solution of equation (1.179) is most readily dealt with using a Green function. In general, the equation that we wish to solve is of the form

$$(\nabla^2 + k^2)p(\boldsymbol{x}) = -Q_{\mathrm{vol}}(\boldsymbol{x}) \tag{1.80}$$

where Q_{vol} represents the source distribution, which in terms of the analysis presented in the last section, may be of monopole or dipole type. We should also note in passing that in analysing the problem of sound generated aerodynamically, Lighthill (1952) showed that the relevant mass and momentum conservation equations may be combined to give an inhomogeneous wave equation with a right-hand side which could be interpreted as a *quadrupole* type source. Irrespective of the nature of the source distribution, the solution of equation (1.176) can be constructed from the pressure field produced by a monopole source at a point. This pressure field is known as the Green function which is denoted by $G(\boldsymbol{x}|\boldsymbol{y})$ and which satisfies the equation

$$(\nabla^2 + k^2)G(\boldsymbol{x}|\boldsymbol{y}) = -\delta(\boldsymbol{x} - \boldsymbol{y}). \tag{1.181}$$

Thus, physically speaking, the Green function is simply proportional to the complex pressure produced at a position $\boldsymbol{x}$ due to a harmonic point monopole source of unit strength at $\boldsymbol{y}$. The presence of a point monopole at $\boldsymbol{y}$ is represented by the three-dimensional Dirac delta function

$$\delta(\boldsymbol{x} - \boldsymbol{y}) = \delta(x_1 - y_1)\delta(x_2 - y_2)\delta(x_3 - y_3).$$

This function has properties which are analogous to the one-dimensional Dirac delta function $\delta(x - y)$. Most importantly, if we have some continuous function $f(\boldsymbol{x})$ in three-dimensional space, then the three-dimensional delta function exhibits the 'sifting' property

$$\int_V f(\boldsymbol{x})\delta(\boldsymbol{x} - \boldsymbol{y})\mathrm{d}V = \begin{cases} f(\boldsymbol{y}) & \boldsymbol{y} \text{ within } V \\ f(\boldsymbol{y})/2 & \boldsymbol{y} \text{ on } S \\ 0 & \boldsymbol{y} \text{ outside } V \end{cases} \tag{1.182}$$

where V is a volume in the medium bounded by the surface S. Note that we can interpret this volume and its bounding surface in the alternative ways depicted in Figs 1.17(a) and 1.17(b). Thus the volume V could simply be enclosed by the surface S as illustrated in Fig. 1.17(a). In the case illustrated in Fig. 1.17(b), the outer part of the bounding surface S may be considered to lie at infinity. The volume V is then of infinite extent but excludes the region bounded by the inner part of the surface S.

Before using the Green function to derive the solution of the inhomogeneous scalar wave equation, it is useful to establish formally that the Green function satisfies the principle of reciprocity. That is, the complex pressure at $\mathbf{y}_1$ (say) that is produced by a point monopole source at $\mathbf{y}_2$ is identical to the complex pressure at $\mathbf{y}_2$ that is produced by a point monopole source at $\mathbf{y}_1$. The two pressure fields due to these sources may be written as $G(\mathbf{x}|\mathbf{y}_1)$ and $G(\mathbf{x}|\mathbf{y}_2)$ respectively and they must satisfy

$$(\nabla^2 + k^2)G(\mathbf{x}|\mathbf{y}_1) = -\delta(\mathbf{x} - \mathbf{y}_1) \tag{1.183}$$

$$(\nabla^2 + k^2)G(\mathbf{x}|\mathbf{y}_2) = -\delta(\mathbf{x} - \mathbf{y}_2). \tag{1.184}$$

If we multiply the first of these equations by $G(\mathbf{x}|\mathbf{y}_2)$, and subtract the second equation multiplied by $G(\mathbf{x}|\mathbf{y}_1)$, then it follows that

$$\begin{aligned} &G(\mathbf{x}|\mathbf{y}_2)\nabla^2 G(\mathbf{x}|\mathbf{y}_1) - G(\mathbf{x}|\mathbf{y}_1)\nabla^2 G(\mathbf{x}|\mathbf{y}_2) \\ &\quad = G(\mathbf{x}|\mathbf{y}_1)\delta(\mathbf{x} - \mathbf{y}_2) - G(\mathbf{x}|\mathbf{y}_2)\delta(\mathbf{x} - \mathbf{y}_1). \end{aligned} \tag{1.185}$$

Now we integrate both sides of equation (1.185) over a volume V that contains the point sources at $\mathbf{y}_1$ and $\mathbf{y}_2$. If S is the surface that encloses the volume V, then Green's theorem states that

$$\int_V (a\nabla^2 b - b\nabla^2 a)\mathrm{d}V = \int_S (a\nabla b - b\nabla a).\mathbf{n}\ \mathrm{d}S \tag{1.186}$$

where a and b are scalar functions and $\mathbf{n}$ is the unit vector normal to the surface S pointing *outwards* from the volume V (see Fig. 1.17). Integrating equation (1.185) over the volume V thus shows that

$$\begin{aligned} &\int_S [G(\mathbf{x}|\mathbf{y}_2)\nabla G(\mathbf{x}|\mathbf{y}_1) - G(\mathbf{x}|\mathbf{y}_1)\nabla G(\mathbf{x}|\mathbf{y}_2)].\mathbf{n}\ \mathrm{d}S \\ &\qquad = G(\mathbf{y}_2|\mathbf{y}_1) - G(\mathbf{y}_1|\mathbf{y}_2) \end{aligned} \tag{1.187}$$

where the property of the delta function (1.182) has been used in integrating the right-hand side of equation (1.185). We can now show that provided certain boundary conditions are satisfied on S, then the left-hand side of equation (1.187) is zero. For example, assume that the pressure and the normal component of particle velocity on S are linearly related by some specific impedance $z(\mathbf{x})$. Remembering that $G(\mathbf{x}|\mathbf{y}_1)$ and $G(\mathbf{x}|\mathbf{y}_2)$ are actually proportional to complex pressures then it follows from the linearized momentum conservation equation that on the surface S

$$\nabla G(\mathbf{x}|\mathbf{y}_1) = -\frac{j\omega\rho_0 G(\mathbf{x}|\mathbf{y}_1)\mathbf{n}}{z(\mathbf{x})}, \quad \nabla G(\mathbf{x}|\mathbf{y}_2) = -\frac{j\omega\rho_0 G(\mathbf{x}|\mathbf{y}_2)\mathbf{n}}{z(\mathbf{x})} \tag{1.188a,b}$$

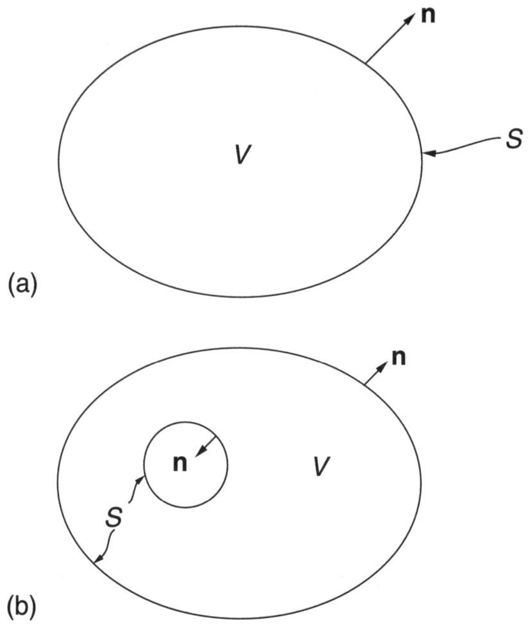

Figure 1.17 (a) The volume V in the medium and its bounding surface S. (b) An alternative interpretation where the outer part of the bounding surface S can be considered to be at infinity.

Substitution of these expressions into equation (1.187) shows that the left-hand side vanishes and thus

$$G(y_2|y_1) = G(y_1|y_2). \tag{1.189}$$

The principle of reciprocity therefore holds for two points inside a volume V in which the Helmholtz equation is satisfied and on whose bounding surface S there is some prescribed acoustic impedance. Conditions of zero normal particle velocity (a rigid surface) or zero pressure (a pressure release surface) are clearly special cases corresponding to an infinite or a zero impedance respectively. Note also that in the case of either a part or a whole of the bounding surface being at infinity (as illustrated in Fig. 1.17(b)), then the relevant impedance is the characteristic acoustic impedance of the medium. This is due to the fact that as the distance from the sources of sound approaches infinity then the waves which they emit will become increasingly like plane waves. This amounts to the application of the Sommerfeld radiation condition, which requires that, as the radial distance r from a harmonic source of sound approaches infinity,

$$\lim_{r\to\infty}[r(p(\boldsymbol{x}) - \rho_0 c_0 u_r(\boldsymbol{x}))] = 0 \tag{1.190}$$

where u_r is the radial component of the particle velocity field. Note also that this condition requires that as $r \to \infty$, then only *outgoing* waves can exist. (See Pierce (1981) or Dowling and Ffowcs Williams (1983) for fuller discussions of both the reciprocity principle and the Sommerfeld radiation condition.)

1.22 THE SOLUTION OF THE INHOMOGENEOUS WAVE EQUATION

Having established the principle of reciprocity, we can now use the Green function to derive the solution to the inhomogeneous Helmholtz equation in the form of an integral equation for the acoustic pressure. In general terms we wish to solve an equation of the form

$$\nabla_y^2 p(\boldsymbol{y}) + k^2 p(\boldsymbol{y}) = -Q_{\mathrm{vol}}(\boldsymbol{y}) \tag{1.191}$$

where, for example, the source strength distribution could consist of a combination of monopole and dipole type sources where $Q_{\mathrm{vol}}(\boldsymbol{y}) = j\omega\rho_0 q_{\mathrm{vol}}(\boldsymbol{y}) - \nabla_y . \boldsymbol{f}_{\mathrm{vol}}(\boldsymbol{y})$. Note that we have used the coordinate $\boldsymbol{y}$ instead of the coordinate $\boldsymbol{x}$ in the wave equation. This, of course, is quite permissible and has been done deliberately to enable the solution to be easily derived. We shall now use the Green function $G(\boldsymbol{y}|\boldsymbol{x})$ which satisfies

$$\nabla_y^2 G(\boldsymbol{y}|\boldsymbol{x}) + k^2 G(\boldsymbol{y}|\boldsymbol{x}) = -\delta(\boldsymbol{y} - \boldsymbol{x}) \tag{1.192}$$

to help us find the solution. Thus $G(\boldsymbol{y}|\boldsymbol{x})$ is the complex pressure produced at $\boldsymbol{y}$ due to a harmonic point source at $\boldsymbol{x}$ and as we have seen, an important property of the Green function is that it satisfies the principle of reciprocity, and therefore that $G(\boldsymbol{y}|\boldsymbol{x}) = G(\boldsymbol{x}|\boldsymbol{y})$. In order to derive the solution to equation (1.191) we firstly multiply the equation by $G(\boldsymbol{y}|\boldsymbol{x})$ and subtract equation (1.192) multiplied by $p(\boldsymbol{y})$. This yields

$$G(\boldsymbol{y}|\boldsymbol{x})\nabla_y^2 p(\boldsymbol{y}) - p(\boldsymbol{y})\nabla_y^2 G(\boldsymbol{y}|\boldsymbol{x}) = -Q_{\mathrm{vol}}(\boldsymbol{y})G(\boldsymbol{y}|\boldsymbol{x}) + p(\boldsymbol{y})\delta(\boldsymbol{y} - \boldsymbol{x}). \tag{1.193}$$

We now integrate both sides of this equation over a volume V that is bounded by the surface S. Using the properties of the delta function specified by equation (1.193) shows that

$$\int_V G(\boldsymbol{y}|\boldsymbol{x})\nabla_y^2 p(\boldsymbol{y}) - p(\boldsymbol{y})\nabla_y^2 G(\boldsymbol{y}|\boldsymbol{x})\mathrm{d}V + \int_V Q_{\mathrm{vol}}(\boldsymbol{y})G(\boldsymbol{y}|\boldsymbol{x})\mathrm{d}V = \begin{cases} p(\boldsymbol{x}) & \boldsymbol{x} \text{ within } V \\ p(\boldsymbol{x})/2 & \boldsymbol{x} \text{ on } S \\ 0 & \boldsymbol{x} \text{ outside } V. \end{cases} \tag{1.194}$$

We now use Green's theorem as specified by equation (1.186). This enables the first volume integral on the left-hand side of equation (1.194) to be transformed into an integral over the surface S. Also using the property $G(\boldsymbol{y}|\boldsymbol{x}) = G(\boldsymbol{x}|\boldsymbol{y})$ then enables equation (1.194) for points $\boldsymbol{x}$ inside V to be written as

$$p(\boldsymbol{x}) = \int_V Q_{\text{vol}}(\boldsymbol{y})G(\boldsymbol{x}|\boldsymbol{y})\mathrm{d}V + \int_S [G(\boldsymbol{x}|\boldsymbol{y})\nabla_y p(\boldsymbol{y}) - p(\boldsymbol{y})\nabla_y G(\boldsymbol{x}|\boldsymbol{y})].\boldsymbol{n}\ \mathrm{d}S. \tag{1.195}$$

Thus, given a distribution of source strength $Q_{\text{vol}}(\boldsymbol{y})$ in the volume V, we can compute the contribution to the complex pressure $p(\boldsymbol{x})$ produced by this source distribution by performing the volume integral on the right-hand side of equation (1.195). However, there will also be a contribution from the surface integral term in equation (1.195). The solution of this equation thus requires a specification of the pressure $p(\boldsymbol{y})$ and pressure gradient $\nabla_y p(\boldsymbol{y})$ on the surface S and a knowledge of the Green function $G(\boldsymbol{x}|\boldsymbol{y})$. Note that there is (in principle) considerable latitude in the choice of Green function that we use to solve a given problem. The only assumptions that we have made is that $G(\boldsymbol{x}|\boldsymbol{y})$ satisfies equation (1.192) and that the principle of reciprocity holds. In certain instances the Green function can be chosen (for example) to satisfy $\nabla_y G(\boldsymbol{x}|\boldsymbol{y}) = 0$, in which case we only need specify $\nabla_y p(\boldsymbol{y})$ in order to compute the resulting pressure field.

1.23 THE SOLUTION OF THE INHOMOGENEOUS WAVE EQUATION IN AN UNBOUNDED MEDIUM

If the medium is unbounded, we can apply the Sommerfeld radiation condition discussed in section 1.21 to show that the surface integral is zero on the right-hand side of equation (1.195). Thus the bounding surface S is at infinity and since we assume that the field there will consist of only *outgoing* plane waves, then the two terms in the surface integral will add to zero at all points on the surface S. The solution of the inhomogeneous wave equation thus becomes

$$p(\boldsymbol{x}) = \int_V Q_{\text{vol}}(\boldsymbol{y})G(\boldsymbol{x}|\boldsymbol{y})\mathrm{d}V. \tag{1.196}$$

In addition, in an unbounded medium, the Green function $G(\boldsymbol{x}|\boldsymbol{y})$ takes a particularly simple form; it is known as the 'free space' Green function $g(\boldsymbol{x}|\boldsymbol{y})$ where

$$G(\boldsymbol{x}|\boldsymbol{y}) = g(\boldsymbol{x}|\boldsymbol{y}) = \frac{\mathrm{e}^{-jk|\boldsymbol{x}-\boldsymbol{y}|}}{4\pi|\boldsymbol{x}-\boldsymbol{y}|}. \tag{1.197}$$

The proof that this represents a solution of equation (1.181) is presented in outline by Morse and Ingard (1968). Now consider the case where the 'source' term on the right-hand side of the inhomogeneous scalar Helmholtz equation (1.179) is due to a distribution of volume sources such that $Q_{\mathrm{vol}}(\boldsymbol{y}) = j\omega\rho_0 q_{\mathrm{vol}}(\boldsymbol{y})$. Then the solution of the equation becomes

$$p(\boldsymbol{x}) = \int_V j\omega\rho_0 q_{\mathrm{vol}}(\boldsymbol{y}) g(\boldsymbol{x}|\boldsymbol{y})\mathrm{d}V. \tag{1.198}$$

The physical interpretation of this equation should now become clear; the volume integral simply represents the superposition of all the contributions of the elemental monopole sources of strength $q_{\mathrm{vol}}(\boldsymbol{y})$ per unit volume. In the particular case of a point monopole source at $\boldsymbol{y}_1$ where $q_{\mathrm{vol}}(\boldsymbol{y}) = q\delta(\boldsymbol{y} - \boldsymbol{y}_1)$, performing the volume integral yields

$$p(\boldsymbol{x}) = \frac{j\omega\rho_0 q \mathrm{e}^{-jk|\boldsymbol{x}-\boldsymbol{y}_1|}}{4\pi|\boldsymbol{x} - \boldsymbol{y}_1|} \tag{1.199}$$

which is the familiar expression for the complex pressure produced by a point monopole source.

In the case of a distribution of *force* per unit volume applied to the medium, then $Q_{\mathrm{vol}}(\boldsymbol{y}) = -\nabla_y.\boldsymbol{f}_{\mathrm{vol}}(\boldsymbol{y})$ and the solution can be written as

$$p(\boldsymbol{x}) = -\int_V \nabla_y.\boldsymbol{f}_{\mathrm{vol}}(\boldsymbol{y}) g(\boldsymbol{x}|\boldsymbol{y})\mathrm{d}V. \tag{1.200}$$

We can use the vector identity $a\nabla_y.\boldsymbol{b} = \nabla_y.(a\boldsymbol{b}) - \nabla_y.a\boldsymbol{b}$ to rewrite this expression as

$$p(\boldsymbol{x}) = \int_V \boldsymbol{f}_{\mathrm{vol}}(\boldsymbol{y}).\nabla_y g(\boldsymbol{x}|\boldsymbol{y}) - \nabla_y.(g(\boldsymbol{x}|\boldsymbol{y})\boldsymbol{f}_{\mathrm{vol}}(\boldsymbol{y}))\mathrm{d}V. \tag{1.201}$$

The divergence theorem can now be used to show that the contribution of the second term in the volume integral vanishes. Thus since

$$\int_V \nabla_y.(g(\boldsymbol{x}|\boldsymbol{y})\boldsymbol{f}_{\mathrm{vol}}(\boldsymbol{y}))\mathrm{d}V = \int_S g(\boldsymbol{x}|\boldsymbol{y})\boldsymbol{f}_{\mathrm{vol}}(\boldsymbol{y}).\boldsymbol{n}\mathrm{d}S \tag{1.202}$$

and assuming that $\boldsymbol{f}_{\mathrm{vol}}(\boldsymbol{y})$ is non-zero over a restricted region of the medium, then if we choose the bounding surface S to be just outside this restricted region, it follows that the surface integral will be zero. Thus the solution for the complex pressure can be written as

$$p(\boldsymbol{x}) = \int_V \boldsymbol{f}_{\mathrm{vol}}(\boldsymbol{y}).\nabla_y g(\boldsymbol{x}|\boldsymbol{y})\mathrm{d}V. \tag{1.203}$$

We can now make explicit that the distribution of force per unit volume acts as a distribution of *dipole* sources. Consider the particular case of a point force applied to the medium at $\boldsymbol{y}_1$ where $\boldsymbol{f}_{\mathrm{vol}}(\boldsymbol{y}) = \boldsymbol{f}\delta(\boldsymbol{y} - \boldsymbol{y}_1)$. In this case, performing the volume integral yields

$$p(\boldsymbol{x}) = \boldsymbol{f}.\nabla_y \left[\frac{\mathrm{e}^{-jk|\boldsymbol{x}-\boldsymbol{y}_1|}}{4\pi|\boldsymbol{x}-\boldsymbol{y}_1|} \right]. \tag{1.204}$$

Comparison of this expression with that given by equation (1.153) demonstrates that the sound field produced by two closely spaced monopoles of strength q and $-q$ and separated by a vector distance $\boldsymbol{d}$ is exactly equivalent to the sound field produced by a point force $\boldsymbol{f}$ where $\boldsymbol{f} = j\omega\rho_0 q\boldsymbol{d}$. The integral over the volume in equation (1.203) thus evaluates the superposition of the fields of a distribution of such sources.

1.24 THE KIRCHHOFF–HELMHOLTZ INTEGRAL EQUATION

If in a given volume V there is zero source strength, $Q_{\text{vol}}(\boldsymbol{y})$, and if we choose the Green function $G(\boldsymbol{x}|\boldsymbol{y})$ to be the free space Green function $g(\boldsymbol{x}|\boldsymbol{y})$, the integral form of solution given by equation (1.195) reduces to the Kirchhoff–Helmholtz integral equation (see Pierce, 1981, Ch. 4). This is given by

$$\int_S [g(\boldsymbol{x}|\boldsymbol{y})\nabla_y p(\boldsymbol{y}) - p(\boldsymbol{y})\nabla_y g(\boldsymbol{x}|\boldsymbol{y})].\boldsymbol{n}\,\mathrm{d}S = \begin{cases} p(\boldsymbol{x}) & \boldsymbol{x} \text{ within } V \\ p(\boldsymbol{x})/2 & \boldsymbol{x} \text{ on } S \\ 0 & \boldsymbol{x} \text{ outside } V. \end{cases} \tag{1.205}$$

Note that the terms appearing on the left-hand side of this equation have a well-defined physical interpretation. Since the equation of conservation of momentum for harmonic excitation shows that $\nabla_y p(\boldsymbol{y}) = -j\omega\rho_0\boldsymbol{u}(\boldsymbol{y})$, then the first term in the integral of the left-hand side of equation (1.205) can be written as

$$-\int_S g(\boldsymbol{x}|\boldsymbol{y})j\omega\rho_0\boldsymbol{u}(\boldsymbol{y}).\boldsymbol{n}\mathrm{d}S = \int_S g(\boldsymbol{x}|\boldsymbol{y})j\omega\rho_0 q_{\text{surf}}\mathrm{d}S \tag{1.206}$$

where $q_{\text{surf}} = -\boldsymbol{u}(\boldsymbol{y}).\boldsymbol{n}$ is a volume velocity per unit surface area, and therefore represents a distribution of monopole source strength. (Note that since $\boldsymbol{n}$ points out from the volume considered, then q_{surf} represents a positive rate of introduction of volume into the volume.) Similarly, the second term in the integral of equation (1.205) can be written as

$$-\int_S p(\boldsymbol{y})\nabla_y g(\boldsymbol{x}|\boldsymbol{y}).\boldsymbol{n}\mathrm{d}S = -\int_S \boldsymbol{f}_{\text{surf}}.\nabla_y g(\boldsymbol{x}|\boldsymbol{y})\mathrm{d}S \tag{1.207}$$

where $\boldsymbol{f}_{\text{surf}} = p(\boldsymbol{y})\boldsymbol{n}$ is a force per unit area acting at the surface S. It can be seen by comparison of this equation with equation (1.203) that this term has a clear interpretation as the sound field generated by a surface distribution of dipole sources. Thus, in the absence of any volume sources, the pressure at a given position for the inside volume V can be specified by the equivalent monopole and dipole source layers on the bounding surface S.

Thus we see that if we distribute elementary plane monopole and dipole sources over the bounding surface S, then provided their relative strengths are appropriately defined, we can produce the sound field $p(\boldsymbol{x})$ within the volume V whilst producing zero radiation outside the volume V. An elementary dipole source specified by the effective force per unit surface area f_{surf} will produce antisymmetric radiation (with the 'cos θ' directivity pattern described in section 1.17) while an elementary monopole source will produce omnidirectional radiation. These elementary sources can be visualized, in the dipole case, as small massless pistons which oscillate to and fro and thus produce force per unit area on the fluid. In the case of an elementary monopole source, two small massless pistons are forced to oscillate apart by the introduction and withdrawal of volume flow. In the particular case of plane wave radiation, when $p(\boldsymbol{y}) = \rho_0 c_0 \boldsymbol{u}(\boldsymbol{y}).\boldsymbol{n}$, then the resulting directivity of the combined radiation from a single elementary monopole–dipole source can be shown to take a Cardioid form. Equation (1.205) shows that the integrated effect of such elementary contributions is again to produce unidirectional radiation. A further description of this principle and details of its application to problems in both acoustic and electromagnetic radiation is given by Baker and Copson (1950) who also describe the relationship of this formulation to Huygen's Principle and the contributions of both Helmholtz and Kirchhoff to its development.

1.25 REFERENCES

Baker, B.B. and Copson, E.T. (1950) *The Mathematical Theory of Huygen's Principle*. Clarendon Press, Oxford.

Doak, P.E. (1973) Fundamentals of aerodynamic sound theory and flow duct acoustics. *Journal of Sound and Vibration* **28**, 527–61.

Dowling, A.P. and Ffowcs-Williams, J.E. (1983) *Sound and Sources of Sound*. Ellis Horwood, Chichester.

Hall, D.E. (1987) *Basic Acoustics*. John Wiley and Sons, New York.

Kinsler, L.E., Frey, A.R., Coppens, A.B. and Sanders, J.V. (1982) *Fundamentals of Acoustics*, 3rd edn, John Wiley, New York.

Lighthill, M.J. (1952) On sound generated aerodynamically: I. General theory. *Proceedings of the Royal Society* **A211**, 564–78.

Lighthill, M.J. (1978) *Waves in Fluids*. Cambridge University Press, Cambridge.

Morse, P.M. and Ingard, K.U. (1968) *Theoretical Acoustics*. McGraw-Hill, New York.

Nelson, P.A. and Elliott, S.J. (1992) *Active Control of Sound*. Academic Press, London.

Pierce, A.D. (1981) *Acoustics: An Introduction to its Physical Properties and Applications*. McGraw-Hill, New York. (also reported in 1989 by the Acoustical Society of America).

Sabine, W.C. (1898) Architectural Acoustics. *Eng. Rec.* **38**, 520–2.

Skudrzyk, E. (1971) *The Foundations of Acoustics*. Springer, New York.

Spencer, A.J.M., Parker, D.F., Berry, D.S., England, A.H., Faulkner, T.R., Green, W.A., Holden, J.T., Middleton, D. and Rogers, T.G. (1977) *Engineering Mathematics*. Van Nostrand Reinhold, New York.

1.26 QUESTIONS

1. A sliding piston in a semi-infinite tube moves with a maxium velocity of 0.1 ms^{-1} for a short duration. Given that the gas in the tube is air with $\rho_0 = 1.2\,kgm^{-3}$, $p_0 = 10^5\,N/m^2$ and $\gamma = 1.4$, calculate the maximum pressure and density changes produced by the motion of the piston.
2. A harmonic plane wave having a pressure amplitude of 5×10^{-3} Nm^{-2} travels in a fluid having an ambient density ρ_0 of 10^3 kgm^{-3} and a sound speed c_0 of 1500 ms^{-1}. Calculate the amplitude of the acoustic particle velocity fluctuation and the value of the time-averaged acoustic intensity in the direction of propagation of the wave.
3. A tube driven by a vibrating piston is terminated by a sample of absorbing material having a specific acoustic impedance which is $4\rho_0 c_0$, where ρ_0 and c_0 denote the ambient density and sound speed of the air in the tube. If the piston vibrates harmonically at a small amplitude and a frequency at which the acoustic wavelength is equal to the length of the tube, calculate the ratio of the amplitude of the pressure on the face of the piston to the amplitude of the velocity of the piston.
4. Two travelling harmonic plane waves propagate in a direction parallel to the x-axis, one in the positive x-direction and the other in the negative x-direction. The waves produce pressure fluctuations that are in phase at $x = 0$. Derive an expression for the modulus and phase of the net pressure fluctuation produced by the interference of the waves as a function of x and as a function of the ratio of the amplitudes of the positive and negative going waves. Illustrate your results graphically with a sketch of the dependence on x of the modulus and phase of the net pressure fluctuation.
5. Two uncorrelated spherically radiating acoustic sources (in air) are respectively 15 m and 5 m from a point at which the sound pressure level is 80 dB re 2×10^{-5} Nm^{-2}. Given that the sound power level of the more distant source is 120 dB re 10^{-2} W, calculate the sound power level of the nearer source.
6. A point monopole source radiating single frequency sound at 100 Hz is placed 1 m above a rigid reflecting surface. The sound pressure is measured at a distance of 10 m from the source at the same height of the source above the plane. Calculate the complex ratio of this pressure to that produced at the same position in the absence of the reflecting plane.
7. (a) A rectangular air-filled enclosure measures 5 m $\times$ 7 m $\times$ 3 m. All the surfaces are covered with plaster having an absorption coefficient of 0.07 at 500 Hz, except the floor (which measures 5 m $\times$ 7 m) which is covered with carpet having an absorption coefficient of 0.2 at 500 Hz. Calculate the average absorption coefficient at 500 Hz.
   (b) A material having an unknown absorption coefficient is used to cover one of the walls (measuring 5 m $\times$ 3 m) of the enclosure as a result of which the reverberation time in the enclosure is reduced to three-quarters of its original value. Estimate the absorption coefficient of the material.
   (c) Explain why this estimate is likely to be unreliable.
8. Derive an expression relating the sound power level of a spherically radiating source to the sound pressure level at a radial distance r from the source. Calculate the sound pressure level at 10 m from such a source of which the sound power is 5 W. Assume the reference level for the measurement of sound pressure level is 2×10^{-5} Nm^{-2}.

9. Two point monopoles of equal and opposite complex strength q lie on the x_1 axis at a distance d apart that is small compared to the wavelength of the sound radiated. Make use of the Taylor series expansion

$$f(y_1 + \varepsilon) = f(y_1) + \varepsilon \frac{\partial}{\partial y_1} + \frac{1}{2!} \varepsilon^2 \frac{\partial^2}{\partial y_1^2} f(y_1) \ldots$$

to show that the complex pressure field radiated can be expressed as

$$p(x) = j\omega\rho_0(qd) \frac{\partial}{\partial y_1} \left[\frac{e^{-jkr}}{4\pi r} \right]$$

where y_1 denotes the coordinate of the midpoint of the two sources and r is the distance from y_1 to the field point x. Show how this result can be reduced to a form which demonstrates a distinctive directivity pattern of the field radiated.

10. The solution to the inhomogeneous Helmholz equation in an unbounded medium driven by a distribution of force per unit volume $\boldsymbol{f}_{vol}(\boldsymbol{y})$ can be written as

$$p(\boldsymbol{x}) = -\int_V \Delta_y \boldsymbol{f}_{\mathrm{vol}}(\boldsymbol{y}) g(\boldsymbol{x}|\boldsymbol{y}) \mathrm{d}V$$

where Δ_y represents the del operator in the $\boldsymbol{y}$ coordinates and $g(\boldsymbol{x}|\boldsymbol{y})$ is the free space Green function. Demonstrate that the field produced by a point force, such that $\boldsymbol{f}_{\mathrm{vol}}(\boldsymbol{y}) = \boldsymbol{f}\delta(\boldsymbol{y} - \boldsymbol{y}_0)$, reduces to that given in Question 9 when the force points in the direction of the x-axis and has a magnitude $j\omega\rho_0 qd$.

2

Fundamentals of vibration

N. LALOR

2.1 INTRODUCTION

Vibration is an integral part of our everyday lives; we experience its results when, for example, we drive to work, operate a domestic appliance or listen to music. Although all these cases involve both the vibration of mechanical components and of particles of air, only the former will be considered in this chapter. Nevertheless, it will be readily appreciated that many of the concepts discussed here have direct parallels in the field of acoustics.

Most people have seen the film of the destruction of the Tacoma Narrows bridge. Here, the vibration was clearly detrimental although this would not be the case with, for example, a musical instrument. However, whether detrimental or not, the phenomenon of vibration is so important and widespread that it is essential that all designers understand it and take account of its consequences in their designs.

Although this chapter covers only the fundamental aspects of vibration theory, the principles discussed here have a much wider application than might at first be thought. This is because nearly all vibration phenomena that occur in practice follow these basic principles so that generally only a simple model is necessary to qualitatively explain an event. Of course, for a fully quantitative simulation, much more detailed models are required, but even these often consist essentially of an assembly of simple models. The size of a model is most conveniently specified by the number of degrees of freedom it possesses, i.e. the number of independent co-ordinates required to uniquely define its configuration at any instant of time. This chapter only deals with systems having either one or two degrees of freedom – the former can be considered as the basic building block and the latter as an example of how such blocks can be joined together.

Two forms of vibration are discussed here, namely transient and periodic. A vibration phenomenon is described as transient if it has a definite beginning and end, particularly if the total time involved is short. If, however, the event repeats itself at regular intervals, it is termed periodic motion. Of course, nearly all phenomena in practice have a beginning and an end but, provided that these are sufficiently well separated, the motion can still be considered periodic. A special form of periodic motion is where the displacement of the vibrating system varies sinusoidally with time. This is termed harmonic motion and is particularly important because all periodic motion can be decomposed into a series of sinusoids. It is for this reason that only this form of periodic motion is considered here.

2.2 SIMPLE FREELY VIBRATING SPRING–MASS SYSTEM

2.2.1 Without damping

Consider the case of a spring with stiffness k rigidly fixed at one end and with a point mass m at the other end (Fig. 2.1). If the mass is displaced downwards from its equilibrium position and is then released, the spring force will accelerate the mass upwards.

At any instant of time following release, when the mass is a distance x from its equilibrium position, the force acting on the mass is

$$-kx.$$

(The sign is negative because the force acts in the opposite direction to the displacement.)

Applying Newton's Second Law of motion gives

$$m\ddot{x} = -kx$$

or

$$m\ddot{x} + kx = 0. \tag{2.1}$$

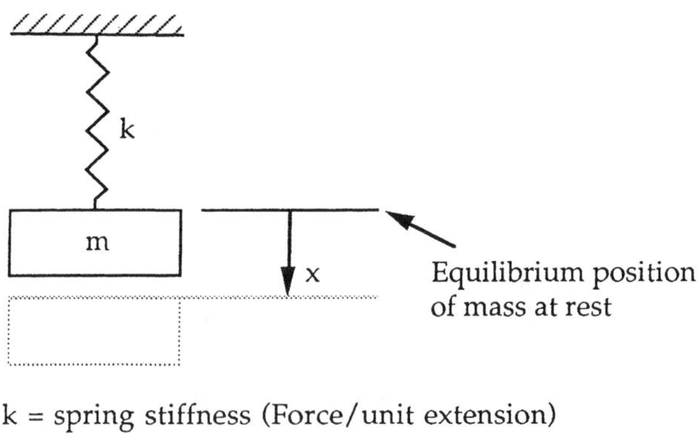

Figure 2.1 Mass–spring system.

The most general solution of this differential equation is

$$x = B_1 \sin \omega t + B_2 \cos \omega t \tag{2.2}$$

where B_1, B_2 are constants determined by the initial conditions, $\omega = \omega_n = 2\pi f_n = \sqrt{\frac{k}{m}}$ and f_n is the natural frequency of the system in Hertz (Hz). ω_n is also often referred to as the undamped natural frequency although, of course, its units are radians per second.

The equation for ω_n can be verified as follows. Differentiating equation (2.2) twice with respect to time gives

$$\ddot{x} = -\omega^2 \{B_1 \sin \omega t + B_2 \cos \omega t\} \tag{2.3}$$

Substituting equations (2.2) and (2.3) into equation (2.1) gives

$$(-\omega^2 m + k)\{B_1 \sin \omega t + B_2 \cos \omega t\} = 0$$

and since $\{B_1 \sin \omega t + B_2 \cos \omega t\} \neq 0, (-\omega^2 m + k) = 0$
Therefore

$$\omega = \sqrt{\frac{k}{m}}. \tag{2.4}$$

The above analysis shows that, unless $B_1 = B_2 = 0$ (a trivial solution indicating that the system is not vibrating), equation (2.2) is a solution of equation (2.1) and that equation (2.4) is valid for all values of x. This is because the system is 'linear' (as are all the other systems analysed in this chapter) which, in this case, is the direct result of the fact that the force in the spring is directly proportional to its extension, i.e. it obeys Hooke's Law. Later on, when we include the effects of damping and exciting forces, we will always find that the natural frequencies and the solutions to equations of motion are independent of the amplitude of vibration and that this amplitude of vibration is proportional to the amplitude of the exciting force.

Because it is often more convenient to use the exponential form for x (i.e. $x = Be^{j\omega t}$) instead of the trigonometric form of equation (2.2), a brief explanation of its use is given below. De Moivre's theorem states that $e^{j\theta} = \cos \theta + j \sin \theta$ also, because B is usually complex, it can be written as $(\alpha - j\beta)$. Therefore

$$\begin{aligned} Be^{j\omega t} &= (\alpha - j\beta)(\cos \omega t + j \sin \omega t) \\ &= (\alpha \cos \omega t + \beta \sin \omega t) \\ &\quad - j(\beta \cos \omega t - \alpha \sin \omega t) \end{aligned}$$

Thus

$$x = \mathrm{Re}\{Be^{j\omega t}\} = \alpha \ \cos \omega t + \beta \sin \omega t$$

where Re{ } denotes the real part of { }

This is of the same form as equation (2.2). Note that we really should write $x = \mathrm{Re}\{Be^{j\omega t}\}$ although the Re{} is usually omitted for convenience. Referring to Fig. 2.2, it can be seen that $\alpha = \sqrt{(\alpha^2 + \beta^2)} \cos \phi$ and $\beta = \sqrt{(\alpha^2 + \beta^2)} \sin \phi$. Thus

$$x = \sqrt{(\alpha^2 + \beta^2)}(\cos \omega t \cos \phi + \sin \omega t \sin \phi).$$

or

$$x = \sqrt{(\alpha^2 + \beta^2)} \cos(\omega t - \phi).$$

The amplitude of x is therefore given by $\sqrt{[(\mathrm{Re}\{Be^{j\omega t}\})^2 + (\mathrm{I_m}\{Be^{j\omega t}\})^2]}$. The significance of the phase angle ϕ will be seen later in section 2.3. An alternative formulation for x is

$$x = Be^{j(\omega t - \phi)}$$

where B is real and not complex. Note that the use of the imaginary exponent of e (i.e. jωt) implies that the motion of x is harmonic.

In the case shown in Fig. 2.1, the mass has its maximum deflection x_0 at the point of release ($t = 0$). Substituting this initial condition gives:

$$x_0 = 0 + B_2 \text{ or } B_2 = x_0$$

Similarly the velocity is zero at $t = 0$. Therefore $0 = B_1 + 0$ or $B_1 = 0$.

Substituting these values into equation (2.2) gives

$$x = x_0 \cos(\omega t). \tag{2.5}$$

A very useful way of expressing the natural frequency of the above system is in terms of its static deflection. Due to the influence of gravity the spring will stretch statically by an amount

$$\delta_{st} = \frac{mg}{k}.$$

Therefore

$$\sqrt{\frac{k}{m}} = \sqrt{\frac{g}{\delta_{st}}}$$

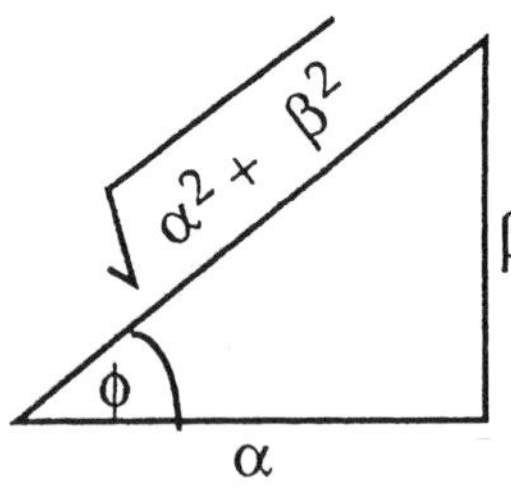

Figure 2.2 Phase triangle.

or

$$f_n = \frac{1}{2\pi}\sqrt{\frac{g}{\delta_{st}}}. \tag{2.6}$$

The natural frequency of this system can also be determined from energy considerations. Figure 2.3 shows the variation of x with time as determined by equation (2.5). At point A the velocity is zero but the deflection from the datum position is a maximum. Therefore,

$$\text{Strain Energy (SE)} = (\text{SE})_{\max} = \tfrac{1}{2}kx_0^2 \text{ and Kinetic Energy (KE)} = 0.$$

At point B the velocity is a maximum and the displacement is zero. Therefore,

$$\text{KE} = (\text{KE})_{\max} = \tfrac{1}{2}mv_0^2 = \tfrac{1}{2}m(\omega x_0)^2 \text{ and SE} = 0.$$

Thus, if total energy is conserved,

$$(\text{KE}) + (\text{SE}) = \text{Constant} = (\text{KE})_{\max} = (\text{SE})_{\max}.$$

Therefore

$$\tfrac{1}{2}kx_0^2 = \tfrac{1}{2}m\omega^2x_0^2$$

so that

$$m\omega^2 = k$$

or

$$\omega = \sqrt{\frac{k}{m}}$$

i.e. the same result as before.

Up to this point the mass of the spring has been neglected. Its effect can be allowed for by using the previously described energy method.
With reference to Fig. 2.4,

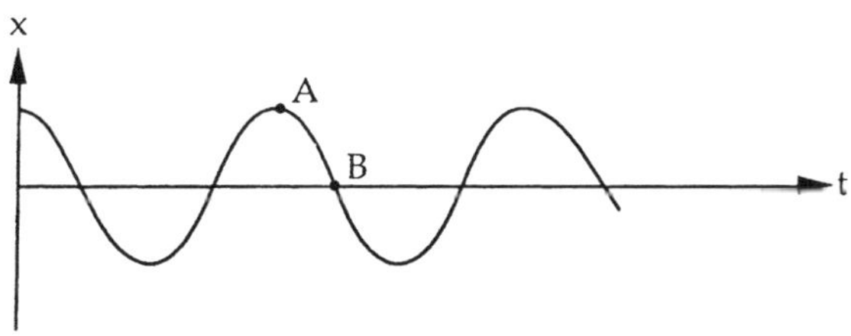

Figure 2.3 Displacement–time history of a vibrating mass.

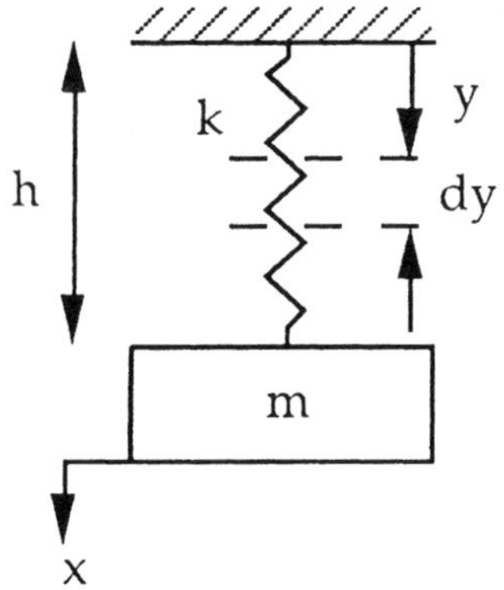

Figure 2.4 System with heavy spring.

h = length of spring (at datum position)
m_s = mass of spring
v_s = amplitude of velocity of element dy of spring $= \omega x_0 \frac{y}{h}$, and
$\frac{m_s}{h}\mathrm{d}y$ = mass of element dy of spring.

Now

$$(\mathrm{SE})_{\max} = \tfrac{1}{2}kx_0^2$$

and

$$(\mathrm{KE_{spring}})_{\max} \simeq \frac{1}{2}\int_0^h \frac{m_s}{h}\omega^2 x_0^2 \frac{y^2}{h^2}\mathrm{d}y = \frac{1}{6}m_s\omega^2 x_0^2.$$

(Note: The KE relationship is an approximation because h actually varies by $\pm x_0$. However, the net error is small.)
As before,

$$(\mathrm{KE_{mass}})_{\max} = \frac{1}{2}m\omega^2 x_0^2.$$

Therefore

$$\frac{1}{2}kx_0^2 = \frac{1}{6}m_s\omega^2 x_0^2 + \frac{1}{2}m\omega^2 x_0^2$$

so that

$$k = \omega^2\left(\frac{m_s}{3} + m\right)$$

or

$$\omega = \sqrt{\frac{k}{m + \frac{m_s}{3}}}. \tag{2.7}$$

2.2.2 With viscous damping

If a viscous damper is placed in parallel with the spring, as shown in Fig. 2.5, the differential equation of motion becomes where c = viscous damping constant (force/unit velocity)

$$m\ddot{x} + c\dot{x} + kx = 0. \tag{2.8}$$

Note that the damping force $c\dot{x}$ also opposes the motion of the mass and therefore has the same sign as the spring force kx.

The most general solution of equation (2.8) is

$$x = B_1 e^{s_1 t} + B_2 e^{s_2 t} \tag{2.9}$$

where B_1 and B_2 are constants determined by the initial conditions.

Each of these components of x are independent solutions. Therefore we can consider a typical solution as

$$x = Be^{st}.$$

Thus,

$$\dot{x} = Bse^{st}$$

and

$$\ddot{x} = Bs^2 e^{st}.$$

Therefore

$$mBs^2 e^{st} + cBse^{st} + kBe^{st} = 0.$$

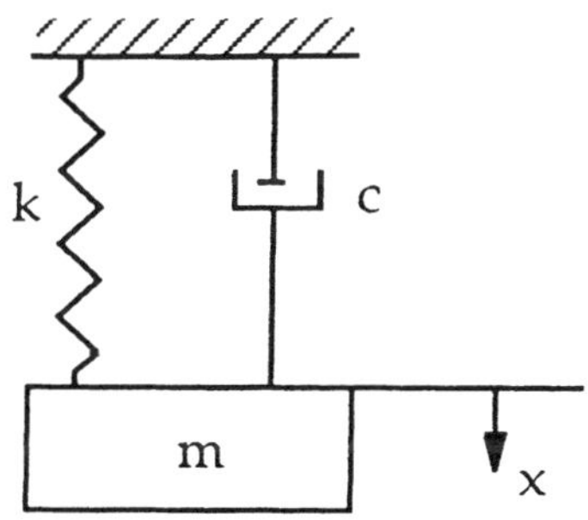

Figure 2.5 Mass–spring system with viscous damping.

For $B \neq 0$ and $\mathrm{e}^{st} \neq 0$

$$ms^2 + cs + k = 0$$

Using the quadratic formula to solve for s we obtain

$$s = \frac{-c \pm \sqrt{c^2 - 4mk}}{2m} = \frac{-c}{2m} \pm \sqrt{\left(\frac{c}{2m}\right)^2 - \frac{k}{m}}.$$

Therefore

$$s_1 = \frac{-c}{2m} + \sqrt{\left(\frac{c}{2m}\right)^2 - \frac{k}{m}}.$$

and

$$s_2 = \frac{-c}{2m} - \sqrt{\left(\frac{c}{2m}\right)^2 - \frac{k}{m}}.$$

This solution has two basic forms. The first is where

$$\left(\frac{c}{2m}\right)^2 > \frac{k}{m}.$$

When this is the case, both s_1 and s_2 are real numbers and, on release, the mass will return exponentially to its equilibrium position without performing any oscillations. If, however,

$$\frac{k}{m} > \left(\frac{c}{2m}\right)^2$$

the quantity inside the square root is negative and s_1, s_2 are given by

$$s_1, s_2 = \frac{-c}{2m} \pm j\sqrt{\frac{k}{m} - \left(\frac{c}{2m}\right)^2}.$$

By substituting these values of s_1 and s_2 into equation (2.9), the solution of the differential equation can then be written in the form

$$x = \mathrm{e}^{-ct/2m}(B_1' \cos(\omega_\mathrm{d} t) + B_2' \sin(\omega_\mathrm{d} t)) \tag{2.10}$$

where B_1' and B_2' are constants determined by the initial conditions and

$$\omega_\mathrm{d} = \sqrt{\frac{k}{m} - \left(\frac{c}{2m}\right)^2}.$$

The way in which equation (2.10) is obtained from equation (2.9) is as follows. After substituting the values of s_1 and s_2, equation (2.9) becomes

$$x = \mathrm{e}^{-ct/2m}\left(B_1 \mathrm{e}^{j\omega_\mathrm{d} t} + B_2 \mathrm{e}^{-j\omega_\mathrm{d} t}\right).$$

Expanding the terms in the brackets using De Moivre's theorem gives

$$x = e^{-ct/2m}\{(B_1 + B_2)\cos(\omega_d t) + j(B_1 - B_2)\sin(\omega_d t)\}.$$

Now B_1 and B_2 are, in general, complex (due to the damping), i.e. $B_1 = \alpha_1 + j\beta_1$ and $B_2 = \alpha_2 + j\beta_2$. Substituting for B_1 and B_2 and taking the real part of the resulting equation gives

$$\mathrm{Re}\{x\} = e^{-ct/2m}(B_1' \cos(\omega_d t) + B_2' \sin(\omega_d t))$$

where $B_1' = \alpha_1 + \alpha_2$ and $B_2' = \beta_2 - \beta_1$.
The borderline between oscillatory and non-oscillatory motion is when

$$\frac{k}{m} = \left(\frac{c}{2m}\right)^2$$

or

$$c = \text{critical damping} = c_c = 2\sqrt{km}.$$

Therefore

$$\frac{c}{c_c} = \zeta = \frac{c}{2\sqrt{km}} \tag{2.11}$$

which is termed the viscous damping ratio. Thus,

$$\omega_d = \omega_n\sqrt{1 - \zeta^2} \tag{2.12}$$

so that

$$\left(\frac{\omega_d}{\omega_n}\right)^2 = 1 - \zeta^2.$$

This is the equation of a circle, or a quadrant of a circle if only positive values of $\frac{\omega_d}{\omega_n}$ and ζ are plotted, as shown in Fig. 2.6. It can be seen that for values of $\zeta < 0.2$ (which is usually the case in most practical applications) the effect of damping on the natural frequency of the system is negligible. Above this value the damped natural frequency becomes increasingly reduced.

Using the same initial conditions as before, the solution of the equation of motion reduces to

$$x = x_0 e^{-\zeta\omega_n t}\left(\cos(\omega_d t) + \frac{\zeta}{\sqrt{1-\zeta}}\sin(\omega_d t)\right). \tag{2.13}$$

Figure 2.7 shows a plot of equation (2.13) and it can be seen that the waveform is no longer sinusoidal but now becomes a damped sine wave under the envelope

$$\pm x_0 e^{-\zeta\omega_n t}$$

Strictly speaking the motion is therefore no longer harmonic (in fact it is not even periodic but is transient). However, provided the damping is not too large, the motion can still be considered harmonic.

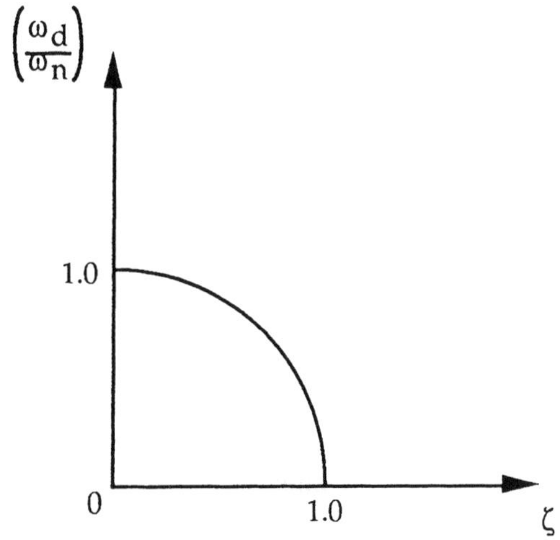

Figure 2.6 Effect of damping on natural frequency.

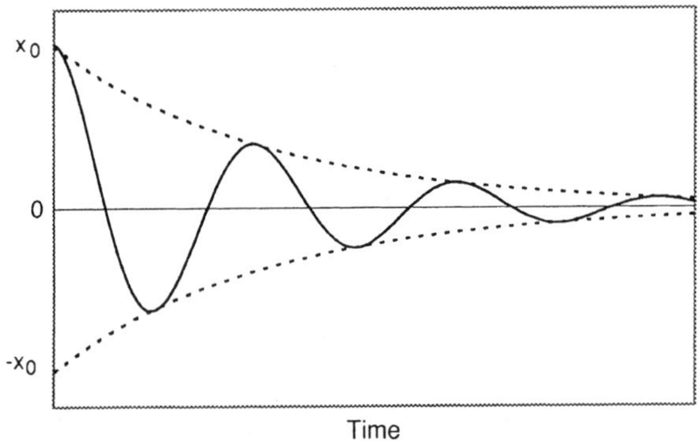

Figure 2.7 Displacement–time history of a damped mass.

2.2.3 Logarithmic decrement

It is often of considerable importance to know quite accurately how much damping is actually present in a system. Considering the simple damped system just analysed, the amount of damping present can be calculated from the damped sine wave time history shown in Fig. 2.7. At any maximum on the curve (i.e. where $\cos(\omega_d t) + \frac{\zeta}{\sqrt{1-\zeta^2}}\sin(\omega_d t) = 1$) the value of x will be given by

$$x(t) = x_0 e^{-\zeta\omega_n t}.$$

The next maximum will occur one period ($T = \frac{2\pi}{\omega_d}$) later and the value of x will then be

$$x(t+T) = x_0 e^{-\zeta\omega_n(t+T)}.$$

Therefore

$$\frac{x(t)}{x(t+T)} = e^{\zeta\omega_n T} \simeq e^{2\pi\zeta}. \tag{2.14}$$

Note that

$$\zeta\omega_n T = \zeta\omega_n \frac{2\pi}{\omega_d} \simeq 2\pi\zeta.$$

Taking natural logarithms of both sides of equation (2.14) gives

$$\log_e\left(\frac{x(t)}{x(t+T)}\right) = \delta = 2\pi\zeta \tag{2.15}$$

where δ is called the logarithmic decrement. Therefore,

$$\zeta = \frac{1}{2\pi}\log_e\left(\frac{x(t)}{x(t+T)}\right). \tag{2.16}$$

If the damping is very small it is more accurate to measure the damping ratio over several cycles rather than just one.

If amplitudes n cycles apart are considered then

$$\frac{1}{n}\log_e\left(\frac{x(t)}{x(t+nT)}\right) = \delta = 2\pi\zeta. \tag{2.17}$$

The damping can also be calculated from the slope of the envelope. The equation of the positive part of the envelope is

$$x(t) = x_0 e^{-\zeta\omega_n t}$$

Therefore

$$\frac{x(t)}{x_{ref}} = \frac{x_0}{x_{ref}} e^{-\zeta\omega_n t}$$

where, x_{ref} = arbitary reference for x.
Therefore

$$\log_e\left(\frac{x(t)}{x_{ref}}\right) = \log_e\left(\frac{x_0}{x_{ref}}\right) - \zeta\omega_n t.$$

Now

$$\log_e(x) = \frac{\log_{10}(x)}{\log_{10}(e)} = \frac{\log_{10}(x)}{0.4343}.$$

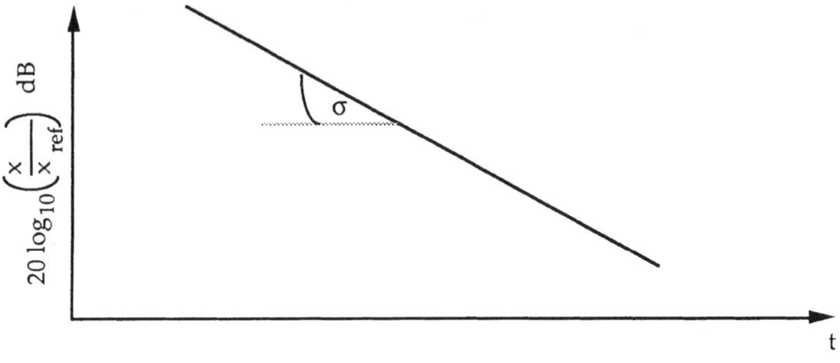

Figure 2.8 Reverberation decay.

Therefore

$$\log_{10}\left(\frac{x(t)}{x_{\text{ref}}}\right) = \log_{10}\left(\frac{x_0}{x_{\text{ref}}}\right) - 0.4343\zeta\omega_{\text{n}}t$$

or

$$20\log_{10}\left(\frac{x(t)}{x_{\text{ref}}}\right) = 20\log_{10}\left(\frac{x_0}{x_{\text{ref}}}\right) - 8.6859\zeta\omega_{\text{n}}t. \tag{2.18}$$

Figure 2.8 shows a plot of equation (2.18), the right-hand side of which is a linear function of time, t. The slope of this graph, $\sigma = 8.6859\ \zeta\omega_{\text{n}} = \frac{60}{T_{60}}$ where T_{60} = reverberation time = time for the vibration to decay by 60 dB. Therefore,

$$\zeta = \frac{60}{\omega_{\text{n}}8.6859T_{60}} = \frac{60}{2\pi f_{\text{n}}8.6859T_{60}} = \frac{1.1}{f_{\text{n}}T_{60}}$$

and the loss factor, (η, defined in section 2.4), is given by

$$\eta = 2\zeta = \frac{2.2}{f_{\text{n}}T_{60}}. \tag{2.19}$$

2.2.4 Other forms of damping

There are two other common forms of damping apart from viscous dămping which are as follows.

2.2.4.1 Coulomb or friction damping

This occurs at joints which rub together and is a major form of damping present in bolted or riveted structures. Here the damping force is constant in magnitude and in-phase with the velocity. However, the analysis of this is quite complex and is therefore beyond the scope of this chapter.

2.2.4.2 Hysteretic damping

This is present when the force versus deflector characteristics of a structure for loading and unloading form a hysteresis loop. It occurs mainly when rubber or similar materials are strained and at joints, particularly where a gasket is used. In this case the damping force is proportional to the displacement but is still in-phase with the velocity. The response of a mass/spring system with this type of damping is analysed later in section 2.4.

For many practical problems, an 'equivalent' viscous damping can be assumed (see section 2.4.1).

2.3 VISCOUSLY DAMPED SINGLE DEGREE-OF-FREEDOM SYSTEM WITH HARMONIC FORCE EXCITATION

Figure 2.9 shows the same model previously considered in section 2.2.2, but in this case excited into vibration by an externally applied force, $F(t)$, i.e. by a force that varies with time. Strictly speaking x should also be written as $x(t)$ because it is also a function of time. However, the (t) is usually omitted in the interest of simplicity of notation.

As before, the motion of this system can be analysed by invoking Newton's Second Law of motion, i.e.

$$m\ddot{x} = -c\dot{x} - kx + F(t)$$

or

$$m\ddot{x} + c\dot{x} + kx = F(t). \tag{2.20}$$

Considering the special case where $F(t)$ is harmonic, equation (2.20) becomes

$$m\ddot{x} + c\dot{x} + kx = F_0 e^{j\omega t} \tag{2.21}$$

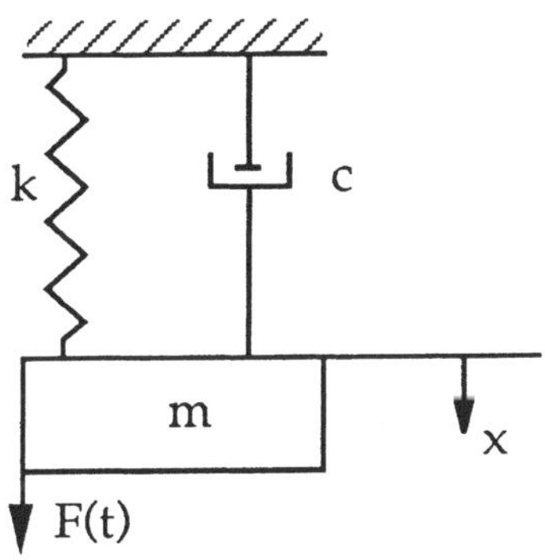

Figure 2.9 Forced mass–spring system with viscous damping.

where F_0 = amplitude of applied force.

The general solution of equation (2.21) is

$$x = e^{-\zeta\omega_n t}\{B_1 \sin(\omega_d t) + B_2 \cos(\omega_d t)\} + B_3 \sin(\omega t) + B_4 \cos(\omega t). \quad (2.22)$$

The first two terms of this solution involving sin $(\omega_d t)$ and $\cos(\omega_d t)$ are due to the fact that equation (2.21) is only valid for positive values of t. Thus the exciting force has to start at $t = 0$ and this causes an initial transient. However, this dies away because of the influence of the damping term $e^{-\zeta\omega_n t}$. Once steady-state conditions have been reached the solution is

$$x = B_3 \sin(\omega t) + B_4 \cos(\omega t) \quad (2.23a)$$

which may be written as

$$x = x_0 \sin(\omega t - \phi) \quad (2.23b)$$

where x_0 = displacement amplitude, and ϕ = phase angle between the applied force and the displacement.

Since the exciting force is harmonic, the displacement must also be harmonic. This is because the system is linear so that the response must always be proportional to the force, as discussed previously in section 2.2.1. Thus

$$x = x_0 e^{j\omega t}. \quad (2.24)$$

It should be noted that, due to the presence of the damping, it is appropriate to make both F_0 and x_0 complex. Nevertheless, it is the normal convention to assume that F_0 is the datum for the phase and is therefore real.

Substituting equation (2.24) into equation (2.21) gives

$$(-\omega^2 m + j\omega c + k)x_0 = F_0. \quad (2.25)$$

Therefore

$$x_0 = \frac{F_0}{\{(k - \omega^2 m) + j\omega c\}}.$$

This can be separated into its real and imaginary parts by multiplying the numerator and denominator by the complex conjugate of the denominator, i.e.

$$x_0 = \frac{F_0\{(k - \omega^2 m) - j\omega c\}}{\{(k - \omega^2 m) + j\omega c\}\{(k - \omega^2 m) - j\omega c\}} = \frac{F_0\{(k - \omega^2 m) - j\omega c\}}{(k - \omega^2 m)^2 + (\omega c)^2}.$$

Dividing numerator and denominator by k^2 gives

$$x_0 = \frac{\frac{F_0}{k}\left\{\left(1 - \frac{\omega^2}{\omega_n^2}\right) - \frac{j\omega c}{k}\right\}}{\left(1 - \frac{\omega^2}{\omega_n^2}\right)^2 + \left(\frac{\omega c}{k}\right)^2}.$$

Now, $(F_0/k) = x_{st}$ = the deflection due to static force of magnitude F_0, and $(c/k) = (2\zeta/\omega_n)$.
Therefore,

$$\frac{x_0}{x_{st}} = \frac{\left(1-\frac{\omega^2}{\omega_n^2}\right)}{\left(1-\frac{\omega^2}{\omega_n^2}\right)^2+\left(2\zeta\frac{\omega}{\omega_n}\right)^2} - j\frac{2\zeta\frac{\omega}{\omega_n}}{\left(1-\frac{\omega^2}{\omega_n^2}\right)^2+\left(2\zeta\frac{\omega}{\omega_n}\right)^2} = \mathrm{Re}\left\{\frac{x_0}{x_{st}}\right\} - j\mathrm{Im}\left\{\frac{x_0}{x_{st}}\right\}. \tag{2.26}$$

The magnitude or 'modulus' of $\frac{x_0}{x_{st}}$, written as $\left|\frac{x_0}{x_{st}}\right|$, is given by

$$\left|\frac{x_0}{x_{st}}\right| = \sqrt{\left(\mathrm{Re}^2\left\{\frac{x_0}{x_{st}}\right\}\right)^2+\left(\mathrm{Im}\left\{\frac{x_0}{x_{st}}\right\}\right)^2} = \frac{\sqrt{\left(1-\frac{\omega^2}{\omega_n^2}\right)^2+\left(2\zeta\frac{\omega}{\omega_n}\right)^2}}{\left(1-\frac{\omega^2}{\omega_n^2}\right)^2+\left(2\zeta\frac{\omega}{\omega_n}\right)^2}.$$

Therefore, the dynamic magnification factor

$$\left|\frac{x_0}{x_{st}}\right| = \frac{1}{\sqrt{\left(1-\frac{\omega^2}{\omega_n^2}\right)^2+\left(2\zeta\frac{\omega}{\omega_n}\right)^2}} \tag{2.27}$$

and the phase angle ϕ between x_0 and F_0 is given by

$$\tan\phi = \frac{\mathrm{Im}\left\{\frac{x_0}{x_{st}}\right\}}{\mathrm{Re}\left\{\frac{x_0}{x_{st}}\right\}} = \frac{-2\zeta\frac{\omega}{\omega_n}}{\left(1-\frac{\omega^2}{\omega_n^2}\right)}. \tag{2.28}$$

Figure 2.10(a) shows how $\left|\frac{x_0}{x_{st}}\right|$ and ϕ vary with frequency, for different amounts of damping. It can be seen that $\left|\frac{x_0}{x_{st}}\right|$ reaches a maximum value or 'resonance' as it is termed, when the frequency of the exciting force is approximately equal to ω_n. The exact frequency for resonance is given below (see equation (2.33) and Fig. 2.12) although, for typical values of damping, the difference between these two frequencies is small. For this reason, ω_n is often called (albeit incorrectly) the resonant frequency. It should also be noted that $\phi = 90°$ when $\omega = \omega_n$. The variation of the real and imaginary parts of $\left(\frac{x_0}{x_{st}}\right)$ with frequency is shown in Fig. 2.10(b). This latter figure also shows how the damping ratio ζ can be estimated from both the modulus and the imaginary response curves.

It is quite common practice to present these results as a polar plot of $\left|\frac{x_0}{x_{st}}\right|$ against ϕ. This is identical to plotting the imaginary part against the real part, on rectangular axes. This Nyquist diagram, as it is called, is shown in Fig. 2.10(c).

It is helpful to represent the various terms in equation (2.25) in the form of a vector diagram, as shown in Fig. 2.11.

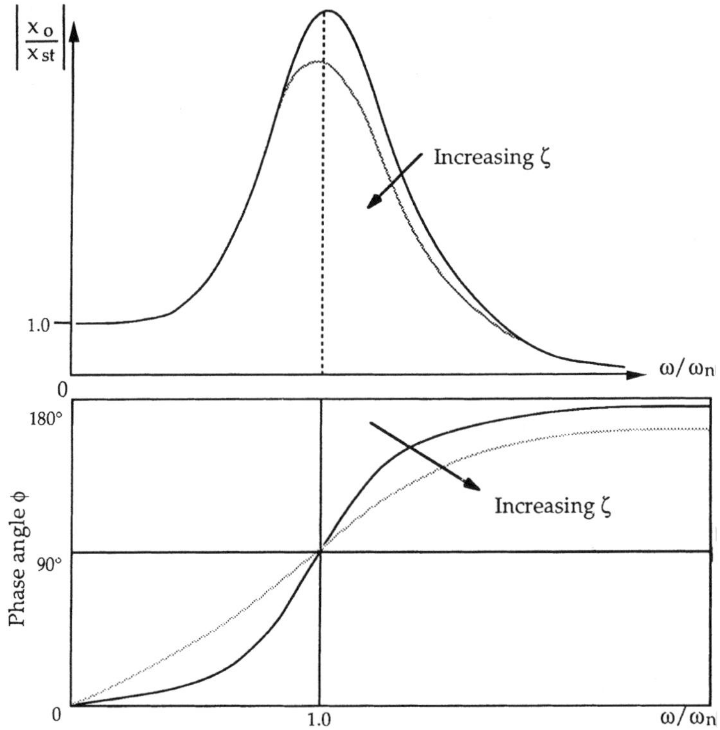

Figure 2.10 (a) Forced response and phase spectra of system with viscous damping.

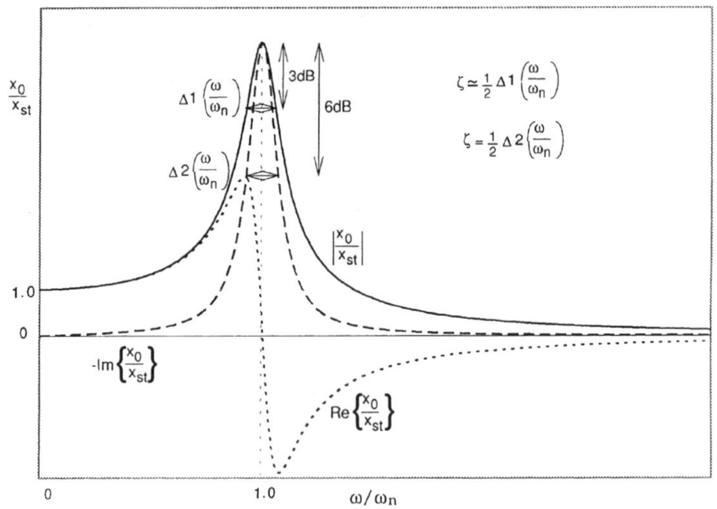

Figure 2.10 (b) Real and imaginary parts of response.

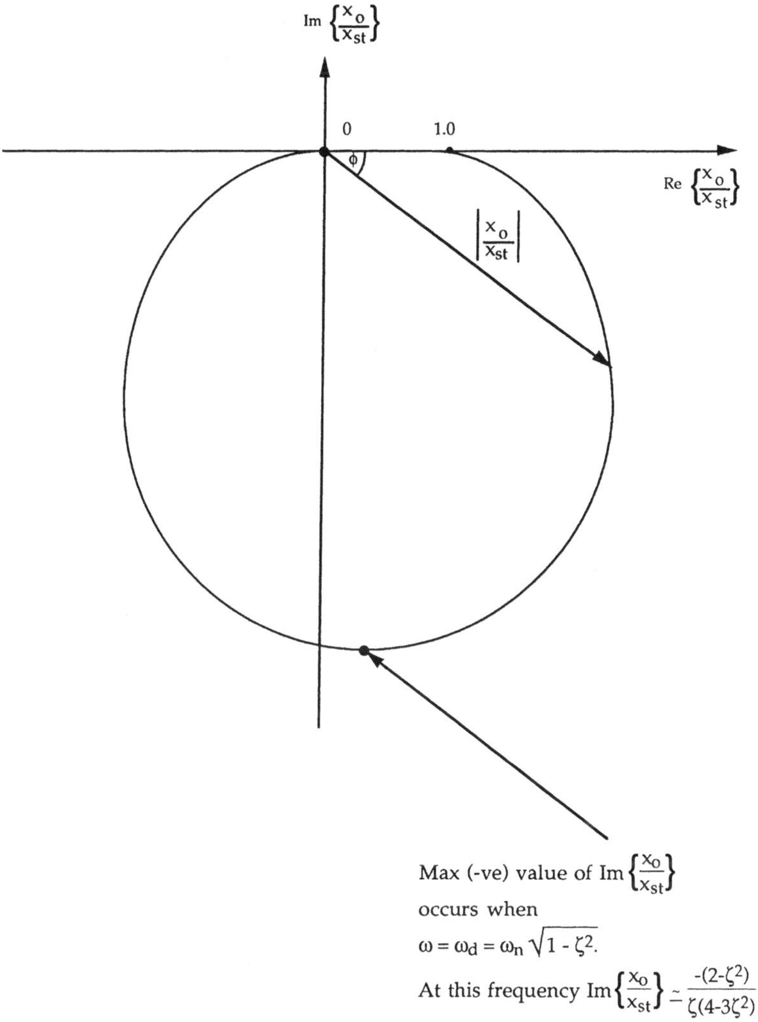

Figure 2.10 (c) Nyquist diagram of system with viscous damping.

When the inertia force is equal to the spring force, $m\omega^2 x_0 = kx_0$, and therefore $\omega = \omega_n = \sqrt{\frac{k}{m}}$. At this frequency the phase angle ϕ is 90° and the resultant force is equal to the damping force. Therefore

$$F_0 = c\omega_n x_0$$

or

$$x_0 = \frac{F_0}{c\omega_n}. \tag{2.29}$$

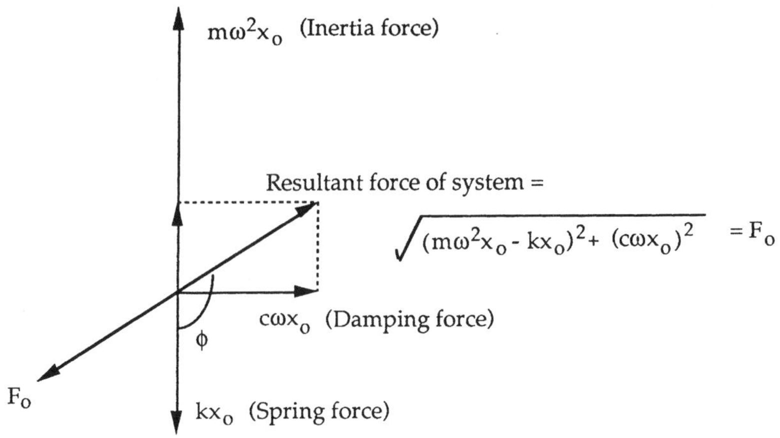

Figure 2.11 Vector diagram of system forces.

In this case the system response is controlled by the damping. This explains why the resonant amplitude approaches infinity when the damping approaches zero. It should also be noted that in this case x_0 is inversely proportional to ω_n.

When $\omega \ll \omega_n : F_0 \simeq kx_0$. Therefore,

$$x_0 \simeq \frac{F_0}{k} \tag{2.30}$$

i.e. the response is controlled by the spring stiffness.

For, $\omega \gg \omega_n : F_0 \simeq \omega^2 m x_0$. Therefore,

$$x_0 \simeq \frac{F_0}{\omega^2 m} \tag{2.31}$$

i.e. the response is controlled by the mass.

The phase angle ϕ is also dependent upon the frequency. Below resonance the displacement of the mass is in phase ($\phi \simeq 0°$) with the exciting force; at resonance, it is at 90° to the force; and above resonance it is out phase ($\phi \simeq 180°$) with the force.

The choice of real and imaginary axes for Fig. 2.11 is, in principle, completely arbitrary. However, as stated previously, it is normal practice to align the real axis with the F_0 vector.

The maximum value of the displacement response amplitude can be obtained by differentiating $\left|\frac{x_0}{x_{st}}\right|$ (equation (2.27)) with respect to ω/ω_n and equating to zero. Choosing the root of the resulting equation for a maximum value gives:

$$\left|\frac{x_0}{x_{st}}\right|_{max} \simeq \frac{1}{2\zeta} = \text{Q factor.} \tag{2.32}$$

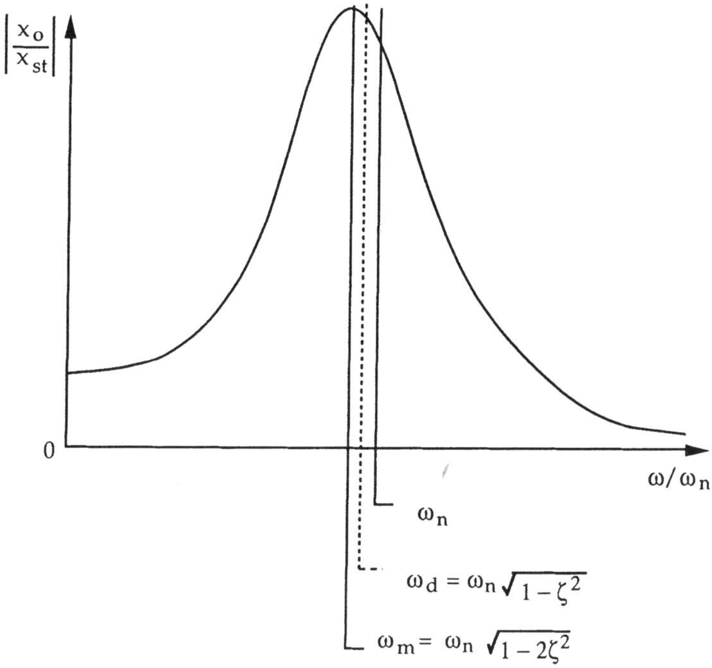

Figure 2.12 Definition of main system frequencies.

This occurs at a frequency ω_m where

$$\omega_m = \omega_n\sqrt{1 - 2\zeta^2}. \tag{2.33}$$

The relative positions of ω_m, ω_d and ω_n are shown in Fig. 2.12, although it should be noted that the differences have been exaggerated, for clarity.

2.3.1 Transmissibility

Suppose the mass in the previous example is a machine mounted on springs to the floor which is assumed to be rigid. If the machine body is excited by, say, a rotating out-of-balance force, it is important to know how much of that force is transmitted to the floor. The ratio of force transmitted to the floor to the force applied to the machine (mass) is called the 'transmissibility', T, and can be calculated as follows:

The force transmitted to the base $= c\dot{x} + kx = (j\omega c + k)x_0 e^{j\omega t}$.

Therefore the modulus of the transmitted force $= x_0\sqrt{(\omega c)^2 + k^2}$

and the modulus of the applied force $= F_0 = x_0\sqrt{(k - \omega^2 m)^2 + (c\omega)^2}$.

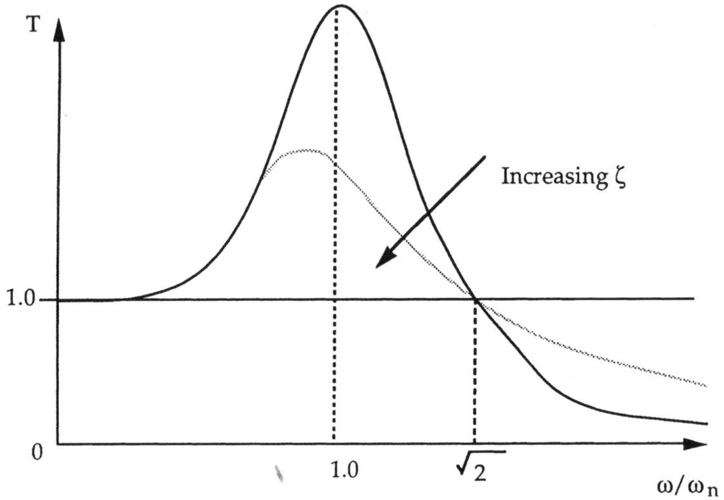

Figure 2.13 Effect of damping on transmissibility.

Therefore

$$T = \frac{|\text{Transmitted Force}|}{|\text{Applied Force}|} = \frac{\sqrt{(\omega c)^2 + k^2}}{\sqrt{(k - \omega^2 m)^2 + (c\omega)^2}} = \frac{\sqrt{1 + \left(\frac{\omega c}{k}\right)^2}}{\sqrt{\left(1 - \frac{\omega^2 m}{k}\right)^2 + \left(\frac{c\omega}{k}\right)^2}}$$

or

$$T = \frac{\sqrt{1 + \left(2\zeta \frac{\omega}{\omega_n}\right)^2}}{\sqrt{\left(1 - \frac{\omega^2}{\omega_n^2}\right)^2 + \left(2\zeta \frac{\omega}{\omega_n}\right)^2}}. \tag{2.34}$$

Figure 2.13 shows the variation of transmissibility with frequency for different values of damping. It will be seen that above $\omega/\omega_n = \sqrt{2}$, damping actually increases the amount of force transmitted to the floor.

2.3.2 Excitation by base motion

Sometimes a system is excited into vibration by base motion e.g. the vibration of a car due to undulations in the road. Referring to Fig. 2.14, the equation of motion is

$$m\ddot{x} = c(\dot{y} - \dot{x}) + k(y - x).$$

Substituting for x and y gives

$$(-\omega^2 m + j\omega c + k)x_0 = (k + j\omega c)y_0$$

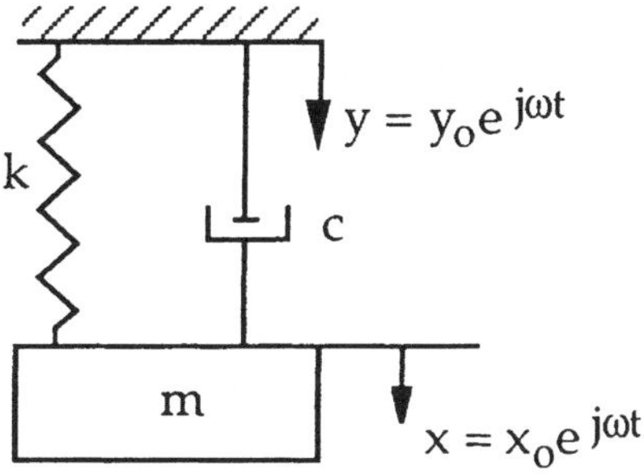

Figure 2.14 Damped system with base motion.

$$\frac{x_0}{y_0} = \frac{(k + j\omega c)}{(k - \omega^2 m) + j\omega c} = \frac{(k + j\omega c)\{(k - \omega^2 m) - j\omega c\}}{(k - \omega^2 m)^2 + (\omega c)^2}$$

$$\frac{x_0}{y_0} = \frac{k(k - \omega^2 m) + (\omega c)^2}{(k - \omega^2 m)^2 + (\omega c)^2} - \frac{j\omega^3 cm}{(k - \omega^2 m)^2 + (\omega c)^2} = \mathrm{Re}\left\{\frac{x_0}{y_0}\right\} + j\mathrm{Im}\left\{\frac{x_0}{y_0}\right\}.$$

Therefore

$$\left|\frac{x_0}{y_0}\right| = \sqrt{\left(\mathrm{Re}\left\{\frac{x_0}{y_0}\right\}\right)^2 + \left(\mathrm{Im}^2\left\{\frac{x_0}{y_0}\right\}\right)^2} = \frac{\sqrt{k^2 + (\omega c)^2}}{\sqrt{(k - \omega^2 m)^2 + (\omega c)^2}}$$

or

$$\left|\frac{x_0}{y_0}\right| = \frac{\sqrt{1 + \left(2\zeta \frac{\omega}{\omega_n}\right)^2}}{\sqrt{\left(1 - \frac{\omega^2}{\omega_n^2}\right)^2 + \left(2\zeta \frac{\omega}{\omega_n}\right)^2}} \tag{2.35}$$

Note that this is exactly the same expression as was previously derived for the transmissibility.

2.3.3 Vibration isolation

The cases just considered in the last two sections were concerned with evaluating the way in which conditions at the base differ from those at the mass when they are connected by a spring and a damper – and there are many situations when this difference is a highly desirable characteristic. For example, it is very important that the large out-of-balance forces on a car engine are not transmitted to the vehicle body so that the passengers are protected from excessive noise and vibration. Similarly, it is important to ensure that a delicate instrument mounted in the nose-cone of a space probe is not subjected to the very high levels of vibration experienced by the rest of the rocket during the launch phase.

In both these examples the transmission of undesirable force or vibration can be reduced by mounting the mass (e.g. engine or instrument) on a suitable spring/damper system. Although a detailed treatment of this so-called 'vibration isolation' concept is beyond the scope of this chapter, there are some general design principles that can be deduced from the previous analysis. It can be seen from Fig. 2.13 that at low frequencies (well below resonance) the value of $|T|$ or $|x_0/y_0|$ is approximately unity, i.e. the spring and damper have no effect so that the mass behaves as though it is connected directly to the base. At resonance, in contrast to this, the spring and damper actually amplify the transmitted force or vibration, i.e. they make the situation worse. However, as frequency is further increased beyond the point where it is $\sqrt{2}$ times ω_n, transmission is progressively reduced. It is therefore clear that, for good isolation, the resonant frequency must be well below the excitation frequency, i.e. the spring should be as soft as possible. It will also be observed that in this high-frequency region the damping actually reduces the isolation effect. Nevertheless, it is always advisable to include some damping to avoid high-resonant amplitudes – a certain amount of low-frequency excitation is inevitably present. e.g. when a car engine is started. The high-frequency penalty of this damping can be mitigated by employing special dampers that are only effective at low frequencies.

2.4 HYSTERETICALLY DAMPED SINGLE DEGREE-OF-FREEDOM SYSTEM WITH HARMONIC FORCE EXCITATION

With hysteretic damping the damping force is in-phase with the velocity, as in the previous case, but here its magnitude is proportional to the displacement.

Considering the same system as shown in Fig. 2.9, the equation of motion is

$$m\ddot{x} + c_h\frac{\dot{x}}{\omega} + kx = F_0 e^{j\omega t} \tag{2.36}$$

where c_h = hysteretic damping constant.

It should be noted that this equation is only valid for harmonic motion since only in this case is the magnitude of the displacement equal to the magnitude of the velocity divided by ω.

Substituting $x = x_0 e^{j\omega t}$ gives

$$(-\omega^2 m + jc_h + k)x_0 = F_0,$$

so that

$$x_0 = \frac{F_0}{(-\omega^2 m + jc_h + k)} = \frac{F_0\{(k - \omega^2 m) - jc_h\}}{\{(k - \omega^2 m) + jc_h\}\{(k - \omega^2 m) - jc_h\}}$$
$$= \frac{F_0(k - \omega^2 m)}{(k - \omega^2 m)^2 + (c_h)^2} - j\frac{F_0 c_h}{(k - \omega^2 m)^2 + (c_h)^2}.$$

Therefore

$$\left|\frac{x_0}{F_0/k}\right| = \frac{k}{\sqrt{(k - \omega^2 m)^2 + (c_h)^2}} = \frac{1}{\sqrt{\left(1 - \frac{\omega^2}{\omega_n^2}\right)^2 + \left(\frac{c_h}{k}\right)^2}} \tag{2.37}$$

and

$$\tan\phi = \frac{c_h/k}{\left(1 - \frac{\omega^2}{\omega_n^2}\right)}. \tag{2.38}$$

The parameter c_h/k is termed 'the loss factor' and is usually denoted by the symbol η. Note that F_0/k is not equal to the static deflection. This can be demonstrated by setting ω to zero in equation (2.37).

Figure 2.15(a) and 2.15(b) shows the variation of $\left|\frac{x_0}{F_0/k}\right|$ and ϕ with frequency. It should be noted here that Q is equal to the value of $\left|\frac{x_0}{F_0/k}\right|$ at frequency ω_n, i.e. $Q = \frac{1}{\eta}$. The equivalent Nyquist plot is shown in Fig. 2.15(c).

Note that maximum response occurs at $\omega = \omega_m = \sqrt{\frac{k}{m}}$, regardless of the amount of damping present.

2.4.1 Equivalent viscous damping

Because of the simplicity of the viscous damping model, it is often convenient to replace the hysteretic damping with an equivalent viscous damping. However, care must be exercised with this substitution. This is because the response curves for the two types of damping are of different shape and therefore the responses can only be made equal at one frequency (see Fig. 2.16). This is usually taken as the resonance frequency.

For viscous damping,

$$\left(\frac{x_0}{F_0/k}\right)_{\max} \simeq \frac{1}{2\zeta}.$$

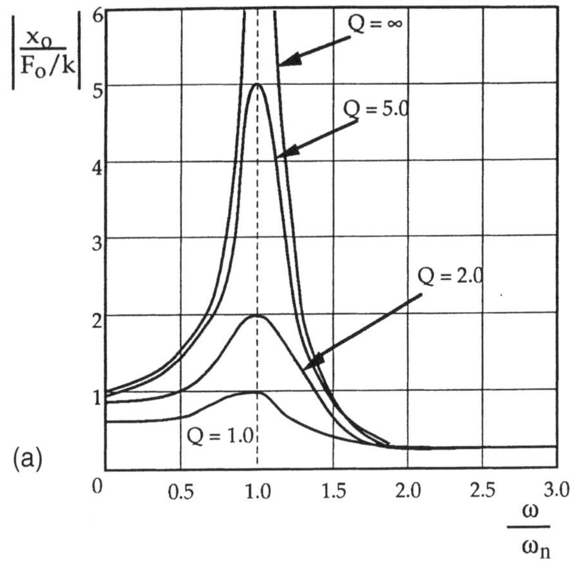

Figure 2.15 (a) Forced response spectrum of system with hysteretic damping.

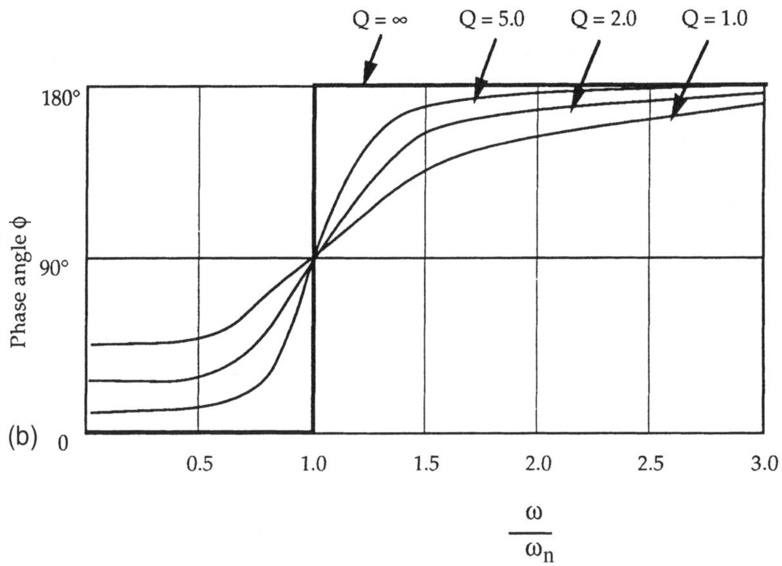

Figure 2.15 (b) Phase spectrum of forced system with hysteretic damping.

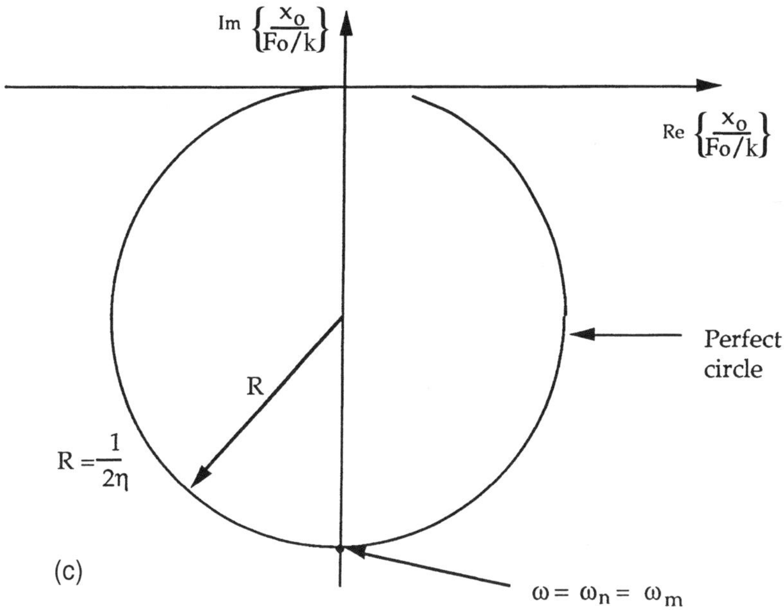

Figure 2.15 (c) Nyquist diagram of system with hysteretic damping.

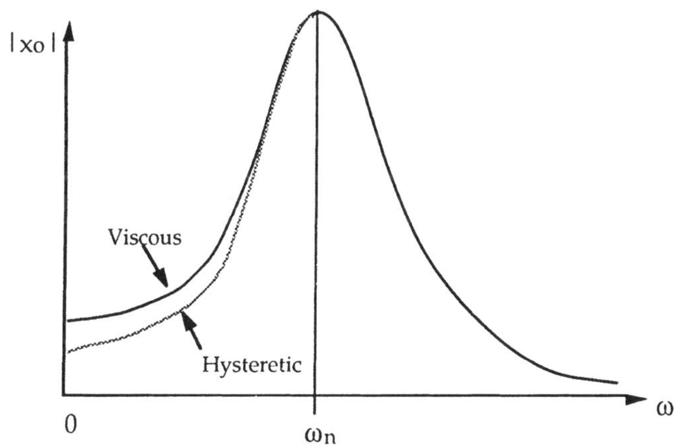

Figure 2.16 Equivalent viscous damping.

For hysteretic damping,

$$\left(\frac{x_0}{F_0/k}\right)_{\max} = \frac{1}{c_h/k}.$$

Therefore, for equivalent damping at resonance

$$\frac{1}{2\zeta_{eq}} = \frac{1}{c_h/k}$$

or

$$2\zeta_{eq} = \eta = c_h/k \tag{2.39}$$

and

$$\zeta_{eq} = \frac{c_h}{2k}.$$

Also

$$\frac{c_{eq}}{2\sqrt{km}} = \frac{c_h}{2k},$$

therefore

$$c_{eq} = \left(\frac{c_h}{\omega_n}\right). \tag{2.40}$$

2.5 TRANSIENT RESPONSE OF SINGLE DEGREE-OF-FREEDOM SYSTEM WITH VISCOUS DAMPING

For a transient, the equation of force equilibrium is the same as that given in section 2.3, viz:

$$m\ddot{x} + c\dot{x} + kx = F(t).$$

However, since the motion is no longer harmonic, we cannot make the substitution $x(t) = x_0 e^{j\omega t}$ and $F(t) = F_0 e^{j\omega t}$. Nevertheless, if $F(t)$ is a simple mathematical function we can still solve the differential equation, as shown in the following sections.

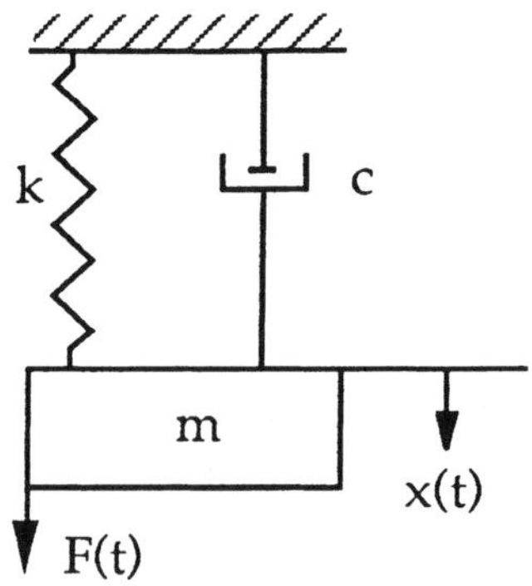

Figure 2.17 Forced mass–spring system with viscous damping.

2.5.1 Response to a step function

For example, if $F(t)$ is a step function (Fig. 2.18) the equation of motion becomes

$$m\ddot{x} + c\dot{x} + kx = a, \quad t \geq 0. \tag{2.41}$$

Assuming that the damping is less than critical, the solution of the complementary function (CF) is

$$x = \mathrm{e}^{-\zeta\omega_{\mathrm{n}}t}(A\cos\omega_{\mathrm{d}}t + B\sin\omega_{\mathrm{d}}t)$$

and the solution of the particular integral (PI) is

$$x = \frac{a}{k}.$$

Therefore the complete solution is

$$x = \mathrm{e}^{-\zeta\omega_{\mathrm{n}}t}(A\cos\omega_{\mathrm{d}}t + B\sin\omega_{\mathrm{d}}t) + \frac{a}{k}. \tag{2.42}$$

In this case, the initial conditions are

$$x = \dot{x} = 0 \text{ at } t = 0$$

where

$$\dot{x} = \mathrm{e}^{-\zeta\omega_{\mathrm{n}}t}\omega_{\mathrm{d}}(-A\sin\omega_{\mathrm{d}}t + B\cos\omega_{\mathrm{d}}t) + (A\cos\omega_{\mathrm{d}}t + B\sin\omega_{\mathrm{d}}t)\left(-\zeta\omega_{\mathrm{n}}\mathrm{e}^{-\zeta\omega_{\mathrm{n}}t}\right) \tag{2.43}$$

Substituting the first initial condition, $x = 0$ at $t = 0$, into equation (2.42) gives

$$0 = 1(A + 0) + \frac{a}{k}.$$

Therefore

$$A = \frac{-a}{k}.$$

Substituting the second initial condition, $\dot{x} = 0$ at $t = 0$, into equation (2.43) gives:

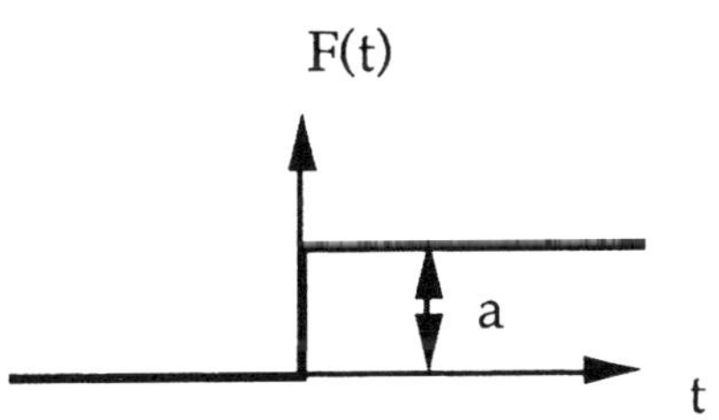

Figure 2.18 Step exciting force.

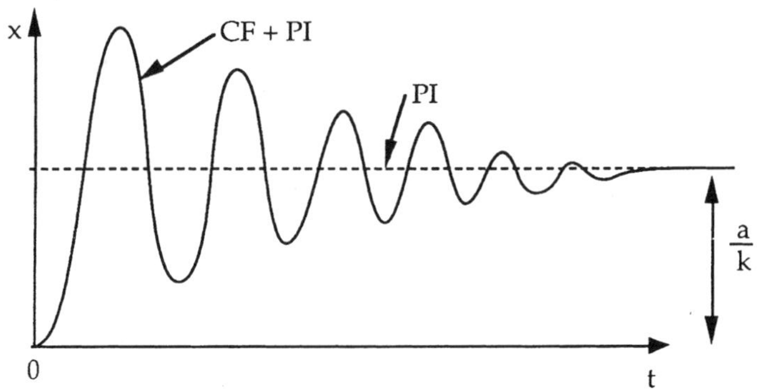

Figure 2.19 Response of system to step force.

$$0 = 1\omega_d(0 + B) + (A + 0)(-\zeta\omega_n).$$

Therefore

$$B = \frac{\zeta\omega_n}{\omega_d}A$$

and

$$x = \frac{a}{k}\left[1 - e^{-\zeta\omega_n t}\left(\cos\omega_d t + \frac{\zeta\omega_n}{\omega_d}\sin\omega_d t\right)\right]. \tag{2.44}$$

Figure 2.19 shows how the system response varies with time. It should be noted that the time history of the applied force can be at least qualitatively obtained by constructing a mean line through the oscillating total response. This can be a very useful way of recognizing the source of a vibration problem.

2.5.2 Response to a rectangular pulse

The above result can be used to determine the response to a more complex force. For example if $F(t)$ is a rectangular pulse, as shown in Fig. 2.20(a), this force time history can be constructed by superimposing two step functions as shown in Fig. 2.20(b).

Since the system is linear, the response to the rectangular pulse can be determined by summing the responses to each step function. This is shown in Fig. 2.20(c).

The response to the first step function is

$$x_1 = \frac{a}{k}\left[1 - e^{-\zeta\omega_n t}\left(\cos\omega_d t + \frac{\zeta\omega_n}{\omega_d}\sin\omega_d t\right)\right] : x_1 = 0 \text{ for } t < 0. \tag{2.45}$$

The response to the second step function is obtained from the first by substituting $(t - \varepsilon)$ for (t) and $(-a)$ for (a), viz:

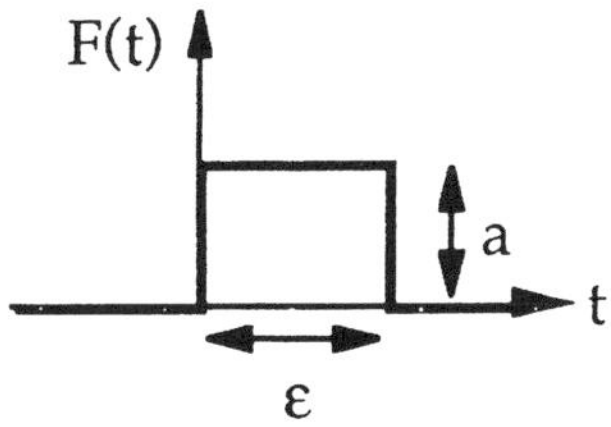

(a)

Figure 2.20 (a) Rectangular force pulse.

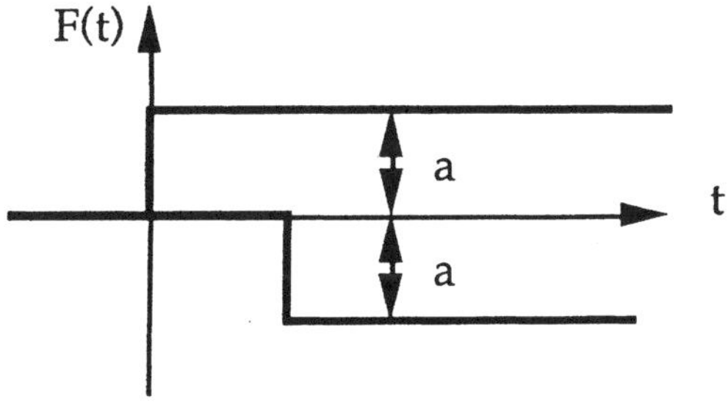

(b)

Figure 2.20 (b) Components of rectangular pulse.

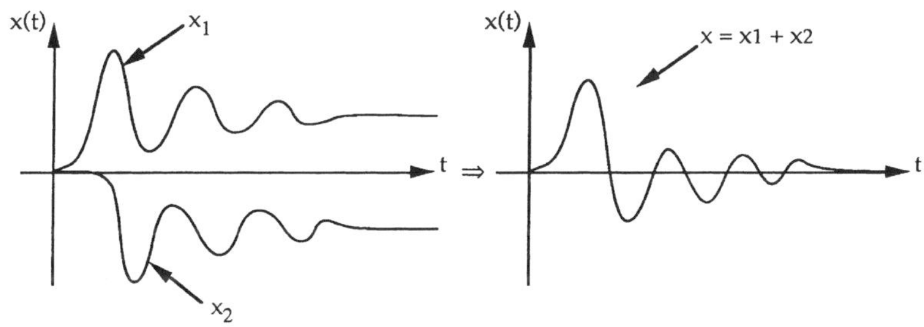

(c)

Figure 2.20 (c) Response of system to rectangular pulse.

$$x_2 = \frac{-a}{k}\left[1 - e^{-\zeta\omega_n(t-\varepsilon)}\left(\cos\omega_d[t-\varepsilon] + \frac{\zeta\omega_n}{\omega_d}\sin\omega_d[t-\varepsilon]\right)\right] : x_2 = 0 \text{ for } t < \varepsilon \tag{2.46}$$

The combined response is

$$x = x_1 + x_2. \tag{2.47}$$

2.5.3 Response to a force impulse

A further extension of this analysis is to consider the special case of a rectangular pulse of infinite magnitude and zero duration. Using the relationships given below in square brackets to expand the above expression for x_2 gives:

$$x_2 = \frac{-a}{k}\left[1 - e^{-\zeta\omega_n t} e^{\zeta\omega_n \varepsilon}(\cos\omega_d t \cos\omega_d \varepsilon + \sin\omega_d t \sin\omega_d \varepsilon \right.$$
$$\left. + \frac{\zeta\omega_n}{\omega_d}\{\sin\omega_d t \cos\omega_d \varepsilon - \cos\omega_d t \sin\omega_d \varepsilon\})\right].$$

$$\begin{bmatrix} \cos(A-B) & = \cos A \cos B + \sin A \sin B \\ \sin(A-B) & = \sin A \cos B - \cos A \sin B \\ e^{(\alpha+\beta)} & = e^{\alpha}.e^{\beta} \\ e^{\theta} & = 1 + \theta + \frac{\theta^2}{2!} + \frac{\theta^3}{3!} + \cdots \end{bmatrix}$$

Note that as ε tends to zero, $\cos\omega_d\varepsilon \simeq 1$, $\sin\omega_d\varepsilon \simeq \omega_d\varepsilon$, $e^{\zeta\omega_n\varepsilon} \simeq (1 + \zeta\omega_n\varepsilon)$. Therefore,

$$x_2 \simeq \frac{-a}{k}\left[1 - e^{\zeta\omega_n t}(1 + \zeta\omega_n\varepsilon)(\cos\omega_d t + \omega_d\varepsilon\sin\omega_d t \right.$$
$$\left. + \frac{\zeta\omega_n}{\omega_d}\{\sin\omega_d t - \omega_d\varepsilon\cos\omega_d t\})\right]$$
$$\simeq \frac{-a}{k}\left[1 - e^{-\zeta\omega_n t}\left(\cos\omega_d t + \frac{\zeta\omega_n}{\omega_d}\sin\omega_d t + \varepsilon\left\{\omega_d\sin\omega_d t + \frac{\zeta^2\omega_n^2}{\omega_d}\sin\omega_d t\right\}\right)\right].$$

Neglecting terms in ε^2 and above gives:

$$x = x_1 + x_2 = \frac{a}{k}\left[e^{-\zeta\omega_n t}\left(\varepsilon\left\{\omega_d\sin\omega_d t \frac{\zeta^2\omega_n^2}{\omega_d}\sin\omega_d t\right\}\right)\right].$$

Therefore

$$x = \frac{(a\varepsilon)}{k\omega_d}e^{-\zeta\omega_n t}(\omega_d^2 + \zeta^2\omega_n^2)\sin\omega_d t.$$

Now,

$$\omega_d = \omega_n\sqrt{1-\zeta^2}$$

or

$$\omega_d^2 + \zeta^2\omega_n^2 = \omega_n^2.$$

Therefore

$$x = \frac{(a\varepsilon)\omega_n^2}{k\omega_d}e^{-\zeta\omega_n t}\sin\omega_d t = \frac{(a\varepsilon)}{m\omega_d}e^{-\zeta\omega_n t}\sin\omega_d t \tag{2.48}$$

where $(a\varepsilon)$ is the area under the force time history of the impulse and is denoted by the symbol $\hat{F} = \int_t^{t+\varepsilon} F\mathrm{d}t$.

Thus the response to a unit impulse (i.e. $\hat{F} = 1$), which is termed the Impulse Response Function, is

$$g(t) = \frac{e^{-\zeta\omega_n t}}{m\omega_d} \sin \omega_d t. \tag{2.49}$$

This function is of considerable importance because it can be used to calculate the responses to any arbitrary force, as will be shown later.

It is of interest to calculate the height of the first peak since this will be the maximum response. Differentiating equation (2.48) with respect to time, equating to zero and choosing the root of the resulting equation for a maximum value gives

$$\omega_d \cos \omega_d t_m = \zeta\omega_n \sin \omega_d t_m.$$

where t_m is the time when this maximum occurs. Therefore

$$\tan \omega_d t_m = \frac{\omega_d}{\zeta\omega_n} = \frac{\sqrt{1-\zeta^2}}{\zeta}.$$

Referring to Fig. 2.21 it can be seen that

$$\sin \omega_d t_m = \sqrt{1-\zeta^2}.$$

Therefore

$$[x(t)]_{max} = \frac{(a\varepsilon)}{m\omega_d} e^{-\zeta\omega_n t_m} \sqrt{1-\zeta^2} = \frac{(a\varepsilon)}{m\omega_n} e^{-\zeta\omega_n t_m} \tag{2.50}$$

It is important to note that the height of first peak $[x(t)]_{max}$ is only slightly affected by damping.

This can be shown by reference to Fig. 2.22. Since

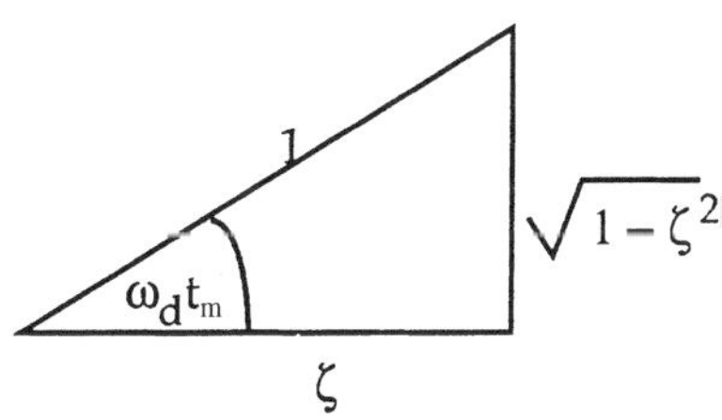

Figure 2.21 Definition of $\sin(\omega_d t_m)$.

$$t_{\mathrm{m}} \simeq \frac{T}{4}$$

$$\zeta\omega_{\mathrm{n}}t_{\mathrm{m}} \simeq \zeta\frac{2\pi}{T}\frac{T}{4} = \frac{\zeta\pi}{2}.$$

For example if $\zeta = 0.05$ (quite high damping), $\mathrm{e}^{-\zeta\pi/2} = 0.92$, i.e. the damping only reduces the peak amplitude by 8%.

This is the justification for applying a series of standard charts termed 'shock spectra' (which assume zero damping) to practical situations which do have damping.

2.5.4 Response to arbitrary force excitation

Figure 2.23 shows an arbitrary transient force time history of total duration t_{p}. This can be split up into a series of impulses each of which occurs at time $t = \psi$ and has a magnitude of $f(\psi)$, as shown cross-hatched in Fig. 2.24.

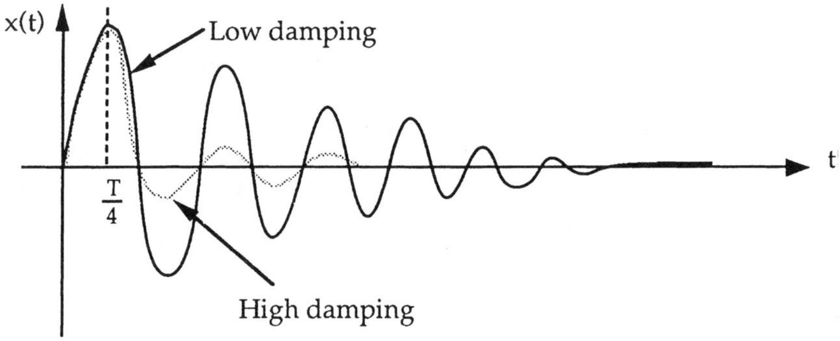

Figure 2.22 Response of system to impulsive force.

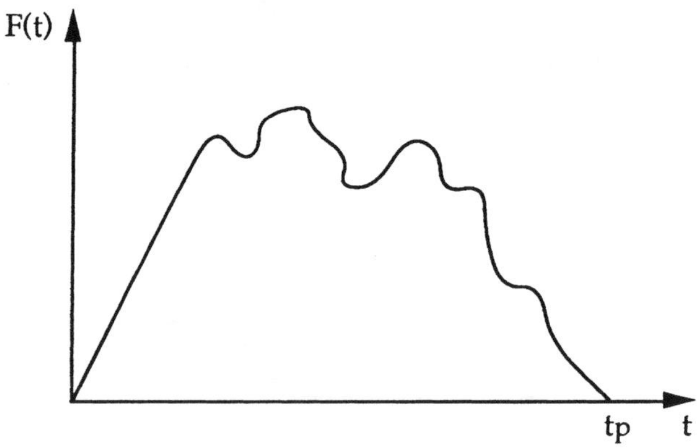

Figure 2.23 Arbitrary force–time history.

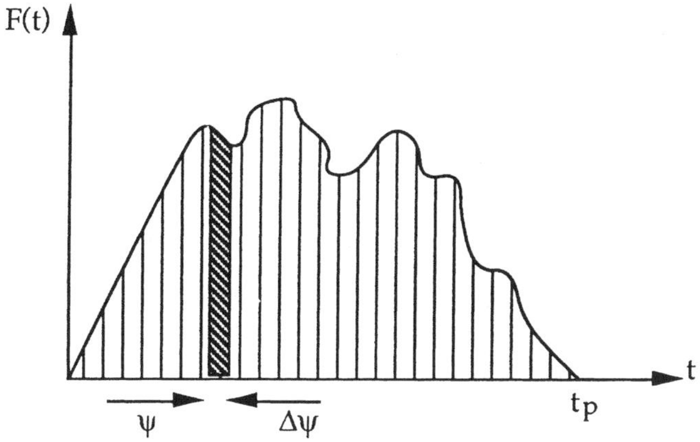

Figure 2.24 Force–time history as series of impulses.

Since the duration of this impulse is $\Delta\psi$, its area, $\hat{F} = f(\psi).\Delta\psi$. The resultant response is thus $\hat{F}$ times the response to a unit impulse, $g(t)$, as shown in Fig. 2.25.

Now because the response due to this impulse does not start at the origin, but, at $t = \psi$, it is expressed as

$$[f(\psi)\Delta\psi]g(t - \psi).$$

Therefore the total response at time t to the complete pulse $F(t)$ is obtained by summing the responses to all the impulses, viz:

$$x(t) = \int_0^t f(\psi)g(t - \psi)\mathrm{d}\psi. \tag{2.51}$$

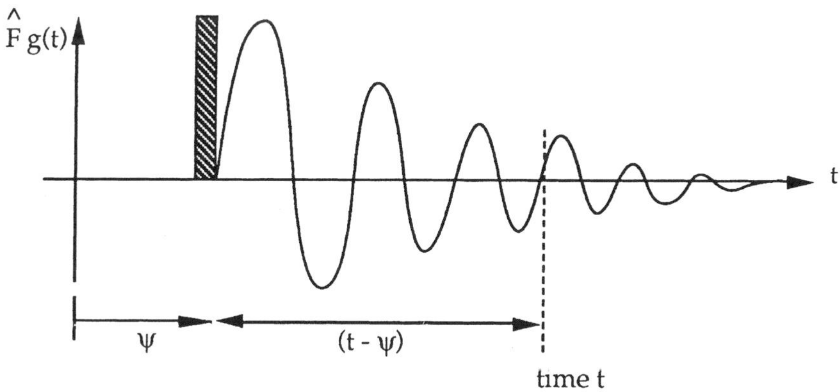

Elapsed time from point of impulse

Figure 2.25 Response of system to arbitrary impulse.

This is known as the Duhamel or Convolution integral.

Now if $t \geq t_p$, then

$$\int_0^t f(\psi)g(t-\psi)\mathrm{d}\psi = \int_0^{t_p} f(\psi)g(t-\psi)\mathrm{d}\psi.$$

This is because the value of the integral is zero for $t \geq t_p$, since $f(\psi)$ is then equal to zero.

2.5.5 Response to arbitrary motion of base

The equation of motion of the system shown in Fig. 2.26 is

$$m\ddot{x} = -k(x-y) - c(\dot{x}-\dot{y}).$$

If we make the substitution $z = x - y$, and note that $\dot{z} = \dot{x} - \dot{y}$ and $\ddot{z} = \ddot{x} - \ddot{y}$ then

$$m(\ddot{z}+\ddot{y}) + c\dot{z} + kz = 0$$

or

$$m\ddot{z} + c\dot{z} + kz = -m\ddot{y}. \tag{2.52}$$

Thus we can use equation (2.51) except that we must substitute z for x and $-m\ddot{y}$ for $f(t)$. The response of the mass to arbitrary motion of the base defined by $\ddot{y}$ is therefore

$$z(t) = \frac{-1}{\omega_\mathrm{d}}\int_0^t (\ddot{y}(\psi))\mathrm{e}^{-\zeta\omega_\mathrm{n}(t-\psi)}\sin\omega_\mathrm{d}(t-\psi)\mathrm{d}\psi. \tag{2.53}$$

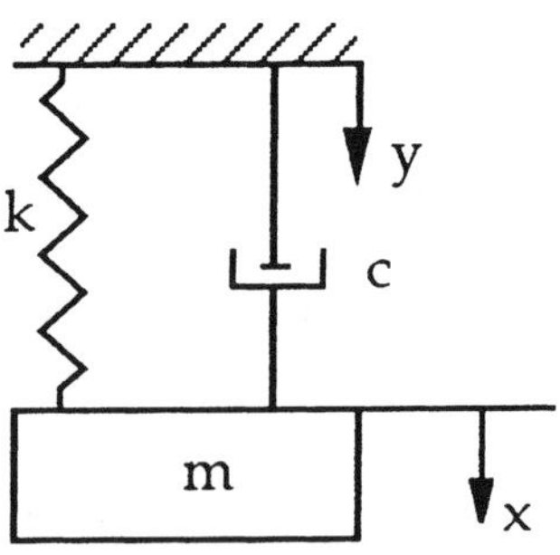

Figure 2.26 Damped system with base motion.

2.6 TWO DEGREE-OF-FREEDOM SYSTEM WITH VISCOUS DAMPING

Figure 2.27 shows the general configuration of a two degree-of-freedom system. It is analysed by applying Newton's Second Law of motion to each mass in turn.

Equation of motion for m_1:

$$m_1\ddot{x}_1 = -k_1x_1 + k_3(x_2 - x_1) - c_1\dot{x}_1 + c_3(\dot{x}_2 - \dot{x}_1) + F_1(t)$$

or

$$m_1\ddot{x}_1 + (c_1 + c_3)\dot{x}_1 - c_3\dot{x}_2 + (k_1 + k_3)x_1 - k_3x_2 = F_1(t). \tag{2.54}$$

Equation of motion for m_2:

$$m_2\ddot{x}_2 = -k_2x_2 - k_3(x_2 - x_1) - c_2\dot{x}_2 - c_3(\dot{x}_2 - \dot{x}_1) + F_2(t).$$

or

$$m_2\ddot{x}_2 + (c_2 + c_3)\dot{x}_2 - c_3\dot{x}_1 + (k_2 + k_3)x_2 - k_3x_1 = F_2(t) \tag{2.55}$$

These two equations can be written in matrix form as

$$\begin{bmatrix} m_1 & 0 \\ 0 & m_2 \end{bmatrix}\begin{Bmatrix} \ddot{x}_1 \\ \ddot{x}_2 \end{Bmatrix} + \begin{bmatrix} (c_1 + c_3) & -c_3 \\ -c_3 & (c_2 + c_3) \end{bmatrix}\begin{Bmatrix} \dot{x}_1 \\ \dot{x}_2 \end{Bmatrix} + \begin{bmatrix} -(k_1 + k_3) & -k_3 \\ -k_3 & (k_2 + k_3) \end{bmatrix}\begin{Bmatrix} x_1 \\ x_2 \end{Bmatrix} = \begin{Bmatrix} F_1(t) \\ F_2(t) \end{Bmatrix}. \tag{2.56}$$

This is a special case of the general equation for all two degree-of-freedom systems, viz

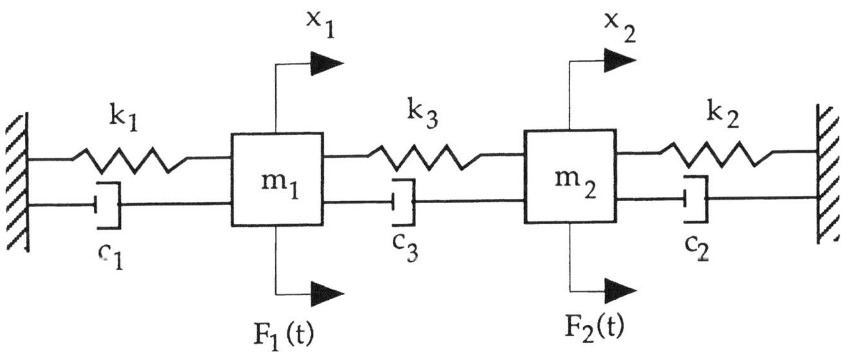

Figure 2.27 General two degree-of-freedom system.

$$\underset{\substack{\uparrow\\ \text{Mass}\\ \text{Matrix}}}{\begin{bmatrix} m_{11} & m_{12} \\ m_{21} & m_{22} \end{bmatrix}} \underset{\substack{\uparrow\\ \text{Acceleration}\\ \text{Vector}}}{\begin{Bmatrix} \ddot{x}_1 \\ \ddot{x}_2 \end{Bmatrix}} + \underset{\substack{\uparrow\\ \text{Damping}\\ \text{Matrix}}}{\begin{bmatrix} c_{11} & c_{12} \\ c_{21} & c_{22} \end{bmatrix}} \underset{\substack{\uparrow\\ \text{Velocity}\\ \text{Vector}}}{\begin{Bmatrix} \dot{x}_1 \\ \dot{x}_2 \end{Bmatrix}}$$

$$+ \underset{\substack{\uparrow\\ \text{Stiffness}\\ \text{Matrix}}}{\begin{bmatrix} k_{11} & k_{12} \\ k_{21} & k_{22} \end{bmatrix}} \underset{\substack{\uparrow\\ \text{Displacement}\\ \text{Vector}}}{\begin{Bmatrix} x_1 \\ x_2 \end{Bmatrix}} = \underset{\substack{\uparrow\\ \text{Force}\\ \text{Vector}}}{\begin{Bmatrix} F_1(t) \\ F_2(t) \end{Bmatrix}} \tag{2.57}$$

which can be written as

$$[M]\{\ddot{X}\} + [C]\{\dot{X}\} + [K]\{X\} = \{F\}. \tag{2.57a}$$

The elements of the above matrices actually have physical significance. For example, with the stiffness matrix:

k_{11} = Force at m_1 to produce unit positive deflection of m_1 $(x_2 = 0)$;
k_{12} = Force at m_1 to keep $x_1 = 0$ when m_2 is given unit positive deflection;
k_{21} = Force at m_2 to keep $x_2 = 0$ when m_1 is given unit positive deflection;
k_{22} = Force at m_2 to produce unit positive deflection of m_2 $(x_1 = 0)$.

The damping and mass matrices can be determined in a similar way by replacing unit deflection in the above by unit velocity or unit acceleration, respectively.

Note that $m_{12} = m_{21}$; $k_{12} = k_{21}$; $c_{12} = c_{21}$, i.e. all the matrices are symmetric because the system is linear.

For harmonic excitation

$$F_1(t) = F_{01}\mathrm{e}^{j\omega t}$$
$$F_2(t) = F_{02}\mathrm{e}^{j\omega t}$$

$$\left.\begin{aligned} x_1 &= a_1\mathrm{e}^{j\omega t} \\ x_2 &= a_2\mathrm{e}^{j\omega t} \end{aligned}\right\} \text{ note that in general } a_1,\ a_2 \text{ are complex quantities}$$

$$\dot{x} = j\omega a\mathrm{e}^{j\omega t} : \ddot{x} = -\omega^2 a\mathrm{e}^{j\omega t}$$

Substituting these relationships into equation (2.56) gives

$$\begin{bmatrix} (k_1 + k_3 + j\omega[c_1 + c_3] - \omega^2 m_1) & (-k_3 - j\omega c_3) \\ (-k_3 - j\omega c_3) & (k_2 + k_3 + j\omega[c_2 + c_3] - \omega^2 m_2) \end{bmatrix} \begin{Bmatrix} a_1 \\ a_2 \end{Bmatrix} = \begin{Bmatrix} F_{01} \\ F_{02} \end{Bmatrix} \tag{2.58}$$

or, in general,

$$[-\omega^2[M] + j\omega[C] + [K]]\{a\} = \{F_0\}. \tag{2.58a}$$

2.6.1 Natural frequencies of the system

For free undamped vibration

$$[-\omega^2 M + K]\{a\} = \{0\}$$

which, when expanded, gives

$$\begin{bmatrix} (-\omega^2 m_1 + k_1 + k_3) & -k_3 \\ -k_3 & (-\omega^2 m_2 + k_2 + k_3) \end{bmatrix} \begin{Bmatrix} a_1 \\ a_2 \end{Bmatrix} = \begin{Bmatrix} 0 \\ 0 \end{Bmatrix} \tag{2.59}$$

or

$$[K_\mathrm{d}]\{a\} = \{0\}$$

where $[K_\mathrm{d}]$ = Dynamic Stiffness Matrix.

Now one possible solution to this equation is for $\{a\} = \{0\}$ but this is the trivial case of zero vibration. The alternative is that the determinant of $[K_\mathrm{d}] = 0$ and this gives the roots (natural frequencies) of the system, viz

$$\omega^4 m_1 m_2 - \omega^2\{m_1(k_2 + k_3) + m_2(k_1 + k_3)\} + k_1 k_2 + k_1 k_3 + k_2 k_3 = 0. \tag{2.60}$$

This is a quadratic equation in ω^2 which has two possible roots (ω_1^2 and ω_2^2) because, in this case, the system has two natural frequencies (a system having N independent degrees of freedom will have N natural frequencies). Strictly speaking equation (2.60) has four roots, the additional two being $-\omega_1$ and $-\omega_2$. However, since these negative frequencies have no physical significance, they are ignored.

2.6.2 Modes of the system

From equation (2.59) we have

$$(-\omega^2 m_1 + k_1 + k_3)a_1 - k_3 a_2 = 0.$$

Therefore

$$\frac{a_1}{a_2} = \frac{k_3}{(-\omega^2 m_1 + k_1 + k_3)}. \tag{2.61}$$

Substituting $\omega^2 = \omega_1^2$ gives $\left(\frac{a_1}{a_2}\right)_{\text{mode 1}}$.

Substituting $\omega^2 = \omega_2^2$ gives $\left(\frac{a_1}{a_2}\right)_{\text{mode 2}}$.

Note that it is only possible to get the ratio of a_1 to a_2 because their absolute values are arbitrary.

2.6.3 Forced response

Writing equation (2.58) for the special case of zero damping (for simplicity of analysis) gives

$$\begin{bmatrix} (-\omega^2 m_1 + k_1 + k_3) & -k_3 \\ -k_3 & (-\omega^2 m_2 + k_2 + k_3) \end{bmatrix} \begin{Bmatrix} a_1 \\ a_2 \end{Bmatrix} = \begin{Bmatrix} F_{01} \\ F_{02} \end{Bmatrix} \tag{2.62}$$

or

$$[K_d]\{a\} = \{F_0\}. \tag{2.62a}$$

Thus

$$\begin{Bmatrix} a_1 \\ a_2 \end{Bmatrix} = [K_\mathrm{d}]^{-1} \begin{Bmatrix} F_{01} \\ F_{02} \end{Bmatrix}$$

or

$$\begin{Bmatrix} a_1 \\ a_2 \end{Bmatrix} = \frac{1}{|K_\mathrm{d}|} \begin{bmatrix} (-\omega^2 m_2 + k_2 + k_3) & k_3 \\ k_3 & (-\omega^2 m_1 + k_1 + k_3) \end{bmatrix} \begin{Bmatrix} F_{01} \\ F_{02} \end{Bmatrix}$$

where, $|K_\mathrm{d}|$ = the determinant value of K_d. This gives

$$a_1 = \frac{(-\omega^2 m_2 + k_2 + k_3)F_{01} + k_3 F_{02}}{|K_\mathrm{d}|} \tag{2.63}$$

$$a_2 = \frac{k_3 F_{01} + (-\omega^2 m_1 + k_1 + k_3)F_{02}}{|K_\mathrm{d}|} . \tag{2.64}$$

The analysis with damping is carried out in exactly the same way except that, because $[K_\mathrm{d}]$ is now complex, its inversion is somewhat more tedious to perform. The resultant values of a_1 and a_2 will also be complex so that their moduli must be obtained by taking the square root of the sum of the squares of the real and imaginary parts. As with the single degree-of-freedom system it is necessary to choose a datum for the phase. It is normal practice to use one of the exciting forces (e.g. F_1) for this purpose. Figure 2.28 shows the typical response of $|a_1|$ and $|a_2|$ with frequency for a damped system.

Note that both $|a_1|$ and $|a_2|$ reach maximum values at the same frequencies (ω_1 and ω_2) because the system is then at resonance, not the individual masses. This can be seen from equations (2.63) and (2.64), which both have the same denominator ($|K_\mathrm{d}|$) and this goes to a minimum value at each resonance of the system.

2.6.4 Illustrative example

The solution of the frequency equation and subsequent determination of the mode shapes is best illustrated by a simple example. Consider the special case where $k_1 = k_2 = k$ and $m_1 = m_2 = m$.

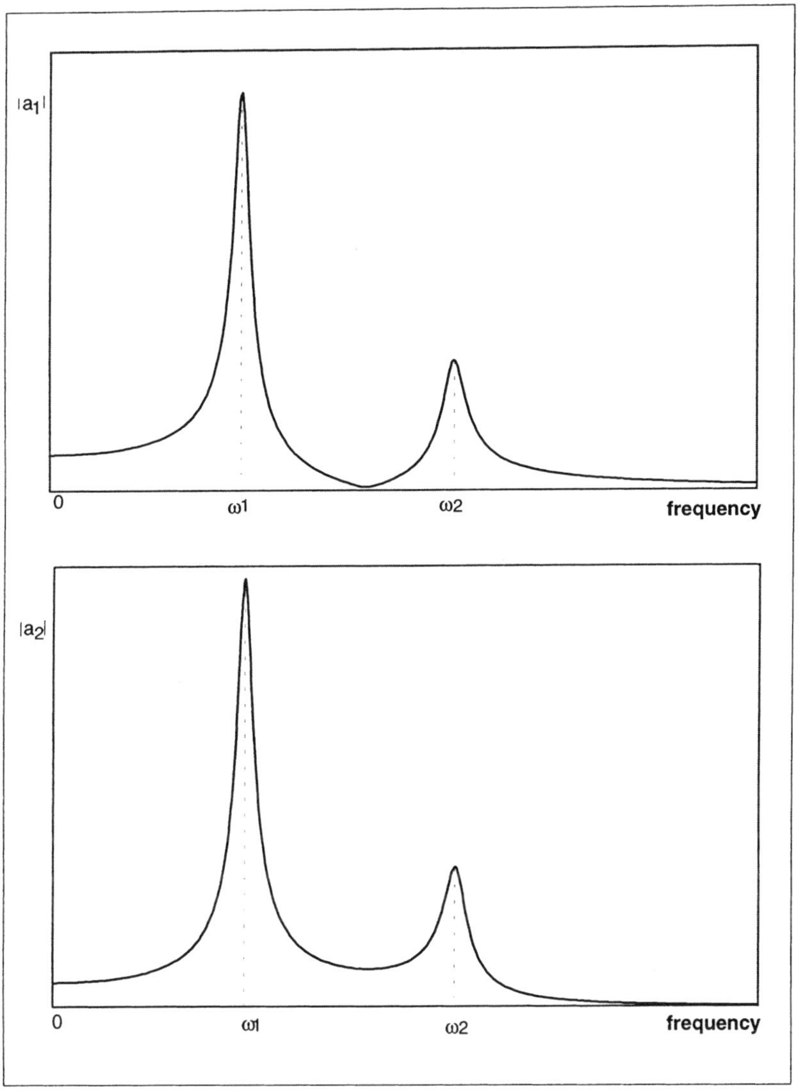

Figure 2.28 Forced response spectra of two degree-of-freedom system.

The frequency equation (2.60) becomes

$$\omega^4 m^2 - \omega^2\{m(k + k_3) + m(k + k_3)\} + k^2 + kk_3 + kk_3 = 0$$

or

$$\omega^4 m^2 - 2\omega^2 m(k + k_3) + k^2 + 2kk_3 = 0.$$

This gives

$$\omega^2 = \frac{2m(k+k_3) \pm \sqrt{4m^2(k+k_3)^2 - 4m^2k(k+2k_3)}}{2m^2}$$

$$= \frac{(k+k_3) \pm \sqrt{k^2 + 2kk_3 + k_3^2 - k^2 - 2kk_3}}{m}$$

or

$$\omega^2 = \frac{(k+k_3) \pm k_3}{m}.$$

Therefore

$$\omega_1^2 = \frac{k}{m} \quad \text{or} \quad \omega_1 = \sqrt{\frac{k}{m}}$$

and

$$\omega_2^2 = \frac{k+2k_3}{m} \quad \text{or} \quad \omega_2 = \sqrt{\frac{k+2k_3}{m}}.$$

Substituting these natural frequencies into equation (2.61) gives:
Mode 1: $\omega^2 = \omega_1^2$
Therefore

$$\left(\frac{a_1}{a_2}\right)_1 = \frac{k_3}{(-(k/m)m + k + k_3)} = 1$$

so that

$$a_1 = a_2 = a.$$

Thus the two masses move in-phase with the same amplitude, as shown in Fig. 2.29. Note that the natural frequency is independent of k_3 because there is no relative motion between the masses.
Mode 2: $\omega^2 = \omega_2^2$
Therefore

$$\left(\frac{a_1}{a_2}\right)_2 = \frac{k_3}{((-(k+2k_3)/m)m + k + k_3)} = -1$$

so that

$$a_1 = -a_2 = a.$$

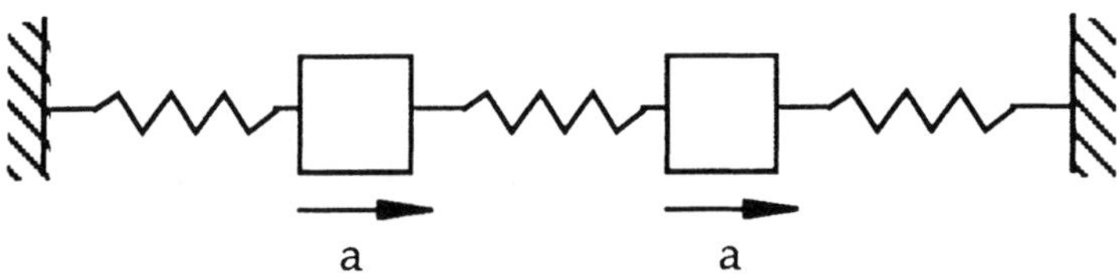

Figure 2.29 First mode of system.

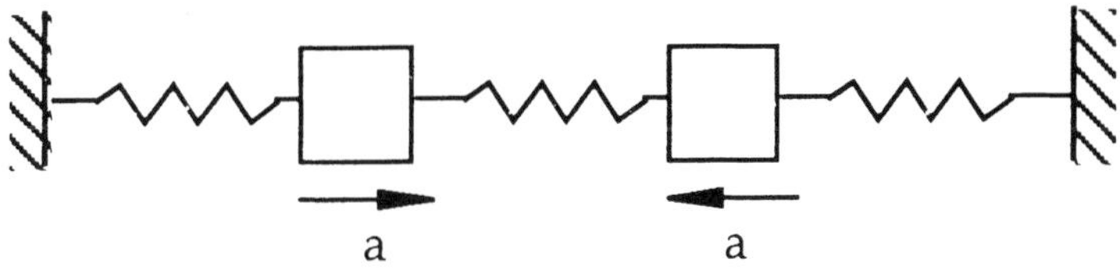

Figure 2.30 Second mode of system.

For the second mode the masses are in anti-phase but still have the same amplitude, as shown in Fig. 2.30.

The orthogonality principle, which states that the motion of one mode is uncoupled from that of another, means that these two modes can exist simultaneously, but independently. This also implies that the vibration time history of the system can always be expressed as the sum of the vibration in each of the two modes.

2.6.5 Co-ordinate coupling

Consider a rigid body mounted on two springs, with centre of gravity at G as shown in Fig. 2.31. Damping has been omitted in order to simplify the analysis. This system has two degrees of freedom because motion is restricted to either vertical (x) or rotation (θ) about an axis normal to the plane of the diagram. Here the deflection vector is $\begin{Bmatrix} x \\ \theta \end{Bmatrix}$.

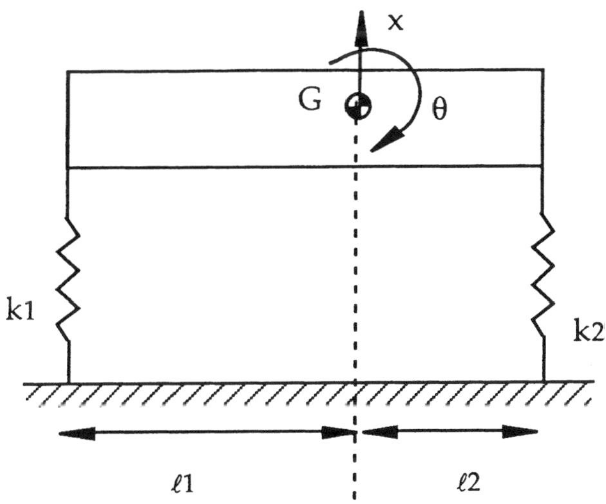

Figure 2.31 Rigid body system: degrees of freedom defined at centroid.

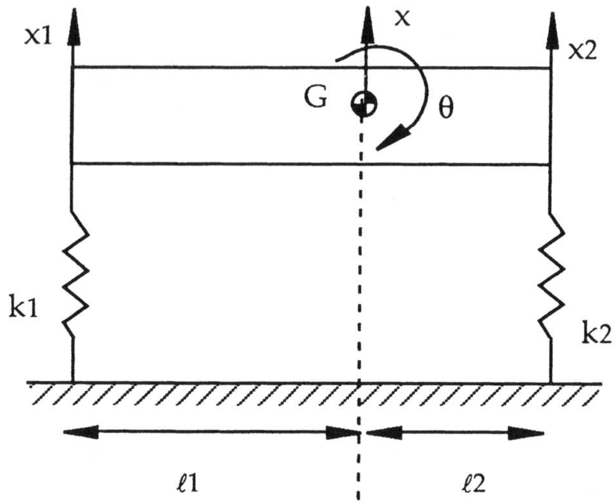

Figure 2.32 Rigid body system: degrees of freedom defined at ends.

Although x and θ turn out to be very convenient, the choice of these two parameters is, in fact, arbitrary and x_1 and x_2 (see Fig. 2.32) are equally valid.

The stiffness and mass matrices are most easily obtained from energy considerations. In the case of the stiffness matrix it is easier to first formulate the problem in terms of the x_1 and x_2 variables and then convert it to x and θ using the identities

$$x_1 = x + \ell_1\theta$$
$$x_2 = x - \ell_2\theta.$$

The Strain Energy, SE, of the system is given by

$$SE = \tfrac{1}{2}\{k_1x_1^2 + k_2x_2^2\}.$$

Substituting for x_1 and x_2 gives

$$= \tfrac{1}{2}\{k_1(x+\ell_1\theta)^2 + k_2(x-\ell_2\theta)^2\} = \tfrac{1}{2}\{k_1(x^2 + 2\ell_1x\theta + \ell_1^2\theta^2) + k_2(x^2 - 2\ell_2x\theta + \ell_2^2\theta^2)\}.$$

Therefore

$$SE = \tfrac{1}{2}\{x^2(k_1+k_2) + \theta^2(k_1\ell_1^2 + k_2\ell_2^2) + 2x\theta(\ell_1k_1 - \ell_2k_2)\}. \qquad (2.65)$$

Considering the part within the {} brackets, it can be seen that there is a x^2 term, a θ^2 term and a $x.\theta$ term. The coefficients of the squared terms represent the leading diagonal values. The coefficient of the $x.\theta$ term represents the value of the off-diagonal term. This coefficient is divided by two because there are always two of these terms in the matrix, i.e. the matrix is symmetric.

Therefore

$$[K] = \begin{bmatrix} (k_1 + k_2) & (\ell_1 k_1 - \ell_2 k_2) \\ (\ell_1 k_1 - \ell_2 k_2) & (k_1 \ell_1^2 + k_2 \ell_2^2) \end{bmatrix}.$$

The mass matrix is determined by consideration of the kinetic energy. In this case it is convenient to work directly in terms of x and θ, i.e.

$$\text{KE} = \tfrac{1}{2}\{M\dot{x}^2 + I_G\dot{\theta}^2\}$$

where M = mass of body, and I_G = moment of inertia about G.

Hence, applying the same principles as before

$$M = \begin{bmatrix} M & 0 \\ 0 & I_G \end{bmatrix}.$$

Note that there is no $\dot{x}\dot{\theta}$ term, so that the off-diagonal terms are zero.

Hence, the equation of motion for free vibration is

$$\begin{bmatrix} M & 0 \\ 0 & I_G \end{bmatrix} \begin{Bmatrix} \ddot{x} \\ \ddot{\theta} \end{Bmatrix} + \begin{bmatrix} (k_1 + k_2) & (k_1\ell_1 - k_2\ell_2) \\ (k_1\ell_1 - k_2\ell_2) & (k_1\ell_1^2 + k_2\ell_2^2) \end{bmatrix} \begin{Bmatrix} x \\ \theta \end{Bmatrix} = \begin{Bmatrix} 0 \\ 0 \end{Bmatrix}. \quad (2.66)$$

Thus the system is dynamically uncoupled because there is no $\ddot{x}.\ddot{\theta}$ term. However, it is statically coupled because there is an $x\theta$ term unless, by chance, $k_1\ell_1 = k_2\ell_2$.

If we now define the displacement about an arbitrary point P, the equations of motion now become

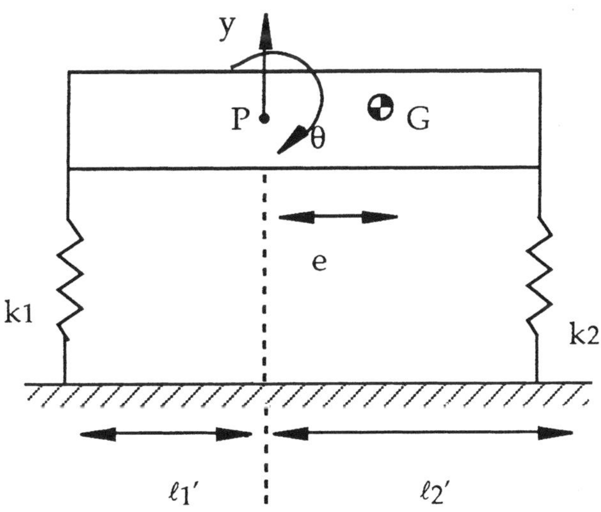

Figure 2.33 Rigid body system: degrees of freedom defined at arbitrary point.

$$\begin{bmatrix} M & -Me \\ -Me & I_P \end{bmatrix} \begin{Bmatrix} \ddot{y} \\ \ddot{\theta} \end{Bmatrix} + \begin{bmatrix} (k_1 + k_2) & (k_1\ell_1' - k_2\ell_2') \\ (k_1\ell_1' - k_2\ell_2') & (k_1\ell_1'^2 + k_2\ell_2'^2) \end{bmatrix} \begin{Bmatrix} y \\ \theta \end{Bmatrix} = \begin{Bmatrix} 0 \\ 0 \end{Bmatrix} \tag{2.67}$$

where, I_P = moment of inertia about $P = I_G + Me^2$.

By a suitable choice of ℓ_1', ℓ_2' the equations can be statically uncoupled, i.e.

$$\ell_1' + \ell_2' = \ell$$

so that

$$k_1\ell_1' = k_2(\ell - \ell_1').$$

Therefore

$$\ell_1' = \frac{k_2\ell}{(k_1 + k_2)} \quad : \ell_2' = \ell - \ell_1' = \frac{k_1\ell}{(k_1 + k_2)}.$$

However, the system is now dynamically coupled.

By using generalized co-ordinates the system can be made to be both statically and dynamically uncoupled. This has the advantage that each of the two separate equations in (2.66) is only a function of one variable so that it can be easily solved independently. Although this does not save much effort for this two degree-of-freedom system, the technique is very useful when the number of degrees of freedom is large. However, such analysis is outside the scope of this chapter so is not included here.

The natural frequencies and modes of the system are obtained from equation (2.66). For harmonic vibration

$$x = x_0 e^{j\omega t} \quad \text{and} \quad \theta = \theta_0 e^{j\omega t}.$$

Therefore

$$\ddot{x} = -\omega^2 x_0 e^{j\omega t} \quad \text{and} \quad \ddot{\theta} = -\omega^2 \theta_0 e^{j\omega t}.$$

Hence the matrix equation (2.66) becomes

$$\begin{bmatrix} (-\omega^2 M + k_1 + k_2) & (k_1\ell_1 - k_2\ell_2) \\ (k_1\ell_1 - k_2\ell_2) & (-\omega^2 I_G + k_1\ell_1^2 + k_2\ell_2^2) \end{bmatrix} \begin{Bmatrix} x_0 \\ \theta_0 \end{Bmatrix} = \begin{Bmatrix} 0 \\ 0 \end{Bmatrix} \tag{2.68}$$

or

$$[K_d]\{x\} = \{0\}.$$

The determinant of $[K_d]$ is equated to zero and the roots of the resultant frequency equation give the two natural frequencies, as before.

The determination of the modes is slightly different. Using the first of the two equations in the above matrix equation (2.68) we obtain:

$$\frac{x_0}{\theta_0} = \frac{(k_1\ell_1 - k_2\ell_2)}{(\omega^2 M - k_1 - k_2)}. \tag{2.69}$$

If $(k_1\ell_1 - k_2\ell_2) \neq 0$ then the two natural frequencies (ω_1 and ω_2) are substituted in turn into this equation, as before. The presence of the coupling term $(k_1\ell_1 - k_2\ell_2)$ means that one of the modes (not necessarily the first) will be predominantly a vertical motion (x_0), but with a small amount of twist (θ_0). The other mode will be predominantly twist, but with a small amount of vertical motion.

However, if $(k_1\ell_1 - k_2\ell_2) = 0$ the modes cannot be obtained from equation (2.69) because the numerator is zero and, in fact, the denominator will also be zero for one of the modes. The physical explanation is that for one mode the motion is entirely vertical (i.e. $\theta_0 = 0$) and for the other the motion is entirely twist (i.e. $x_0 = 0$). This occurs because x_0 and θ_0 are completely uncoupled in the equations. Fortunately, this means that x_0 can be found directly from the first equation in (2.68) and θ_0 directly from the second.

2.7 VIBRATION ABSORBER (NEUTRALIZER)

Consider the following two degree-of-freedom system shown in Fig. 2.34. In the analysis of this system damping will be neglected as it mainly affects the amplitude at resonance and has only a small effect otherwise.

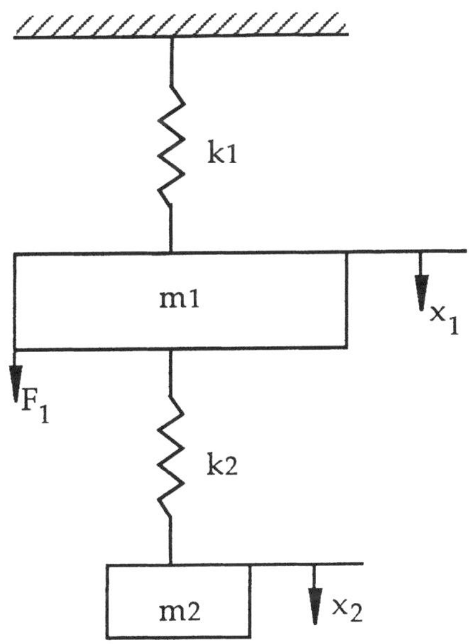

Figure 2.34 Mass–spring system with vibration absorber.

Suppose the main mass m_1 is excited by a harmonic force, $F_1 = F_0 e^{j\omega t}$, where ω is near to the resonant frequency of the main system. In the absence of the secondary system m_2, k_2, large amplitudes of vibration will result. If the frequency of the exciting force is fixed the problem can be cured by adding the secondary system.

For the combined system

$$M = \begin{bmatrix} m_1 & 0 \\ 0 & m_2 \end{bmatrix} : K = \begin{bmatrix} (k_1 + k_2) & -k_2 \\ -k_2 & k_2 \end{bmatrix}.$$

Thus the equations of motion become

$$\begin{bmatrix} (k_1 + k_2 - \omega^2 m_1) & -k_2 \\ -k_2 & (k_2 - \omega^2 m_2) \end{bmatrix} \begin{Bmatrix} a_1 \\ a_2 \end{Bmatrix} = \begin{Bmatrix} F_0 \\ 0 \end{Bmatrix}$$

where $x_1 = a_1 e^{j\omega t}$ and $x_2 = a_2 e^{j\omega t}$. Therefore

$$\begin{Bmatrix} a_1 \\ a_2 \end{Bmatrix} = \frac{1}{|K_d|} \begin{bmatrix} (k_2 - \omega^2 m_2) & k_2 \\ k_2 & (k_1 + k_2 - \omega^2 m_1) \end{bmatrix} \begin{Bmatrix} F_0 \\ 0 \end{Bmatrix}$$

where

$$|K_d| = (k_2 - \omega^2 m_2)(k_1 + k_2 - \omega^2 m_1) - k_2^2.$$

This gives

$$a_1 = \frac{(k_2 - \omega^2 m_2) F_0}{(k_2 - \omega^2 m_2)(k_1 + k_2 - \omega^2 m_1) - k_2^2}$$

$$a_2 = \frac{k_2 F_0}{(k_2 - \omega^2 m_2)(k_1 + k_2 - \omega^2 m_1) - k_2^2}.$$

We divide the numerator and denominator by $k_1 k_2$ and define

$$\omega_a^2 = \frac{k_2}{m_2}$$

$$\Omega_n^2 = \frac{k_1}{m_1}.$$

Therefore,

$$a_1 = \frac{(F_0/k_1)(1 - (\omega^2/\omega_a^2))}{(1 - (\omega^2/\omega_a^2))(1 + (k_2/k_1) - (\omega^2/\Omega_n^2)) - (k_2/k_1)}$$

and

$$a_2 = \frac{(F_0/k_1)}{(1 - (\omega^2/\omega_a^2))(1 + (k_2/k_1) - (\omega^2/\Omega_n^2)) - (k_2/k_1)}.$$

Let

$$\left(\frac{F_0}{k_1}\right) = x_{\text{st}}.$$

Therefore,

$$\frac{a_1}{x_{\text{st}}} = \frac{1 - (\omega^2/\omega_{\text{a}}^2)}{\left(1 - (\omega^2/\omega_{\text{a}}^2)\right)\left(1 + (k_2/k_1) - (\omega^2/\Omega_{\text{n}}^2)\right) - (k_2/k_1)} \tag{2.70}$$

and

$$\frac{a_2}{x_{\text{st}}} = \frac{1}{(1 - (\omega^2/\omega_{\text{a}}^2))(1 + (k_2/k_1) - (\omega^2/\Omega_{\text{n}}^2)) - (k_2/k_1)}. \tag{2.71}$$

At resonance, the amplitudes (a_1 and a_2) of both masses will go to infinity (no damping) and for this to happen $|K_{\text{d}}| = 0$. This gives the frequency equation as

$$\omega^4 - \omega^2\left(\frac{k_1}{m_1} + \frac{k_2}{m_1} + \frac{k_2}{m_2}\right) + \frac{k_1 k_2}{m_1 m_2} = 0. \tag{2.72}$$

It will be seen that the numerator of equation (2.70) is equal to $(1 - \omega^2/\omega_{\text{a}}^2)$. This will become zero when $\omega = \omega_{\text{a}}$. Thus the main mass, m_1 will stop moving when the frequency of the exciting force equals the natural frequency of the secondary mass–spring system. At this frequency the secondary mass will have an amplitude of

$$a_2 = \frac{-F_0}{k_2}.$$

The negative sign is because the secondary mass is exactly 180° out-of-phase with the force applied to the primary mass.

It must be stressed, however, that this technique only works when the excitation is at a fixed frequency. Of course, if this single-frequency excitation is well removed from the resonant frequency of the system, amplitudes will be low anyway and it is hardly worth taking remedial action. However, if the excitation is at or close to resonance, large amplitudes of the main mass will result and a simple solution to the problem is to attach a second mass–spring system, as described. Under these circumstances, the secondary mass–spring system is known as a dynamic vibration absorber. It should be pointed out that the name 'absorber' is rather misleading because the secondary system does not absorb any energy (indeed, it cannot do so in the undamped system described). The vibration reduction is because the exciting force F_0 is cancelled (neutralized) by an equal and opposite force from the secondary spring. It must be remembered that when such an absorber is fitted, the whole system will then have two natural frequencies (one either side of the excitation frequency) so that any drift in the excitation frequency will again produce large amplitudes of the main mass.

The size of the vibration absorber mass will be mainly dependent upon the amount of vibration amplitude of the absorber that can be tolerated – the lower the mass (and hence the lower k_2) the greater will be the amplitude of vibration ($a_2 = -F_0/k_2$.)

The distance between the two resonant peaks is also dependent upon the absorber mass. If we define the mass ratio as

$$\mu = \frac{m_2}{m_1}$$

and consider the case where the main system (m_1, k_1) is excited at resonance. i.e.

$$\omega = \omega_a = \Omega_n.$$

Then the roots of the frequency equation (2.72) become:

$$\left(\frac{\omega_1}{\omega_a}\right)^2, \left(\frac{\omega_2}{\omega_a}\right)^2 = \left(1 + \frac{\mu}{2}\right) \pm \sqrt{\mu + \frac{\mu^2}{4}}. \tag{2.73}$$

Figure 2.35 shows a plot of the variation of ω_1/ω_a and ω_2/ω_a with μ, from equation (2.73). It can be seen that the resonant peaks become further apart as the absorber mass (μ) increases. Conversely, when the absorber mass is zero (i.e. there is no absorber), there is a single peak, as would be expected, at the natural frequency of the original system (Ω_n).

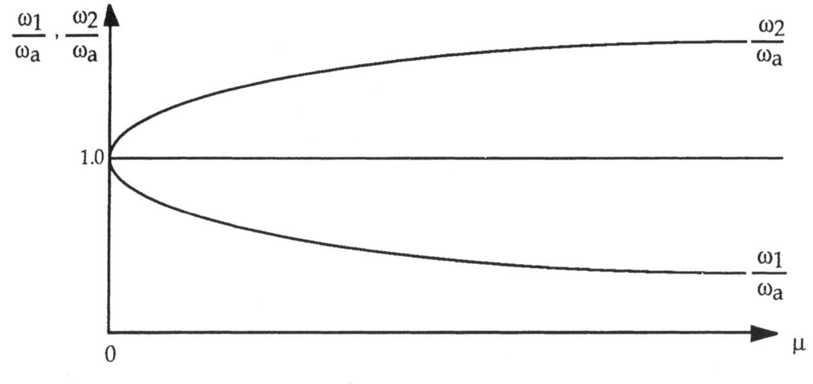

Figure 2.35 Effect of mass ratio on natural frequencies.

2.8 LAGRANGE EQUATIONS

Up until this point Newton's Second Law has been invoked to derive the equations of motion. Because vector quantities are involved, extreme care must be taken to ensure strict adherence to the sign convertion. A way out of this difficulty is to formulate the equations of motion from energy considerations. Largrange, a French mathematician, was first to achieve this and the resulting equations are named after him.

2.8.1 Single degree-of-freedom system

In this case the Lagrange equation is

$$\frac{\mathrm{d}}{\mathrm{d}t}\left(\frac{\partial T}{\partial \dot{x}}\right)+\frac{\partial D}{\partial \dot{x}}+\frac{\partial U}{\partial x}=f \tag{2.74}$$

where

T = kinetic energy = $\frac{1}{2}m\dot{x}^2$
U = strain energy = $\frac{1}{2}kx^2$
D = dissipation function = $\frac{1}{2}c\dot{x}^2$(viscous damping)
f = exciting force.

Hence,

$$\left(\frac{\partial T}{\partial \dot{x}}\right)=m\dot{x}$$

$$\frac{\mathrm{d}}{\mathrm{d}t}\left(\frac{\partial T}{\partial \dot{x}}\right)=m\ddot{x} \tag{2.75}$$

and

$$\frac{\partial D}{\partial \dot{x}}=c\dot{x} \tag{2.76}$$

$$\frac{\partial U}{\partial x}=kx. \tag{2.77}$$

Substituting equations (2.75), (2.76) and (2.77) into (2.74) gives

$$m\ddot{x}+c\dot{x}+kx=f$$

which is the result previously determined for a single-degree-of-freedom system.

2.8.2 Multi-degree-of-freedom system

Although the previous case is a trivial one and is included only to introduce the technique, the Lagrange equations are very useful for determining the equations of motion of complex systems.

For more than one degree of freedom the equations are

$$\frac{\mathrm{d}}{\mathrm{d}t}\left(\frac{\partial T}{\partial \dot{\boldsymbol{q}}_j}\right)+\frac{\partial D}{\partial \dot{\boldsymbol{q}}_j}+\frac{\partial U}{\partial \boldsymbol{q}_j}=\boldsymbol{Q}_j \qquad j=1,2,\ldots \tag{2.78}$$

where

$$\{\boldsymbol{q}\}=\begin{Bmatrix} x_1 \\ \vdots \\ x_n \end{Bmatrix} : \boldsymbol{Q}=\begin{Bmatrix} f_1 \\ \vdots \\ f_n \end{Bmatrix}$$

and

$$T=\frac{1}{2}\{\dot{\boldsymbol{q}}\}^T[M]\{\dot{\boldsymbol{q}}\} : \frac{\mathrm{d}}{\mathrm{d}t}\left(\frac{\partial T}{\partial \dot{\boldsymbol{q}}}\right)=[M]\{\ddot{\boldsymbol{q}}\} \tag{2.79}$$

$$D=\frac{1}{2}\{\dot{\boldsymbol{q}}\}^T[C]\{\dot{\boldsymbol{q}}\} : \frac{\partial D}{\partial \dot{\boldsymbol{q}}}=[C]\{\dot{\boldsymbol{q}}\} \tag{2.80}$$

$$U=\frac{1}{2}\{\boldsymbol{q}\}^T[K]\{\boldsymbol{q}\} : \frac{\partial U}{\partial \boldsymbol{q}}=[K]\{\boldsymbol{q}\}. \tag{2.81}$$

Substituting equations (2.79), (2.80) and (2.81) into (2.78) gives

$$[M]\{\ddot{\boldsymbol{q}}\}+[C]\{\dot{\boldsymbol{q}}\}+[K]\{\boldsymbol{q}\}=\boldsymbol{Q}. \tag{2.82}$$

It can be seen that equation (2.82) is of exactly the same form as equation (2.57a).

Of course, the above analysis assumes a knowledge of M, C and K. However, for a lumped parameter system (i.e. with discrete masses, dampers and springs) the values of T, D and U can be determined directly in terms of the various parameters.

Consider, as an example, the two degree-of-freedom system shown in Fig. 2.36. In this case

$$T=\tfrac{1}{2}m_1\dot{x}_1^2+\tfrac{1}{2}m_2\dot{x}_2^2$$

$$D=\tfrac{1}{2}c_1\dot{x}_1^2+\tfrac{1}{2}c_2(\dot{x}_2-\dot{x}_1)^2=\tfrac{1}{2}(c_1+c_2)\dot{x}_1^2-c_2\dot{x}_1\dot{x}_2+\tfrac{1}{2}c_2\dot{x}_2^2.$$

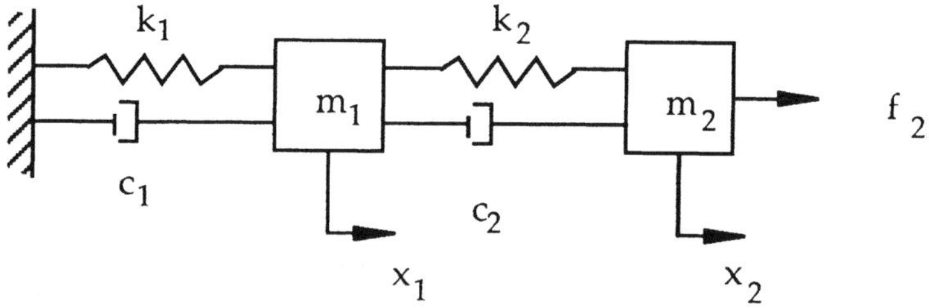

Figure 2.36 Forced two degree-of-freedom system.

$$U = \tfrac{1}{2}k_1x_1^2 + \tfrac{1}{2}k_2(x_2 - x_1)^2 = \tfrac{1}{2}(k_1 + k_2)x_1^2 - k_2x_1x_2 + \tfrac{1}{2}k_2x_2^2$$

Taking the values of j in turn:
$j = 1$:

$$\frac{\partial T}{\partial \dot{x}_1} = m_1\dot{x}_1 : \frac{\mathrm{d}}{\mathrm{d}t}\left(\frac{\partial T}{\partial \dot{x}_1}\right) = m_1\ddot{x}_1$$

$$\frac{\partial D}{\partial \dot{x}_1} = (c_1 + c_2)\dot{x}_1 - c_2\dot{x}_2$$

$$\frac{\partial U}{\partial x_1} = (k_1 + k_2)x_1 - k_2x_2$$

$$\boldsymbol{Q}_1 = f_1 = 0.$$

Therefore

$$m_1\ddot{x}_1 + (c_1 + c_2)\dot{x}_1 - c_2\dot{x}_2 + (k_1 + k_2)x_1 - k_2x_2 = 0.$$

$j = 2$:

$$\frac{\mathrm{d}}{\mathrm{d}t}\left(\frac{\partial T}{\partial \dot{x}_2}\right) = m_2\ddot{x}_2$$

$$\frac{\partial D}{\partial \dot{x}_2} = -c_2\dot{x}_1 + c_2\dot{x}_2$$

$$\frac{\partial U}{\partial x_2} = -k_2x_1 + k_2x_2$$

$$\boldsymbol{Q}_2 = f_2.$$

Therefore

$$m_2\ddot{x}_2 - c_2\dot{x}_1 + c_2\dot{x}_2 - k_2x_1 + k_2x_2 = f_2.$$

Arranging these two equations in matrix form gives exactly the same mass and stiffness matrices as were obtained for the same system in section 2.7.

Note also that

$$T = \frac{1}{2}\{\dot{\boldsymbol{q}}\}^T[M]\{\dot{\boldsymbol{q}}\} = \frac{1}{2}\{\dot{x}_1, \dot{x}_2\}\begin{bmatrix} m_{11} & m_{12} \\ m_{21} & m_{22} \end{bmatrix}\begin{Bmatrix} \dot{x}_1 \\ \dot{x}_2 \end{Bmatrix}$$
$$= \frac{1}{2}\{(\dot{x}_1m_{11} + \dot{x}_2m_{21}), (\dot{x}_1m_{12} + \dot{x}_2m_{22})\}\begin{Bmatrix} \dot{x}_1 \\ \dot{x}_2 \end{Bmatrix}.$$

Therefore,

$$T = \tfrac{1}{2}\{\dot{x}_1^2m_{11} + \dot{x}_1\dot{x}_2m_{21} + \dot{x}_1\dot{x}_2m_{12} + \dot{x}_2^2m_{22}\}.$$

Comparing this with

$$T = \tfrac{1}{2}m_1\dot{x}_1^2 + \tfrac{1}{2}m_2\dot{x}_2^2,$$

it can be seen that

$$m_{11} = m_1$$
$$m_{22} = m_2$$
$$m_{12} = m_{21} = 0.$$

Therefore

$$[M] = \begin{bmatrix} m_1 & 0 \\ 0 & m_2 \end{bmatrix}.$$

By a similar process

$$[K] = \begin{bmatrix} (k_1 + k_2) & -k_2 \\ -k_2 & k_2 \end{bmatrix} : [C] = \begin{bmatrix} (c_1 + c_2) & -c_2 \\ -c_2 & c_2 \end{bmatrix}.$$

This is the origin of the energy method for determining $[K]$ and $[M]$ used previously in section 2.6.5.

2.9 QUESTIONS

1. Figure Q2.1 shows the ground level overpressure (i.e. in excess of atmospheric pressure) due to an aircraft flying overhead at supersonic speed.

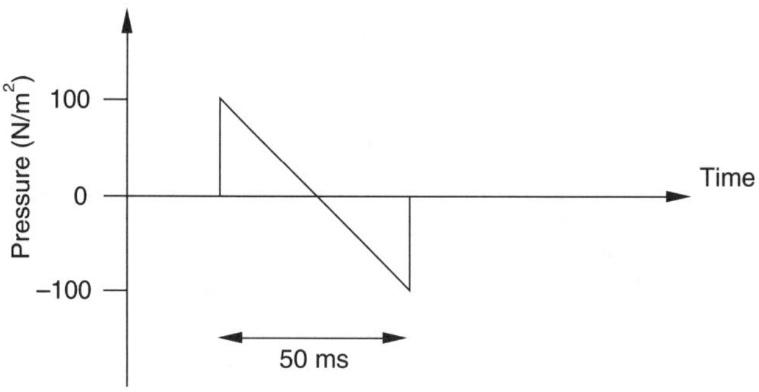

Figure Q2.1.

A window pane, having a natural frequency of 20 Hz, experiences this pressure shock and is set into vibration. Neglecting the effect of damping, calculate the amplitude of vibration of the window at this frequency, after the shock has passed.

Assume that the effective force on the window is equal to its surface area times the overpressure.

Relevant details of window are:

Surface area $= 1.0\,\mathrm{m}^2$
Modal mass $= 12.0\,\mathrm{kg}$.

The convolution integral is:

$$x(t) = \int_0^t f(\psi) \cdot g(t - \psi)\mathrm{d}\psi.$$

2. A generator set, which has a mass of 800 kg, is mounted on the floor of a building. Unfortunately it is found that, when the set is running at its constant design speed of 1500 rev/min, the out-of-balance force at this frequency excites the mass (generator set)–spring (floor) system exactly at resonance. In order to solve the problem it is decided to fit a vibration absorber.

   Calculate the mass of the absorber if its amplitude of vibration at the design condition is to be 1 cm when the amplitude of the out-of-balance generator force is 250 N.

   The following assumptions can be made:

   - the generator set acts as a point mass;
   - the floor acts as a massless, linear spring;
   - the effect of damping is negligible.

3. Figure Q2.2 shows a simple lumped parameter model of a torsional system. All the inertias in the figure are constrained such that their only freedom is rotation about the axis AB.

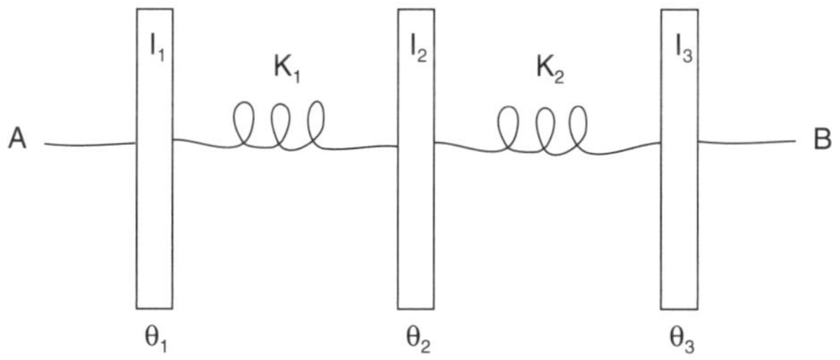

Figure Q2.2.

   Calculate the natural frequencies of the system and the relative amplitudes of θ_1, θ_2, and θ_3 at each mode of vibration, using the data given below.

   Sketch the three mode shapes.

$I_1 = I_2 = I_3 = 2.0 \times 10^{-2}\,\mathrm{kg\,m}^2$
$K_1 = K_2 = 1.25 \times 105\,\mathrm{Nm/rad}$.

4. A small electric motor driven pump unit, which has a mass of 5 kg, is supportted on a vibration isolator. It is observed that the isolator spring statically deflects 1 cm under the weight of the pump.

   (a) Calculate the natural frequency of the pump on its isolator.
   (b) The pump is deflected by 8 mm and is then released. After 10 cycles of oscillation its amplitude of vibration has reduced to 1 mm. What is the viscous damping ratio of the system?
   (c) The speed of the pump is then varied until its vibration amplitude reaches a maximum of 2 mm. What is the amplitude of the out-of-balance exciting force, at this speed?

Assume that the pump is a point mass and that the isolator is a simple linear, massless spring.

3

Fundamentals of human response to sound

I.H. FLINDELL

3.1 INTRODUCTION

In physical acoustics, the word 'sound' describes the transmission of vibrational waves through the atmosphere. In this chapter, however, we really need to think of 'sound' as the human perceptions of those same vibrational waves, and not the waves themselves. This distinction is important when we consider noise pollution, since sound only becomes noise when there is someone there to object to it. The highest levels of community noise only very rarely exceed 80 $L_{Aeq,16h}$ or 120 L_{Amax} (see Section 3.5.3 for an explanation of these noise units) and even at these high sound levels there are no permanent effects on the environment because the amount of physical energy involved is too small. In addition, even at these high sound levels, there are no direct effects on the body due to physical disruption or heating effects in the tissues. Transient sounds with peak overpressures reaching a fraction of one atmosphere within a few milliseconds can cause direct physical disruption, particularly to sensitive tissues such as the eardrum and the lungs, but these sounds only occur in association with extremely powerful sources such as gunfire, explosives, and rocket motors, not normal community noise.

At more typical levels of community noise caused by road traffic or noisy neighbours (normally in the range 45 $L_{Aeq,16h}$ to 75 $L_{Aeq,16h}$), the acoustic stimulation leads to a direct response in the auditory nerve and this may or may not lead to higher level nervous and endocrine (hormone) responses. These responses are essentially temporary adaptations by the body to meet changes in the external environment. It is by no means clear by what biological mechanisms, if any, these temporary (acute) adaptations might in turn lead to more long-lasting (chronic) indirect effects on the body, such as raised blood pressure or mental health problems. This is clearly not the same as the cumulative chemical poisoning caused by agricultural pesticides or similar chemical agents, where excess exposure leads to identifiable fatalities and where residues can be found months or years later in contaminated ground. As a result of this difference, it is the mainly indirect effects of noise and vibration on people such as long-term annoyance, interference with various activities, and possible health effects that are important, and not the actual levels of physical energy *per se*. The indirect effects are important because they form the only real basis for assessing the need for, and effectiveness of, noise control action.

The main objectives of this chapter are to show how human response to sound is dependant on auditory capabilities and on attitudes, memory, and expectation; to describe the function of the auditory system in relation to the underlying anatomy and physiology; and to show how current objective measurement practice can be related to these auditory and perceptual capabilities. The relevance to engineering acoustics is that when considering noise control action there are three different types of situation, as follows:

(a) where it is possible to eliminate an objectionable noise by completely effective noise control action;
(b) where the objectionable noise cannot be completely eliminated, but where residual nuisance can be balanced against the costs and inconvenience of noise control action;
(c) where, irrespective of noise control action, the objectionable noise remains completely unacceptable.

The likely human response to the noise is an important consideration where noise control action can have a material effect (situation (b) above), but whether the likely human response matters or not in situations (a) and (c) is debatable. Under situation (a), where the objectionable noise cannot be heard, then the noise cannot be a problem, by definition. Under situation (c), where the noise is completely unacceptable regardless of any noise control action, the only realistic solution is to keep people away from the noise. On the other hand, the criteria for acceptability and unacceptability should nevertheless be determined in the light of the likely human response.

A good example of situation (b) above is the current compromise which has been reached between the costs and inconvenience of controlling road traffic noise and the resulting noise nuisance. Road traffic is the most dominant source of community noise in England and Wales and in all other similarly developed countries of the world. Road traffic noise adversely affects a significant proportion of the population and, notwithstanding the fact there is no convincing scientific evidence in support, there is concern (Passchier-Vermeer *et al.*, 1996) that it could directly contribute to cardiovascular disease in a very small proportion of the population.

In Europe, the noise from individual vehicles has been reduced in recent years by successive reductions in new vehicle type approval noise limits (European Commission Green Paper, 1996). Mitigation in the form of noise barriers and sound insulating windows has undoubtedly helped in the case of new roads. In addition, new housing adjacent to main roads is often designed with the most noise-sensitive rooms facing away from the road. On the other hand, new roads continue to be built and the total amount of road traffic increases year on year. There is some evidence that, while the small proportion of the population exposed at the highest levels of road traffic noise has reduced over the past few years, the much larger proportions of the population exposed at successively lower levels of road traffic noise have increased. Data from two national noise surveys carried out in 1972 (Harland and Abbott, 1977) and in 1990 (Sargent and Fothergill, 1993) are plotted in Fig. 3.1 below to illustrate this.

The reduction in the proportion of the population exposed at the highest noise levels results from the introduction of reduced new vehicle type approval noise limits over the past few years. The increase in the proportion of the population exposed at lower noise levels results from the general increase in road traffic and from its wider distribution across the country as a result of people taking increasing advantage of the individual opportunities for travel provided by road transport. There is a trade-off here between the economic and opportunity benefits of road transport and the

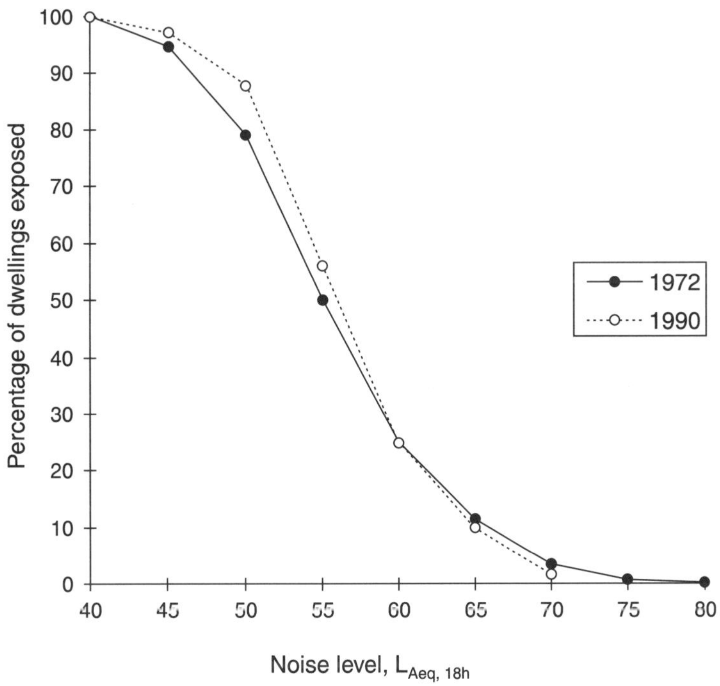

Figure 3.1 Distribution of noise exposure outside dwellings.

costs in air pollution and noise effects. While small improvements can and will continue to be made through the development of new technology, some of these improvements might be opposed by the population at large if they increased the costs of road transport significantly. In addition, the most effective way to make further significant improvements at a stroke would be to impose drastic restrictions on the individual freedom to own and use motor vehicles or to impose economic penalties on their use.

At the time of writing, it is clear that the environmental effects of unrestricted road traffic are not generally perceived as outweighing the benefits to society at large, although there are indications that the balance may be changing with increasing public awareness of the various issues involved. The precise nature of human responses to road traffic noise at different levels of exposure is an important part of the equation for future generations to consider.

3.2 NOISE EFFECTS

3.2.1 The range of effects of noise

Noise can have a wide range of effects on people, depending on the level and on the situation in which people are exposed. High-level sounds occurring in heavy industry and similar situations can damage hearing, lower level sounds occurring in a community context cause significant annoyance and can potentially damage non-auditory health, and even the lowest level sounds can have significant effects on the perceived quality of the environment.

Transient sounds reaching peak overpressures of a fraction of one bar within a few milliseconds caused by explosions, gunfire, or impulsive industrial processes can damage the ear drum and other parts of the ear through direct physical disruption. The peak action level of 200 pascals (140 dB peak unweighted SPL) included as part of the current Noise at Work Regulations 1989 (made under the Health and Safety at Work Act, 1974) is intended to reduce the risk of traumatic damage to hearing as far as possible. In practice, 200 pascals is only one 500th of one bar and the risk of traumatic damage is probably quite low at this level. A generous handclap or an enthusiastic door slam close to the ear can generate peak sound levels of 140 dB peak without too much difficulty. On the other hand, it is clear that regular exposure to transient sounds of this magnitude is something to be avoided.

Short-term exposure to continuous sounds at lower absolute sound levels (than 140 dB rms) can lead to a temporary loss of hearing sensitivity. This is known as a temporary threshold shift (TTS) and is often associated with tinnitus. Tinnitus, or '*ringing in the ears*' is an audible sensation occurring when there is no actual sound present. One of the possible causes of TTS appears to be fatigue of the sensitive hair cells in the inner ear. Hearing sensitivity begins to return immediately the noise exposure is ceased, although it may take several hours for full recovery to take place, depending on the degree of threshold shift.

Prolonged exposure at similar absolute sound levels to those that can cause TTS after short-term exposure can lead to a permanent loss of hearing sensitivity associated with progressive atrophy of the sensitive hair cells in the inner ear. The three most important factors are the noise level, the total duration of exposure to the noise and individual susceptibility, although the absence of a full recovery period between periods of noise exposure may also be important. The Noise at Work Regulations 1989 require both the employer and the employee to carry out a number of preventive actions where individual noise exposure exceeds 85 $L_{Aeq,8h}$ to reduce the risk of permanent hearing loss and there is a second over-riding limit of 90 $L_{Aeq,8h}$ which must not be exceeded. It is generally believed that the risk of noise induced hearing loss is negligible where individual noise exposure does not exceed 75 $L_{Aeq,8h}$ (Berglund and Lindvall, 1995). The various biological mechanisms leading to noise-induced hearing loss and the available preventive methods and systems are covered more extensively in Axelsson *et al.*, (1996).

Moving down the scale, we know from the survey carried out by the Building Research Establishment in 1990 (Sargent and Fothergill, 1993) that the main source of community noise outside homes in England and Wales is road traffic. An earlier survey (Morton-Williams, Hedges and Fernando, 1978) found that around 16% of the population (of England) described road traffic noise as the 'biggest nuisance' of those noise sources audible from outside their homes. The 1990 survey found that around 60% of the population of England and Wales were exposed to community noise levels outside their houses of between 50 and 60 $L_{Aeq,16h}$, and that the proportion exposed to outside noise levels exceeding 70 $L_{Aeq,16h}$ was around 2%.

At these noise levels, the amount of physical energy involved is microscopic. A sound pressure level (SPL) of 70 dB is only 0.06 N/m^2, and corresponds to an acoustic intensity in a free field (see Chapter 1) of only 10 $\mu W/m^2$. There is clearly no possibility of any direct disruptive effects on the tissues caused by such small amounts of physical energy. On the other hand, there can be a considerable amount of annoyance, particularly in the upper parts of the range of exposure. Noise at night can contribute to sleep disturbance, particularly where the noise source is relatively unfamiliar. Outdoors, community noise levels in the upper part of the range can easily interfere with speech. This can be particularly important where the 16-hour daytime average noise level comprises a sequence of short aircraft or railway noise events at higher absolute sound levels. Effective speech communications might be completely masked for relatively short periods of time, although the much longer periods at lower noise levels making up the long-term average must also be borne in mind. The various surveys of community noise and effects which have been carried out around the world over the last 30 to 40 years are covered more extensively in the chapter on 'Environmental and Transportation Noise' in Volume 2.

There is concern in the literature that community noise exposure at the higher levels (around 70 $L_{Aeq,16h}$ and above) has been associated with non-auditory health effects such as hypertension (high blood pressure), myocardial infarction (heart attacks), low birthweight babies, psychiatric problems and similar problems. None of the available data is conclusive and this might be either because there are no such effects, or because there are too many methodological problems to be able to design fully satisfactory research studies. However, the possibility of excessive noise leading to annoyance leading to some non-specified stress which eventually leads to non-auditory health effects in susceptible individuals is sufficiently plausible to justify continuing research in this area.

Other than in the field of noise annoyance, the most convincing research results to date suggest that aircraft noise can reduce scholastic achievement, although the available data is still open to different interpretations. A major problem with most noise and non-auditory health effects research is the difficulty of proving cause and effect just from an observed statistical association, particularly where there is no discussion of a plausible causative mechanism for the observed effect (Ludlow and Flindell, 1997).

Finally, there are many situations where the absolute noise level is less important than the character of the noise or even some completely non-acoustical aspect of the situation in determining the response. Examples of these situations might include complaints of low-level noise from a neighbour's TV or hi-fi stereo system, or from an unattended barking dog. The extent to which early morning wake-up calls made by a particularly keen cockerel should be considered as an implicit part of the rural environment has been debated a number of times in the courts. The noise complainant's perception of the degree of consideration shown by the person making the noise is often very important. In addition, it is often the case that particular acoustic features (Porter, Flindell and Berry, 1993) such as tonality or impulsivity which may be present are more important than the overall level. Acoustic features are particularly important in sound-quality considerations, where manufacturing industry is becoming increasingly concerned with the character of sound generated by their products. The precise character of the sounds generated by a particular type of tyre tread pattern or piston sidewall configuration can affect the outcome of competitive tenders to supply components to major vehicle manufacturers, and literally thousands of jobs can be affected if a component is perceived as making the wrong contribution to the overall sound quality of the vehicle.

3.2.2 Noise and perception

The human ear is obviously extremely sensitive to very small amounts of acoustic energy. Why is this? The information which the auditory system is able to extract from environmental sounds contributes to the competitive survival of the organism as a whole. We must not forget that the auditory system in humans has evolved to meet basic survival needs in pre-industrial settings and might not be particularly well adapted to modern requirements.

To understand auditory perception, we must understand the ways in which auditory information is processed at a number of different levels within the auditory system leading to successively more complex responses as the information progresses up the chain to the auditory cortex.

First, an incoming sound wave enters the ear canal leading to corresponding deflections of the ear drum. A system of small bones transmits these vibrations across the middle-ear cavity into the fluid filled inner-ear cavity known as the cochlea. The cochlea is a spiral organ located within the lower part of the skull which is divided into upper and lower chambers by the basilar membrane. The basilar membrane contains many thousands of sensitive hair cells disposed along its length. Different hair cells are stimulated by different frequencies of sound to transmit nerve impulses along the auditory nerve to the brain. There are numerous reflex connections along the way and there is also considerable evidence that the basilar membrane can 'tune in' to certain sounds through a feedback mechanism involving some of the hair cells. The information progresses along the auditory pathways as more or less complex patterns of nerve impulses until it reaches the auditory cortex where the highest level of perceptual processing takes place.

The two main functions of auditory perception are as follows:

(a) to assist in constructing an internal model, or perceptual construct, of the external surroundings;
(b) to understand speech.

The precise term 'perceptual construct' is used because the main focus of perception is to make sense of a constantly changing pattern of sensory stimulation by constructing a stable internal model of the external surroundings. As a person moves around within a complex acoustic environment, the balance between direct and reverberant sounds is constantly changing and certain wanted sounds may even be temporarily masked from time to time. The precise pattern of stimulation available at the different sense organs is likely to give conflicting information from one moment to the next. The perceptual processor has evolved in such a way that it attempts to sort everything out on the basis of previous experience of similar patterns of stimulation. For example, the first time one hears a foreign language spoken it may be impossible to make any sense of the utterances at all, yet after a period of familiarity, certain key words will begin to be recognized and eventually, most people will be able to repeat a selection of words with some fluency, even if they do not know precisely what the words actually mean. As another example, the loudness of a familiar sound as heard at the ear gives some clue as to the relative distance of the source, but this does not apply in the case of an unfamiliar sound whose relative distance might be completely misjudged, particularly where the surroundings are unfamiliar as well.

It is important to understand that because auditory perception is focused on representing the distant external surroundings, the auditory system is not well-adapted to making precise judgements of the level of stimulation at the ear. When asked to judge the loudness of a particular noise source, most people will consider the relative loudness of that particular source as compared to similar sources from the same class of object with due regard to the apparent distance of the source and will not pay much attention to the actual sound level at the ear. For example, a distant aircraft might be judged as noisier than a nearby electric-powered milk delivery vehicle even though the sound level generated by the milk float might actually be higher. This is because, when asked to judge subjective 'noisiness', people often tend towards describing their impression of the sound power of the source, not the sound level at the ear.

Perceptual constructs are formed on the basis of the evidence available to the organism at the time. This evidence comprises the acoustic stimuli, memory, and the evidence from the other senses. People do not pay attention to all the available evidence all the time and indeed, there is a mechanism of 'selective attention' where information deemed to be of no relevance is positively disregarded. People ignore acoustic stimuli while asleep, but the lower levels of neural processing within the auditory system are still fully operational and will rapidly cause awakening where there is an acoustic stimulus which either requires, or is deemed to require, some form of action. An unfamiliar sound is more likely to cause awakening than a familiar sound which can be easily categorized by some lower level of neural processing.

A good example of the hierarchy of lower and higher level processes which are carried out within the auditory system as a whole is the acoustic startle effect. Most people will have experienced this effect caused by unexpected sounds with a rapid onset rate. There is an entirely involuntary reflex action comprising a sequence of protective flexor muscle contractions which serve to make the body as small and inconspicuous as possible, and simultaneously to arrest whatever activity is in progress at the time. The heart rate and blood pressure are temporarily increased and there will usually be a number of slower acting endocrine system responses. These activities are all part of a sequence of preparing the body for action, if required. At the same time, a process of cognitive appraisal is initiated, characterized by a brief period of disorientation until the nature of the threat has been evaluated and appropriate action taken. The acoustic startle effect is stronger with higher-level sounds and the most rapid onset rates, but the main determinants are the unexpectedness of the sound and the degree of familiarity. Anecdotal evidence suggests that people can get used to the acoustic startle effect to some extent with familiar sounds, depending on the magnitude of the initial response.

The acoustic startle effect is just one of the many ways in which the body can react to sound. Since it is an entirely natural sequence of events, there is no *a priori* reason to suppose that it is inherently harmful, although there is an implied linkage between the resulting short-term physiological changes and possible long-term cardiovascular and endocrine effects in much of the literature on noise and non-auditory health. Probably the most important reason to suppose that it might be harmful under certain circumstances is where it causes annoyance. The concept of annoyance forms the key to understanding much of what is written about noise. Noise annoyance is effectively a higher-order perceptual construct which is applied to a particular source by an individual listener, and to describe it as an emotional state of the listener can be misleading. For example, a person can describe a particular noise source as being annoying although they display no outward signs of actually being annoyed. Probably, such a person is really saying that they would be annoyed by the noise if they were unable to keep their emotions under control. The whole concept of noise annoyance as being a property of the noise source is consistent with the general focus of perception on the external or distal environment.

3.2.3 Noise and speech interference

Returning to the understanding of speech, it is clear that the mechanisms within the auditory system which contribute to this have evolved on top of the underlying perceptual processors concerned with the extraction of more general information from the auditory environment. It appears that there are specialized systems within the auditory cortex which are particularly well adapted to the identification of speech features within sounds, but these must have evolved in parallel with the development of different speech sounds by the vocal tract. A basic understanding of the fundamental processes of auditory perception defines the building blocks on which any more detailed understanding of the mechanisms of speech perception must be based. Of course, speech perception also involves memory and expectation such that the same utterances can be perceived with different meanings in different linguistic contexts. In addition, visual perception (lip reading) can also play an important role. It is beyond the scope of this chapter to go into the complex biological mechanisms of speech perception in any more detail.

The speech signal can be partially or completely masked wherever there is an interfering noise or a reverberant sound field present. Speech intelligibility is defined as the proportion of words or sentences which are correctly understood and obviously varies across different talkers and listeners and under different listening situations. Most talkers will automatically compensate for a certain amount of background noise by using a raised voice level, but this does not happen where the speech sounds are reproduced through an electro-acoustic communications system. In addition, where the masking noise mainly comprises reverberation, using a raised voice level will merely raise the reverberant field level in the same proportion, without achieving any net benefit.

The loss of intelligibility under standardized conditions can be calculated from the ratio of the speech frequency spectrum to the masking noise frequency spectrum by taking the assumed frequency and time resolving powers of the peripheral auditory system into account. The classic Articulation Index (AI) calculation procedure (ANSI, S3.5, 1969) compares the long-term rms average one-third octave frequency band spectrum of the wanted speech signal against the long-term rms average one-third octave frequency band spectrum of the masking noise and then aggregates the separate contributions to overall intelligibility in each one-third octave frequency band according to the standardized weighting scheme which is given in Table 3.1.

Of course, the instantaneous level of any typical speech signal varies significantly over time. It is difficult to define any standardized procedure for measuring the true level of the speech peaks which contribute most to the

Table 3.1 Weightings for each one-third frequency band

Band centre frequency (Hz)	*Weighting*
200	0.0004
250	0.0010
315	0.0010
400	0.0014
500	0.0014
630	0.0020
800	0.0020
1000	0.0024
1250	0.0030
1600	0.0037
2000	0.0038
2500	0.0034
3150	0.0034
4000	0.0024
5000	0.0020

overall intelligibility, because of this inherent variability. In the AI calculation procedure, the speech peaks are assumed to rise 12 dB above the long-term rms average level, which is a much easier quantity to define. This is a reasonable assumption for normal speech signals, but might not apply in the case of shouting or whispering or where the dynamics of the speech signal have been compressed by either intentional audio processing or unintentional non-linearities and distortions in an electro-acoustic speech communications system.

In the AI calculation procedure, the assumed contribution to overall intelligibility made by each one-third octave frequency band varies from a maximum of 30 dB in each band where the long-term rms speech signal to masking noise ratio equals or exceeds plus 18 dB (the speech peaks are then assumed to be 30 dB above the long-term rms level of the masking noise), down to a minimum of zero where the long-term rms speech signal to masking noise ratio is equal to or less than −12 dB (the speech peaks are then assumed to be at the same level as the long-term rms level of the masking noise). The weightings for each one-third octave frequency band are as shown in Table 3.1.

Note that the 1600 Hz to 3150 Hz one-third octave frequency bands are assumed to contribute the most to overall intelligibility, and that the 200 Hz one-third octave frequency band is assumed to contribute the least. In addition, note that a wideband 30 dB speech peak to long-term rms masking noise ratio multiplied by the weightings in each of the 16 one-third octave bands adds to 1. A wideband 15 dB speech peak to long-term rms masking noise ratio multiplied by the weightings in each of the 16 one-third octave bands adds to 0.5 and a wideband 0 dB speech peak to long-term rms masking ratio in each of the 16 one-third octave bands adds to 0.

The resulting AI score varies from 1 to 0 where an AI score of 0.5 is usually considered acceptable for an electroacoustic communications system, and an AI score or 0.7 is considered desirable where the communications system is 'to be used under a variety of stress conditions and by a large number of different talkers and listeners having varying degrees of skill'.

BS 7443: 1991 requires a Speech Transmission Index (STI) value of greater than or equal to 0.5 for sound systems used for emergency purposes. The STI measurement is very similar to the AI, but a special test signal is used to allow standardization of the measurement procedure. The STI test signal (and the simplified Rapid Speech Transmission Index, RASTI, test signal (BS6840: Part 16: 1989)) is made up from amplitude modulated narrow band noise signals to simulate typical speech, and the speech signal to noise ratios are then inferred from the measured modulation transfer functions in each frequency band. The AI calculation procedure requires that the speech signal (which is not precisely defined) and the masking noise should be measured separately, which is not always practical, particularly where the masking noise results mainly from reverberant sound following on from previous utterances in the speech signal. The STI measures the effect of both reverberation and steady masking noise at the same time in terms of the resulting reductions in the modulation transfer functions through the speech communications system as a whole. The modulation transfer functions can also be calculated from the measured system impulse response functions, and this can be very useful in system design where the impulse response can be predicted in advance using image source and ray tracing models.

Note that the median AI score of 0.5 can be obtained where the long-term rms speech signal to masking noise ratio averages only 3 dB across the one-third octave frequency bands from 200 Hz to 5000 Hz. Experimental data reported in the ANSI standard shows that over 95% of sentences will be correctly understood where the AI score is only 0.5, although the percentage of phonetically balanced isolated words heard correctly with this degree of masking noise will fall to around 75%. These results demonstrate how robust speech communications is against interfering masking noise, and also show how dependent the auditory processor is on the contextual cues present in sentences which are not present in random isolated test words.

In practice, there will always be variability above and below the calculated loss of speech intelligibility caused by masking noise due to individual differences in talking and listening skill and also in the level of motivation among those being tested. Selective attention and other higher-order perceptual processes also play an important part, such that the actual loss of intelligibility in background noise will depend on the individual listener and on the degree of listener effort applied. All objective speech intelligibility calculation or measurement procedures are intended to apply to the ideal, motivated listener case and thereby take no account of individual variability. In addition, real speech is far more complex than the artificial cosine modulated test signals involved in STI measurements, which cannot therefore represent this complexity. Real speech comprises a sequence of phonemes, each of which is made up from a series of harmonically related voice frequencies generated by the vocal cords and a series of mostly high-frequency broad-band components generated by turbulence within the vocal tract. The overall speech frequency spectrum is then filtered through the variable formant frequency bands generated by changing the articulation of the vocal tract. The transitions and silences between one phoneme and the next can also be important. This means that the AI and STI procedures should be used more for relative comparisons of the overall performance of similar speech communications environments than for absolute assessments where any degree of precision is required. The only way to determine true speech intelligibility with precision is to carry out a behavioural measurement using real talkers and real listeners and even this will be subject to considerable variability if the talkers, listeners or word lists and test procedures are changed.

3.2.4 Noise and sleep disturbance

Sleep disturbance is an interesting effect which can be quite variable depending on the circumstances. Everyone seems to need sleep, although people can often get by with less sleep than they would prefer without apparent ill-effects. Research has not provided a convincing explanation as to why sleep appears to be necessary for health and well-being and it is even possible that sleep is merely a biological adaptation to the day–night cycle. There is presumably a survival advantage to having a built-in mechanism which encourages people to curl up and hide away from sources of danger at night. There is also a survival advantage in waking up briefly at regular intervals throughout the night just to check on the surroundings, and this is in fact what happens, although most people rarely remember any of this in the morning. In fact, remaining completely still for any length of time can damage the tissues due to local compression effects and it is necessary for people to wake up to some extent just to move about and change their position while still 'sleeping'.

Research shows that people go through a sequence of different sleep stages several times during each sleep period as a whole. There are certain patterns of neural activity in the brain which can be detected using electrodes fitted to the scalp. The electroencephalograph (EEG) records these brain wave patterns, which can be correlated with observed behaviour and subjective reports (people can be woken immediately particular EEG patterns are observed and asked to describe their experiences at that time, i.e. nothing, or a particular type of dream). The typical pattern of rapid irregular waves changes to a regular sequence known as alpha waves during relaxation while people are beginning to drift off to sleep. There are then four discernible stages during a typical sleep cycle lasting 1–1.5 hours as follows:

Stage 1 prolonged reductions in wave amplitude and frequency
Stage 2 bursts of 'spindle waves' mixed with single slow large amplitude waves
Stage 3 periods of slow high amplitude delta waves
Stage 4 delta waves present for more than 50% of the time

The cycle returns to stage 1 at the end, which is then accompanied by the rapid eye movements (REM sleep) stage most often associated with dreaming.

The peripheral auditory system remains fully active during sleep, but the higher-level processing systems shut down and pay no attention to the incoming sensory input for most of the time. Any transient sound will cause some neural response which may or may not lead to a partial or complete arousal depending on the nature of the stimulus and on the sleep stage at the time. The normal sleep stage cycle can be disrupted or delayed and other reflex responses can occur such as body movements or changes in autonomic functions (heart rate, vasoconstriction, etc.), and a sufficiently intense or unfamiliar signal will cause behavioural awakening. Habituation to familiar sounds seems to reduce the likelihood of behavioural awakening without significantly affecting the likelihood of changes in sleep stage or other reflex responses.

There is some debate in scientific circles as to the relative importance of these measurable reflex responses as compared to actual behavioural awakening (Institute for Environment and Health, 1997). Measurable reflex responses have been observed in sleep laboratories at quite modest sound levels, down to around 45–55 L_{Amax} indoors, and the latest revision of the 1980 World Health Organisation Environmental Health Criteria Document 12 on Noise (WHO, 1980) prepared by Berglund and Lindvall in 1995 (Berglund and Lindvall, 1995) recommends guideline values of 30 L_{Aeq} and 45 L_{Amax} indoors to avoid negative effects on sleep. Berglund and Lindvall further suggest that 'to protect sensitive persons a still lower guideline value would be preferred'. On the other hand, there is no real evidence that any such reflex responses at low sound levels are anything other than normal adaptations by the body to changes in the external environment and there is considerable evidence that people can habituate to much higher levels of environmental and transportation noise without behavioural awakening taking place. At the time of writing, the Berglund and Lindvall revision of the WHO noise guidelines has not been officially adopted by the WHO.

A recent study of noise and sleep carried out around Heathrow, Gatwick, Manchester, and Stansted airports was carried out by the UK Department of Transport as part of a review of night time aircraft noise controls (Ollerhead *et al.*, 1992). The main difference between this type of study and previous sleep laboratory type studies was that all measurements were carried out in people's own homes where they were presumably habituated to the typical night time noise environments. Wrist-worn actimeters recorded upper arm movements at night using a research protocol which was separately shown to correlate well with the more complex EEG recording technique. The main advantage of actimetry is that the subject is not required to wear scalp electrodes which are then connected to a recording machine by flexible cables and hence take some time to get used to. In all, some 400 subjects were monitored for a total of 5742 subject-nights making a very large database. There were around 40,000 subject hours, broken down into more than 4.5 million 30 second brief time periods for correlation against external aircraft noise events.

The results showed that, for an average sleeping period of 7.25 hours, there are typically around 45 arousals per night, with around 18 corresponding to awakenings of 10–15 seconds or more. There was no measurable increase in the rate of sleep disturbance where outdoor event levels were below 90 dBA SEL (80 L_{Amax}), and the chance of the average person being awakened at higher levels (80–95 L_{Amax} outdoors) was only about 1 in 75. There were, however, large differences in individual sensitivity. Most people were least sensitive to aircraft noise disturbance during the first 45 minutes of sleep.

There was generally poor agreement between the objective actimetry measures of sleep disturbance and self-ratings of sleep quality. However, the report mentioned further analyses of aircraft noise disturbance in the early mornings, the relationship with reported annoyance, and the effects of sleep disturbance on daytime sleepiness which do not appear to have been published. The study generally indicated that aircraft noise at night at the four airports was not a major cause of sleep disturbance, and this encouraged the Department of Transport to introduce a revised night time aircraft noise management regime to allow for an increased throughput of passengers at night while still encouraging the use of quieter aircraft. The revised schemes were hotly contested through the courts by the local authorities concerned.

The continuing controversy over aircraft noise management at night at the four airports is a good example of the clear difference which can exist between objective scientific facts and public perceptions in the general field of human response to noise. It is public perceptions that determine community action and individual complaints, and these need not be entirely consistent with scientific facts. Evidently, a significant minority of the populations exposed to aircraft noise at night strongly object to this exposure, and this is irrespective of whether it causes actual sleep disturbance or not. As in many similar fields, the actual results of research studies seem to be open to different interpretations depending on the interpreter's point of view.

How can we summarize what is known about noise and sleep disturbance? First, the sound suspected of causing sleep disturbance must intrude above the background noise level to a sufficient extent that it would be clearly identifiable if the person was awake. Next, it must be either unfamiliar or relatively sudden or both, so that the various intermediate level perceptual processors which continue to function while the person is asleep are unable to appraise the sound as one which can be safely ignored. Finally, the increased neural activity resulting from the stimulus might contribute to increased arousal of the central nervous system, which, depending on the immediately preceding state of arousal, might then lead to awakening. Since people tend to wake up and then immediately fall asleep again many times during the night without necessarily having any recollection of these awakenings in the morning, then it is quite possible that actual disturbance caused by noise might also not be remembered in the morning. It is also possible that particular noise events might be wrongly attributed as causing sleep disturbance. People can wake up for other reasons just before a noise event occurs, which is then attributed as the reason for the awakening. On the other hand, noise might cause difficulty in going back to sleep once awakened by some other cause and this would also be perceived as a negative effect of the noise. On balance, it seems probable that until sleep research can precisely quantify the long-term negative effects (other than subjective reports of general undesirability) of sleep disturbance caused by noise, then it might be difficult to justify noise acceptability criteria based on objective measures of sleep disturbance.

3.2.5 Activity interference

Noise clearly causes direct interference with speech communications and can therefore have disruptive effects in the office and in the home. Listening to TV and radio and telephone conversations can all be affected. However, effective outdoor to indoor attenuation can successfully mitigate this type of interference in most community situations. For example, even where outdoor noise exposure exceeds 75 $L_{Aeq,16h}$, which is quite rare, high-performance acoustic double glazing or secondary glazing (with appropriate attention to ventilation) can attenuate sound levels down to around 40 $L_{Aeq,16h}$, where direct interference with speech communications is likely to be minimal. Average speech levels at a distance of 1 m range from 60 to 65 dBA, which exceeds the 40 L_{Aeq} masking noise by a more than sufficient margin to avoid significant interference. Outdoors, of course, considerable interference with speech communications is likely, and there can be a particular problem with schools and other institutional buildings such as hospitals which might be difficult to insulate effectively.

The average noise level, L_{Aeq}, does not differentiate between generally steady sounds such as distant road traffic noise and intermittent sounds such as aircraft noise. Intermittent sounds at the same L_{Aeq} will have higher levels when present compensated by periods of relative silence when not present. Aircraft and railway noise, for example, could cause greater (but intermittent) speech interference than road traffic noise present at the same L_{Aeq}. For steady sounds, the overall amount of speech interference is proportional to level, but for intermittent sounds the overall amount of speech interference might be more related to the proportion of time that the sound is present. There is no empirical data available to quantify these effects in respect of typical transportation noise.

Most people have experience of noise interference with some other activity such as concentrated study or relaxation. People differ in their preferences for either complete silence or for some form of innocuous masking noise which might prevent otherwise distracting noise from being heard. Experiments on noise and mental task performance show that noise can have both positive and negative effects, tending to maintain alertness when tired. Experiments on the combined effect of noise and the common cold show that, where noise can enhance performance in healthy individuals, it can lower performance in people with a cold. There is very little research data on noise and performance effects in the field, although practical experience shows that work productivity is often higher in open-plan offices despite the increased levels of potentially distracting noise. Many people will have experience of teenagers who cannot complete their homework without music which older people find objectionable.

Finally, it is important to mention that background music through the PA system is sometimes provided in public places such as railway station concourses and shopping malls, chiefly because the people in charge perceive it to be beneficial. The provision of military band style music during the late morning rush hour at a busy railway station can assist in encouraging people to disperse rapidly, while more restful music at less busy periods can help to discourage people from complaining about late trains. Different types of music at the supermarket checkout can have quite significant effects, but the problem is that these effects can be quite different depending on personality. If people were all the same, then they would all like to listen to the same things, which is clearly not the case. It is hardly possible to treat personal preference as an engineering quantity in this respect, and measured sound levels are not usually a useful guide to this type of effect.

3.2.6 Noise annoyance

As implied above, noise annoyance is a general term used to describe the degree of unwantedness of a particular sound in a particular situation. It is essentially a property of the noise source, but it can vary when that noise source is heard in different listening situations. It is not usually a description of the emotional state of the person describing the noise. It necessarily depends on higher-order auditory processing, memory, expectations and attitudes in addition to noise level alone.

There is no statutory definition of noise annoyance, and this can lead to confusion. Noise complainants may describe themselves as being highly annoyed by a particular noise, but this can mean very different things to different people. There have been a large number of noise annoyance surveys carried out over the last 30 years and more in which researchers have attempted to establish engineering relationships between the amount of noise as measured using sound-level meters and the amount of community annoyance response as measured using questionnaires. While these surveys have shown that average noise annoyance increases with noise level, various attempts (Schultz, 1978; Fidell, Barber and Schultz, 1991) to find a common relationship across all noise sources and listening situations have largely foundered on the difficulties of comparing different surveys using different measuring instruments (Fields, 1994).

The problem here is that the precise form of noise annoyance measured in any questionnaire survey can only be defined by the precise wording of the questionnaire actually used. Different researchers have invariably used different questionnaire wording based on 'bother', 'dissatisfaction', 'noisiness', etc., as well as plain and simple 'annoyance'. There are also problems in comparing the precise meanings of different annoyance descriptors in different languages. In some cases researchers have even constructed artificial annoyance scales from a number of activity interference components (see for example, McKennell, 1963) where the contribution of each of those components to overall annoyance has been assumed by definition in advance.

A specific example of the type of problem that this uncertainty over the precise definition of noise annoyance can cause is the current research interest in the possibility of different exposure response relationships for different transportation noise sources. The European Commission have proposed a system of Trans-European-Transport-Networks (TETNs) to encourage economic development and European cohesion. It is becoming increasingly recognized that the environmental consequences of alternative policy decisions to achieve the general aims of the TETNs should be assessed at a strategic level, which, in noise terms, means comparing the relative noise impacts of developing the different transport modes (road, rail, air and water) to different extents. In the absence of any definitive data to the contrary, it is likely that any such relative assessments will use a common noise scale (L_{Aeq}) that does not differentiate between the different noise sources, although there is a considerable amount of research evidence which suggests that there are important differences between people's responses to different sources of noise at the same L_{Aeq}.

Noise Team 6 of the International Commission of the Biological Effects of Noise is currently (1996–98) working on the problem of inter-comparability of different noise survey measurement techniques (Fields, 1996), but even if the necessary standardized questionnaire items can be introduced, it is still not clear that all possible problems of interpretation can be overcome. Some researchers prefer to use a numeric scale of annoyance, others prefer to use a semantic scale and these scales can be both bipolar (extending from positive to negative) and unipolar (extending from neutral to negative).

There is also debate as to whether annoyance scales can be treated as either relative or absolute. A relative scale allows for comparison between two or more different situations, whereas an absolute scale implies discrete meaning. The problem here is that nearly all field research into noise annoyance only asks respondents about the situation in which they personally live and the only comparisons which can then be made are those between different respondents living in different situations. It could be simply a matter of wishful thinking to assume that respondents living in noisy situations interpret noise annoyance scales in the same way as respondents living in quiet situations, and there does not seem to be any method available for calibrating these scales. Laboratory studies where volunteer listeners report their subjective impressions can overcome this particular problem and, incidentally, tend to be much more efficient than field studies in terms of the resources required. On the other hand, no laboratory study can be fully representative of real life.

3.2.7 Noise and non-auditory health

The 'stress hypothesis' suggests that sustained or repeated exposure could lead to persistent changes in the neurophysiological, endocrine, sensory, digestive and cardiovascular systems. If proven, any of these chronic effects could be classified as a deterioration in health, possibly leading to significant adverse health outcomes. The main problem with the stress hypothesis is that it does not differentiate between beneficial and harmful stress. A certain amount of 'stress' is desirable to maintain arousal in the short term and basic fitness in the longer term. There is no clear mechanism by which the observable short-term physiological responses to noise (which can by no means be considered directly harmful if they are within the normal repertoire of bodily responses) might be assumed to translate into potentially damaging chronic changes in the level of basic physiological functions. On the other hand, increasing public concerns that such chronic health effects of environmental and community noise might exist cannot be dismissed out of hand. The fact that no satisfactory theoretical mechanism has yet been proposed is clearly not a proof that such effects do not exist.

Laboratory studies using animals have shown adverse health effects of prolonged exposure to very high noise levels. However, field studies using real people at existing levels of noise are much less convincing, either showing no effects at all, or merely demonstrating non-significant trends. There are essentially five serious methodological problems which need to be overcome before definitive data can be obtained. First, any such relationships between community noise and long-term health outcomes (other than annoyance) must be quite weak, if they exist at all. This is not to say that they are unimportant as even weak effects could have significant consequences when aggregated over the population as a whole. However, weak relationships are much harder to detect, requiring very large sample sizes and extremely careful matching between experimental (noise exposed) and control (non-noise exposed) groups of subjects. This leads on to the second problem in that individual susceptibility might well vary significantly. Suppose that only one person in twenty was actually susceptible to this type of effect at typical levels of noise exposure in the community. Experimental sample sizes would then have to be very large to have a reasonable chance of including susceptible individuals within any research study. The way around this problem is to identify susceptible individuals in advance, but this is rather a 'chicken and egg' situation at the present time. Some researchers advocate the concentration of research resources onto so-called vulnerable groups, suggesting children and old people as candidates. However, there do not seem to be any data to show that such groups are actually any more susceptible than anyone else so this does not really solve the problem.

The third problem is that of confounding factors. There are many individual and community lifestyle factors which probably have greater effects on long-term health than noise. Smoking and diet are obviously important, but there are likely to be many others. Experimental and control groups need to be carefully matched across all such potential confounding factors, and in many cases this simply will not be possible without experimental manipulation of noise exposure which is often not practical and might even be unethical. For example, road traffic noise levels are often highest outside small houses directly fronting onto main roads and lower outside big houses which tend to be set back in their own grounds. The differences in likely socio-economic backgrounds of residents in the differently noise exposed properties might well be important in terms of long-term health outcomes. One of the main theoretical issues in the stress hypothesis is the degree of personal control, and assuming that, all other things being equal, houses in noisy areas are less desirable than houses in quieter areas, then people with more money presumably have more freedom to choose quieter houses in which to live. In addition, since the majority of noise sources (main roads, airports, railways) have been around for many years, it seems likely that the most heavily noise exposed residents are to some extent self-selected as those residents who are most sensitive to noise will probably have moved away by now unless there are other factors present which outweigh the perceived disadvantages of the noise. As another example, consider any research study of the effects of road traffic noise. Here, the main independent variable could alternatively be thought of as distance from a main road, or the amount of traffic on the road outside the house and not just the external road traffic noise level because this essentially depends on these other two variables. There are two types of question here: (a) on what theoretical basis should the distance from a main road or the amount of traffic on it have adverse health effects for nearby residents? (b) if there are adverse health effects, what is (are) the intervening causal factor(s)? It might be air pollution, perceived danger, difficulty in crossing over the road to the local shop or any of a dozen or more other factors.

The fourth and potentially the most serious problem is the difficulty of measuring individual exposure to noise. Virtually all community noise surveys classify individual noise exposure on the basis of noise levels measured outside the house. Sometimes measurements outside the noisiest facade are used to represent noise energy incident on the external walls of the house, and sometimes free field measurements or calculations are used to represent the original noise exposure at that place with the effect of reflecting vertical walls taken out of consideration. Neither of these measures provides a very good description of individual noise exposure. Individual noise exposure due to externally measured community noise sources is invariably lower than that measured due to the outdoor to indoor noise attenuation of the house, and due to the fact that people are unlikely to be present at their house for 24 hours a day. Most people go out to work, to school, and for shopping, social and leisure activities and their subsequent degree of noise exposure is then completely unknown. For many people, noise exposure while travelling to and from work is probably more significant than noise exposure due to outside community noise sources while indoors at home. In addition, many other people attend discotheques, motor racing events, football matches and other noisy leisure activities which might lead to significant personal noise exposure.

Finally, most research studies which have been carried out to date are essentially cross-sectional. This means that differently noise exposed groups of residents are compared over a short period of time and there is no longer term follow-up. Cross-sectional studies can detect statistically significant associations between variables but cannot be used to establish cause and effect. Cause and effect can only really be established by means of longitudinal studies where there is a significant change in noise exposure over time. Very few of these studies have been done.

3.2.8 Sound quality

Sound quality is a general term which describes those features of a sound which contribute to its subjective appraisal which are not well described by simple objective measures such as A-weighted sound level. Sounds can have almost infinite variation in spectral and temporal characteristics and many of these differences can be separately perceived, such as the degree of tonality or impulsivity. In practice, this means that some sounds can be rated as noisier or more annoying than others even where the overall level or the amount of physical energy present is actually lower. This means that any product labelling scheme based on A-weighted sound level or sound power level alone will not tell the whole story as far as consumers are concerned.

Sound quality is becoming increasingly important as a part of the design of manufactured products such as motor vehicles, washing machines, and office machinery (Porter and Berry, 1997). At the time of writing, large contracts for the supply of components to equipment manufacturers can be won or lost on the sound quality aspects of the component, which, unfortunately, cannot always be described objectively in a way which allows both parties to quantify the precise differences between what is being offered and what is actually required.

The main focus of sound quality research in the past has been to contribute to the development of more sophisticated single-figure objective rating schemes that more properly reflect relative noisiness or annoyance judgements. For example, early work in the aircraft noise field very quickly established that the different frequency components present in turbojet engine noise were not well represented by any of the standard A, B, or C frequency weighting curves based on equal loudness contours (see below). The aircraft industry quite quickly standardized on the Effective Perceived Noise Level (EPNL) scale (ISO 3891, 1978) which is a relatively complex method of aggregating the contributions to overall noisiness from each one-third octave frequency band measured separately. There are additional corrections for prominent individual one-third octave frequency bands where present and for the duration of a flyover event. With the benefit of hindsight, it now seems doubtful whether the significantly increased complication of measurement as compared to a simple overall A-weighted measurement is fully justified, although the EPNdB scale is too entrenched into the established methods of aircraft type approval noise certification procedures to be able to contemplate any change.

There are a number of 'loudness' calculation schemes which follow the same general principle of aggregating the contributions to overall subjective impressions from each one-third octave band frequency band measured separately. For example, the ISO standard method for calculating loudness (ISO 532, 1975) takes into account the upward spread of masking by which low-frequency components present in the sound are assumed to mask (render less audible) higher frequency components to an extent determined from psychoacoustics experiments in the research laboratory. Another approach has been to add subjective corrections or penalties to the measured overall A-weighted level to account for any particularly noticeable characteristic features present in the sound. None of these approaches has been completely successful because of their focus on single-number descriptors.

More recent work (Porter, Flindell and Berry, 1993; Flindell and Lewis, 1996) has recognized the importance of different individual preferences in concentrating on the separate assessment of acoustic features where present, without attempting to combine the measurement of those features according to some essentially arbitrary scheme which is unlikely to work in all situations. Instead, the acoustic feature approach concentrates on providing an objective description of all relevant features present but does not even attempt to prescribe which one of those features might be the most significant in any particular case. As an example, noise complaints attributable to mechanical air handling plant in commercial premises are relatively common. In many cases, the main problem often seems to be discrete frequency components and harmonics generated by fans which are perceived as tonal. This situation can easily happen where the fan designers ignore noise considerations. BS 4142: 1990 (BSI, 1990) paragraph 7.2 specifies a 5 dB correction factor which can be added to the A-weighted measurement of the noise to account for this tonal feature. The presence or absence of a tonal feature is left as a matter for subjective judgement. The latest revision of the equivalent American National Standard (ANSI S12.9, 1997) specifies a similar correction.

There is a problem, however, with the method of applying correction factors in that it often does not properly represent subjective response. If the reason for the complaint is the noticeability or prominence of a subjectively perceived tonal component, then the overall A-weighted sound level might not be relevant at all. In addition, the 5 dB tonal correction factor implies that removal of the tone will be subjectively equivalent to a reduction in overall A-weighted sound level by 5 dB. In some situations, this might be true, but in other situations it is quite possible that a tonal component, once identified and forming the subject of complaints, might have to be completely eliminated to satisfy complainants, irrespective of any effects on the overall A-weighted level.

To give another example, recent work at ISVR on sound quality inside motor vehicles for commercial sponsors has shown that individuals clearly differ in their personal preferences for different degrees of irregular impulsive noise caused by a diesel engine at idle. The acoustic feature approach recognizes this problem by attempting to quantify the magnitude of each acoustic feature separately in terms of sensation level, the amount in decibels by which the feature exceeds the just noticeable threshold for that feature. The feature sensation level is not intended to correlate directly with the degree of subjective noisiness or annoyance, but it very usefully shows the amount by which the feature would have to be reduced to eliminate the subjective problem, and this is directly relevant to the definition of the most appropriate noise control action. Where there are two or more features present, the approach does not attempt to define which is the more objectionable as this is often different for different listeners. Finally, the overall A-weighted level could be a feature in its own right, so the new approach is not exclusive of the more classical methods of assessment.

Where sound quality work becomes most difficult is in the area of hi-fi sound. Modern hi-fi equipment has tended to reach a plateau in terms of reproduced sound quality with the advent of the Compact Disc using 16 bit PCM (Pulse Code Modulation) digital sound. In fact, the potential reproduced sound quality of CD systems is so good that many of the later derivatives such as mini-disc and digital compact cassette have traded audio quality for playing time. Carefully controlled laboratory tests have shown that listeners often cannot reliably distinguish between competitive high-quality audio systems under blind test conditions where they are prevented from distinguishing between the two systems by any means other than reproduced sound quality alone. In addition, auditory memory is such that small differences which are just detectable under strict A–B comparison conditions with no time interval between presentations might be completely undetectable where there is a time gap of minutes of more between presentations.

To set against these scientific 'facts', there are those who consider that a short A–B comparison does not allow listeners to fully appreciate subtle but important differences. These 'subjectivist' reviewers argue that there are subtle aspects of the quality of audio reproduction which might only become apparent after several days or more of careful listening, and that it is clearly impractical to be able to do comparative A–B tests on this basis. Unfortunately, there does not seem to be any way of testing this hypothesis by any controlled scientific experiment, which means that the subjectivists must be allowed to have their cake and eat it. Since the reproduced sound quality differences between competitive high-quality audio systems are purely a matter of personal preference anyway, perhaps none of this really matters.

In addition, there are certainly aspects of stereo audio reproduction which appear to require a few minutes or so to fully appreciate. An example of this is a little-known technical defect of old fashioned vinyl LP records which has been overcome by the modern CD audio system but only at the cost of degraded stereo images. Vinyl LPs suffer from significant high-frequency cross-talk between left and right stereo channels due to the mechanical nature of the system. This fortuitously appears to compensate for a subjective widening of the stereo image at high-frequencies due to the head shadow effect (Behrens, Flindell and Watanabe, 1995) which would otherwise occur. The CD audio system does not suffer from high-frequency cross-talk and this leads to the high-frequency components of a stereo image being displaced relative to the low-frequency components, in other words, a slightly blurred stereo image as compared to when a vinyl LP is used. There are undoubtedly other subtle differences between competing systems which at present appear to have no rational explanation in terms of existing psychoacoustic knowledge, but which might well be explained in the future. On the other hand, it is likely that there are also subtle differences reported by reviewers which are no more or less than pure fantasy. If it helps to sell more audio products, then why not?

3.3 AUDITORY ANATOMY

All of the various human responses to sound and noise described above have their roots in the way in which the peripheral auditory system is initially stimulated by the incident sound, and by the various ways in which higher-order neural and endocrine systems respond to these sounds.

Most of what we know about the auditory system comes from two main research methodologies. The first is an understanding of the capabilities of the ear, as determined by personal observation and by what are termed psychophysical methods. This means investigating the time, frequency, and sound level resolving powers of the ear by measuring just detectable thresholds, and by comparing subjective judgements of comparative magnitudes. The second method is to correlate observations of the anatomy of the auditory system against measurable loss of capabilities associated with some departure from the normal anatomy caused by disease or trauma. Neither research methodology on its own can provide a complete understanding. It is also important to point out that while many of the current theories of how the system works in detail seem to be largely correct, all such theories are subject to modification and experimental verification at some stage in the future. It is very difficult to study the micro-anatomy of the ear under *in-situ* conditions without causing interference with normal function and the dissection of cadaver ears cannot provide all the information required to understand normal operation while alive. A third possible method of investigating the various functions of the auditory system would be to use classical methods of introspection. These might have a role in contributing to understanding the mechanisms of perception, but cannot really assist with understanding any of the lower-level processes on which perception depends.

A cross-section of the ear is given at Fig. 3.2. Humans have two ears, located one to either side of the head. The possession of two ears allows for the sampling of complex sound fields at two different points in space at the same time which also allows for much quicker sound source localization than would be possible with only one ear. The ear canal protects the sensitive ear drum from damage and contributes to high-frequency sensitivity where the length approximates to odd multiples of quarter wavelength.

The external ear, or pinna, contributes to the localization of high-frequency sounds by providing a small amount of shielding against sounds from the rear. The various folds in the pinna cause differential reflection of very high-frequency sounds into the ear canal at different angles of elevation and this also contributes to localization in height. It has been found to be important to include some representation of the pinna in dummy heads used for binaural sound recording techniques, as there is otherwise a loss of discrimination between direct and reverberant sound. The reader is recommended to try cupping the hands as extensions behind each ear so as to simulate the possession of abnormally large ears as an experiment. An immediate improvement in the subjective impression of the direct to reverberant sound ratio for sound sources directly in front should result. The same technique will usually be found to provide an immediate improvement in stereo image quality when listening to an audio hi-fi system, but, and perhaps surprisingly, artificial Mickey mouse ears for the purpose of stereo image enhancement are rarely found in specialist hi-fi stores, presumably because it would be difficult to make them expensive enough to qualify as hi-fi.

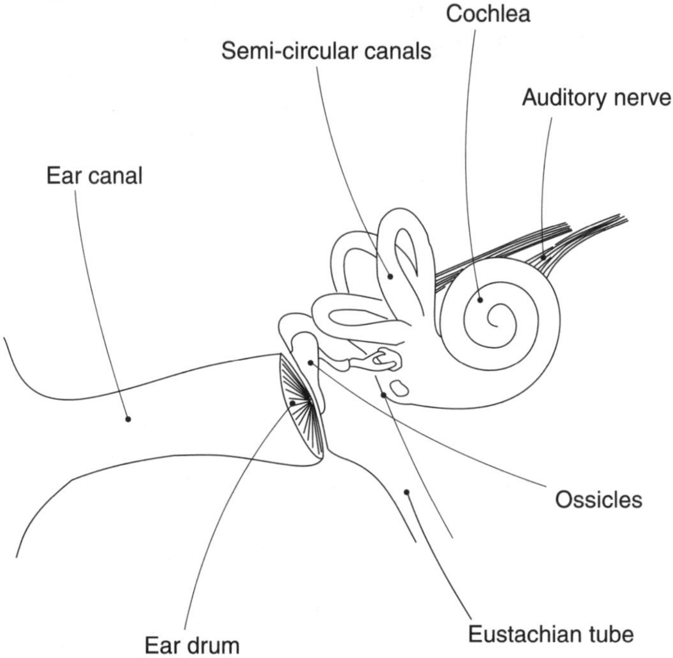

Figure 3.2 Cross-section of the ear.

The ear drum vibrates in response to sound pressure in the ear canal. The sound pressures in the ear canal are modified by the acoustics of the head and shoulders and are significantly different at high frequencies from the sound pressures that would have been present at the centre of the listener's head with the listener absent. The real-ear transfer function varies for different angles of azimuth and elevation relative to the head. This is an important consideration from the point of view of relating objective measurements of sound to subjective impressions. Some researchers have suggested standardization using dummy head measurements to represent sound pressures at a standard ear drum rather than at a standard measuring microphone positioned at the same place. The logic behind this suggestion is subject to some doubt, because the auditory system already comes pre-calibrated for its own real-ear transfer functions as based on a lifetime of listening experience. These real-ear transfer functions vary considerably from one individual to the next. What is important is the pre-existing sound field present before either a listener's head or a measuring microphone is inserted. Of course, there are special cases where a measurement in the ear canal is appropriate, such as where hearing aids or headphones are used, but in most cases it is useful to be able to relate the measurements back to an equivalent free-field sound spectrum whenever comparisons must be made.

There is an air-filled middle-ear cavity behind the ear drum. The primary purpose of the middle-ear cavity is to provide a movement space behind the eardrum. Vibrational motion of the ear drum is transmitted across the middle-ear cavity by a chain of three small bones (the ossicles) into the fluid filled inner-ear chambers known as the cochlea. The primary function of the ossicles is simply to transmit the vibrational motion across the middle-ear cavity, but they also provide additional leverage to convert the motion from the relatively low mechanical impedance of the ear drum to the relatively high mechanical impedance of the fluid filled inner-ear cavities.

There are two flexible membranes in the dividing wall between the cochlea and the middle-area cavity, the upper membrane (the oval window) accepting movement via the middle-ear bones, and the lower membrane (the round window) vibrating in sympathy as a result of the vibrational waves transmitted through the cochlea. The cochlea is coiled through two and a half turns in a spiral. The basilar membrane divides the upper and lower chambers of the cochlea throughout its length as shown in cross-section at Fig. 3.3.

The basilar membrane supports inner and outer rows of sensitive hair cells which are stimulated by the shearing action of the tectorial membrane across their tips whenever the basilar membrane as a whole is deflected by vibrational waves in the fluid contained within the cochlea. The pattern of vibration along the basilar membrane represents the time varying frequency content of the acoustic stimulus in the following way:

1. Incoming positive acoustic sound pressures cause inward deflection of the eardrum and middle-ear bones, eventually causing an inward displacement of the oval window, creating a positive pressure in the upper chamber of the cochlea.
2. This causes downward deflection of the nearest part of the basilar membrane immediately adjacent to the oval window and there is a corresponding outward deflection of the round window directly beneath.
3. The initial downward deflection progresses along the basilar membrane from the basal end towards the apex at the other end of the spiral as a travelling wave.
4. The travelling wave reaches a maximum amplitude at a point along the basilar membrane which depends on the frequency content of the acoustic stimulus and is then rapidly attenuated further along.
5. Low-frequency signals penetrate to the apex of the basilar membrane. The point of maximum amplitude moves closer to the basal end (near to the oval and round windows) as the frequency increases. High-frequency signals might only penetrate a very short distance along the cochlea.

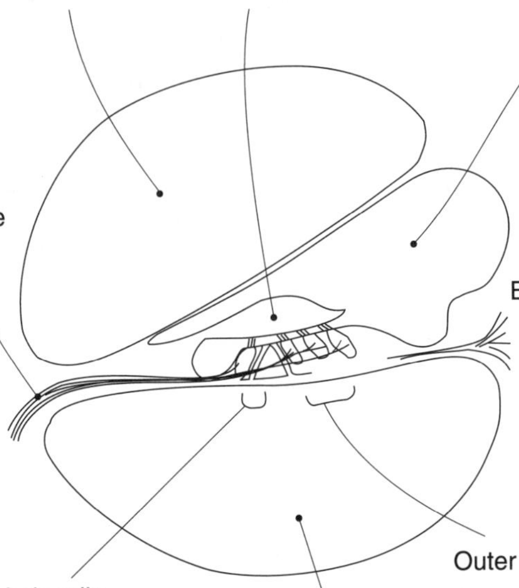

Figure 3.3 Cross-section of the cochlea.

The distribution of the travelling wave along the basilar membrane acts as a mechanical frequency analyser. There is a tonotopic organization of the sensitive hair cells along the basilar membrane whereby auditory nerve impulses originating from different parts of the basilar membrane become associated with different frequencies of sound. It is important to note that the travelling wave passes through, and therefore stimulates, higher frequency regions before reaching its point of maximum amplitude. Incoming sounds of any particular frequency stimulate all higher frequency parts of the basilar membrane (but not the lower frequency parts) to some extent before reaching the point of maximum amplitude. This is reflected in the upward spread of masking curves whereby the masking effect of a narrow band noise signal spreads to higher but not lower frequencies.

If and when they have been sufficiently stimulated by the shearing action of the tectorial membrane across them, the inner rows of hair cells send nerve impulses to the brain. Each nerve impulse is a single electro-chemical discharge which travels along the nerve cell membrane from the cell body to the tips of the associated nerve fibres. Each nerve impulse involves a temporary depletion of the nerve cell's biochemical resources which must then be replenished through the normal processes of metabolism. Consequently, there is a maximum firing rate for each nerve cell and an associated much lower resting firing rate, even when the nerve cell is not being stimulated. Each nerve impulse results from a build-up of stimulation and the precise time at which it is most likely to occur is essentially unpredictable, since it also depends on the elapsed time since the last firing and on the metabolic condition of the nerve cell. Larger-amplitude audio signals generate more vibratory excitation of the hair cells and thus a higher firing rate for each associated nerve fibre. The dynamic range of individual hair cells is limited and the most sensitive nerve fibres become saturated at the higher sound levels. It is thought that other less sensitive combinations of hair cells and nerve fibres take over at higher sound levels to overcome the problems of the limited dynamic range of each hair cell and nerve fibre combination considered separately.

The pattern of nerve impulses in the auditory nerve does not represent the incident acoustic waveform as such, but rather it represents the resulting pattern of vibratory excitation along the basilar membrane, as modified by the current metabolic condition of the various hair cells and nerve cells involved in the communications chain. In other words, it does not act as a linear transducer, but rather it detects the presence or absence of acoustic signals of particular amplitude and frequency content occurring at different times.

A wide-band transient signal is likely to excite the basilar membrane throughout its length, causing a sequential series of nerve impulses to spread 'tonotopically' across the auditory nerve, with the high-frequency connected nerve fibres excited first. A continuous pure tone will generate a continuous series of nerve impulses in those fibres most closely associated with the part of the basilar membrane which is most strongly excited. The ability of the nerve impulses to convey relative phase information is limited because of the general unpredictability of the precise time of firing.

The peripheral auditory system can be modelled as a bank of parallel band-pass filters operating in real time. This model is sufficient for most purposes, but it must be borne in mind that there is no point in the system at which each band-pass filter output signal could be separately obtained. A better model might be a bank of feature detectors based on band-pass filters and operating in parallel.

There is some evidence that the frequency analysis functions of the basilar membrane vary with time over short fractions of a second. The outer rows of hair cells are believed to vibrate in sympathy with an incoming acoustic signal and thus assist in 'tuning' in the basilar membrane to discreet frequency components. This enhances both frequency selectivity and sensitivity to weak sounds. It has been difficult to study this 'active process' *in vivo* because access to the systems involved cannot be obtained without drastically altering or even destroying them. The active process presumably assists the ear in adopting the best compromise solution between time and frequency resolution that can be achieved. In a purely physical sense, it is not possible to have fine frequency resolution and fine temporal resolution simultaneously. The accuracy to which any discrete frequency component can be described depends on the time (and hence the number of cycles) for which it is observed.

It is probably quite important for the peripheral auditory system to have good temporal resolution for the detection of transient sounds. The frequency content of transient sound is unimportant for the first few milliseconds, when the only response of the organism might be the early stages of some reflex activity. Rather more gradually, the auditory system will begin to process the incoming signal, at which stage information regarding frequency content becomes potentially more useful. Providing that the signal is still present, it is then advantageous to be able to tune in to specific frequency components to assist in discrimination between other possible signals. When we look at the construction of typical speech signals we can see that the patterns of temporal and spectral variation appear to match quite closely the known temporal and spectral resolving powers of the auditory system as a whole. This matching helps provide the most efficient means of communicating the maximum amount of information in the shortest possible time within the other biological and environmental constraints that apply to the overall system design.

The final link in the chain is the system of auditory nerves which transmit nerve impulses from the basilar membrane to the brain. The auditory nerve pathways pass through a number of interconnection points where nerves branch off to different parts of the brain and finally reach the auditory cortex where the most complex processing is believed to take place. There is a considerable amount of evidence that different nerve cells in the auditory cortex are wired as specific 'feature detectors' and respond only to those features when present in the external stimuli.

There are multiple interconnections between the left and right auditory nerves which presumably support binaural processing of signals. Small differences in the time of arrival of signals between the left and right auditory nerves are associated with different azimuth angles of the source relative to the head. In addition, and at high frequencies only, different azimuth angles lead to quite large differences in the relative amplitude of the signals at the two ears. It is likely that there are specialized nerve cells located in the auditory cortex and possibly at lower level interconnection points which can respond to these differences. The existence of feature detector nerve cells reinforces the general concept that the auditory system as a whole appears to be mostly organized as a very sophisticated auditory pattern identification and recognition device and not at all as a simple transducer from acoustic to nervous energy as might appear at first sight.

3.4 AUDITORY RESPONSE

3.4.1 Auditory capabilities

The normal range of hearing is shown at Fig. 3.4. The normal threshold of hearing curve at the bottom of the chart shows that the human ear is most sensitive at around 3–4 kHz. This is consistent with the first quarter wave resonance in the ear canal. In this frequency range, and for healthy young adults with no hearing loss, (this is defined as normal hearing), the minimum sound pressure in a diffuse field that can be just detected when listening with both ears (the minimum audible field) is around 10–20 μPa corresponding to sound levels of from −6 to 0 dB. 'Just detected' means that the sound present condition can be correctly discriminated from the sound not present condition at a better than chance level and does not necessarily mean that sounds at this level would be of any significance in the environment. Individual thresholds vary quite a lot above and below the mean at different frequencies, even within a sample where everyone has so-called normal hearing. It is not clear whether this variability represents fundamental

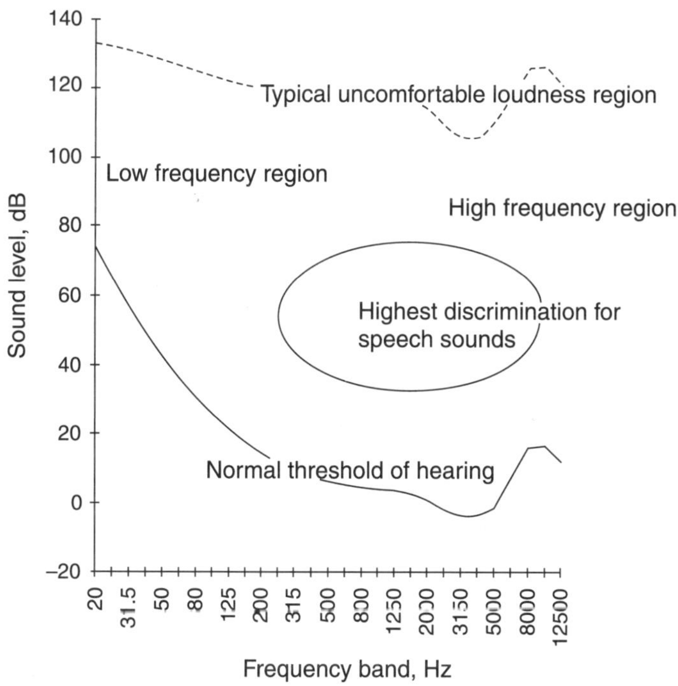

Figure 3.4 The normal range of hearing.

differences in hearing sensitivity or whether it merely results from measurement uncertainty.

The normal hearing thresholds increase below 300 Hz and above 8 kHz, such that much higher sound levels are required for them to be audible. Children may be able to hear higher level sounds up to 20 kHz and beyond, but 15 kHz is a more practical limit for most adults.

High-frequency sounds with wavelengths that are usually short compared to the dimensions of the source are consequently quite directional compared to lower frequency sounds. This directionality probably means that the detection of high-frequency sounds in the environment takes a lower priority for the organism than lower frequency, more omni-directional sounds. In addition, high frequencies are less useful for speech sounds which are intended for wide broadcasting in social animals. High-frequency hearing, because of its directionality, is much more useful for animals such as bats which use high-frequency sound for echo-location.

People are very insensitive to low-frequency sounds below 50 Hz, but there is probably no very good reason why being able to hear very low-frequency sound would provide survival advantage. There is no information content in speech sounds below 300 Hz. Most low-frequency environmental sounds come from relatively remote sources which are often of less importance to the individual than nearer sources.

The normal hearing threshold curve can alternatively be measured in the ear canal as a minimum audible pressure. This is generally higher than the minimum audible field, and does not exhibit the same peak sensitivity at around 3 kHz. The differences between the minimum audible field and the minimum audible pressure must be taken into account by the designers of headphones, earpieces, and audiometric devices, but are not otherwise of any great concern.

The upper limit of hearing is often quoted as being somewhere around 120 dB SPL, but this can be misleading. In the mid-range of frequencies from around 1–5 kHz, a continuous pure tone at 120 dB is very unpleasant, but it could be increased to 140 dB and not go above the range of audibility. Broad-band signals at these levels tend to be less unpleasant than pure tones. This is possibly because the vibratory excitation pattern on the basilar membrane is not as concentrated at one point. High-level pure tones tend to become increasingly unpleasant with frequency until they eventually exceed the limits of audibility and then become inconsequential. Because the ear is relatively insensitive to low-frequency sounds, it is not meaningful to quote an upper limit of hearing at the very lowest just audible frequencies, which are more likely to be felt as a resonance in some part of the body than to be heard as such.

The smallest just noticeable differences between similar pure tones are in the middle range of sound levels from 40 dB SPL to 80 dB SPL and frequencies from 300 Hz to 5 kHz. Within this range, it is quite possible (under laboratory conditions) to correctly identify the louder of two sounds which differ in level by less than 1 dB or the higher pitched of two sounds which differ in frequency by less than 1%. A significant time gap between listening to the two sounds results in an apparent loss of intensity or frequency discrimination which might be more properly described as a failure of auditory memory. In addition, the just noticeable differences for sound level and frequency become progressively wider outside of this middle range of sound levels and frequencies.

In practice, the just noticeable difference in sound level or frequency content between two real-life sounds which are heard several hours or even days apart will be much greater than 1 dB or 1%. It is likely that at least a 3 dB difference in sound level would be required before most people would be able to correctly identify the louder of two aircraft overflight events heard on separate days, and even larger differences would be required where there is a time difference of months or years. This auditory memory effect can often lead to confusion where the minimum audible difference in average sound levels due to some new environmental noise source is being debated. On average, there might be a just detectable increase in the proportion of people describing themselves as 'highly annoyed' where average noise levels have increased by just a few decibels but, on the other hand, it is likely that averaging might conceal the fact that only a minority would have noticed the difference.

The smallest just noticeable differences for frequency are concentrated in the same frequency range where the ear has its highest sensitivity. It is presumably just as advantageous to be able to discriminate between similar frequencies in this range as it is to be able to hear them in the first place. The smallest just noticeable differences for sound level seem to be concentrated in the middle region between very soft and very loud sounds, which is again probably the most useful area. Finally, the smallest just noticeable differences are concentrated in the same sound level and frequency ranges as normal conversational speech. This means that the human ear tends to be well adapted to discrimination between the different sounds involved in human speech. This of course begs the question as to which evolved first, the sound level and frequency discrimination patterns of the human ear, or the sound level and frequency differences involved in differentiating between different speech sounds. Probably, both systems evolved in tandem.

There is an implicit assumption here that no aspect of the auditory system would have evolved in the way that it has unless there is a plausible survival advantage. The localization gap in the mid-frequency range shows that this is not always a clear cut issue. The localization gap refers to the mid-frequency range between around 750 Hz and around 2 kHz where the auditory system is relatively poor at determining the azimuth angle of a direct sound source, despite the fact that it would probably have been useful to possess better directional discrimination for speech sounds in this frequency region. The localization gap results from the wavelength of sound at different frequencies compared to the dimensions of the head and could have conceivable been shifted to a much lower frequency range by adopting much wider spacing between the ears, or alternatively, it could have been shifted to a much higher frequency range by adopting much narrower spacing between the ears. Neither solution fits in very well with the general architecture of the human body.

3.4.2 Frequency response of the ear

Figure 3.5 below shows empirically derived standard curves of equal loudness across different frequencies. Listeners are asked to adjust the level of a variable frequency pure tone until it matches the subjective loudness of a reference pure tone, normally at 1 kHz. A complete set of iso-loudness curves are then plotted out for different levels of the reference pure tone (BS 3383: 1988, ISO 226, 1987) Different listeners often give quite different results, so the composite equal loudness contours represent an average. There has been some concern recently (Lydolf, Frandsen and Moller, 1996) that the standardized equal loudness contours may have been significantly affected by the psychometric methods used to determine the point of subjective equality for each combination of test tone frequency and reference tone level, and it is possible that the ISO standard curves might be updated to reflect this at some stage in the future. However, it is probably more important to note that the method of deriving the equal loudness contours represents an artificial laboratory situation. There is no guarantee that the standard equal loudness contours really represent the way that the auditory system responds to the more complex wide-band sounds that are representative of real-life experience.

On the other hand, since the general form of the equal loudness contours is reasonably consistent, it seems reasonable to suppose that they represent certain fundamental properties of the peripheral auditory system as follows:

1. The raised thresholds at very low and very high frequencies (about which there are no reasonable doubts) are reflected in reduced loudness of very low and very high frequency pure tones as compared to mid frequency band pure tones.
2. The equivalent frequency response of the ear appears to get flatter as the sound level increases, so that the relative loudness of low frequency sounds at higher sound levels is proportionately greater.
3. The growth of loudness with increasing sound level is much steeper at low and very high frequencies than it is in the mid-range of frequencies. This is also associated with reduced sound level and frequency discrimination at these extreme frequencies as mentioned above. This means that at 1 kHz, the average listener might be able to

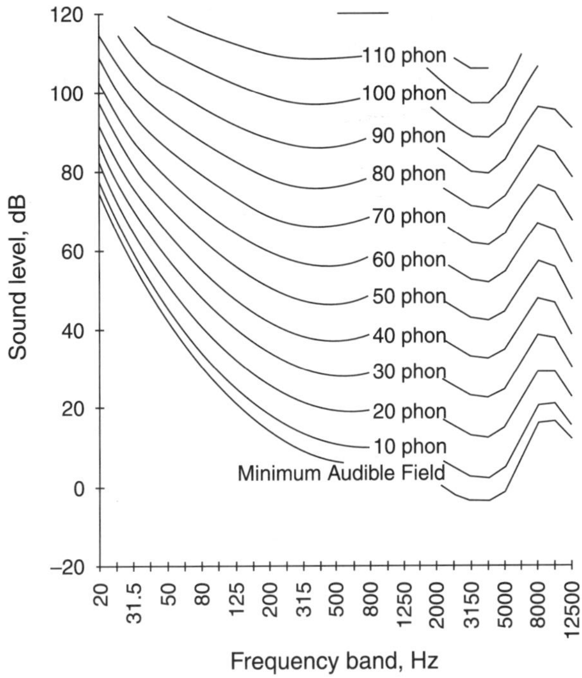

Figure 3.5 Standard equal loudness contours.

discriminate between more than 100 different sound levels between the auditory threshold at around 0 dB SPL and the uncomfortable loudness level at around 120 dB SPL. At 50 Hz the total number of discriminable sound levels is squashed into a much narrower range and is also very much smaller because the just noticeable differences are much greater.

3.4.3 Time response of the ear

The ear requires a finite amount of time to register a sound. First, the various mechanical parts of the ear must be set in motion. Secondly, there must be sufficient mechanical stimulation of the sensitive hair cells to cause nerve impulses to be sent off along the auditory nerve in some quantity. It is doubtful whether any single nerve impulse would be separately distinguishable from the random resting pattern of nerve impulses. Finally, the resulting increase in neural firing rates in the auditory nerve must be sufficient over a long enough period of time for the central nervous system to become aware of the change in sensory input.

The time response of the ear is best described by an integration time of between 100 and 200 ms. Transient sounds of shorter duration are harder to detect and do not register the same loudness as if they were to continue for one-fifth of a second or more. Figure 3.6 (based on data given by Stephens, 1974) shows a generic curve of constant detectability for transient sounds with constant overall energy (sound pressure squared times time) and increasing duration. For constant overall energy, the instantaneous sound level decreases proportionately to increases in duration.

Over the middle range of durations from about 15 ms to about 150 ms, detectability is a function of energy. As the duration increases beyond around 150 ms, detectability decreases because the instantaneous sound level continues to decrease, yet the increasing duration no longer contributes to loudness. At very short durations below around 15 ms, detectability also decreases because time frequency uncertainty means that the energy in the toneburst becomes spread over an increasingly wider range of frequencies. Consequently, there is less energy available for detection in each separate part of the basilar membrane.

The range of constant detectability with duration varies with frequency to some extent, being shorter for high frequencies and longer for low frequencies. It is possible to plot similar curves for loudness against

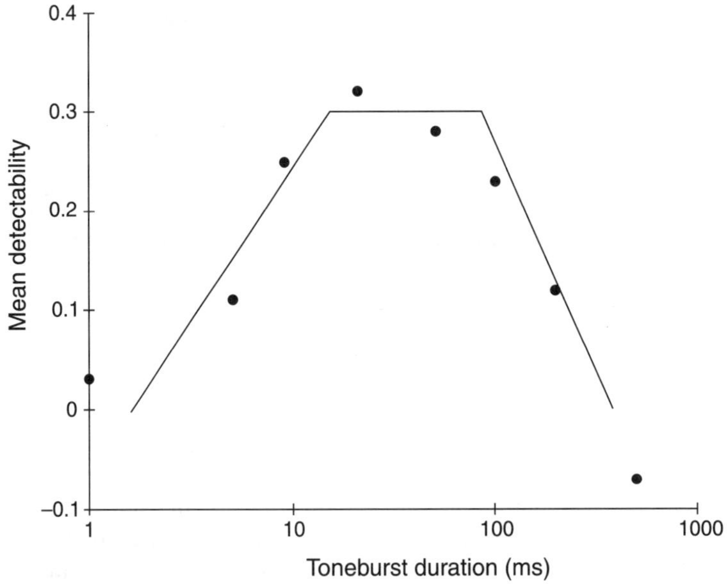

Figure 3.6 Detectability of constant energy tonebursts: effect of increasing duration.

duration for constant overall energy, but the precise shape is dependant on the experimental method used. Comparative loudness is essentially subjective and different listener's criteria vary. Short transient sounds with different durations vary in sound quality or character in addition to loudness, and it is easy to confuse differences in quality with loudness. This is not the case for detectability which is much more dependant on the fundamental capabilities of the auditory system, although listener effort can still play a part.

One practical consequence of the finite integration time of the ear is that the subjectively perceived loudness of very short transient sounds might not be commensurate with the true peak level. This means that the potentially damaging effects of very short high-energy transient sound might be underestimated, and is a good reason for setting the Noise at Work Regulations 1989 Peak Action Level conservatively.

3.4.4 Masking effects

Most of the fundamental capabilities of the auditory system to detect faint sounds of different temporal and frequency characteristics can be explained by the properties of the travelling wave along the basilar membrane, in conjunction with the properties of the inner and outer rows of hair cells and subsequent neural processing. The upward spread of masking curve shown at Fig. 3.7 below (based on data given in Ehmer, 1959) describes the influence of a narrow band masking signal at higher but not at lower frequencies on the detection threshold of a pure tone target signal (see section 3.3 above).

This phenomenon results from the spread of the travelling wave through the high-frequency regions of the basilar membrane only as far as the point of maximum amplitude. There is little or no excitation of the lower frequency parts of the basilar membrane beyond the point of maximum amplitude. Thus, a high-frequency target signal might not be detected where the excitation pattern on the basilar membrane is submerged below the high-frequency tail of the excitation pattern due to the masker. A low-frequency target signal is much more easily detected because the low-frequency tail of the excitation pattern due to the masker is much lower in amplitude.

What happens when the target signal is within the same frequency band occupied by the masker? The most important consideration here is the effective bandwidth of the parallel auditory filters comprised of both the travelling wave mechanical filtering and the subsequent tuning-in process attributable to the outer rows of hair cells. These auditory filters are known as 'critical bands' in the psychoacoustic literature. They are approximately one-third octave band wide throughout most of the auditory frequency range except at very low frequencies where they are wider. Masking signals

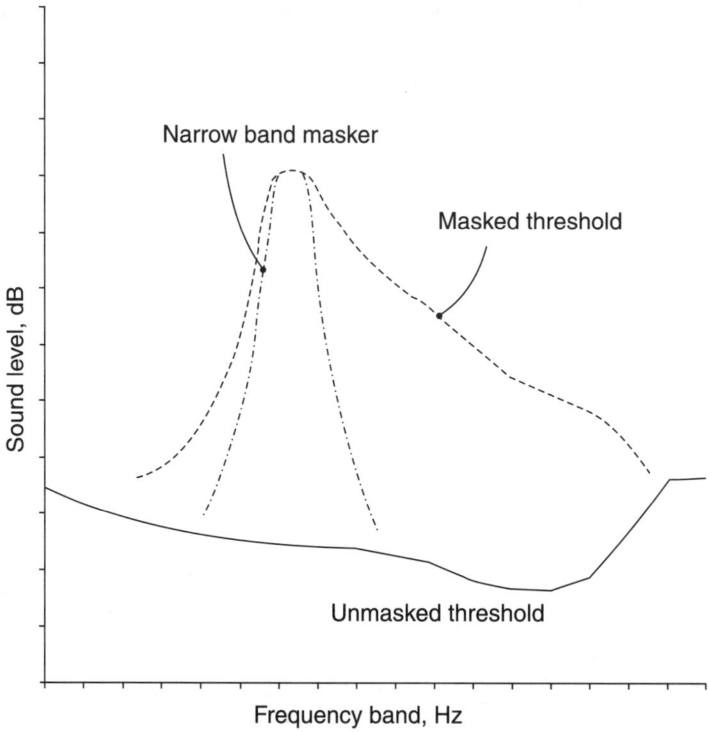

Figure 3.7 The upward spread of masking.

contained within the auditory filter bandwidth centred on the target signal frequency contribute equally to the masking effect, whereas masking signals outside of this bandwidth do not. In the middle range of sound levels and frequencies, the detection threshold of a pure tone target signal is at a level roughly 4 dB below the level of the masking signal measured over the surrounding one-third octave bandwidth. Increasing the bandwidth of the masking signal beyond the surrounding one-third octave bandwidth has very little effect on the masked threshold of the pure tone target signal, even though the overall sound level of the masker will increase. Decreasing the bandwidth of the masking signal within the surrounding one-third octave bandwidth reduces the masked threshold of the pure tone target signal commensurate with the reduction of overall sound level of the masker.

This restriction of the masking effect of a broad band noise on a pure tone target signal to the one-third octave band surrounding the tone is responsible for the commonly heard assertion that the threshold of audibility of a pure tone can be as much as 20 dB below the overall sound level of a broad band masker. In these cases, the contributions to the overall sound level of the broad band masker made by frequency components outside of the one-third octave band surrounding the pure tone target are irrelevant to the masking effect. When considering auditory masking effects, it is important to bear in mind that the sound levels in each separate frequency band add up to the overall sound level measured broad band. Overall subjective loudness is likely to be proportional to the overall broad band level, whereas the audibility of separate pure tone components is only affected by the surrounding one-third octave bandwidth of any masker present.

There is another common misconception that the auditory filter bandwidths as measured by masking phenomena ought to be somehow the same as the frequency resolution of the human ear, which is much narrower. In fact, there are a large number of overlapping one-third octave bandwidth auditory filters with much closer frequency spacing than one-third octaves. The auditory filter centred on a pure tone target signal might not be very sensitive to small differences in the frequency of that tone, but there are numerous overlapping auditory filters present which are likely to be more sensitive as the pure tone target tracks up and down the off-frequency filter slopes. In addition, there is some evidence that the frequency of continuous repetitive signals can be coded by successive 'volleys' of nerve impulses which tend to be bunched together at successive repetitions of the wave. This phenomenon is known as 'phase locking' and is limited at high frequencies by the inherent uncertainty, or jitter, of several milliseconds in the precise time at which nerve cells actually fire (or not) in response to sufficient stimulation.

Non-simultaneous masking represents masking effects before and after the masker sound. Masking effects can precede the masker (backward or pre-stimulatory masking, or follow on after the masker (forward or post-stimulatory masking). High-level maskers lead to temporary threshold shifts (fatigue). These noise-induced temporary threshold shifts can be considered as more extreme examples of forward or post-stimulatory masking. Temporary threshold shifts can turn into noise-induced hearing loss in extreme cases.

Most non-simultaneous masking effects can be explained, at least in qualitative terms, by the excitation patterns of different test signals along the basilar membrane which vary according to the time varying frequency content of the signal. Considered overall, masking effects are important because, while possessing exquisite sensitivity to low-level continuous sounds, the auditory system has limited time and frequency resolution and might not be able to detect individual components within composite signals than can easily be detected by an appropriate instrument using a sufficiently long averaging time.

3.4.5 Spatial capabilities

There are complex variations in the detailed pattern of sound fields at the two ears which are always correlated (to a greater or lesser extent) with the nature of the surroundings, particularly the spatial disposition of sound sources and associated reflecting surfaces. Two extreme examples of this variation are: a single omni-directional point source positioned directly in front of a listener in a non-reverberant free field (i.e. in an anechoic room or outdoors in a very quiet place with a very absorbent ground surface); or a multiplicity of incoherent point sources widely distributed throughout a highly reverberant enclosed space. The auditory system has evolved in such a way that it is capable of extracting useful information from the nature of this variability.

There is normally one ear located on each side of the head, making two in all. This confers a number of advantages. First, each ear samples the sound field in a different place. Multiple sound waves bouncing around inside any confined space interfere with one another both constructively and destructively to produce a multiplicity of nodes and anti-nodes at different frequencies. Clearly, any pressure sensor (such as the human ear) positioned at a node will be at a serious disadvantage in terms of being able to detect a sound wave at that frequency. It is relatively unlikely that two spaced pressure sensors would both be positioned at nodes at the same time, meaning that one or the other is likely to be able to detect the signal.

A good example of this situation is stereophonic sound reproduction equipment, where a listener sits equidistant from and in front of two loudspeakers reproducing the left and right audio channels. Signals which are 'panned' to the centre of the sound stage by the recording engineer will normally be reproduced with equal strength (and in phase) by both loudspeakers. This will produce a complex interference pattern between the two loudspeaker signals at all listening positions which are not precisely equidistant from both loudspeakers. The combination of constructive and destructive interference between the two loudspeaker signals produces a comb-filtering effect with harmonic node frequencies dependant on the difference in distances from the two loudspeakers. This comb-filtering effect is rarely heard by an average listener, although it can easily be demonstrated by measuring the actual frequency response of the stereo system at off-centre positions. The reason it is not heard is because the auditory system is able to average across the sounds detected at the two ears located at physically different positions. Head movement can provide the same benefits as having two ears in terms of spatial sampling, although the necessary information to allow the internal perceptual processor to average across the spatial anomalies in the sound field takes longer to accumulate and requires more complex processing.

The second main advantage of having two ears results from the additional opportunities this provides for direct comparison of the left and right auditory nerve signals. In essence, sounds coming from any point on the median plane through the centre of the head should be effectively identical in both phase and intensity at the two ears. Sounds coming from 90 degrees left or right of the median plane (i.e. from any point to either side of the head on the extended centre line orthogonal to the median plane) should have the maximum difference between left and right ears. At low frequencies where the head diameter is small compared to a wavelength, the main difference between signals at the ears is the relative phase. The relative phase difference is proportional to the angle that the source direction is displaced away from the median plane and increases with frequency. Above around 700 Hz, the relative phase difference can exceed π radians for the more extreme angles off to the side, and it then becomes ambiguous as to whether the signal in the left or right ear is leading. At higher frequencies above about 1.5–2 kHz, the resulting relative phase differences contribute very little to source localization accuracy, although the relative time of arrival of short transient sounds at the left and right ears might still provide useful source localization cues up to quite high frequencies.

The wavelength of these higher frequency sounds becomes increasingly comparable with and then smaller than the head diameter. A significant shielding effect begins to develop which can reach 15 or 20 dB at even higher frequencies. Thus source localization with the head fixed in position relative to the source can be achieved by comparing the left and right auditory nerve signals as follows:

1. low-frequency phase differences for continuous tones;
2. high-frequency intensity differences for continuous tones;
3. time of arrival differences for transients (wideband).

These between-the-ear differences with the head fixed in position relative to the source provide a great deal of useful information but can still be ambiguous. Particularly in the middle range of frequencies, neither the relative phase difference or the relative amplitude difference provide particularly useful source localization cues. The best audio signal to provide good source localization accuracy would seem to be pink noise, because with all frequencies present this provides the maximum redundancy of information.

As in the case of spatial sampling, head movement is also important in localizing the azimuth and elevation of sound sources. The differences in relative phase and intensity between the two ears can be used to define the angle of the source relative to the median plane, but this does not help in resolving front from back or up from down. However, the ways in which the between-the-ear differences change when the head is rotated relative to the source (or when the source moves relative to the head) provides a considerable amount of extra information which allows front-or-back and up-or-down confusions to be otherwise resolved. Consider the case of a source located at 45 degrees off to the side in front. With the head fixed, it is difficult to tell whether the source is actually in front or behind. As soon as the head is rotated towards to the side on which the source appears to lie, then if the source is in front, the differences between the left and right ear signals reduce. If the source is behind, then the differences between the left and right ear signals increase.

Finally, we have to consider the role of the pinna, or external ear. It obviously has no acoustic function at low frequencies where it is small compared to the wavelength of sound. First, the pinna provides some screening against high-frequency sounds coming from the rear. The relative balance between high- and low-frequency components present in a particular sound should be subtly different dependent on whether the source is in front of or behind the listener. In addition, there is some evidence that the various folds can reflect high frequency sounds into the ear canal to produce a comb filtering effect through destructive interference with the direct sound. The comb filter null frequencies vary depending on the angle of elevation. Since the detailed shape of the pinna varies between different individuals, it seems likely that the precise association between the angle of azimuth and the amount of high frequencies present, and the angle of elevation and the resulting comb filter nulls would have to be learnt on the basis of experience. In addition, the listener must be familiar with a particular sound source to be able to detect the effect of angle-dependent filtering.

3.5 MEASUREMENT OF SOUND

3.5.1 Sound level meter measurements

The main function of a sound level meter is to measure sound levels as accurately as possible. This requirement is met by providing a precision condenser microphone, accurate and stable electronic systems and appropriate and regular calibration. However, there is a wide range of possible noise units which could be used, each of which would give different results under different circumstances. A great deal of research effort has been expended over the last 30–40 years in various attempts to perfect the correlation between sound level meter readings and subjective impressions of a noise by an appropriate choice of a more or less complex noise unit. This goal now seems increasingly unrealistic, because subjective impressions often depend as much on expectation and so-called non-acoustic factors (which cannot be measured by a sound-level meter) as on physical sound levels. On the other hand, the A frequency weighting and the F, S, and longer time weightings are now in general use and it is important to understand how these can be related to known auditory capabilities.

For maximum objectivity (which may be required for research purposes), it is desirable that the sound level meter should not impose any unrepresentative time or frequency weighting or other non-linearity on the measurement. Most sound-level meters are provided with a 'Lin' or unweighted frequency response setting which can be used in these circumstances. The precise frequency limits for the Lin frequency weighting are not defined in International Standards and this means that the user may have to approach the question of calibration with some care. A precision grade microphone can be specified with a flat frequency response over the entire audio range of interest (typically 30 Hz to 15 kHz), but the microphone directivity may become important at high frequencies where the diameter becomes comparable with a wavelength. At these frequencies, it is important to select carefully between a free-field or pressure corrected microphone and also to orientate the microphone correctly in relation to the sound source to obtain the maximum objectivity. Clearly, neither the microphone frequency response nor its directivity will match that of the human ear.

3.5.2 Frequency weightings

Virtually all commercially produced precision sound level meters are provided with an A frequency weighting setting in addition to an unweighted or Lin frequency response setting. The A frequency weighting is now used for nearly all outdoor community noise measurements. The A frequency weighting reduces the sensitivity of the sound level meter to low and high frequency sounds as compared with sounds in the upper mid-band of frequencies from around 1 kHz up to around 4 kHz, and approximately follows the shape (inverted) of the 40 dB sound level at 1 kHz equal loudness contour. Some older sound level meters are fitted with B and C frequency weightings in addition, and these approximately follow the shape (inverted) of the 70 and 100 dB sound level at 1 kHz equal loudness contours. The A, B and C frequency weighting curves are shown at Figure 3.8.

The original intention at the time when the A, B and C frequency weighting curves were standardized was to roughly match the assumed frequency response of the ear as implied by the standard equal loudness contours. Three different frequency weightings were required because the equal loudness contours become flatter (wider) at increasing sound levels. The A frequency weighting was intended for use at around 40 dB sound level, the B frequency weighting at around 70 dB sound level and the C frequency weighting at around 100 dB sound level. This well-intentioned scheme suffered from two major defects. The most important defect is that simply approximately matching the frequency response of the sound-level meter to the assumed frequency response of the ear (as determined by equal loudness contours for pure tones) does not account for all the other acoustic

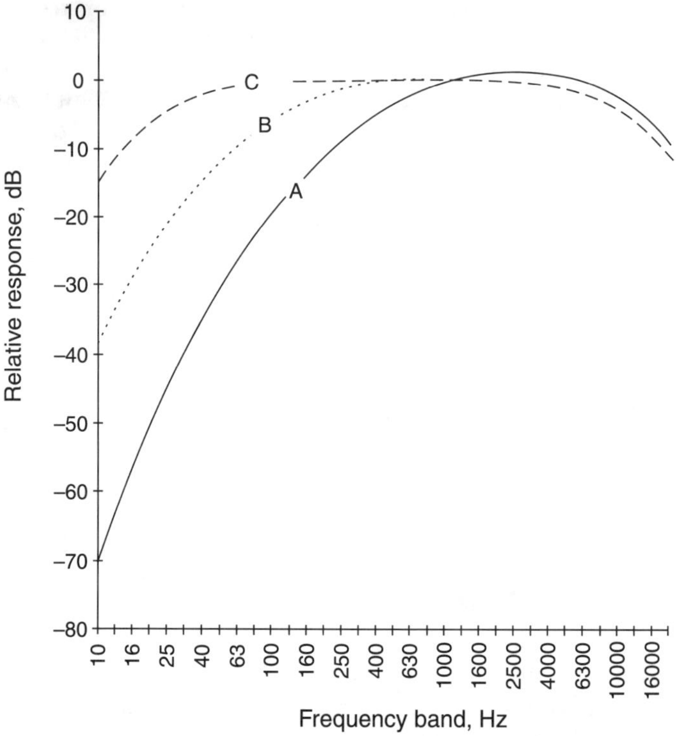

Figure 3.8 The A, B, and C frequency weighting curves.

and non-acoustic factors that might affect subjective opinions of the noise being measured. The second defect is that for sounds with prominent low- or high-frequency content, switching the frequency weightings can have a considerable effect on the overall reading. Switching between frequency weightings would therefore have to follow precisely specified rules depending on overall sound level, but no such rules were ever agreed upon.

Practical experience shows that the A frequency weighting provides some immunity against wind noise generated at the microphone windshield when taking measurements outdoors, and also discriminates against predominately low frequency noise from distant sources. These factors alone would be a sufficient reason for its widespread use. A large number of noise levels and annoyance response surveys have been carried out at community sites over the last few years. In most, if not all, of these surveys nearly all noise measurements were carried out using the A frequency weighting. Reasonable correlations have been obtained between A frequency weighted measures of sound level and reported annoyance or bother, etc., obtained by using a questionnaire. These reasonable correlations have then been taken as sufficient evidence that the A frequency weighting is the 'best' frequency weighting curve that could be devised. The B and C frequency weightings have largely fallen out of use.

On the other hand, although there have been a number of laboratory type studies looking at the superiority or otherwise of the A frequency weighting, most field studies have only used the A frequency weighting, i.e. there is no data against which to compare different frequency weightings. This means that the field evidence that it is the 'best' frequency weighting to use is somewhat lacking. Technically, the A frequency weighting becomes increasingly incorrect at higher sound levels where the low frequencies are downweighted too much to comply with the inverse of the equal loudness contours.

The A frequency weighting has been criticised for indoor measurements, where typical facade attenuation suppresses mid- and high-frequency components, emphasizing the low-frequency components within a room. This could be significant for premises adjacent to busy urban roads where there can be a lot of low-frequency noise generated by bus and lorry engines accelerating from rest and changing gears.

The A frequency weighting has also been criticized on the basis that people lose high-frequency hearing as they get older, and that low frequencies might then become more important than implied by the weighting curve. Although this suggestion appears plausible, there is no evidence that the A frequency weighting provides a better match to older people's hearing.

The Noise Rating (NR) frequency weighting curves (see Fig. 3.9) also represent equal loudness contours, but in this case by using octave band frequency analysis of the sound being measured. Many types of commercial sound-level meters can be fitted with switchable octave or one-third octave filters to facilitate this type of measurement. NR curves were devised in Europe, and the very similar Noise Criteria (NC) and Preferred Noise Criteria (PNC) curves were devised in the USA.

NR curves are widely used in building acoustics to define noise limits for building services or for acceptable noise break-in from outside sources. Typically, NR 10–20 is specified for audiometry, NR 20–30 for classrooms, NR 30–40 for quiet offices, and NR 60–70 for industry. The NR 70 suggestion for industry is based on avoiding interference with speech communication and is often highly optimistic in terms of what can be achieved in practice.

Finally, there are a number of more complex frequency weighting schemes, such as the Effective Perceived Noise Level (EPNdB) scheme and the Zwicker and Stevens Loudness Level schemes which involve more or less complex calculations to aggregate the loudness or noisiness contributions of each one-third octave frequency bandwidth present. There is some evidence

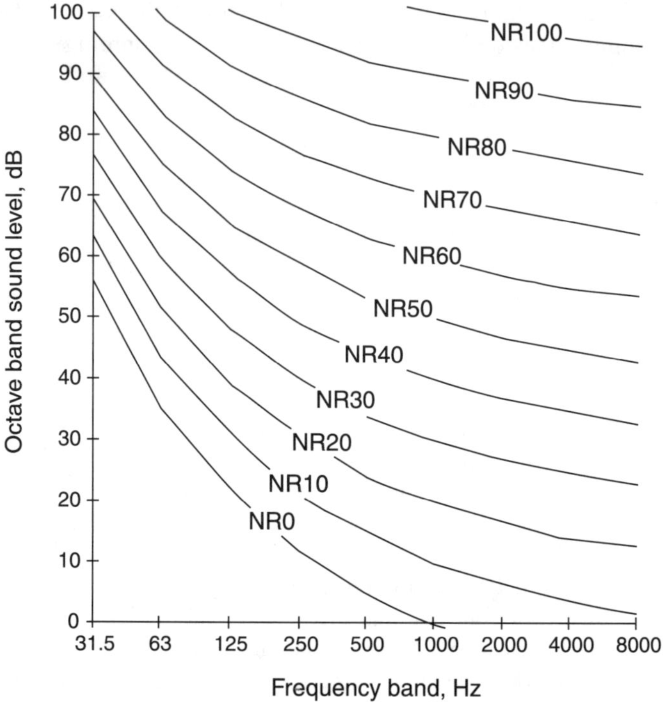

Figure 3.9 The Noise Rating (NR) frequency weighting curves.

obtained in controlled laboratory studies that the use of these complex frequency band addition schemes can lead to higher correlations with subjective impressions of the noise, but there is very little evidence obtained in field studies that the additional complexity is either worthwhile or even justified by greater accuracy.

3.5.3 Time weightings

We know from the experimental data relating to the integration time of the ear that transient sounds of less than about do 125 ms duration do not register their full loudness. Virtually all commercially produced precision sound-level meters are provided with an F time weighting setting in addition to one or more of the following time weightings; Peak, S (slow), and long-term linear averaging (L_{Aeq}).

The F (fast) time weighting has an exponential time constant of 125 ms, to correspond approximately to the ear integration time. The intention is that the sound-level meter readout will then respond to changes in the level of a sound at about the same speed as the ear. Any transient sound with duration less than 125 ms will give a lower reading than if it lasted for 125 ms or longer. However, if a sound-level meter with an F time weighting is used to monitor the fluctuating sound levels in a rapidly varying sound such as normal speech, then the variation in the sound-level meter readout will be too fast for the eye to follow. Many sound-level meters are fitted with an S (slow) time weighting setting, which with an exponential time constant of 1 s, slows down the sound-level meter display to allow for the average level to be estimated with greater precision. Of course, digital readout displays which are updated only once or twice per second can create a false impression of how rapidly the underlying instantaneous sound level is changing.

Two other time weighting settings are sometimes provided for the measurement of individual transient sounds. The I (impulse) time weighting has a fast rise (35 ms) and slow decay and is meant to mirror the ear's response to impulse sounds, although there is some doubt as to whether it actually does this very well; it is hardly used in the UK. The Peak time weighting has a time constant of below 100 microseconds to respond as quickly as possible to the true peak level of transient sounds. The Peak time weighting is specified for the measurement of transient sounds to determine compliance with the Peak Action Level under the Noise at Work Regulations, 1989, and in this case the Lin frequency weighting must be used.

Visual estimation of the average sound level from a continually fluctuating sound-level meter display is not the best method of obtaining a longer term average. Integrating sound-level meters are fitted with a long-term averaging circuit which generates an L_{Aeq} readout, which is defined as the equivalent continuous sound level, i.e. that steady level that would have the same energy as that averaged over a stated measurement duration. The 'A' in L_{Aeq} signifies that the A frequency weighting is used. For historical reasons, the Department of Environment, Transport and Regions in the UK use the term L_{eq} instead of L_{Aeq} to describe aircraft noise exposure, but apart from the implication that a duration of 16 hours from 0700 to 2300 hrs is used, it is otherwise exactly the same.

In the older type of analogue sound-level meter, L_{Aeq} is calculated by integrating an electronic signal which is proportional to the instantaneous A-weighted sound pressure squared, and then dividing by the total measurement duration and then converting back to a decibel equivalent by using a logarithmic amplifier. The result is effectively a long-term root mean square (LTRMS) average of the instantaneous sound pressure, and is often described as an energy average because of the use of sound pressure squared. Of course, without the squaring process, the average sound pressure would tend to zero regardless of the magnitude of the sound. In a modern digital sound-level meter, this is all done in software.

The measurement duration must be stated or assumed because a 3 dB change in L_{Aeq} is equivalent to a factor of two in the number of events included in the measurement, in the number of noise sources included in the measurement, or in the aggregate event duration within a defined overall measurement duration. Doubling the measurement duration to include periods of relative silence before and after a single noise event reduces the overall L_{Aeq} by 3 dB. The duration is normally added as a suffix such as $L_{Aeq,16h}$ or $L_{Aeq,5min}$, and it may even be important to state the start and end times of the measurement. For example, it is becoming a *de facto* standard in the UK to report average daytime community noise levels as $L_{Aeq,16h}$ from 0700 hrs to 2300 hrs local time. Average night-time community noise levels ($L_{Aeq,8h}$ from 2300 hrs to 0700 hrs local time) would normally be considerably lower in the UK.

The Sound Exposure Level (SEL), (the acronym is sometimes more easily remembered as the Single Event Level), is a specific measure applicable to separately identifiable noise events. Obviously, the maximum F time weighted level (known as L_{Amax}) of an event cannot convey any impression of the duration or relative frequency content of the sound. SEL is effectively a measure of the total acoustic energy in the event scaled back to a standard duration of 1 s and is calculated in the same way as L_{Aeq} except that the integrated A-weighted sound pressure squared signal is not divided by the measurement duration.

The best way to appreciate the difference between SEL and L_{Aeq} is to consider two extreme examples, a completely steady sound and a single transient event surrounded by extended periods of silence. For a completely steady sound, the L_{Aeq} does not change with different measurement durations as the average sound level does not change. The SEL, however, increases logarithmically as the total acoustic energy received at the measurement microphone increases with measurement duration. For a single transient sound surrounded by extended periods of silence, the L_{Aeq} reduces logarithmically as the average acoustic energy received at the measurement microphone reduces with increases in the measurement duration. For the same situation, the SEL does not change with different measurement durations as the total acoustic energy received at the measurement microphone does not change. There is no difficulty in translating between measurements of SEL and L_{Aeq} provided that the relevant measurement durations are known. For the measurement of transient events, SEL has an advantage as the onset and offset times of the measurement are not critical.

The two acoustic energy based measures (L_{Aeq} and SEL) have come into increasingly widespread use in recent years and have largely superseded a number of older measures. There are two main reasons for this: lower cost and greater convenience. First, integrating sound-level meters which are capable of measuring L_{Aeq} and SEL are now only a little more costly than non-integrating instruments, whereas only a few years ago the technology did not exist to allow for this type of measurement in a simple hand-held instrument. Bearing in mind the labour and overhead costs of noise measurement, the additional cost of using an integrating sound-level meter is usually negligible.

In terms of convenience, L_{Aeq} and SEL measurements are simple and robust. Calculations of the effect of changes in the noise environment due to, for example, the introduction of a new noise source, will agree quite closely with before and after measurements. Perhaps surprisingly, this physical simplicity does not apply in the case of some of the more esoteric noise units and noise indices which have been used in the past. On the other hand, the very fact that L_{Aeq} and SEL measurements are consistent with physical reality does not automatically imply that they are consistent with subjective impressions. For example, doubling the number of events within a defined measurement period will double the average acoustic energy received over that period and increase the L_{Aeq} by 3 dB. The same effect could be achieved by increasing the sound level of the individual events by 3 dB. It seems likely that many people would not notice a 3 dB change in sound level of the individual events, particularly if the change happens gradually. On the other hand, the increase in the number of events might be more easily noticed. This means that the interpretation of measures such as L_{Aeq} and SEL still requires some caution. The most extreme example is the effect of adding a new noise source with the same L_{Aeq} as the existing noise. The overall L_{Aeq} will increase by 3 dB which implies that the psychological effect of adding a completely new noise (which is unlikely to be missed) is equivalent to an increase in sound level due to the existing noise alone of only 3 dB, which would not be noticed by a large proportion of residents exposed to the noise.

Another way of describing complex time varying noise environments is by the use of what are known as the statistical levels, L_n. Any particular L_n is the sound level which is just exceeded for n% of the defined measurement duration. In the days before integrating sound-level meters became widespread, L_n values were calculated by using a statistical distribution analyser. A time-varying sequence of instantaneous sound-level readings would be fed into an array of counters, each of which was set up to count the amount of time the sound-level reading was between two narrow limits, say from 52 dB to 54 dB. The array of counters recorded the probability distribution of different sound levels within the overall measurement duration. Any desired value of L_n could then be read off from a plot of the measured probability distribution on standard chart paper. Of course, modern digital sound-level meters can be programmed to perform the same functions in software.

The following specific L_n values remain in reasonably widespread use in the UK, although their usage in other parts of the world seems to be generally declining. The L_{01} is associated with the maximum level occurring during a measurement sample period, but it is clearly not the actual maximum level which might be higher for a long duration measurement. The L_{10} is specified for road traffic noise measurement and assessment in the

UK. Unattended measurements of L_{10} are relatively insensitive to extraneous noise sources which are not road traffic unless they are present for more than 10% of the time. There have been attempts to apply noise standards written in L_{10} to other noise sources such as aircraft and railway noise. A moment's thought will show the folly of this approach, since the sound level which is just exceeded for 10% of the time will be completely unaffected by any noise source which is present for a smaller proportion of the overall time. The choice between an L_n or an L_{Aeq} type measure tends to depend more on practical considerations such as these than on any correlations with subjective impressions of the noise. In terms of reported noise annoyance studies, it is often unproductive to attempt any differentiation between L_n and L_{Aeq} type measures on a statistical basis because the correlation between the different types of measures is often too high. This is another way of saying that they both measure largely the same thing except where there are detailed differences which make one of the scales (usually L_n) completely inappropriate.

The L_{50} is sometimes used as a descriptor of the average noise level. The L_{90} is used in BS 4142: 1990 (BSI, 1990) as a descriptor of the steady background noise level, and the L_{99} is associated with the minimum level during the measurement sample period. The steady background noise level during a measurement sample period might be important for certain types of noise assessment. An alternative way of describing the steady background noise level is to use the concept of residual noise level. This is defined as the average noise level or L_{Aeq} of what is left when the contributions from all specific noise sources present have been removed. The residual noise level can be easily calculated from a detailed sound-level time history, provided that the contributions from all specific noise sources present can be properly identified. This can involve an element of subjectivity, and therefore for administrative and regulatory purposes the L_{90} provides a more robust descriptor.

The most comprehensive descriptions of environmental noise levels which are generally available short of a full bandwidth audio recording are provided by data logging sound level meters which record a time history of short L_{Aeq} measurements. The duration of each short L_{Aeq} in the sequence can be selected by the user, but is normally either 1 s or 125 ms to approximately correspond with the S and F time weightings. Instruments which can record faster short L_{Aeq} sequences down to 10 ms are available for the measurement of impulsive noise, but it is doubtful whether this provides any real advantages because of the finite integration time of the human ear. One advantage of the short L_{Aeq} sequence is that any other A-weighted noise level index can be calculated from it, but since the international community is now increasingly standardizing on the use of L_{Aeq} as a general purpose noise unit the need for this type of flexibility is somewhat reduced.

3.5.4 Detailed frequency analysis

The basic function of a sound-level meter measurement is to provide a single number description of the overall magnitude of the noise. This is obviously very important where noise exposure is compared against some defined standard and is additionally useful where one noise environment is to be compared against another. However, as pointed out in the discussion of sound quality above, there are often individual acoustic features present which are likely to contribute to subjective impressions in matters of detail which are not capable of being represented by any overall single-number descriptor.

There are really two approaches which could be adopted to deal with these more complex situations. First, a full frequency range recording can be made using precision microphones and a DAT recorder (or any alternative system of providing equivalent audio quality using uncompressed Pulse Code Modulation digital audio at 16 bits, 44.1 kHz sample rate, or better). The recordings are then reproduced in completely subjective listening tests. Secondly, a more detailed analysis of the noise can be carried out using various forms of more or less sophisticated frequency analysis instruments to identify and quantify acoustic features of interest in objective terms. Subjective listening tests can be robust and repeatable, but it can be difficult to interpret the results in terms of contributing to policy decisions. As a result, many acoustical engineers involved in noise assessment in general and in sound-quality work in particular tend to concentrate on objective measures wherever possible. There are a number of sound quality analyser systems becoming more widely available at the present time. Most of these instruments can provide a whole range of detailed analyses of the time varying frequency content of the signals and can display the results of these analyses in different ways.

There are two generic types of frequency analyser system in common use. These have separate frequency bands distributed either linearly or logarithmically across the auditory range. The Fast Fourier Transform (FFT) frequency analyser generates linearly spaced frequency analysis bands and tends to be the most commonly encountered of the two generic types. The other type uses a sequence of constant percentage bandwidth filters which are logarithmically spaced by definition. Both types of frequency analyser can use either logarithmically or linearly spaced frequency axes in any subsequent output displays, but this does not affect the fundamental mode of operation. Frequency analysers can also operate in real-time or sequentially and this can be important for transient signal analysis. The simplest type of switched one-third octave band analyser attached to a sound level meter can only measure one frequency band at a time. A sophisticated real-time frequency analyser can measure all frequency bands simultaneously and continuously. There are a number of halfway houses where the different frequency bands are shared out sequentially and not all are measured both simultaneously and continuously. In any digital instrument, this is simply a matter of processing power and therefore cost.

The logarithmic frequency band spacing of the constant percentage bandwidth analyser corresponds more closely to the essentially logarithmically spaced auditory filters than the linear frequency band spacing of the FFT type analyser. By comparison, the frequency resolution of an FFT analyser often appears too coarse at low frequencies and much too detailed at the high frequencies. A typical 1024 point FFT analyser running at, say, 25 kHz audio sample rate might display 400 lines of frequency resolution at 25 Hz bandwidth from DC up to 10 kHz; 25 Hz bandwidth at 100 Hz is much coarser (25%) than the frequency resolution of the human ear, whereas 25 Hz bandwidth at 10 kHz is much finer (0.25%) than the frequency resolution of the human ear.

The most common type of real-time constant percentage bandwidth frequency analyser has a series of parallel one-third octave band filters. This frequency resolution is approximately proportional to the frequency resolution of the ear throughout most of the audio range, but it is not sufficient to detect fine differences in frequency content down to 1% resolution which might be detected by the ear. A constant percentage frequency analyser with parallel 1/24th octave band filters comes much closer to the frequency resolution of the ear, but even this has only half semi-tone frequency resolution.

The main disadvantage of constant percentage bandwidth frequency analysers (apart from their usually significantly greater cost) is that they are not as good at displaying harmonic relationships between discrete frequency components when present as the linearly spaced frequency analysers. Harmonic relationships between discrete frequency components can be important for sound quality. However, one advantage of constant percentage bandwidth frequency analysers, provided they use either analogue or digital narrow band filter systems, is that each filter operates continuously which allows the various averaging times in each band to be selected to most closely correspond to the averaging time of the ear in each frequency band. FFT analysers operate by taking discrete blocks of data for subsequent processing and this means that the averaging times can only be increased in discrete steps by averaging across more than one block. For the example of the 1024 point analyser working at 25 kHz audio sample rate discussed above, each block of 1024 data points takes approximately 40 ms, and so the overall averaging time can only be set up in multiples of 40 ms across the whole audio band. In addition, the true frequency resolution in any FFT analyser can often be quite a bit coarser than the line spacing because the effective time duration of each sample block varies when the necessary time and frequency windowing effects are taken into account. Other more sophisticated methods of frequency analysis have been developed but these are beyond the scope of this chapter. The reader is referred to Chapter 6 for further details of some of these methods.

3.5.5 Spatial analysis

The auditory system has some quite powerful spatial capabilities whereas the simple sound level meter fitted with a pressure sensitive microphone cannot discriminate between sounds coming from different directions at all. This difference is not relevant for measuring compliance with noise standards and regulations, but it may be important for sound quality work and it is clearly very important for reproduced sound system development. There are a number of commercially available dummy head sound recording systems which go some way to being able to reproduce the spatial capabilities of the human auditory system although they are all subject to a number of limitations. Dummy-head sound recording systems offer obvious benefits as compared to a single pressure sensitive microphone for measurements of the electroacoustic properties of headphone and earpiece systems.

No dummy head based sound recording systems are capable of generating an objective spatial analysis of the sound field that truly correlates with human spatial perceptions. This is because human spatial hearing depends to a considerable extent on higher-order neural processing which is insufficiently understood at the time of writing for it to be copied by any measurement system. The dummy-head-based sound recording system merely reproduces the acoustically relevant parts of the head, pinna and ear canal and does not simulate any of the internal systems. The main function of a dummy-head sound recording system is to be able to reproduce the same ear canal signals at a listener's ear that would have been received had the listener actually been present at the recording situation. The dummy head recording system can in theory provide a more realistic representation of the original (recorded) sound field than any single microphone or conventional stereo microphone recording. There is of course another side to the coin here, in that the dummy head system imposes a frequency-dependent directivity characteristic on any measurement that, while it is representative of the physical characteristics of the human ear, is one stage removed from the simple omni-directional measurement that all existing noise standards and regulations assume. Some manufacturers of commercial dummy head sound recording systems provide an electronic filter system that is intended to return the frequency response characteristics of the recorded signal to those of a single omni-directional microphone, but it is clearly technically impossible to cancel out the directivity effects of the dummy head electronically.

Secondly, no dummy head system yet developed can properly represent the head movement cues that can resolve front-or-back and other ambiguities in real situations. In addition, and particularly when using headphones, the listener will probably use head movement without necessarily being aware of it in an attempt to resolve such front-or-back ambiguities. Head position tracking systems have been developed where the frequency response to the left and right ears is made to depend on the azimuth and elevation of the sound source relative to the head. This is all very well for panning a single synthesised signal source around the head, but no system yet devised is capable of separating out all the individual directional components of a three-dimensional sound field, applying spatial filtering separately to each of these components in response to head position information and then recombining all components into a coherent whole. Without appropriate head position compensation, every time the listener's head is rotated, the apparent sound field rotates with it, which gives a totally false impression. Dummy head sound recording systems are used quite widely in the automotive industry for interior vehicle sound quality work. In this case, users must learn to disregard head rotation cues when listening, but this means that front-or-back confusions will necessarily be quite common when attempting to identify sound sources within the vehicle from recordings.

The third main problem with dummy head sound recording systems is that of individual differences in ear canal acoustics. The dummy head system necessarily imposes its own acoustic characteristics on the recorded signal, which are then filtered again by the transfer function of the headphone system on the individual listener's ear canals. For dummy head systems with full ear canal simulators, there are effectively two sets of ear canal resonances in series, one of which must be electronically filtered out to maintain accurate reproduction. This is always something of a compromise because of these individual differences in ear canal acoustics. In addition, the essence of any dummy head recording system is that an artificial head is substituted for the listener's real head. To the extent that the artificial head cannot match each listener's head dimensions exactly, the spatial impression given will be in error. Current dummy head sound recording systems are generally capable of producing highly convincing spatial impressions even though the spatial accuracy can be somewhat dubious. It would probably be better to use these systems for the subjective assessment of the relative differences between different sound field situations than for any objective analysis purposes, which might be better dealt with by using a more conventional single omni-directional microphone based measurement.

More recently, a number of systems have been developed to allow dummy head recording systems to be operated using loudspeakers instead of headphones for listening. Using headphones simplifies the task of ensuring that only left channel signals go the left ear and only right channel signals go to the right ear. A similar effect can be achieved using loudspeakers if special cross-talk cancellation filters are fitted. Appropriate anti-phase signals are radiated by the left loudspeaker to cancel out the cross-talk radiation that would otherwise have been received at the left ear from the right loudspeaker and vice versa. It is not quite as simple as this, however, as there has to be further anti-phase signals to cancel out cross-talk from the original cancellation signals and so on.

One advantage of using a cross-talk cancellation system to listen to dummy head recordings using loudspeakers is that the sound field does not move when the listener's head is rotated. However, there is a finite limit to the area within which effective cross-talk cancellation can be achieved and so head movement must still be restricted. In addition, the differences between the left and right signals recorded with any standard two-channel dummy head system effectively code the angle of the source relative to the median plane through the centre of the head and do not code front-or-back or up-or-down. This means that the cross-talk cancellation system does not overcome the fundamental difficulty that all two-channel dummy head systems have in discriminating between front-or-back or up-or-down. More audio channels are required to code this additional information, but it is difficult to see where any such additional microphones could be placed about the dummy head without completely changing the fundamental operating philosophy of the system. The unaided human auditory system is relatively poor at resolving front-or-back without using head movement in addition to the two ears, so it is probably rather pointless to expect an electro-acoustic system to be able to do better while not supporting listener head movement as an integral part of the system. Finally, the use of cross-talk cancellation systems does not overcome the fundamental objections to the use of dummy-head recording systems for objective measurement purposes.

3.6 COMMUNITY NOISE

3.6.1 State of the art on community noise

The Wilson Committee on the Problem of Noise was appointed in April 1960 to 'examine the nature, sources and effects of the problem of noise and to advise what further measures can be taken to mitigate it'. The general emphasis at the time was on describing the various sources of noise and assessing the resulting effects in terms of what practical steps could be taken to reduce noise annoyance. Despite the best efforts of the committee to advise on appropriate action, the problem of noise has not gone away. The final report (Wilson *et al.*, 1963) noted (paragraph 20) 'There is a considerable amount of evidence that, as living standards rise, people are less inclined to tolerate noise'. In comparison with 1963, there are two issues here: (a) has the total amount of noise increased, decreased, or stayed about the same? and (b), has people's sensitivity to noise, or their tolerance of noise effects, changed?

The World Health Organisation set up a study group on noise in November 1973. This was eventually followed by the publication in 1980 of the WHO Environmental Health Criteria Report 12 on Noise (WHO, 1980). The report identified the need for noise control and hearing conservation programmes in industry. Road traffic noise was singled out as the main source of community noise, but aircraft noise was also identified as an issue of world-wide concern. The possibility of adverse health effects of noise was clearly identified, but (paragraph 1.1.1) 'because of their complexity, their variability, and the interaction of noise with other environmental factors, the adverse health effects of noise do not lend themselves to a straightforward analysis'. Noise obviously continued to be a problem at the time (1980), but the possibility of adverse health effects was increasingly being discussed in addition to the annoyance affects of noise.

After wide-ranging international consultation on earlier drafts, Berglund and Lindvall prepared a revised version of the 1980 WHO document for the World Health Organisation in 1995 (Berglund and Lindvall, 1995). The 1995 revision included contributions from a number of specialists in particular fields and the noise guideline values suggested were based on the discussions held at a WHO Task Force Meeting in Germany in 1992. The 1995 revision does not have any official WHO status at the time of writing, although it is increasingly being used as an authoritative source by legislators and administrators in the field of noise regulation and control. In the foreword, the editors state: 'the complex and extended work process makes it necessary to declare that the editors are solely responsible for the present text of the document'. It is evident that the noise problem is continuing to increase in importance, as can be seen from the increasing volume of research which has been carried out. On the other hand, it is also becoming increasingly difficult to reach a consensus on the precise magnitude and relative importance of the problem, and the practical steps that can be taken to alleviate it.

Most recently, we have the European Commission Green Paper on Future Noise Policy (European Commission, 1996), which starts with the statement: 'Environmental noise, caused by traffic, industrial and recreational activities, is one of the main local environmental problems in Europe and the source of an increasing number of complaints from the public'. The EC Green Paper follows on from night time noise targets established by the European Commission's Fifth Environmental Action Programme to limit further increases in night noise. The European Commission has recently begun consultation on limiting the impact of noise from air transport (European Commission 1996, 2). The EC Green Paper and other initiatives are welcome in the sense that they have continued to raise the profile of the noise problem, but the practical solutions to the problem remain as difficult issues.

Over the years since the Wilson Committee was appointed there has been a considerable amount of research and development work carried out into the problem of noise. Noise at source has been reduced, in some cases by considerable amounts. For example, modern high bypass ratio aircraft engines are much quieter than the low by-pass ratio aircraft engines used in the Boeing 707s and Comet aircraft used at the time of the Wilson Report, and there are similar improvements in the noise generated by motor vehicle engines. Buildings can now be designed to exclude noise at relatively low cost, and there are numerous standards and regulations all designed to control or limit noise.

On the other hand, the volume of traffic, people's expectations for peace and quiet, and their general willingness to complain have all increased. The noise problem is fundamentally a matter of public perception, and it can be of little use to point out to people that certain aspects of the noise problem have improved if they are unhappy and complaining here and now. It might even be the case that the more effort is spent on noise, the more people might be pre-disposed to complain. There has been a considerable amount of research carried out to better quantify the noise exposure/community response relationship, but this has largely concentrated on static situations. The available scientific information is quite limited on what actually happens in practical situations where a new noise source is introduced or new houses are built in an existing noisy environment, and although planning officers, environmental health officers and acoustics consultants have accumulated a considerable amount of anecdotal experience of these real-life problems, there has been no real effort by government to formalize the different approaches adopted in these practical situations.

3.6.2 Practical solutions

Out of all the possible effects of noise on people, noise annoyance, perhaps by default, seems to provide the best basis for the definition of noise acceptability criteria. The main problem is that there is no single noise/annoyance relationship which applies to all situations and this can make generalization difficult. The scientific basis for the development of universal or generalized noise acceptability criteria based on noise annoyance is weak because so much depends on the circumstances of each individual case. On the other hand, some objectivity is required as otherwise there would be no technical basis at all to underlie the settlement of disputes involving noise.

In most practical situations where a decision has to be made balancing the costs of noise control action against the environmental consequences of the noise, it is often debatable whether any better defined but uniform noise/annoyance relationship would actually help. Each situation is different, with a different history, different people are involved, and the best compromise solution might well be different as well. Sometimes it is the introduction of a new noise source which needs to be assessed, sometimes it is the development of new housing in an existing noisy area which is at issue and sometimes it is an existing noise source in an existing residential area which is the cause of complaint. Each of these situations is different and may need to be treated differently.

3.6.3 New noise sources

Considering first the new noise source, the most important aspect to consider is how will the environment be changed? The standard method of road traffic noise assessment used in the UK (DOT, 1993) requires a comparison between the do-something and do-minimum situations. Where noise levels are not increased by new development, there is clearly no noise impact. There is no need to consider noise/annoyance relationships for this comparison as the only item of interest is the change in the level of noise, as measured using some appropriate physical descriptor. There are technical debates as to which noise descriptor should be used (L_{Aeq}, L_{A10}, etc.) but it probably does not make any real difference providing that the same noise descriptor is used throughout. Most residential environments are affected by more than one noise source and it is important to decide whether the new noise source should either be considered in isolation, or in the context of the contribution it makes to the overall or ambient noise environment. Considering any new noise source in isolation will generally be more conservative than considering the effect that it has on overall noise levels, which depends on the extent to which the new noise source is dominant over other noise sources present, or not.

What degree of change is important? Increases (or decreases) in the level of a steady noise of 1 dB or less will not be detectable except under laboratory conditions. An increase (or decrease) of 5 dB or more might be required for the change to be significant under real-life conditions, particularly when the change happens gradually over a period of months or years (as is often the case where new transport infrastructure is under development). A small increase in the noise level due to a separate noise source which is not the most dominant noise source will have an even smaller effect on the overall noise level comprised of contributions from all sources present. This means that an increase of less than 1 dB in the overall noise level might be easily detected where it results from a much larger increase in the noise level contribution made by a less dominant but nevertheless separately identifiable new noise source. We can see that the question of what degree of change is important can be quite complicated, and is certainly affected by the different contributions made by other noise sources present.

What happens where the noise sources are not steady, such as aircraft or railway noise or even certain classes of road traffic noise where the noise peaks attributable to individual vehicles are separately identifiable? Different noise sources will be audible at different times, and the significance of any change will depend on everything else which is happening at the time. Finally, what is really happening when the noise level due to a separate noise source changes? In the case of aircraft noise, for example, the change could be due to some combination of the individual aircraft becoming quieter as older noisy aircraft are phased out, of the numbers of aircraft increasing to accommodate increasing numbers of passengers, and possibly due to some change in the pattern of operations such as a change of route or a change of timings. All of these factors could have an effect on the average noise levels as represented by L_{Aeq} or some other such noise level descriptor or indicator, but it has not been proven by research that the different effects of each factor on the community would be equally represented by the equivalent changes in L_{Aeq} that would result for each change separately.

In the case of road traffic noise, all other things being equal, a doubling of the vehicle flow rates gives a 3 dB increase in average noise level. The increase in noise level might not be very noticeable, but the increase in vehicle flow rate which causes it might be very noticeable. This is another area of uncertainty in terms of the results of noise attitude research. Further unresolved questions are: Is it better to have noise concentrated in certain parts of the day, or spread out more evenly? Which sources of noise are more or less annoying than others? The noise attitude research to answer these practical questions has not been done. There are also judgements to be made which cannot be based on the results of research. For example, assuming that a new transport infrastructure development has to be accommodated somewhere, is it better to concentrate new noise in existing noisy areas, where it might only result in a small increase on top of what are already high noise levels, or to position the new noise source in a completely new area, which might thereby suffer a large increase in noise levels which nevertheless do not approach the existing higher noise levels in the existing noisy area? It is difficult to imagine any kind of noise attitude research which could effectively answer such essentially political questions.

On a practical level, however, there are two further steps which can be taken when assessing the contribution made by a new noise source. The first step, where noise levels are increased, is to compare noise levels made by the new noise source against the existing background noise level, and the second is to compare noise levels made by the new noise source against absolute noise level criteria derived from regulations or international standards. The purpose of comparing the new source noise levels against the existing background noise level is to eliminate from further consideration those situations where the new source noise levels, even if they have been increased, are still below existing background noise levels and are therefore inconsequential. The same increase in separate source noise levels which applies close to a new noise source will continue to apply at increasing distances from that source, but eventually the distance will be such that even the increased noise levels are inconsequential.

Acoustic features such as tonality or impulsivity might be important in particular cases and some allowance can always be made for these when comparing noise levels (Porter *et al.*, 1993). It is sometimes tempting to apply a test of 'audibility' as an alternative to this test, on the basis that an inaudible noise cannot be a noise problem. On the other hand, inaudibility can be very difficult to define and can vary under different meteorological conditions. A simple comparison against the existing background noise level as a test of significance is much more straightforward. The existing background noise level can be defined as the L_{A90} (the A-weighted sound level exceeded for 90% of the time) (BS 4142:1990) or as the residual L_{Aeq} (the average sound level remaining when the effects of all other specific sound sources have been removed). Alternatively, the specific noise level can be compared against the existing overall or ambient sound level. This test is much less conservative where the ambient noise level is higher than the background noise level, which is normally the case.

The second step, where noise levels are increased, is to compare noise levels made by the new noise source against absolute noise level criteria derived from standards and regulations. Such criteria fall into three distinct classes as follows (Flindell, 1997). First there are criteria which define the lowest noise levels below which noise can be considered effectively irrelevant. In general terms, long-term average outdoor noise levels below 55 L_{Aeq} during the day and 45 L_{Aeq} during the night are not usually considered significant for residents indoors, although the method of averaging across intermittent noise events may be important here. Noise complaints can arise at lower noise levels, but it is harder to justify such complaints on noise level grounds alone as compared to complaints at higher noise levels. The second class of criteria should be considered as a range within which an increasing emphasis should be placed on noise control or mitigation to be set against the practicalities of any particular situation. This range will start at around 55 L_{Aeq} during the day and 45 L_{Aeq} at night and progress up to the third class of criteria which define an upper limit which should not be exceeded under any circumstances. Noise insulation of houses will not necessarily be justified at the lower end of this range, but will certainly be justified at the higher end of the range.

Community noise levels which contribute to noise-induced hearing loss or non-auditory health effects might be used as a basis for setting upper noise limits, but there are problems with this approach. First, as explained above, community noise levels which are high enough to contribute to noise-induced hearing loss are unusual. It may be difficult to argue that criteria based on this approach would actually offer any useful protection against noise to the public, and some lower criterion might be justified. Possible noise and non-auditory health effects might be used as a basis for upper limit criteria, but the evidence to support any particular value is weak (Passchier-Vermeer *et al.*, 1996).

3.6.4 New houses

The development of new houses in existing noisy areas is perhaps more straightforward. There is a 'market forces' argument which suggests that, providing noise levels do not exceed those which will contribute to adverse health effects (the upper limit absolute criteria discussed above), then there is no *a priori* reason to prevent people building houses in noisy areas provided that they have a free choice as to whether to live there or not. On the other hand, people might not have a free choice and they may need to be protected against noise exposure in a new house which they might only perceive as a nuisance after some time living there, when it is too late for them to back out. New housing developments rapidly become established residential areas, and it is debatable whether we want to increase the stock of noisy housing areas as a whole. Finally, new housing developments in existing noisy areas will ultimately restrict the freedom of the existing noise source for further expansion and runs counter to any land use planning policy based on common sense considerations.

In practical terms, noise levels must always be considered against other planning needs, such as a shortage of space for building elsewhere and some compromise regarding noise insulation will usually be justified. Current UK planning guidance is helpful in this respect (DoE, 1994), because it suggests guideline values in different noise exposure categories of increasing severity, but is not overly prescriptive.

3.6.5 Noise complaints

Noise complaints can be straightforward to deal with where there is an easy solution and intractable where there is not. In some cases, noise complaints can be dealt with simply by turning off the noise source at times when it is not required to operate for any useful purpose. Regulations and standards are irrelevant, it is just a matter of common sense. Where there is no simple solution and some compromise must be reached, it is again unlikely that noise standards and regulations alone can be relied upon to provide the best compromise solution. In practice, these more difficult noise complaint situations arise where the complainant perceives a nuisance that the operator or noise producer cannot easily reduce. The best compromise solution will be one that does not unduly inconvenience the operator while still offering a sufficient degree of noise control to satisfy the complainant. It is not always possible to achieve a satisfactory compromise in which case one party at least will not receive satisfaction. It is unlikely that a satisfactory compromise can be reached without negotiation, during which a too rigid reliance on noise regulations and standards by either party might well be counter-productive. The final decision might only be reached in the courts, in which case the concepts of reasonable expectations must be set against best practicable means at economic cost. What is reasonable, practicable, and economic are all essentially value judgements which do not depend on the results of noise/annoyance research.

3.7 CONCLUSIONS

It is clear that the problem of noise has not been solved in the years since the Wilson Committee was appointed. There has been a great deal of progress in noise control at source, but much of this progress appears to have been at least partially cancelled out by the considerable increases in traffic since that date. Objective noise measurement clearly relates to standards and regulations and might not be particularly well based on actual human response to noise. Research into noise/annoyance relationships has clearly demonstrated that reported annoyance generally increases with noise level, but it has left a number of important questions unanswered. Noise assessment continues to be based on largely pragmatic considerations. There are a number of methods of assessment available, and it is important that these should be carefully selected to match the precise objectives of the assessment in each particular case, so as to support the best compromise solutions that can be achieved.

3.8 REFERENCES

American National Standards Inc. (1969) *American National Standard Methods for the Calculation of the Articulation Index*, S3.5.

Acoustical Society of America (1997) *Quantities and Procedures for Description and Measurement of Environmental Sound – Part 4: Noise assessment and prediction of long term community response*, ANSI S12.9.

Axelsson, A. *et al.* (1996) *Scientific Basis of Noise-induced Hearing Loss*, Thieme, New York.

Behrens, T., Watanabe, Y. and Flindell, I.H. (1996) *Der Kammfilter-Effekt in der Stereophonie*, DAGA 96, 22 Deutsche Jahrestagung fur Akustick, Bonn.

Berglund, B. and Lindvall, T. (1995) *Community Noise*, Archives of the Center for Sensory Research, Vol. 2, Issue 1, Stockholm University and Karolinska Institute.

BS 3383: 1988, ISO 226: 1987 (1988) *Normal Equal Loudness Level Contours for Pure Tones under Free-field Listening Conditions*, British Standards Institute.

BS 4142: 1990 (1990). *Method for Rating Industrial Noise Affecting Mixed Residential and Industrial Areas*, British Standards Institution.

BS 6840: Part 16: 1989 (1989) *Sound System Equipment – Part 16. Guide to the RASTI method for the objective rating of speech intelligibility in auditoria*, British Standards Institution.

BS 7443: 1991 (1991). *Specification for Sound Systems for Emergency Purposes*, British Standards Institution.

Commission of the European Communities (1996a) *Future Noise Policy*, EC Green Paper, COM(96) 540 final.

Commission of the European Communities (1996b) *Consultation Paper on the Limitation of the Impact of Noise from Air Transport*, European Commission Directorate-General for Transport, 11 November.

Dept of Environment (1994) *PPG24: Planning and Noise*, HMSO, London.

Dept of Transport (1993) *Design Manual for Roads and Bridges – Volume 11*, HMSO, London.

Ehmer, R. H. (1959) Masking by tones vs. noise bands, *Journal of the Acoustical Society of America*, **31**, 1253–6.

Fidell, S., Barber, D. S. and Schultz, T. J. (1991) Updating a dosage-effect relationship for the prevalence of annoyance due to general transportation noise, *Journal of the Acoustical Society of America*, **89**, 221–33.

Fields, J. M. (1994) *A Review of an Updated Synthesis of Noise/annoyance Relationships*, NASA Contractor Report 194950.

Fields, J. M. (1996) Progress towards the use of shared noise reaction questions, *Proceedings Internoise* 96, 2389–94.

Flindell, I. H. (1997) Noise assessment methodology in multi-source environments, *Proceedings InterNoise 97*, Budapest.

Flindell, I. H. and Lewis, C. (1996) Acoustic features in vehicle interior noise quality, *Proceedings Internoise* 96, Liverpool Institute of Acoustics, 2479–84.

Harland, D. G. and Abbott, P. G. (1977) *Noise and Road Traffic outside Homes in England*, Transport and Road Research Laboratory Report 770.

Institute for Environment and Health (1997) *IEH Report on the Non-auditory Effects of Noise*, MRC Institute for Environment and Health, London.

ISO 226 (1996). *Acoustics – Normal Equal Loudness Level Contours*, International Organization for Standardization.

ISO 532 (1975) *Method for Calculating Loudness Level*, International Organization for Standardization.

ISO 3891 (1978) *Acoustics – Procedure for Describing Aircraft Noise Heard on the Ground*, International Organization for Standardization.

Ludlow, B. P. and Flindell, I. H. (1997) An overview of noise and health effects – one way forward?, *Proceedings InterNoise 97*, Budapest.

Lydolf, M., Frandsen, P. C. and Moller, H. (1996) Psychometric procedures for measurements of equal loudness level contours, *Proceedings InterNoise 96*, Liverpool, Institute of Acoustics, 3221–4.

McKennell, A. C. (1963) *Aircraft Noise Annoyance around London Heathrow Airport*, UK Central Office of Information.

Morton Williams, J., Hedges, B. and Fernando, E. (1978) *Road Traffic and the Environment*, Social and Community Planning Research.

Ollerhead, J. B. *et al.* (1992) *Report of a Field Study of Aircraft Noise and Sleep Disturbance*, Dept of Transport, London.

Passchier-Vermeer, W. *et al.* (1996) Effects of noise on health, *Noise/News International*, I-INCE, New York, 137–50.

Porter, N. D. and Berry, B. F. (1997) *A Study of Standard Methods for Measuring the Sound Quality of Industrial Products: Final Review*, NPL report CIRA(EXT) 021.

Porter, N. D., Flindell, I. H. and Berry, B. F. (1993) An acoustic feature model for the assessment of environmental noise, *Acoustics Bulletin*, Institute of Acoustics, 40–8.

Sargent, J. W. and Fothergill, L. C. (1993) *The Noise Climate around our Homes*, BRE Information Paper IP21/93, London.

Schultz, T. J. (1978) Synthesis of social surveys on noise annoyance, *Journal of the Acoustical Society of America*, **64**(2), 377–405.

Statutory Instrument (1974) *Health and Safety at Work etc. Act, 1974*, HMSO, London.

Statutory Instrument (1989) *The Noise at Work Regulations 1989*, No. 1790, HMSO, London.

Stephens, S. D. G. (1974) Methodological factors influencing loudness of short duration sounds, *Journal of Sound and Vibration*, **37**, 235–46.

WHO (1980). *Environmental Health Criteria 12 – Noise*, World Health Organisation and United Nations Environment Programme, Geneva.

Wilson, A. H. *et al.* (1963) *Committee on the Problem of Noise – Final Report*, Cmnd 2056, HMSO, London.

4

Fundamentals of human responses to vibration

M. J. GRIFFIN

4.1 INTRODUCTION

The high sensitivity of the human body to oscillatory motion is such that human responses to vibration determine the acceptability of vibration in many environments. For example, vibration will usually be felt and become annoying before it damages a building, the vibration in transport may cause discomfort or interfere with activities when it does not cause damage to the vehicle, the vibration of tools and machines produces injuries and disease without breaking the tool or machinery.

It is convenient to consider human exposure to oscillatory motion in three categories.

Whole-body vibration occurs when the body is supported on a surface which is vibrating (e.g. sitting on a seat which vibrates, standing on a vibrating floor or lying on a vibrating surface). Whole-body vibration occurs in transport (e.g. road, off-road, rail, air and marine transport) and when near some machinery.

Motion sickness can occur when real or illusory movements of the body or the environment lead to ambiguous inferences as to the movement or orientation of the human body. The movements associated with motion sickness are always of very low frequency, usually below 1 Hz.

Hand-transmitted vibration is caused by various processes in industry, agriculture, mining, construction and transport where vibrating tools or workpieces are grasped or pushed by the hands or fingers.

There are many different effects of oscillatory motion on the body and many variables influencing each effect. The variables may be categorized as extrinsic variables (those occurring outside the human body) and intrinsic variables (the variability that occurs between and within people), as in Table 4.1.

This chapter introduces the principal human responses to vibration, summarizes current methods of evaluating exposures to vibration, and

Table 4.1 Variables influencing human responses to oscillatory motion

Extrinsic variables
Vibration variables:
vibration magnitude
vibration frequency
vibration direction
vibration input positions
vibration duration
Other variables:
other stressors (noise, temperature, etc.)
seat dynamics
Intrinsic variables
Intra-subject variability:
body posture
body position
body orientation (sitting, standing, recumbent)
Inter-subject variability:
body size and weight
body dynamic response
age
gender
experience, expectation, attitude and personality
fitness

identifies methods of minimizing unwanted effects of vibration (see also Griffin, 1990).

4.2 MEASUREMENT OF VIBRATION

The magnitudes of vibration to which the human body is exposed are usually measured with accelerometers and expressed in terms of the acceleration, in metres per second per second (i.e. ms^{-2}, or m/s^2).

The frequency of vibration is expressed in cycles per second using the SI unit, hertz (Hz). The frequency of vibration influences the extent to which vibration is transmitted to the surface of the body (e.g. through seating), the extent to which it is transmitted through the body (e.g. from seat to head), and the responses to vibration within the body. The three principal categories of human response to oscillatory motion are currently associated with three different ranges of frequency: whole-body vibration (from about 0.5 to 80 Hz), motion sickness (at frequencies below 1 Hz), hand-transmitted vibration (from about 8 to 1000 Hz).

The responses of the body differ according to the direction of the motion (i.e. axes of vibration). The three principal directions of whole-body

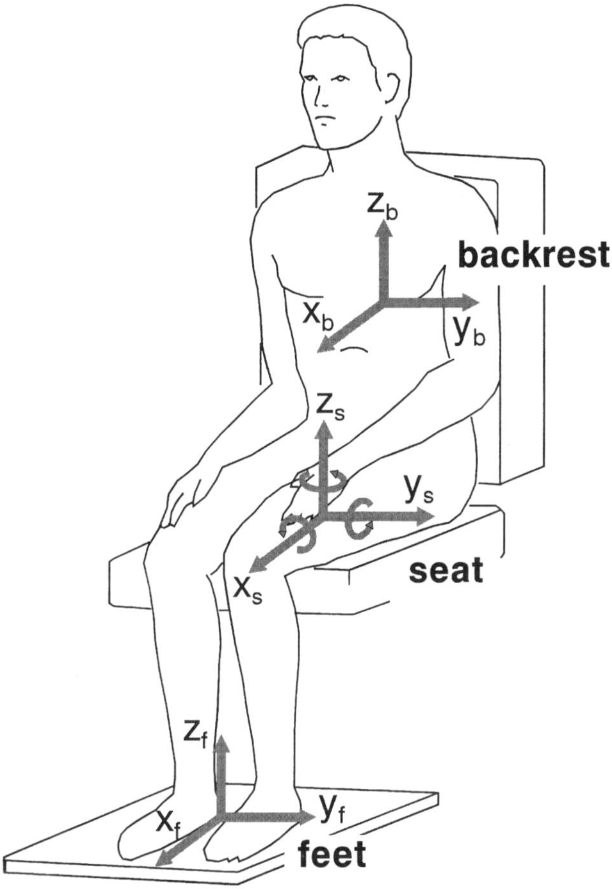

Figure 4.1 Axes of vibration used to measure exposures to whole-body vibration.

vibration for seated and standing persons are: fore-and-aft (x-axis), lateral (y-axis) and vertical (z-axis). The vibration is measured at the interface between the body and the surface supporting the body (e.g. on the seat beneath the ischial tuberosities for a seated person or beneath the feet for a standing person). Figure 4.1 illustrates the translational and rotational axes for an origin on a seat at the ischial tuberosities of a seated person. A similar set of axes is used for describing the directions of vibration at the backrest and feet of seated persons. Figure 4.2 shows a co-ordinate system used when measuring vibration in contact with a hand holding a tool.

Some human responses to vibration depend on the duration of exposure. The duration of measurement of vibration may affect the measured

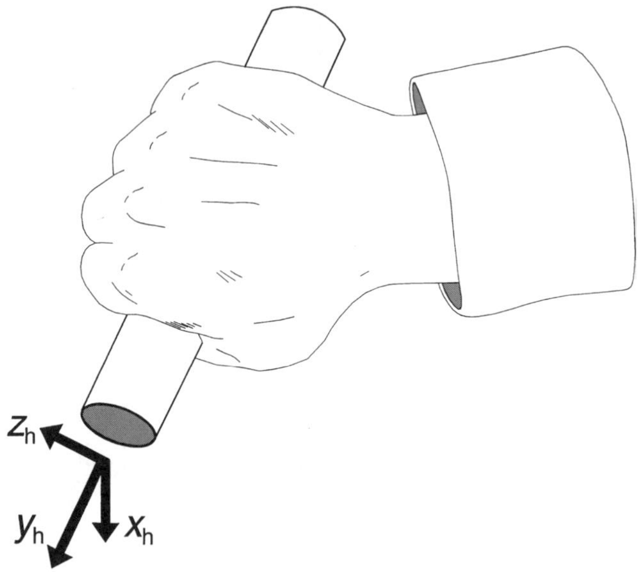

Figure 4.2 Axes of vibration used to measure exposures to hand-transmitted vibration.

magnitude of the vibration when the conditions are not statistically stationary. The root-mean-square (i.e. rms) acceleration is often used, but it may not provide a good indication of vibration severity if the vibration is intermittent, contains shocks, or otherwise varies in magnitude from time-to-time (see section 4.3.4.1).

4.3 WHOLE-BODY VIBRATION

Whole-body vibration can affect human comfort, and the performance of activities and health. Vibration of the whole body is produced by all forms of transport (including road, off-road, rail, sea and air transport) and by various types of industrial machinery.

4.3.1 Biodynamics

Biodynamics is the science of the physical responses of the body to force or motion. The human body is a complex mechanical system which does not, in general, respond to vibration in the same manner as a rigid mass; there are relative motions between the body parts which vary with the frequency and the direction of the applied vibration. Although there are resonances in the body, it is over-simplistic to summarize the dynamic response of the body by merely mentioning one or two resonance frequencies. The biodynamics of the body affect all human responses to vibration, but the discomfort, the interference with activities and the health effects of vibration cannot be predicted solely by considering the body as a mechanical system.

4.3.1.1 Transmissibility of the human body

The extent to which the vibration at the input to the body (e.g. the vertical vibration at a seat) is transmitted to a part of the body (e.g. vertical vibration at the head or the hand) is described by the transmissibility. At low frequencies of oscillation (e.g. below about 1 Hz), the vertical oscillations of a seat and the body are very similar and so the transmissibility is approximately unity. With increasing frequency of oscillation the motions of the body increase above that measured at the seat and the ratio of the motion of the body to the motion of the seat will reach a peak at one or more frequencies (i.e. resonance frequencies). At high frequencies the body motion will be less than that at the seat.

The resonance frequencies, and the transmissibilities at resonance, vary according to the direction (i.e. axis) of vibration, according to where the vibration is measured on the body and according to the posture of the body. There can be large differences between subjects. For seated persons, there may be resonances to the head and the hand at frequencies in the range 4 to 12 Hz for vertical vibration, below 4 Hz with x-axis vibration and below 2 Hz with lateral vibration (see Fig. 4.3 and Paddan and Griffin, 1988a, b). A seat back can greatly increase the transmission of x-axis vibration to the head and upper body of seated people, and bending of the legs can greatly affect the transmission of vertical vibration to the heads of standing persons.

4.3.1.2 Mechanical impedance of the human body

Mechanical impedance reflects the relation between the driving force at the input to the body and the resultant movement of the body. If the human body were rigid, the ratio of force to acceleration applied to the body would be constant and indicate the mass of the subject. Because the body is not rigid, the ratio of force to acceleration is only close to the body mass at very low frequencies (below about 2 Hz with vertical vibration; below about 1 Hz with horizontal vibration).

Measures of mechanical impedance usually show a principal resonance for vertical vibration of seated subjects at about 5 Hz, and sometimes a second resonance in the range 7 to 12 Hz (see Fig. 4.4 and Fairley and Griffin, 1989). Unlike some of the resonances affecting the transmissibility of the body, these resonance are only influenced by movement of large masses close to the input of vibration to the body. The large difference in impedance between that of a rigid mass and that of the human body means that the

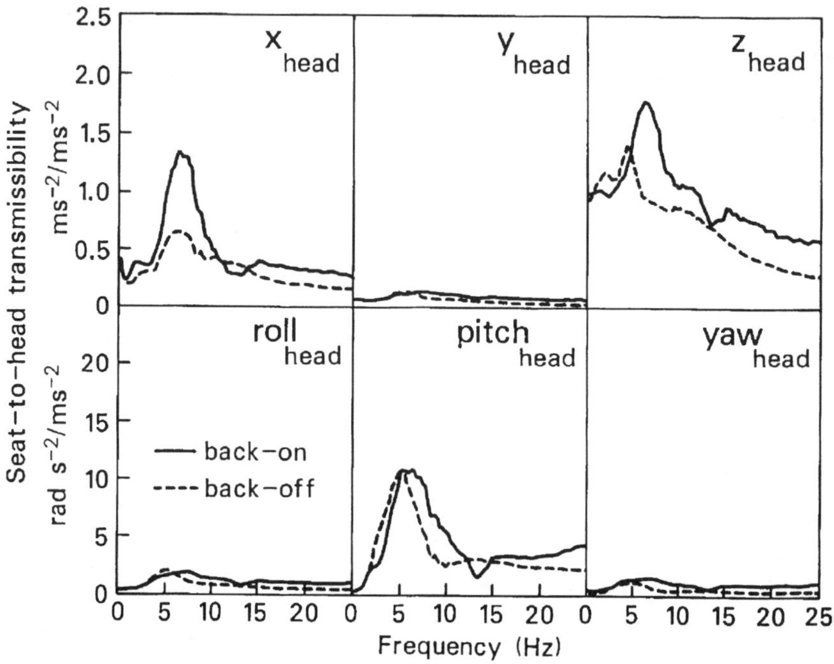

Figure 4.3 Transmissibility of vertical seat vibration to six axes of head vibration (median data from 12 males show effect of contact with seat backrest; data from Paddan and Griffin, 1988a).

body cannot usually be represented by a rigid mass when measuring the vibration transmitted through seats (see section 4.3.6). The mechanical impedance of the body is generally non-linear: the resonance frequency decreases when the vibration magnitude increases (see Fig. 4.5).

4.3.1.3 Biodynamic models

Various mathematical models of the responses of the body to vibration have been developed. A simple model with one or two degrees-of-freedom can represent the point mechanical impedance of the body and a dummy might be constructed to represent this impedance for seat testing (see Wei and Griffin, 1997; Mansfield and Griffin, 1997). Compared with mechanical impedance, the transmissibility of the body is affected by many more variables and so requires a more complex model reflecting the posture of the body and the translation and rotation associated with the various modes of vibration: a finite-element model seems necessary (see Kitazaki and Griffin, 1997).

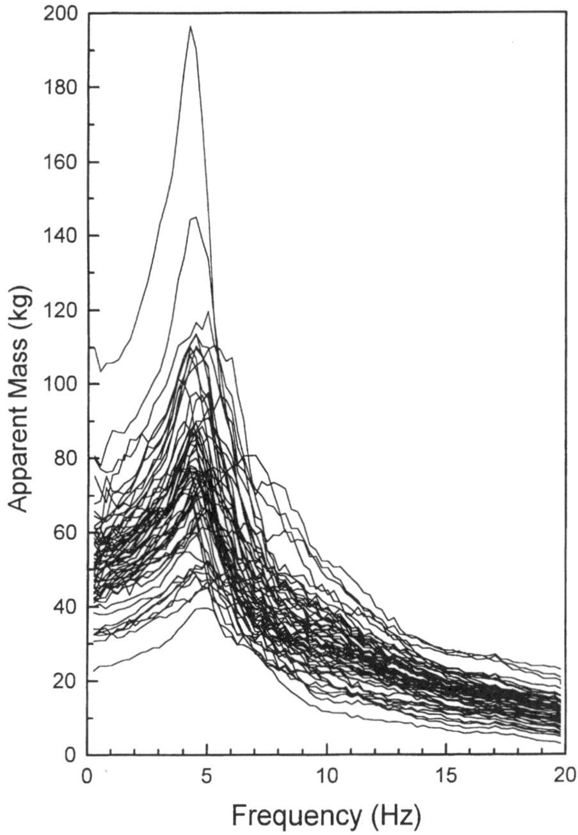

Figure 4.4 Apparent masses of 60 seated subjects (24 males, 24 females, 12 children) in the vertical axis measured during 64 s exposures to 1.0 ms^{-2} rms random vibration with simultaneous vibration of seat and feet (data from Fairley and Griffin, 1989).

4.3.2 Vibration discomfort

For very low magnitudes, the percentage of persons who will be able to feel vibration and the percentage who will not be able to feel the vibration may be estimated. For higher vibration magnitudes, an indication of the extent of subjective reactions is available in a semantic scale of discomfort. The relative discomfort caused by two or more alternative motions is often of greater interest and can be predicted from measurements of the vibration. Vibration limits to prevent discomfort vary between different environments (e.g. between buildings and transport) and between different types of transport (e.g. between cars and trucks) and within types of vehicle (e.g. between sports cars and limousines). The design limit depends on external

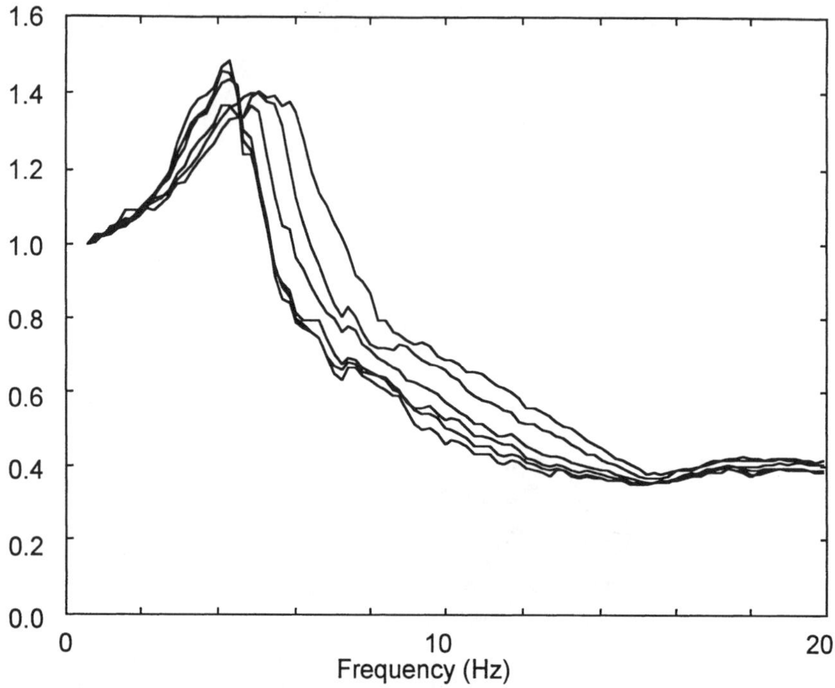

Figure 4.5 Median normalized apparent mass of 12 subjects measured at 0.25, 0.5, 1.0, 1.0, 2.0, 2.5 ms^{-2} rms. (Resonance frequencies decrease with increasing vibration magnitude; from Mansfield and Griffin, 1997).

factors (e.g. cost and speed) and the comfort in alternative environments (e.g. competitive vehicles).

4.3.2.1 Effects of vibration magnitude

The absolute threshold for the perception of vertical whole-body vibration in the frequency range 1 to 100 Hz is, very approximately, 0.01 ms^{-2} rms; a magnitude of 0.1 ms^{-2} will be easily noticeable; magnitudes around 1 ms^{-2} rms are usually considered uncomfortable; magnitudes of 10 ms^{-2} rms are usually dangerous. The precise values depend on vibration frequency and the exposure duration and they are different for other axes of vibration (Griffin, 1990).

Doubling the vibration magnitude (expressed in ms^{-2}) produces an approximate doubling of the sensation of discomfort. A halving of vibration magnitude can therefore produce a very considerable improvement in comfort. For some types of whole-body vibration, differences as small as 10% in the vibration magnitude will be detected by the majority of people.

4.3.2.2 *Effects of vibration frequency and direction*

The extent to which a given vibration acceleration will cause a larger or smaller effect on the body at different frequencies is reflected in 'frequency weightings'; frequencies capable of causing the greatest effect are given the greatest weight and others are attenuated in accord with their decreased importance. Two different frequency weightings (one for vertical and one for horizontal vibration of seated or standing persons) were presented in International Standard 2631 (1974, 1985) and have been reproduced in other relevant standards. However, ISO 2631 has recently undergone major revision (International Organization for Standardization, 1997). The revision of International Standard has a form broadly similar to British Standard 6841 (1987), but it contains various errors and inconsistencies that make it difficult to use.

Frequency weightings for human response to vibration have been derived from laboratory experiments in which volunteer subjects have been exposed to a set of motions having different frequencies. The subjects' responses are used to determine 'equivalent comfort contours'. The reciprocal of such a curve forms the shape of the frequency weighting; greatest 'weight' is given to those frequencies that cause discomfort with the lowest vibration magnitudes. Figure 4.6 shows frequency weightings W_b to W_f defined in British Standard 6841 (1987) as they may be implemented by analogue or digital filters (International Standard 2631 (1997) defines similar, though not identical, weightings). Table 4.2 defines simple asymptotic (i.e. straight line) approximations to these weightings and Table 4.3 shows how the weightings should be applied to the 12 axes of vibration illustrated in Fig. 4.1. (The weightings W_g and W_f are not required to predict vibration discomfort. W_g has been used for assessing interference with activities and is similar to the weighting for vertical vibration in ISO 2631 (1974, 1985); W_f is used to predict motion sickness caused by vertical oscillation, see section 4.4.2.)

Some frequency weightings are used for more than one axis of vibration, with different 'multiplying factors' allowing for overall differences in sensitivity between axes (see Table 4.3). The frequency-weighted acceleration should be multiplied by the multiplying factor before the component is compared with components in other axes, or included in any summation over axes. The rms value of this acceleration (i.e. after frequency weighting and after being multiplied by the multiplying factor) is sometimes called a 'component ride value'. Figure 4.7 illustrates the component ride values for a small car. It may be seen that greatest discomfort may be expected in this car from the vertical vibration on the seat, the vertical vibration at the feet and the fore-and-aft vibration on the backrest. (The old ISO 2631

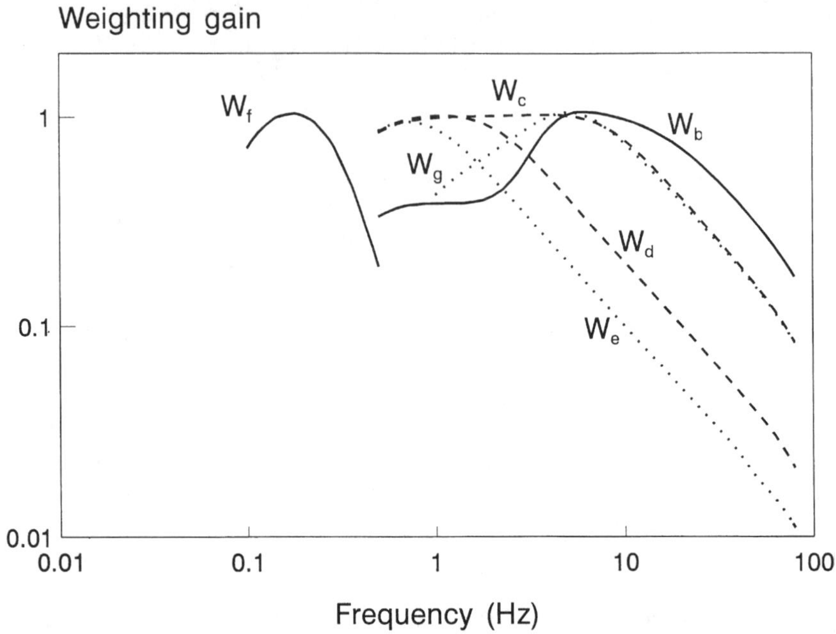

Figure 4.6 Acceleration frequency weightings for whole-body vibration and motion sickness (as defined in British Standard 6841, 1987).

was limited to the evaluation of vibration in the three directions on the seat.)

Vibration occurring in several axes is more uncomfortable than vibration occurring in a single axis. In order to obtain an 'overall ride value', the 'root-sums-of-squares' of the component ride values is calculated:

$$\text{Overall ride value} = (\Sigma(\text{component ride values})^2)^{1/2}$$

Overall ride values from different environments can be compared; a vehicle having the highest overall ride value would be expected to be the most uncomfortable with respect to vibration. The overall ride values can also be compared with the discomfort scale shown in Table 4.4. This scale indicates the approximate range of vibration magnitudes which are significant in relation to the range of vibration discomfort that might be experienced in vehicles.

4.3.2.3 Effects of vibration duration

Vibration discomfort tends to increase with increasing duration of exposure to vibration. The rate of increase may depend on many factors but a simple

Table 4.2 Asymptotic approximations to frequency weightings, $W(f)$, in BS 6841 (1987) for comfort, health, activities and motion sickness (f = frequency, Hz; $W(f) = 0$ where not defined)

Weighting name	*Weighting definition*	
W_b	$0.5 < f < 2.0$	$W(f) = 0.4$
	$2.0 < f < 5.0$	$W(f) = f/5.0$
	$5.0 < f < 16.0$	$W(f) = 1.00$
	$16.0 < f < 80.0$	$W(f) = 16.0/f$
W_c	$0.5 < f < 8.0$	$W(f) = 1.0$
	$8.0 < f < 80.0$	$W(f) = 8.0/f$
W_d	$0.5 < f < 2.0$	$W(f) = 1.00$
	$2.0 < f < 80.0$	$W(f) = 2.0/f$
W_e	$0.5 < f < 1.0$	$W(f) = 1.00$
	$1.0 < f < 20.0$	$W(f) = 1.00/f$
W_f	$0.100 < f < 0.125$	$W(f) = f/0.125$
	$0.125 < f < 0.250$	$W(f) = 1.0$
	$0.250 < f < 0.500$	$W(f) = (0.25/f)^2$
W_g	$1.0 < f < 4.0$	$W(f) = (f/4)^{1/2}$
	$4.0 < f < 8.0$	$W(f) = 1.00$
	$8.0 < f < 80.0$	$W(f) = 8.0/f$

Table 4.3 Application of frequency weightings for the evaluation of vibration with respect to discomfort

Input position	*Axis*	*Frequency weighting*	*Axis multiplying factor*
Seat	x	W_d	1.0
	y	W_d	1.0
	z	W_b	1.0
	r_x (roll)	W_c	0.63
	r_y (pitch)	W_c	0.40
	r_z (yaw)	W_c	0.20
Seat back	x	W_c	0.80
	y	W_d	0.50
	z	W_d	0.40
Feet	x	W_b	0.25
	y	W_b	0.25
	z	W_b	0.40

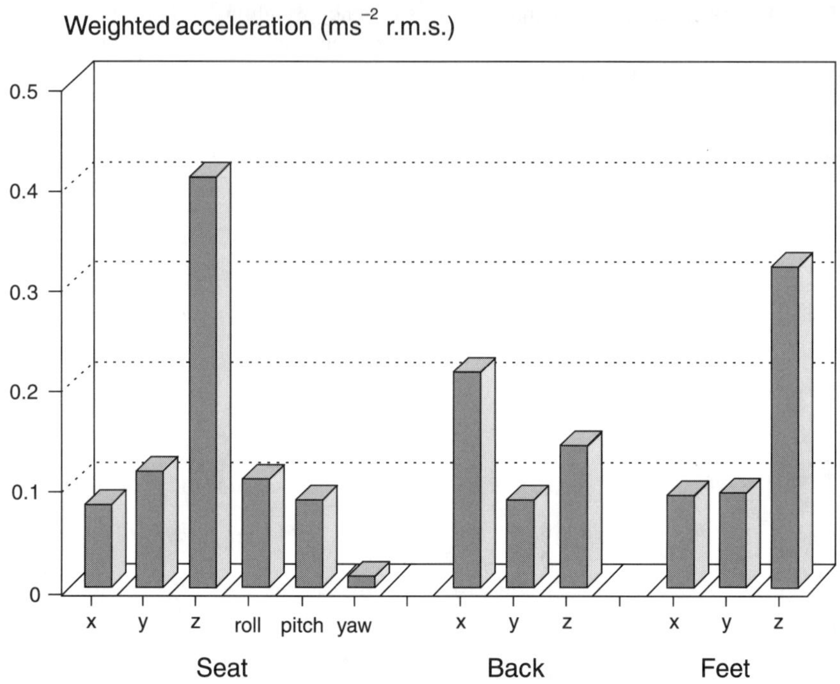

Figure 4.7 Component ride values for 12 axes of vibration for the driver of a small car (values shown after the application of appropriate frequency weighting and axis multiplying factor from BS 6841, 1987).

Table 4.4 Scale of vibration discomfort from British Standard 6841 (1987) and International Standard 2631 (1997)

	Rms weighted acceleration (ms^{-2})	
Extremely uncomfortable	3.15	
	2.5	
	2.0	Very
	1.6	uncomfortable
Uncomfortable	1.25	
	1.0	
	0.8	Fairly
	0.63	uncomfortable
A little uncomfortable	0.5	
	0.4	
	0.315	
	0.25	Not uncomfortable

'fourth power' time-dependency is used to approximate how discomfort varies with duration of exposure from the shortest possible shock to a full day of vibration exposure (i.e. $(\text{acceleration})^4 \times \text{duration} = \text{constant}$, see section 4.3.4.1).

4.3.3 Interference with activities

Vibration can interfere with the acquisition of information (e.g. by the eyes), the output of information (e.g. by hand or foot movements) or the complex central processes that relate input to output (e.g. learning, memory, decision-making).

The greatest and best understood effects of whole-body vibration are on input processes (mainly vision) and output processes (mainly continuous hand control). In both cases, there may be a disturbance occurring entirely outside the body (e.g. vibration of a viewed display or vibration of a hand-held control), a disturbance at the input or output (e.g. movement of the eye or hand), and a disturbance within the body affecting the peripheral nervous system (i.e. afferent or efferent system). Central processes may also be affected by vibration but understanding is currently too limited to make confident generalized statements.

Effects of vibration on vision and manual control are most usually caused by the movement of the affected part of the body (i.e. eye or hand). The effects may be decreased by reducing the transmission of vibration to the eye or to the hand, or by making the task less susceptible to disturbance (e.g. increasing the size of a display or reducing the sensitivity of a control). Consequently, the effects of vibration on vision and manual control can often be reduced by redesign of the task.

4.3.3.1 Effects of vibration on vision

Reading a book or newspaper in a vehicle may be difficult because the paper is moving, the eye is moving, or both the paper and the eye are moving. There are many variables which affect visual performance in these conditions; it is not possible to represent adequately the effects of vibration on vision without considering the effects of these variables (Griffin and Lewis, 1978).

When a stationary observer views a moving display the eye may be able to track the position of the display using 'pursuit eye movements' if the display is moving at frequencies below about 1 Hz and with a low velocity. At slightly higher frequencies of oscillation, the precise value depending on the predictability of the motion waveform, the eye will make 'saccadic' eye movements to redirect the eye with small jumps. At frequencies above about 3 Hz, the eye will be directed to one extreme of the oscillation so as to view the image as it is temporarily stationary while reversing the direction of movement (i.e. at the 'nodes' of the motion). The absolute threshold for the visual detection of the vibration of an object by a stationary observer occurs when the peak-to-peak oscillatory motion gives an angular displacement at the eye of approximately 1 minute of arc. The acceleration required to achieve this threshold is very low at low frequencies but increases in proportion to the square of the frequency to become very high at high frequencies. If the vibration displacement is above the visual detection threshold there will be perceptible blur; the effects of vibration on visual performance (e.g. effects on reading speed and reading accuracy) may then be estimated from the maximum time that the image spends over some small area of the retina (e.g. the period of time spent near the nodes of the motion with sinusoidal vibration). For sinusoidal vibration, this time decreases (and so reading errors increase) in linear proportion to the frequency of vibration and in proportion to the square root of the displacement of vibration (O'Hanlon and Griffin, 1971). With dual-axis vibration (e.g. combined vertical and lateral vibration of a display), this time is greatly reduced and reading performance drops greatly (Meddick and Griffin, 1976). With narrow-band random vibration, there is a greater probability of low image velocity than with sinusoidal vibration of the same magnitude and predominant frequency, so reading performance tends to be less affected by random vibration than by sinusoidal vibration (Moseley *et al.*, 1982). Display vibration reduces the ability to see fine detail in displays while having little effect on the clarity of larger forms.

If an observer sits or stands on a vibrating surface, the effects of vibration on vision depend on the extent to which the vibration is transmitted to the eye. The head is likely to move in both translational axes (i.e. in the x-axis, y-axis and z-axis) and rotational axes (i.e. in the roll, pitch and yaw). The head motions most affecting vision are often the vertical and pitch movements of the head. The pitch motion of the head is well compensated by the 'vestibulo-ocular reflex' which serves to help stabilize the line of sight of the eyes at frequencies below about 10 Hz (e.g. Benson and Barnes, 1978). Hence, although there is often much pitch motion of the head at 5 Hz, there is less pitch motion of the eyes at this frequency. Pitch motion of the head, therefore, has a less than expected effect on vision – unless the display is attached to the head, as with a helmet-mounted display (see Wells and Griffin, 1984). The effects on vision of translational motion of the head depend on viewing distance; the effects are greatest when close to a display. As the viewing distance increases, the retinal image motions produced by translational displacements of the head decrease until, when viewing an object at infinite distance, there is no retinal image motion produced by translational head displacement (Griffin, 1976). For a vibrating observer, there may be little difficulty with low-frequency pitch-head motions when viewing a fixed display and no difficulty with translational head motions when viewing a distant display. The greatest problems occur with pitch-head motion when the display is attached to the head and with translational head motion when viewing near displays. Additionally, there may be resonances of the eye within the head – but these are highly variable between individuals and often occur at high frequencies (e.g. 30 Hz and greater) where it is often possible to attenuate the vibration entering the body.

When an observer and a display oscillate together, in phase, at low frequencies, the retinal image motions (and decrements in visual performance) are less than when either the observer or the display oscillate separately (Moseley and Griffin, 1986a). However, the advantage is lost as the vibration frequency is increased since there is then an increasing phase difference between the motion of the head and the motion of the display. At frequencies around 5 Hz, the phase lags between seat motion and head motion may be 90 degrees or more (depending on seating conditions) and sufficient to eliminate any advantage of moving the seat and the display together.

When reading a newspaper on a train, the motion of the arms may result in the motion of the paper being different in magnitude and phase from the motions of both the seat and the head of the observer. The dominant axis of motion of the newspaper may be different from the dominant axis of motion of the person (Griffin and Hayward, 1994). Increasing the size of detail in a display will often greatly reduce adverse effects of vibration on vision (Lewis and Griffin, 1979). Increasing the spacing between rows of letters and choosing appropriate character fonts can also be beneficial. The contrast of the display, or other reading material, also has an effect but maximum performance may not occur with maximum contrast. The known influence of all such factors are summarized in a design guide for visual displays to be used in vibration environments (Moseley and Griffin, 1986b).

4.3.3.2 Manual control

Writing, and other complex control tasks involving hand control activities, can also be impeded by vibration (see Lewis and Griffin, 1978; McLeod and Griffin, 1989). The characteristics of the task and the characteristics of the vibration combine to determine effects of vibration on performance; a given vibration may greatly affect one type of tracking task but have little effect on another.

The direct mechanical jostling of the hand by vibration that causes unwanted movement of a control is called 'breakthrough' or 'feedthrough' or 'vibration-correlated error'. The inadvertent movement of the pencil caused by 'jostling' while writing in a vehicle is a form of 'vibration-correlated error'. In a simple tracking task, where the operator is required to follow movements of a target, some of the error will also be correlated with the target movements. This is called 'input-correlated error' and often mainly reflects the inability of an operator to follow the target without delays inherent in visual, cognitive and motor activity. The part of the tracking error which is not correlated with either the vibration or the tracking task is called the 'remnant'. This includes operator generated 'noise' and any source of non-linearity; drawing a freehand 'straight line' does not result in a perfect straight line even in the absence of environmental vibration. The effects of vibration on vision can result in increased remnant with some tracking tasks and some studies show that vibration, usually at frequencies above about 20 Hz, interferes with neuromuscular processes (e.g. Goodwin *et al.*, 1972; Martin *et al.*, 1984; Ribot *et al.*, 1986) which may be expected to result in increased remnant. The causes of the three components of the tracking error are shown in the model presented as Fig. 4.8.

The gain (i.e. sensitivity) of a control determines the control output corresponding to a given force, or displacement, applied to the control by the operator. The optimum gain in static conditions (high enough to not cause fatigue but low enough to prevent inadvertent movement) is likely to be too great during exposure to vibration where inadvertent movement is more likely (Lewis and Griffin, 1977).

Vibration may affect the performance of tracking tasks by reducing the visual performance of the operator. Wilson (1974) and McLeod and Griffin (1990) have shown that collimating a display by means of a lens so that it appears to be at infinity can reduce, or even eliminate, errors with some tasks. It is possible that visual disruption has played a significant part in the performance decrements reported in other experimental studies of the effects of vibration on manual control.

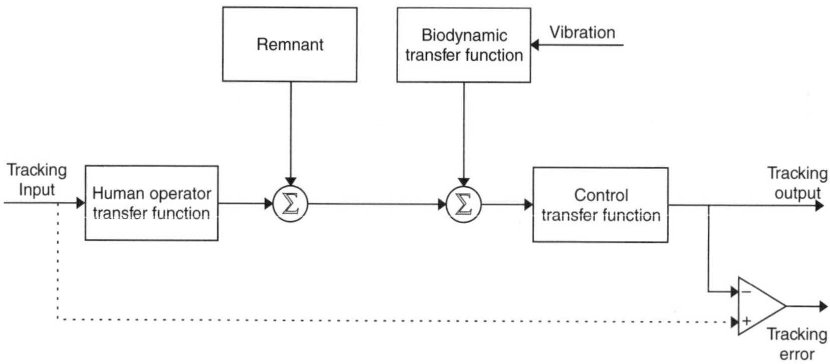

Figure 4.8 Linear model of a pursuit manual control system showing how tracking errors may be caused by the vibration (vibration-correlated error), the task (input correlated error) or some other cause (remnant).

The vibration transmissibility of the body is approximately linear (i.e. doubling the magnitude of vibration at the seat may be expected to approximately double the magnitude of vibration at the head or at the hand). Vibration-correlated error may therefore increase in approximately linear proportion to vibration magnitude. There is no simple relation between the frequency of vibration and its effects on control performance. The effects of frequency depend on the control order (which varies between tasks) and the biodynamic response of the body (which varies with posture and between operators). With zero-order tasks and the same magnitude of acceleration at each frequency, the effects of vertical seat vibration may be greatest in the range 3–8 Hz since transmissibility to the shoulders is greatest in this range (see McLeod and Griffin, 1989). In the horizontal axes (i.e. the x- and y-axes of the seated body), the greatest effects appear to occur at lower frequencies, around 2 Hz or below. Again, this corresponds to the frequencies at which there is greatest transmission of vibration to the shoulders. The axis of the control task most affected by vibration may not be the same axis as that in which most vibration occurs at the seat. Often, fore-and-aft movements of the control (which generally correspond to vertical movements on the display) are most affected by vertical whole-body vibration.

Multiple frequency vibration causes more disruption to performance than the presentation of any one of the constituent single frequencies alone. Similarly, the effects of multiple axis vibration are greater than the effects of any one of the single axes alone.

Few investigations of the effects of vibration have considered everyday tasks. Corbridge and Griffin (1991) showed that the effects of vertical whole-body vibration on spilling liquid from a hand-held cup were greatest close to 4 Hz. They also found that the effects of vibration on writing speed and subjective estimates of writing difficulty were most affected by vertical vibration in the range 4–8 Hz. Although 4 Hz was a sensitive frequency for both the 'drinking' and the writing task, the dependence on frequency of the effects of vibration were different for the two activities.

International Standard 2631 offered a 'fatigue-decreased proficiency' boundary as a means of predicting the effects of vibration on activities (International Organization for Standardization, 1974, 1985). A complex time-dependent magnitude of vibration was said to be 'a limit beyond which exposure to vibration can be regarded as carrying a significant risk of impaired working efficiency in many kinds of tasks, particularly those in which time-dependent effects ('fatigue') are known to worsen performance as, for example, in vehicle driving'. There are no data justifying the complex fatigue-decreased proficiency boundary presented in this standard; among researchers it has long been discarded as unfounded committee speculation. Vibration may influence 'fatigue' but there is little relevant scientific evidence.

4.3.3.3 Cognitive performance

Few investigators have addressed the possibility of vibration affecting cognitive performance. Simple cognitive tasks (e.g. simple reaction time) appear to be unaffected by vibration, other than by changes in arousal or motivation or by direct effects on input and output processes. This may also be true for some complex cognitive tasks. However, the scarcity and diversity of experimental studies allows the possibility of real and significant cognitive effects of vibration (see Sherwood and Griffin, 1990, 1992).

4.3.4 Health effects

Occupational, sport and leisure activities involving exposure to whole-body vibration have been reported to cause various disorders (see Dupuis and Zerlett, 1986; Hulshof and van Zanten, 1987; Bongers and Boshuizen, 1990; Griffin, 1990; Bovenzi and Zadini, 1992). The studies do not all agree on either the type or the extent of disorders and rarely have the findings been related to measurements of the vibration exposures. However, it is widely believed that disorders of the back (back pain, displacement of intervertebral discs, degeneration of spinal vertebrae, osteoarthritis, etc.) may be associated with vibration exposure. There may be several alternative causes of an increase in disorders of the back among persons exposed to vibration (e.g. poor sitting postures, heavy lifting). It is not always possible to conclude confidently that a back disorder is solely, or primarily, caused by vibration.

Other disorders which have been claimed to be due to occupational exposures to whole-body vibration include abdominal pain, digestive disorders, urinary frequency, prostatitis, haemorrhoids, balance and visual disorders, headaches and sleeplessness. Further research is required to confirm whether these signs and symptoms are causally related to exposure to vibration.

4.3.4.1 Method of vibration evaluation

The manner in which the health effects of oscillatory motions depend upon the frequency, direction and duration of motion is currently assumed to be similar to that for vibration discomfort (see section 4.3.2). However, it is assumed that the 'total' exposure, rather than the 'average' exposure, is important and so a 'dose' measure is used.

International Standard 2631 (1974, 1985) defined exposure limits which were 'set at approximately half the level considered to be the threshold of pain (or limit of voluntary tolerance) for healthy human subjects' (see Fig. 4.9). British Standard 6841 (1987) defined an 'action level' for vertical vibration based 'vibration dose values'. The *vibration dose value* can be

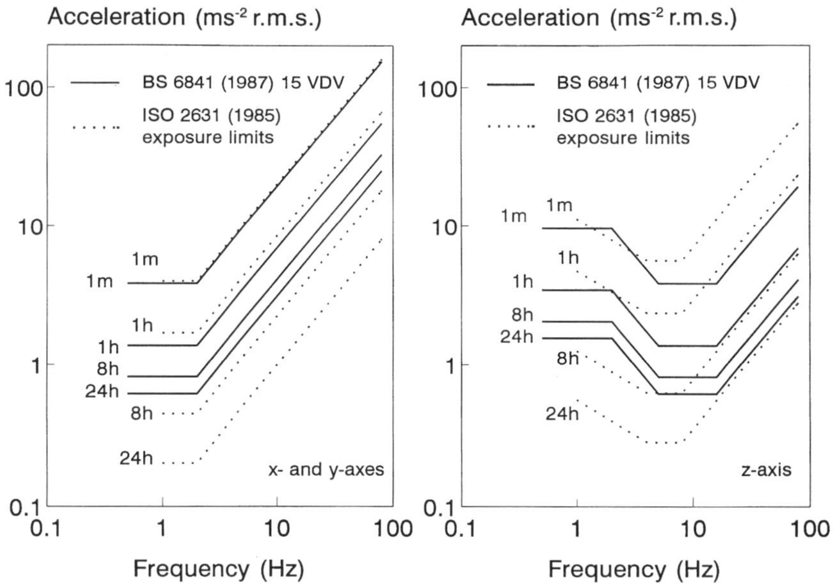

Figure 4.9 Comparison of International Standard 2631 (1985) exposure limits with an 'action level' based on a vibration dose value (VDV) of 15 $ms^{-1.75}$ from British Standard 6841 (1987). When seated: *x*-axis = fore-and-aft; *y*-axis = lateral; *z*-axis = vertical). (From Griffin, 1990).

considered to be the magnitude of a one second duration of vibration which will be equally severe to the measured vibration. The vibration dose value uses a 'fourth power' time dependency to accumulate vibration severity over the exposure period from the shortest possible shock to a full day of vibration (see BS 6841, 1987; ISO 2631, 1997):

$$\text{Vibration dose value} = \left[\int_{t=0}^{t=T} a^4(t)\, dt \right]^{1/4}$$

If the exposure duration (t, seconds) and the frequency-weighted rms acceleration (a_{rms}, ms^{-2} rms) are known for conditions in which the vibration characteristics are statistically stationary, it can be useful to calculate the 'estimated vibration dose value', eVDV:

$$\text{Estimated vibration dose value} = 1.4 a_{rms} t^{1/4}$$

The eVDV is not applicable to transients, shocks and repeated shock motions in which the crest factor (peak value divided by the rms value) is high. The vibration dose value (VDV) can be used to evaluate the severity of transients, shocks and repeated shock motions.

No precise limit can be offered to prevent disorders caused by whole-body vibration, but standards define useful methods of quantifying vibration severity. British Standard 6841 (1987) offers the following guidance.

> High vibration dose values will cause severe discomfort, pain and injury. Vibration dose values also indicate, in a general way, the severity of the vibration exposures which caused them. However there is currently no consensus of opinion on the precise relation between vibration dose values and the risk of injury. It is known that vibration magnitudes and durations which produce vibration dose values in the region of 15 $ms^{-1.75}$ will usually cause severe discomfort. It is reasonable to assume that increased exposure to vibration will be accompanied by increased risk of injury.

An action level might be set higher or lower than 15 $ms^{-1.75}$. In International Standard 2631 (1997), two boundaries are defined and it is claimed that below a boundary corresponding to 8.5 $ms^{-1.75}$ 'health risks have not been objectively observed', between 8.5 and 17 $ms^{-1.75}$ 'caution with respect to health risks is indicated' and above 17 $ms^{-1.75}$ 'health risks are likely'. At high-vibration dose values, prior consideration of the fitness of the exposed persons and the design of adequate safety precautions may be required. The need for regular checks on the health of routinely exposed persons may also be considered.

The vibration dose value provides a robust measure by which highly variable and complex exposures can be compared. The tentative action level merely serves to indicate the approximate values which might be excessive. Figure 4.10 illustrates the root-mean-square accelerations corresponding to a vibration dose values of 8.5, 15.0 and 17 $ms^{-1.75}$ for exposures between 1 and 24 hours.

Unlike the 'exposure limit' in the old International Standard 2631 (1974, 1985), a vibration dose value 'action level' does not appear to allow very high magnitudes at short durations (a few minutes) or require very low magnitudes for long duration exposures (many hours). Any exposure to continuous vibration, or intermittent vibration, or repeated shock may be compared with the action level by calculating the vibration dose value. It would be unwise to exceed the action level without consideration of the possible health effects of an exposure to vibration or shock.

4.3.4.2 EU Machinery Safety Directive

The Machinery Safety Directive of the European Community (89/392/EEC) states: 'machinery must be so designed and constructed that risks resulting from vibrations produced by the machinery are reduced to the lowest level, taking account of technical progress and the availability of means of reducing vibration, in particular at source' (Council of the European

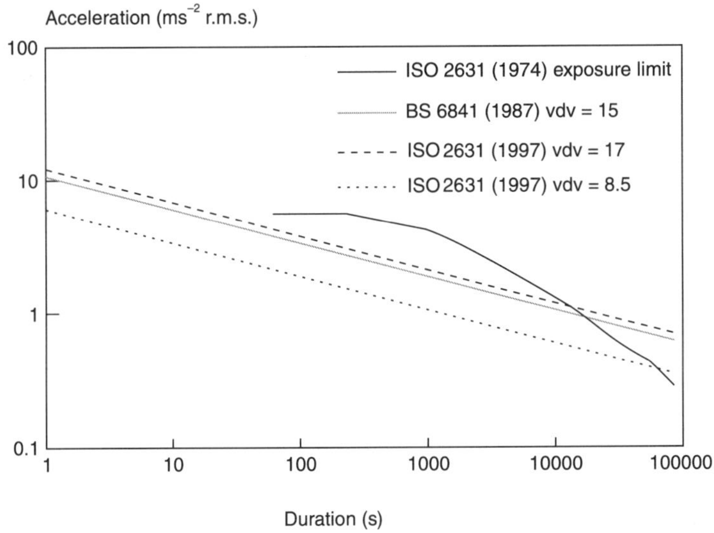

Figure 4.10 Action level corresponding to a vibration dose value (VDV) of 1.5 ms$^{-1.75}$ (see British Standard 6841, 1987) compared with 'exposure limit' from International Standard 2631 (1974, 1985) and 8.5 and 17 ms$^{-1.75}$ boundaries suggested in ISO 2631 (1997).

Communities, 1989). Instruction handbooks for machinery causing whole-body vibration should specify the frequency-weighted acceleration if it exceeds a stated value (currently a frequency-weighted acceleration of 0.5 ms^{-2} rms). The vibration magnitude will often depend on the test conditions and, although the test procedures must be specified, their selection leaves much control over the measured value to the manufacturer. Standardized test procedures are being prepared but the values currently quoted may not always be representative of the operating conditions in the work for which the machinery is used. Additionally, the real risks from whole-body vibration are dependent on the duration of exposure and not solely the magnitude of the vibration.

4.3.4.3 Proposed EU Exposure Directive

The opening principles of a proposed Council Directive on the minimum health and safety requirements regarding vibration are: 'Taking account of technical progress and of the availability of measures to control the physical agent at source, the risks arising from exposure to the physical agent must be reduced to the lowest achievable level, with the aim of reducing exposure to below the threshold level' (Commission of the European Communities, 1994).

In respect of whole-body vibration, the proposed EU Directive currently bases the vibration magnitudes on 8-hour energy-equivalent acceleration magnitudes (called A(8) values). The proposed Directive identifies a threshold level (A(8) = 0.25 ms^{-2} rms), an action level (A(8) = 0.5 ms^{-2} rms), and an exposure limit value (A(8) = 0.7 ms^{-2} rms). When exposures exceed the 'threshold level', it is proposed that workers must receive information concerning the potential risk of exposure to whole-body vibration. The 'action level' is intended to identify the conditions in which training in precautionary measures is required, an assessment of the vibration is to be made, and a programme of preventative measures is to be instituted. The proposed Directive also indicates that when the action level is exceeded workers shall have the right to regular health surveillance, including routine examinations designed for the early detection of disorders caused by whole-body vibration. If the 'exposure limit value' is exceeded, health surveillance must be carried out and Member States of the Community will be expected to control the harmful effects.

This proposed EU Directive does not appear to reflect a mature understanding of human responses to whole-body vibration. The manner in which the 8-hour equivalent magnitudes correspond to higher magnitudes at shorter durations makes them an unsuitable basis for estimating the severity of vibration exposures. The problem is partially addressed by two additional clauses: "Where whole-body exposure involves shocks or other vibration at high magnitudes during low durations, the corresponding action level is established as the hazard due to exposure during a period of one hour to a constant amplitude of 1.25 ms^{-2}" and 'For applying the provisions of this section, it is considered that A(8) is likely to reach 0.5 ms^{-2} when the work equipment used transmits to the whole body a short term (a few minutes) equivalent acceleration equal to that numerical value'. These clauses are obviously insufficient and before the proposed Directive can be enforced it will have to be re-drafted. Similarly, there is a need to reconsider the threshold level, action level and exposure limit value and the appropriate actions to be taken when each is exceeded.

4.3.5 Disturbance in buildings

Acceptable magnitudes of vibration in buildings are close to vibration perception thresholds. The effects of vibration in buildings are assumed to depend on the use of the building in addition to the vibration frequency, direction and duration. Guidance is given in various standards (e.g. International Standard 2631 part 2, 1989; British Standard 6472, 1992). Using the guidance contained in ISO 2631 Part 2 (1989), it is possible to summarize the acceptability of vibration in different types of building in a

Table 4.5 Vibration dose values at which various degrees of adverse comment may be expected in buildings (based on International Standard 2631 Part 2 (1989) and British Standard 6472 (1992) (see Griffin, 1990))

Place	*Low probability of adverse comment*	*Adverse comment possible*	*Adverse comment probable*
Critical working areas	0.1	0.2	0.4
Residential	0.2–0.4	0.4–0.8	0.8–1.6
Office	0.4	0.8	1.6
Workshops	0.8	1.6	3.2

single table of vibration dose values (see Table 4.5 and British Standard 6472, 1992). The vibration dose values in Table 4.5 are applicable irrespective of whether the vibration occurs as a continuous vibration, intermittent vibration or repeated shocks.

4.3.6 Seating dynamics

Ideally, vibration should be reduced at source. This may involve minimizing undulations of the terrain, reducing the speed of travel of vehicles, or improving the balance of rotating parts. However, irrespective of the vibration input, seating dynamics will influence the vibration responsible for discomfort, interference with activities and injury. Most seats exhibit a resonance at low frequencies which results in higher magnitudes of vertical vibration occurring on the seat than on the floor. At high frequencies there is usually attenuation of vibration. The resonance frequencies of common seats are usually in the region of 4 Hz (see Fig. 4.11). The amplification at resonance is partially determined by the 'damping' in the seat. The variations in transmissibility between seats are sufficient to result in significant differences in the vibration experienced by people supported by different seats.

The transmissibility of a seat (i.e. the non-dimensional ratio of the vibration magnitude at the seat-subject interface to the vibration magnitude entering the seat at the floor, expressed as a function of vibration frequency) can be measured in various ways. The transmissibility is highly dependent on the mechanical impedance of the human body, so the transmissibility of a seat measured with a mass supported on the seat will be different from that with a human body sitting in the seat. The non-linearity of the body mechanical impedance results in seat transmissibilities which vary with changes in vibration magnitude and vibration spectra entering the seat. Measurements of seat transmissibility may be undertaken on laboratory simulators with volunteer subjects, but precautions are required to protect subjects from injury. Measurements may also be performed with drivers or

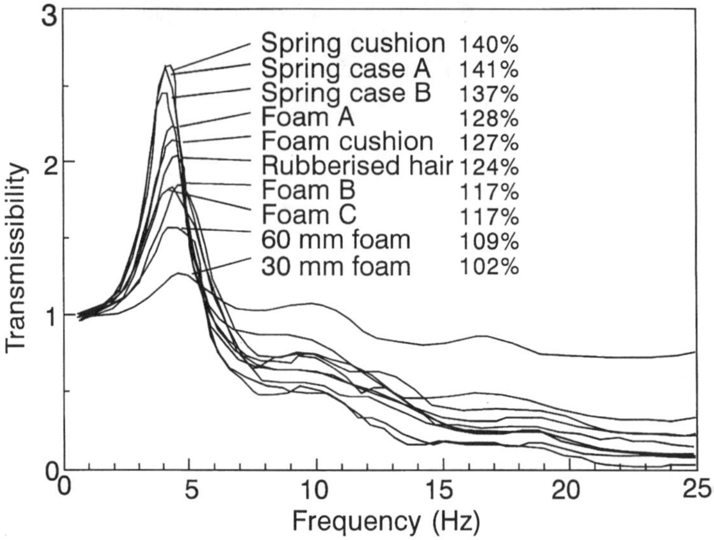

Figure 4.11 Comparison of the vertical transmissibilities and SEAT values for ten alternative cushions of passenger railway seats (data from Corbridge *et al.*, 1989).

passengers in vehicles. Anthropodynamic dummies are being developed to represent the average mechanical impedance of the human body so that laboratory and field studies can be performed without exposing people to vibration. This reduces the costs and inherent risks and should provide a measurement of seat transmissibility unaffected by inter-subject variability in mechanical impedance (e.g. Mansfield and Griffin, 1996). Seat transmissibility may also be predicted using measurements of the impedance of a seat and the known mechanical impedance of the human body (Wei and Griffin, 1998).

The suitability of a seat for a specific vibration environment depends on: (i) the vibration spectra present in the environment; (ii) the transmissibility of the seat; and (iii) the sensitivity of the human body to the different frequencies of vibration. These three functions of frequency are contained within a simple numerical indication of the isolation efficiency of a seat called the 'seat effective amplitude transmissibility', SEAT (Griffin, 1990). In concept, the SEAT value compares the vibration severity on a seat with the vibration severity on the floor beneath the seat:

$$\mathrm{SEAT}(\%) = \frac{\text{ride comfort on seat}}{\text{ride comfort on floor}} \times 100$$

A SEAT value greater than 100% indicates that, overall, the vibration on the seat is 'worse' than the vibration on the floor beneath the seat; SEAT values below 100% indicate that the seat has provided some useful attenuation. Seats should be designed to have the lowest SEAT value compatible with other constraints. In common cars, SEAT values are often in the range 60–80%. In railways carriages the SEAT value for vertical vibration is likely to be greater than 100% because conventional seats cannot provide any attenuation of the low-frequency vibration that is normally present in such vehicles. Consequently, Fig. 4.11 shows high SEAT values in which the vibration discomfort experienced by the occupants of the seats will be greatly degraded by the seat response. Selecting a seat so as to minimize the SEAT value in these circumstances may not result in lower vibration on the seat than on the floor, but should prevent unnecessary exposure to vibration. The optimization of the seating dynamics can be both the cheapest and the most effective method of improving vehicle ride.

The SEAT value is determined using a mathematical procedure. The 'ride comfort' that would result from sitting on the seat or on the floor can be predicted using the frequency-weightings in the appropriate standard. The SEAT value may be calculated from either the frequency-weighted rms values (if the vibration does not contain transients) or the vibration dose values of the frequency-weighted acceleration on the seat and the floor:

$$\mathrm{SEAT}(\%) = \frac{\text{vibration dose value on seat}}{\text{vibration dose value on floor}} \times 100$$

The SEAT value is influenced by the vibration input and not merely by the dynamics of the seat; different values are obtained with the same seat in different vehicles. The SEAT value indicates the suitability of a seat for attenuating a particular type of vibration.

Conventional seating (comprising some combination of foam, rubber or metal springing) usually has a resonance at about 4 Hz and therefore provides no attenuation at frequencies below about 6 Hz. Attenuation can be provided at frequencies down to about 2 Hz using a separate suspension mechanism beneath the seat pan. In such 'suspension seats', used in some off-road vehicles, trucks and coaches, there are low resonance frequencies (often below about 2 Hz). This is beneficial if the dominant vibration is at higher frequencies, but of no value if the dominant motion is at very low frequencies. Standards for the testing of the suitability of suspension seats to specific classes of work vehicle have been prepared (see International Standards, ISO 5007, 1990; ISO 7096, 1994; ISO 10326-1, 1992). One problem with suspension seats is that the suspension mechanism (comprising a spring and damper mechanism) has a limited travel, often 50–100 mm. If the relative motion of the suspension reaches this limit there will be an impact which can be a greater cause of discomfort and hazard than would have been present with a conventional seat. Suspension seats have been marketed showing the benefits of the suspension while assuming a linear single-degree-of-freedom mechanical system. In practice, suspensions are non-linear (having friction affecting response with low magnitude motions, and hitting end-stops with high magnitudes) and their full response requires consideration of the dynamic response of the cushion and the impedance of the human body in addition to the idealized response of the damper and spring.

4.4 MOTION SICKNESS

Sickness (e.g. vomiting, nausea, sweating, colour changes, dizziness, headaches and drowsiness) are normal responses to motion which are experienced by fit and healthy people. Many different motions can cause sickness and reduce the comfort, impede the activities and degrade the well-being of both those directly affected and those associated with the motion sick (Griffin, 1991).

4.4.1 Causes of motion sickness

Translational and rotational oscillation, constant speed rotation about an off-vertical axis, Coriolis stimulation, movements of the visual scene, and various other stimuli producing sensations associated with movement of the body can cause sickness. Although some motions can be reliably predicted as being more nauseogenic than others, motion sickness is neither explained nor predicted solely by the physical characteristics of motion.

Motions of the body are detected by three sensory systems: the vestibular system, the visual system and the somatosensory system. The vestibular system is located in the inner ear and comprises the semi-circular canals, which respond to the rotation of the head, and the otoliths which respond to translational forces (either translational acceleration or rotation of the head relative to an acceleration field, such as the force of gravity). The eyes may detect relative motion between the head and the environment, caused by either head movements (in translation or rotation) or movements of the environment, or a combination of the movements of the head and the environment. The somatosensory systems respond to force and displacement of parts of the body and give rise to sensations of body movement, or force.

In 'normal' environments, the movements of the body are detected by all three sensory systems and this leads to a sufficiently unambiguous indication of the movements of the body in space. In some other environments, the three sensory systems can simultaneously give signals corresponding to different motions (or motions which are not realistic) and lead to some form of conflict. This leads to the idea of a 'sensory conflict theory' of motion sickness in which sickness occurs when the sensory systems disagree on the motions which are occurring. However, this implies some absolute significance to sensory information, whereas the 'meaning' of the information is probably learned. This led to the 'sensory rearrangement theory' of motion sickness which states that: 'all situations which provoke motion sickness are characterised by a condition of sensory rearrangement in which the motion signals transmitted by the eyes, the vestibular system and the non-vestibular proprioceptors are at variance either with one another or with what is expected from previous experience' (Reason, 1970, 1978). Reason and Brand (1975) suggest that the conflict may be sufficiently considered in two categories: inter-modality (between vision and the vestibular receptors) and intra-modality (between the semi-circular canals and the otoliths within the vestibular system). For both categories, it is possible to identify three types of situation in which conflict can occur (see Griffin, 1990, Ch. 7).

There is evidence that the average susceptibility to sickness among males is less than that among females, and susceptibility decreases with increased age among both males and females (e.g. Lawther and Griffin, 1988a). However, there are larger individual differences within any group of either gender at any age; some people are easily made ill by motions that can be endured indefinitely by others. The reasons for these differences are not properly understood.

4.4.2 Sickness caused by oscillatory motion

Motion sickness is not caused by oscillation (however violent) at frequencies much above about 1 Hz; the phenomenon arises from motions at the low frequencies associated with normal postural control of the body. Various experimental investigations have explored the extent to which vertical oscillation causes sickness at different frequencies. These studies have allowed the formulation of a frequency weighting, W_f (see Fig. 4.6), and the definition of a 'motion sickness dose value'. The frequency weighting W_f reflects greatest sensitivity to acceleration in the range 0.125–0.25 Hz, with a rapid reduction in sensitivity at higher frequencies. The motion sickness dose value predicts the probability of sickness from a knowledge of the frequency and magnitude of vertical oscillation (see Lawther and Griffin, 1987; International Standard 2631, 1997):

$$\text{Motion sickness dose value, MSDV} = a_{\text{rms}}t^{1/2}$$

where a_{rms} is the root-mean-square value of the frequency-weighted acceleration (ms^{-2}) and t is the exposure period (seconds). The percentage of unadapted adults who are expected to vomit is given by 1/3 MSDV. (These relationships have been derived from exposures in which up to 70% of persons vomited during exposures lasting between 20 minutes and 6 hours.)

The motion sickness dose value has been used for the prediction of sickness on various marine craft (ships, hovercraft and hydrofoil) in which vertical oscillation has been shown to be the prime cause of sickness (Lawther and Griffin, 1988b). Vertical oscillation is not the principal cause of sickness in many road vehicles and some other environments. The above expression should not be assumed to be applicable to the prediction of sickness in all environments.

4.5 HAND-TRANSMITTED VIBRATION

Exposure of the fingers or the hands to vibration or repeated shock can give rise to various signs and symptoms of disorder. The extent and inter-relation between the various signs and symptoms are not fully understood but five types of disorder may be identified (see Table 4.6). More than one disorder can affect a person at the same time and it is possible that the presence of one disorder facilitates the appearance of another. The onset of each disorder is dependent on several variables, such as the vibration characteristics, the dynamic response of the fingers or hand, individual susceptibility to damage, and other aspects of the environment. Confusingly, the terms 'vibration syndrome', or 'hand–arm vibration syndrome', HAVS, are sometimes used to refer to an unspecified combination of one or more of the effects listed in Table 4.6.

4.5.1 Sources of hand-transmitted vibration

The vibration on tools varies greatly depending on tool design and method of use, so it is not possible to categorize individual tool types as 'safe' or 'dangerous'. However, Table 4.7 lists tools and processes which are sometimes a cause for concern. Figure 4.12 shows vibration spectra measured on 24 tools that might sometimes be associated with injury.

4.5.2 Effects of hand-transmitted vibration

4.5.2.1 Vascular disorders

In the United Kingdom, the possibility of 'neurosis due to vibration' was considered by the Departmental Committee on Compensation for Industrial Diseases in 1907. However, the first published cases of the condition now

Table 4.6 Five types of disorder associated with hand-transmitted vibration exposures (from Griffin, 1990). (Some combination of these disorders is sometimes referred to as the 'hand–arm vibration syndrome', HAVS)

Type	*Disorder*
Type A	Circulatory disorders
Type B	Bone and joint disorders
Type C	Neurological disorders
Type D	Muscle disorders
Type E	Other general disorders (e.g. central nervous system)

Table 4.7 Some tools and processes potentially associated with vibration injuries (standards identifying type tests indicated in parentheses)

Type of tool	*Examples of tool type*
Percussive metal-working tools	Riveting tools (ISO 8662-2) Caulking tools (ISO 8662-5) Chipping hammers (ISO 8662-2) Clinching and flanging tools (ISO 8662-10) Impact wrenches (ISO 8662-7) Impact screw drivers (ISO 8662-7) Nut runners (ISO 8662-7) Scaling hammers Needle guns (ISO 8662-14) Nibbling machines and shears (ISO 8662-10) Swaging
Grinders and other rotary tools	Pedestal grinders Hand-held grinders (ISO 8662-4; ISO 8662-8) Hand-held sanders (ISO 8662-4; ISO 8662-8) Hand-held polishers (ISO 8662-4; ISO 8662-8) Flex-driven grinders/polishers Rotary burring tools Files
Percussive hammers and drills used in mining, demolition, road construction and stone working	Hammer drill (ISO 8662-3) Rock drills (ISO 8662-3) Tampers and rammers (ISO 8662-9) Road breakers (ISO 8662-5) Stone working tools (ISO 8662-14)
Forest and garden machinery	Chain saws (ISO 7505) Anti-vibration chain saws (ISO 7505) Brush saws Mowers (ISO 5395) Hedge cutters and trimmers Barking machines Stump grinders
Other processes and tools	Nailing gun (ISO 8662-14) Stapling gun (ISO 8662-11) Pad saws (ISO 8662-12) Circular saws (ISO 8662-12) Scabblers Engraving pens

Table 4.7 (*Continued*)

Type of tool	*Examples of tool type*
	Shoe-pounding-up machines Vibratory rollers Concrete vibro-thickeners Concrete levelling vibrotables Motorcycle handle bars Pedestrian-controlled machines

most commonly known as 'vibration-induced white finger', VWF, are generally acknowledged to be those reported in Italy by Loriga in 1911. Vibration-induced white finger has subsequently been reported to occur in many other widely varied occupations in which there is exposure of the fingers to vibration (see Taylor and Pelmear, 1975; Griffin, 1990; Gemne *et al.*, 1993).

Vibration-induced white finger (VWF) is characterized by intermittent whitening (i.e. blanching) of the fingers. The finger tips are usually the first to blanch but the affected area may extend to all of one or more fingers with continued vibration exposure. Attacks of blanching are precipitated by cold and therefore usually occur in cold conditions or when handling cold objects. The blanching lasts until the fingers are rewarmed and vasodilation allows the return of the blood circulation. Many years of vibration exposure often occur before the first attack of blanching is noticed. Affected persons often have other signs and symptoms, such as numbness and tingling. Cyanosis and, rarely, gangrene, have also been reported. It is not yet clear to what extent these other signs and symptoms are causes of, caused by, or unrelated to, attacks of 'white finger'.

Vibration-induced white finger cannot be assumed to be present merely because there are attacks of blanching. It will be necessary to exclude other

→

Figure 4.12 One-third octave band spectra on 20 powered hand tools:
—— unweighted spectra;
- - - - weighted spectra;
(a: pneumatic rock drill; b: pneumatic road breaker; c: petrol-driven Wacker compressing road surface after mending; d: a non-anti-vibration chain saw; e: an anti-vibration chain saw; f: a pneumatic metal-chipping hammer; g: pole scabbler; h: needle gun; i: random orbital sander; j: impact wrench; o: riveting gun; l: dolly used with riveting gun; m: nutrunner; n: metal drill; o: wire swaging; p: etching pen; q: electric 9-inch angle grinder; r: pneumatic rotary file; s: pneumatic 5-inch straight grinder; t: pneumatic 7-inch vertical grinder. All spectra from the axis giving the highest weighted acceleration.) (Data from Griffin, 1997).

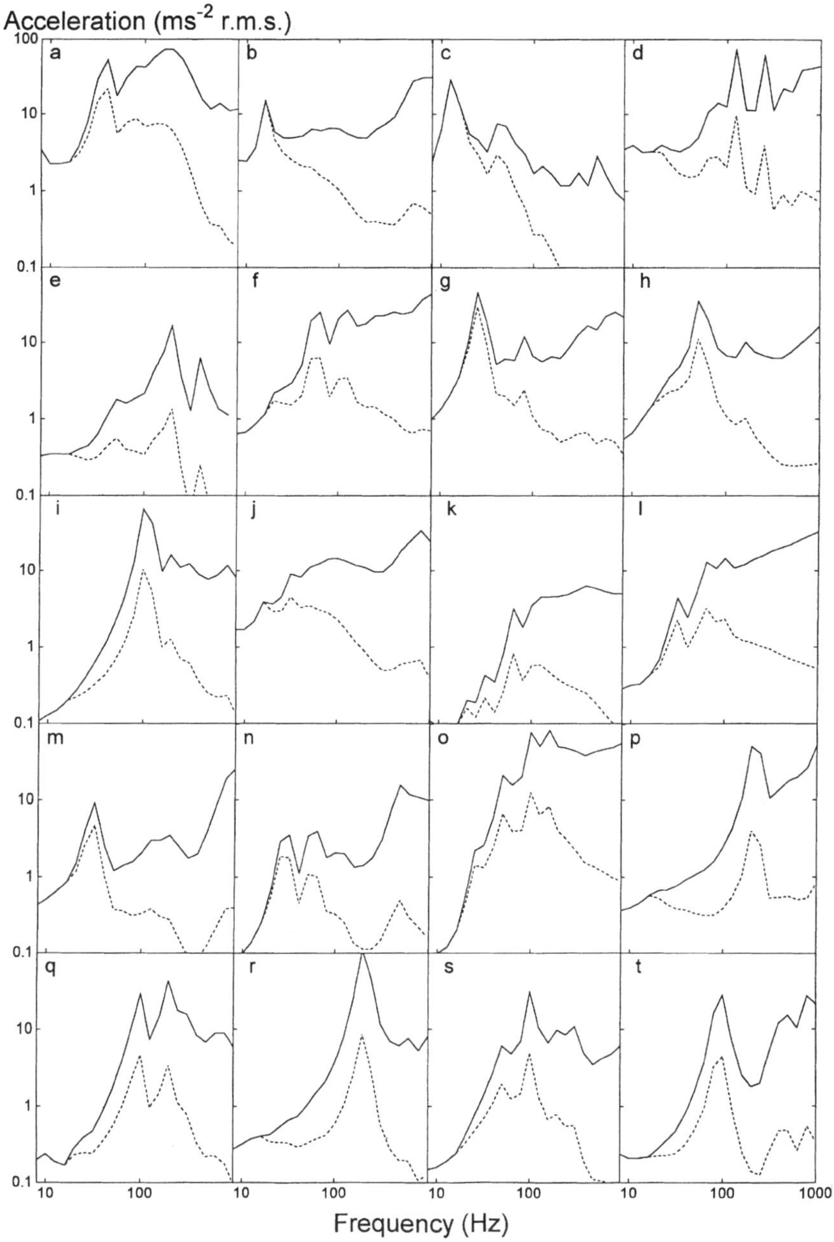

Acceleration (ms⁻² r.m.s.)
100
10
1
0.1
a
b
c
d
e
f
g
h
i
j
k
l
m
n
o
p
q
r
s
t
10
100
1000
Frequency (Hz)

known causes of similar symptoms (by medical examination) and also necessary to exclude so-called Primary Raynaud's disease (also called 'constitutional white finger'). If there is no family history of the symptoms, if the symptoms did not occur before the first significant exposure to vibration, and if the symptoms and signs are confined to areas in contact with the vibration (e.g. the fingers, not the ears etc.), they will often be assumed to indicate vibration-induced white finger. Diagnostic tests for vibration-induced white finger are not infallible indicators of the disease. However, the measurement of finger systolic blood pressure following finger cooling and the measurement of finger rewarming times following cooling can be useful, but many others tests are in use (see Neilsen and Lassen, 1977; Bovenzi, 1993).

The severity of the effects of vibration are sometimes recorded by reference to the 'stage' of the disorder. The staging of vibration-induced white finger is based on verbal statements made by the affected person. In the Stockholm Workshop staging system, the staging is influenced by both the frequency of attacks of blanching and the areas of the digits affected by blanching (see Table 4.8).

A 'scoring system' is used to record the areas of the digits affected by blanching (see Fig. 4.13). The scores correspond to areas of blanching on the digits commencing with the thumb. On the fingers, a score of 1 is given for blanching on the distal phalanx, a score of 2 for blanching on the middle phalanx and a score of 3 for blanching on the proximal phalanx. On the thumbs the scores are 4 for the distal phalanx and 5 for the proximal

Table 4.8 Stockholm Workshop scale for the classification of vibration-induced white finger (Gemne *et al.*, 1987). (If a person has stage 2 in two fingers of the left hand and stage 1 in a finger on the right hand the condition may be reported as 2L(2)/1R(1). There is no defined means of reporting the condition of digits when this varies between digits on the same hand. The scoring system is more helpful when the extent of blanching is to be recorded)

Stage	*Grade*	*Description*
0	–	No attacks
1	Mild	Occasional attacks affecting only the tips of one or more fingers
2	Moderate	Occasional attacks affecting distal and middle (rarely also proximal) phalanges of one or more fingers
3	Severe	Frequent attacks affecting all phalanges of most fingers
4	Very severe	As in stage 3, with trophic skin changes in the finger tips

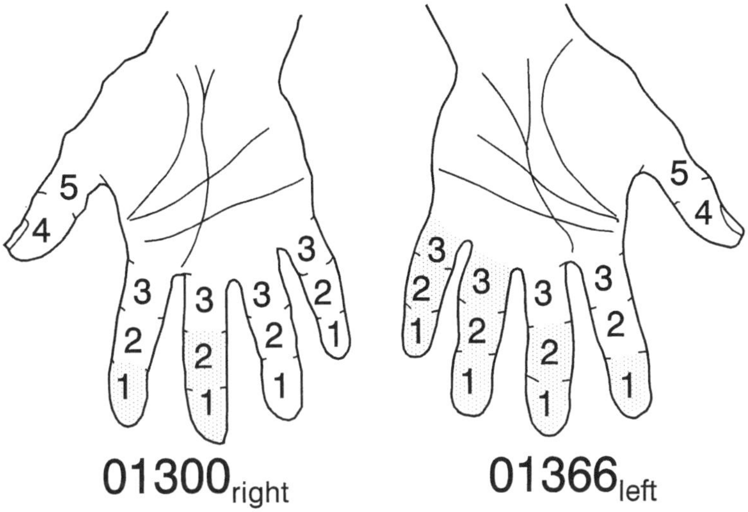

Figure 4.13 Method of scoring the areas of the digits affected by blanching (from Griffin, 1990). The blanching scores for the hands shown are 01300_{right}, 01366_{left}.

phalanx. The blanching score may be based on statements from the affected person or on the visual observations of a designated observer.

4.5.2.2 Neurological disorders

Numbness, tingling, elevated sensory thresholds for touch, vibration, temperature and pain, and reduced nerve conduction velocity are now considered to be separate effects of vibration, and not merely symptoms and signs of vibration-induced white finger. A method of reporting the extent of vibration-induced neurological effects of vibration has been proposed (see Table 4.9). This staging is not currently related to the results of any specific

Table 4.9 Proposed 'sensorineural stages' of the effects of hand-transmitted vibration (Brammer *et al.*, 1987)

Stage	*Symptoms*
0_{SN}	Exposed to vibration but no symptoms
1_{SN}	Intermittent numbness with or without tingling
2_{SN}	Intermittent or persistent numbness, reduced sensory perception
3_{SN}	Intermittent or persistent numbness, reduced tactile discrimination and/or manipulative dexterity

objective test; the 'sensorineural stage' is a subjective impression of a physician based on the statements of the affected person or the results of any available clinical or scientific testing. Neurological disorders are sometimes identified by screening tests using measures of sensory function, such as the thresholds for feeling vibration, heat or warmth on the fingers.

4.5.2.3 Muscular effects

Workers exposed to hand-transmitted vibration sometimes report difficulty with their grip, including reduced dexterity, reduced grip strength and locked grip. Many of the reports are derived from symptoms reported by exposed persons, rather than signs detected by physicians, and could be a reflection of neurological problems. Muscle activity may be of great importance to tool users since a secure grip can be essential to the performance of the job and the safe control of the tool. The presence of vibration on a handle may encourage the adoption of a tighter grip than would otherwise occur and a tight grip may increase the transmission of vibration to the hand. If the chronic effects of vibration result in reduced grip this may help to protect operators from further effects of vibration, but interfere with both work and leisure activities.

4.5.2.4 Articular disorders

Surveys of the users of hand-held tools have found evidence of bone and joint problems, most often among men operating percussive tools such as those used in metal-working jobs and mining and quarrying. It is speculated that some characteristic of such tools, possibly the low-frequency shocks, is responsible. Some of the reported injuries relate to specific bones and suggest the existence of cysts, vacuoles, decalcification, or other osteolysis, degeneration or deformity of the carpal, metacarpal or phalangeal bones. Osteoarthrosis and olecranon spurs at the elbow, and other problems at the wrist and shoulder are also documented (Griffin, 1990). There is not universal acceptance that vibration is the cause of articular problems and there is currently no dose-effect relation which predicts their occurrence. In the absence of specific information, it seems that adherence to current guidance for the prevention of vibration-induced white finger may provide reasonable protection.

4.5.2.5 Other effects

Hand-transmitted vibration may not only affect the fingers, hands and arms. Studies have found a high incidence of problems such as headaches and sleeplessness among tool users and have concluded that these symptoms are caused by hand-transmitted vibration. Although these are real problems to those affected, they are 'subjective' effects which are not accepted as real by all researchers. Some current research is seeking a physiological basis for such symptoms. It would appear that caution is appropriate, but it currently appears reasonable to assume that the adoption of the modern guidance to prevent vibration-induced white finger will also provide some protection from any other effects of hand-transmitted vibration within, or distant from, the hand.

4.5.3 Preventative measures

Table 4.10 summarizes some of the actions which may be appropriate from management, tool manufacturers, technicians and physicians at the workplace and from tool users. When there is reason to suspect that hand-transmitted vibration may cause injury, the vibration at tool–hand interfaces should be measured. It will then be possible to predict whether the tool or process is likely to cause injury and whether any other tool or process could give a lower vibration severity. The duration of exposure to vibration should also be quantified. Reduction of exposure time may include the provision of exposure breaks during the day and, if possible, prolonged periods away from vibration exposure. For any tool or process having a vibration magnitude sufficient to cause injury, there should be a system to quantify and control the maximum daily duration of exposure of any individual.

Gloves are sometimes recommended as a means of reducing the adverse effects of vibration on the hands. When using the frequency weightings in current standards, most commonly available gloves do *not* normally provide effective attenuation of the vibration on most tools (Griffin, 1998). Gloves and 'cushioned' handles may reduce the transmission of high frequencies of vibration but current standards imply that these frequencies are not usually the primary cause of disorders. Gloves may protect the hand from other forms of mechanical injury (e.g. cuts and scratches) and protect the fingers from temperature extremes. Warm hands are less likely to suffer an attack of finger blanching and some consider that maintaining warm hands while exposed to vibration may also lessen the damage caused by the vibration.

Workers who are exposed to vibration magnitudes sufficient to cause injury should be warned of the possibility of vibration injuries and educated on the ways of reducing the severity of their vibration exposures. They should be advised of the symptoms to look out for and told to seek medical attention if the symptoms appear. There should be pre-employment medical screening wherever a subsequent exposure to hand-transmitted vibration may reasonably be expected to cause vibration injury. Medical supervision of each exposed person should continue throughout employment at suitable intervals, possibly annually.

Table 4.10 Some preventative measures to consider when persons are exposed to hand-transmitted vibration (adapted from Chapter 19 of *The Handbook of Human Vibration,* Griffin, 1990)

Group	*Action*
Management	Seek technical advice Seek medical advice Warn exposed persons Train exposed persons Review exposure times Policy on removal from work
Tool manufacturers	Measure tool vibration Design tools to minimize vibration Ergonomic design to reduce grip force etc. Design to keep hands warm Provide guidance on tool maintenance Provide warning of dangerous vibration
Technical at workplace	Measure vibration exposure Provide appropriate tools Maintain tools Inform management
Medical	Pre-employment screening Routine medical checks Record all signs and reported symptoms Warn workers with predisposition Advise on consequences of exposure Inform management
Tool user	Use tool properly Avoid unnecessary vibration exposure Minimise grip and push forces Check condition of tool Inform supervisor of tool problems Keep warm Wear gloves when safe to do so Minimise smoking Seek medical advice if symptoms appear Inform employer of relevant disorders

4.5.4 Standards for the evaluation of hand-transmitted vibration

4.5.4.1 International and national standards

International Standard 5349 (1986) and many national standards use a frequency weighting (called W_h in British Standard 6842) to quantify the severity of hand-transmitted vibration over the frequency range from 8 to 1000 Hz (see Fig. 4.14). This weighting is applied to measurements of vibration acceleration in each of the three axes of vibration at the point of entry of vibration to the hand (see Fig. 4.2). Measurements of tool vibration should be obtained with representative operating conditions. The standards imply that if two tools expose the hand to vibration for the same period of time, the tool having the lowest frequency-weighted acceleration will be least likely to cause injury or disease. Figure 4.12 shows the vibration spectra on 24 tools before and after the application of the frequency weighting W_h.

Current standards for the evaluation of hand-transmitted vibration use the concept called 'equal energy' so that a complex exposure pattern of any duration during the day can be represented by the equivalent value for an exposure of 8 hours (4 hours in ISO 5349, 1986). For an exposure of duration, t, to a frequency-weighted rms acceleration, a_{hw}, the 8-hour energy-equivalent acceleration, $a_{hw(eq,8h)}$, is given by:

$$a_{hw(eq,8h)} = a_{hw}\sqrt{(t/T_{(8)})} \tag{4.1}$$

where $T_{(8)}$ is 8 hours (in the same units as t). The value of $a_{hw(eq,8h)}$ is sometimes denoted by $A(8)$.

A relation between years of vibration exposure, E, the 4-hour energy-equivalent frequency-weighted acceleration, $a_{hw(eq,4h)}$, and the predicted prevalence of finger blanching, C, was proposed in an Annex to International Standard 5349 (1986) (see Table 4.11 and Fig. 4.15):

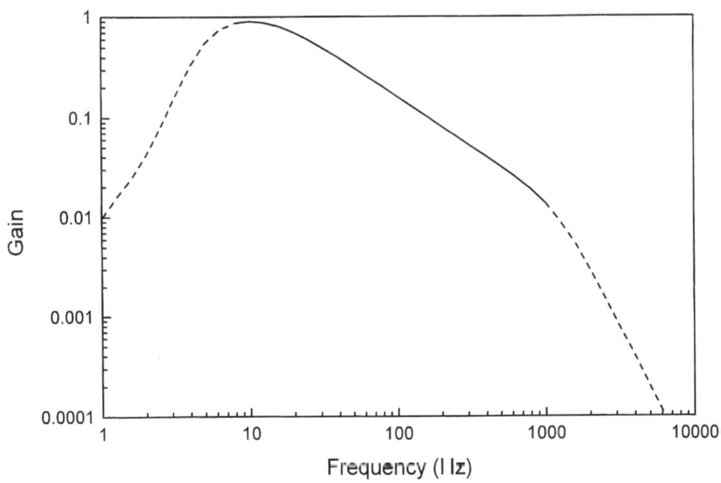

Figure 4.14 Frequency weighting, W_h, (shown here between 8 and 1000 Hz; when evaluating one-third octave band spectra the frequency range is specified as 6.3–1250 Hz).

Table 4.11 Number of years before blanching develops in 10–50% of vibration-exposed persons according to International Standard 5349 (1986)

Weighted acceleration, $a_{hw(eq,\,4h)}$ (ms^{-2} rms)	*Percentage of population affected by finger blanching*				
	10%	*20%*	*30%*	*40%*	*50%*
2	15	23	>25	>25	>25
5	6	9	11	12	14
10	3	4	5	6	7
20	1	2	2	3	3
50	<1	<1	<1	1	1

$$C = 100\left[\frac{a_{\mathrm{hw(eq,4h)}} \cdot E}{95}\right]^2. \qquad (4.2)$$

The values in Table 4.11 and Fig. 4.15 refer to frequency-weighted acceleration (i.e. referenced to the frequency range 8–16 Hz after passing through the filter shown in Fig. 4.14). British Standard 6842 (1987) offers similar guidance, but is restricted to a 10% prevalence of vibration-induced white finger. Figure 4.16 shows how the magnitudes required for a predicted

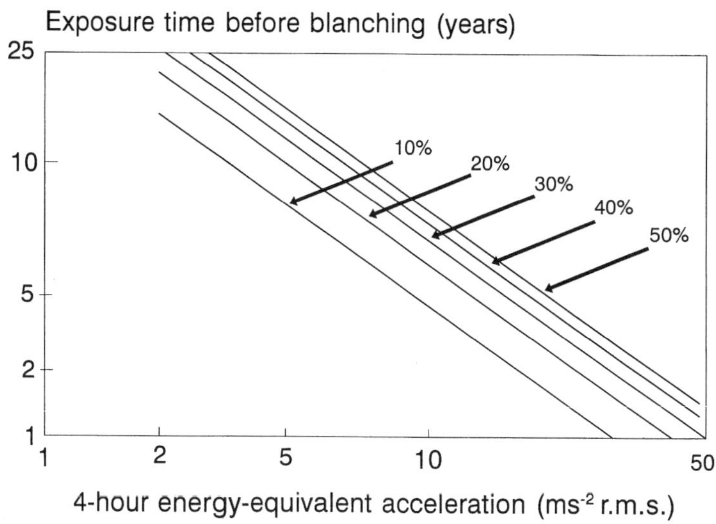

Figure 4.15 Years of exposure to various magnitudes of 4-hour energy-equivalent frequency-weighted hand-transmitted vibration after which finger blanching in 10%–50% of exposed persons should be expected according to International Standard 5349 (1986).

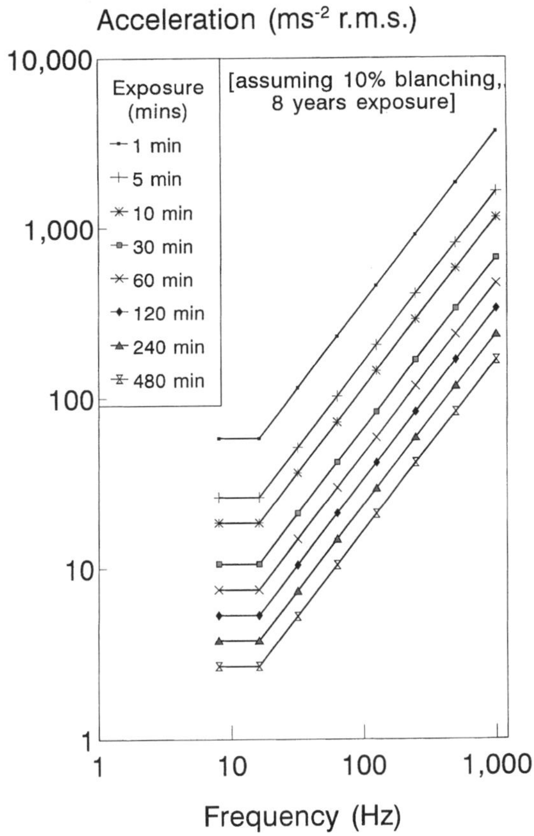

Figure 4.16 Acceleration magnitudes predicted to give 10% prevalence of vibration-induced white finger after 8 years for daily exposure durations from 1 minute to 8 hours (according to International Standard 5349, 1986).

prevalence of 10% vibration-induced white finger after 8 years are assumed to depend on vibration frequency from 8 to 1000 Hz for exposure durations from 1 minute to 8 hours per day.

The frequency weighting, the time-dependency and the dose-effect information used in ISO 5349 (1986) are based on less than complete information, and such information as exists has been interpolated, extrapolated and simplified for practical convenience (see Griffin, 1997). Consequently, the percentage of affected persons in any group of exposed persons will not always closely match the values shown in Table 4.11 or Figs 4.15 and 4.16. Additionally, the number of persons affected by vibration will depend on the rate at which persons enter and leave a group exposed to vibration.

4.5.4.2 EU Machinery Safety Directive

The Machinery Safety Directive of the European Community (89/392/EEC) states: 'machinery must be so designed and constructed that risks resulting from vibrations produced by the machinery are reduced to the lowest level, taking account of technical progress and the availability of means of reducing vibration, in particular at source' (Council of the European Communities, 1989). Instruction handbooks for hand-held and hand-guided machinery should specify the equivalent acceleration to which the hands or arms are subjected where this exceeds a stated value (currently a frequency-weighted acceleration of 2.5 ms^{-2} rms). The relevance of any such value will depend on the test conditions to be specified in 'type testing' standards (see International Standard 8662 Part 1 (1988); Table 4.7 provides a guide to other parts of ISO 8662). In use, very many hand-held vibrating tools can exceed 2.5 ms^{-2} rms. The specified tests do not encompass all realistic exposures and so will not represent the exposures received in some jobs. Additionally, the different exposure durations and other variables involved in different jobs carry different degrees of risk.

4.5.4.3 Proposed EU Exposure Directive

The opening principles of a proposed Council Directive on the minimum health and safety requirements regarding hand-transmitted vibration are: 'Taking account of technical progress and of the availability of measures to control the physical agent at source, the risks arising from exposure to the physical agent must be reduced to the lowest achievable level, with the aim of reducing exposure to below the threshold level' (Commission of the European Communities, 1994).

In respect of hand-transmitted vibration, the proposed EU Directive currently identifies a threshold level ($a_{hw(eq,8h)} = 1.0\,ms^{-2}$ rms), an action level ($a_{hw(eq,8h)} = 2.5\,ms^{-2}$ rms), and an exposure limit value ($a_{hw(eq,8h)} = 5.0\,ms^{-2}$ rms). The manner in which these 8-hour equivalent magnitudes correspond to higher magnitudes at shorter durations is shown in Fig. 4.17. When exposures exceed the 'threshold level', it is proposed that workers must receive information concerning the potential risk of exposure to hand-transmitted vibration. The 'action level' is intended to identify the conditions in which training in precautionary measures is required, an assessment of the vibration is to be made, and a programme of preventative measures is to be instituted. The proposed Directive also indicates that when the action level is exceeded workers shall have the right to regular health surveillance, including routine examinations designed for the early detection of disorders caused by hand-transmitted vibration. If the 'exposure limit value' is exceeded, health surveillance must be carried out and Member States of the Community will be expected to control the harmful effects.

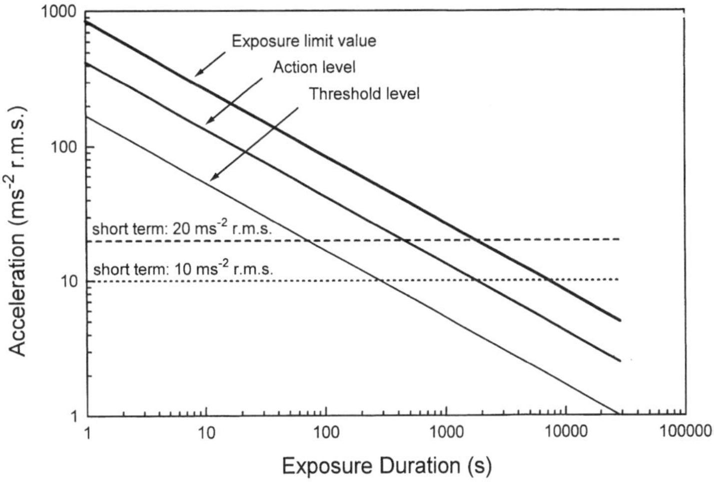

Figure 4.17 Threshold level, action level and exposure limit proposed in a draft EC Directive on Physical Agents (Commission of the European Communities, 1994). In addition to these three criteria, effort shall be made to reduce the hazard when a short-term (a few minutes) equivalent acceleration exceeds 10 ms^{-2} rms. Exposure to a short-term (a few minutes) equivalent acceleration equal to or greater than 20 ms^{-2} rms is considered a hazardous activity.

The extremely high magnitudes for short durations shown in Fig. 4.17 should not be considered to be representative of their safety; these high magnitudes are a consequence of the 'equal energy' concept described in section 4.5.4.1. The proposed Directive identifies exposures 'for a short term (a few minutes) equivalent acceleration equal to or greater than 20 ms^{-2}', as 'activities with increased risk' which must be declared to the 'authority responsible'. Member States would be required to ensure that appropriate measures are taken in order to control the risks associated with these activities. Equipment which 'transmits to the hand–arm system' a short-term (a few minutes) equivalent acceleration equal to or greater than 20 ms^{-2} rms must be marked. For lower magnitudes of vibration, the proposed Directive says:

> 'Where the activity involves the use of work equipment which transmits to the hand–arm system a short-term (a few minutes) equivalent acceleration exceeding 10 ms^{-2}, increased efforts shall be made to reduce the hazard, with priority to the use of low-vibration equipment and processes, including the revision of product design and work practice. Pending the effective implementation the duration of continuous exposure shall be reduced.'

The proposed Directive is based on the 'vector sum' (more correctly the root-sums-of-squares, rss) of the frequency-weighted accelerations in three axes, although an axis can be omitted if the values are less than 50% of the values in another axis at the same location. This differs from International Standard 5349 (1986) which is based on the axis having the highest weighted acceleration, however a forthcoming revision of ISO 5349 is expected to change the evaluation method to the use of rss.

Current standards for evaluating hand-transmitted vibration provide guidance which cannot be ignored. However, those concerned with the design or evaluation of situations involving hand-transmitted vibration should anticipate that further understanding of the relevant pathology, physiology and biodynamics may yield improvements to the methods of assessing the safety of exposures to hand-transmitted vibration.

4.6 REFERENCES

Benson, A. J. and Barnes, G. R. (1978) Vision during angular oscillation: the dynamic interaction of visual and vestibular mechanisms, *Aviation, Space and Environmental Medicine*, **49**(1) Section II, 340–5.

Bongers, P. M. and Boshuizen, H. C. (1990) Back disorders and whole-body vibration at work, Thesis, University of Amsterdam, ISBN 90-9003668-7.

Bovenzi, M. (1993) Digital arterial responsiveness to cold in healthy men, vibration white finger and primary Raynaud's phenomenon. *Scandinavian Journal of Work, Environment and Health*, **19**(4), 271–6.

Bovenzi, M. and Zadini, A. (1992) Self-reported back symptoms in urban bus drivers exposed to whole-body vibration, *Spine*, **17**(9), 1048–59.

Brammer, A. J., Taylor, W. and Lundborg, G. (1987) Sensorineural stages of the hand-arm vibration syndrome, *Scandinavian Journal of Work, Environment and Health*, **13**(4), 279–83.

British Standards Institute (1987a) *Measurement and evaluation of human exposure to whole-body mechanical vibration and repeated shock*, British Standard, BS 6841.

British Standards Institute (1987b) *Measurement and evaluation of human exposure to vibration transmitted to the hand*, British Standard, BS 6842.

British Standards Institute (1992) *Evaluation of human exposure to vibration in buildings (1 Hz to 80 Hz)*, British Standard, BS 6472.

Commission of the European Communities (1994) Amended proposal for a Council Directive on the minimum Health and Safety requirements regarding the exposure of workers to the risks arising from physical agents – Individual Directive in relation to Article 16 of Directive 89/391/EEC. *Official Journal of the European Communities*, No. C 230, 19.8.94, 3–29.

Corbridge, C. and Griffin, M. J. (1991) Effects of vertical vibration on passenger activities: writing and drinking, *Ergonomics*, **34**(10), 1313–32.

Corbridge, C., Griffin, M. J. and Harborough, P. (1989) Seat dynamics and passenger comfort, *Proceedings of the Institution of Mechanical Engineers*, **203**, 57–64.

Council of the European Communities (Brussels) (1989) On the approximation of the laws of the member states relating to machinery. Council Directive (89/392/EEC), *Official Journal of the European Communities*, June, 9–32.

Departmental Committee on Compensation for Industrial Diseases (1907) *Report of the departmental committee on compensation for industrial diseases*. Cd.3495, HMSO.

Dupuis, H. and Zerlett, G. (1986) *The Effects of Whole-body Vibration*, Springer-Verlag, Berlin, ISBN 0-387-16584-3.

Fairley, T. E. and Griffin, M. J. (1989) The apparent mass of the seated human body: vertical vibration, *Journal of Biomechanics*, **22**(2), 81–94.

Gemne, G., Lundström, R. and Hansson, J-E. (1993) Disorders induced by work with hand-held vibrating tools. A review of current knowledge for criteria documentation, *Arbete Och Halsa*, 1933:6, Arbets Miljo Institutet, National Institute of Occupational Health, ISBN 91-7045-209-1, ISSN 0346-7821.

Gemne, G., Pyykko, I., Taylor, W. and Pelmear, P. (1987) The Stockholm Workshop scale for the classification of cold-induced Raynaud's phenomenon in the hand-arm vibration syndrome (revision of the Taylor-Pelmear scale), *Scandinavian Journal of Work, Environment and Health*, **13**(4), 275–8.

Goodwin, G. M., McCloskey, D. I. and Matthews, P. B. C. (1972) The contribution of muscle afferents to kinaesthesia shown by vibration induced illusions of movement and by the effects of paralysing joint afferents, *Brain*, **95**, 705–48.

Griffin, M. J. (1976) Eye motion during whole-body vertical vibration, *Human Factors*, **18**(6), 601–6.

Griffin, M. J. (1990) *Handbook of Human Vibration*. Academic Press, London, ISBN: 0-12-303040-4.

Griffin, M. J. (1991) Physical characteristics of stimuli provoking motion sickness, Paper 3. In *Motion Sickness: Significance in Aerospace Operations and Prophylaxis*. AGARD Lecture Series LS – 175, ISBN 92-835-0634-0.

Griffin, M. J. (1997) Measurement, evaluation and assessment of occupational exposures to hand-transmitted vibration. *Occupational and Environmental Medicine*, **54**(2), 73–89.

Griffin, M. J. (1998) Evaluating the effectiveness of gloves in reducing the hazards of hand-transmitted vibration. Awaiting publication.

Griffin, M. J. and Hayward, R. A. (1994) Effects of horizontal whole-body vibration on reading, *Applied Ergonomics*, **25**(3), 165–9.

Griffin, M. J. and Lewis, C. H. (1978) A review of the effects of vibration on visual acuity and continuous manual control, Part I: Visual acuity. *Journal of Sound and Vibration*, **56**(3), 383–413.

Hulshof, C. and Zanten, B. V. van. (1987) Whole-body vibration and low-back pain, *International Archives of Occupational and Environmental Health*, **59**, 205–20.

International Organization for Standardization (1974) Guide for the evaluation of human exposure to whole-body vibration, *International Standard*, ISO 2631 (E).

International Organization for Standardization (1983) Acoustics – preferred reference quantities for acoustic levels, *International Standard*, ISO 1683.

International Organization for Standardization (1985) Evaluation of human exposure to whole-body vibration – Part 1: General requirements., *International Standard* 2631/1.

International Organization for Standardization (1986) Mechanical vibration – Guidelines for the measurement and the assessment of human exposure to hand-transmitted vibration, *International Standard* ISO 5349.

International Organization for Standardization (1988) Hand-held portable tools – Measurement of vibration at the handle – Part 1: General, *International Standard* ISO 8662-1.

International Organization for Standardization (1989) Evaluation of human exposure to whole-body vibration – Part 2: Continuous and shock-induced vibration in buildings, *International Standard* ISO 2631-2.

International Organization for Standardization (1990) Agricultural wheeled tractors – operator's seat – laboratory measurement of transmitted vibration, *International Standard*, ISO 5007.

International Organization for Standardization (1992) Mechanical vibration – Laboratory method for evaluating vehicle seat vibration – Part 1: Basic requirements, *International Standard*, ISO 10326-1.

International Organization for Standardization (1994) Earth-moving machinery – laboratory evaluation of operator seat vibration, *International Standard*, ISO 7096.

International Organization for Standardization (1997) Mechanical vibration and shock – evaluation of human exposure to whole-body vibration – Part 1: General requirements, *International Standard*, ISO 2631-1.

Kitazaki, S. and Griffin, M. J. (1997) A modal analysis of whole-body vertical vibration, using a finite element model of the human body, *Journal of Sound and Vibration*, **200**(1), 83–103.

Lawther, A. and Griffin, M. J. (1987) Prediction of the incidence of motion sickness from the magnitude, frequency, and duration of vertical oscillation, *Journal of the Acoustical Society of America*, **82**(3), 957–66.

Lawther, A. and Griffin, M. J. (1988a) A survey of the occurrence of motion sickness amongst passengers at sea, *Aviation, Space and Environmental Medicine*, **59**(5), 399–406.

Lawther, A. and Griffin, M. J. (1988b) Motion sickness and motion characteristics of vessels at sea, *Ergonomics*, **31**(10), 1373–94.

Lewis, C. H. and Griffin, M. J. (1977) The interaction of control gain and vibration with continuous manual control performance, *Journal of Sound and Vibration*, **55**(4), 553–62.

Lewis, C. H. and Griffin, M. J. (1978) A review of the effects of vibration on visual acuity and continuous manual control, Part II: continuous manual control, *Journal of Sound and Vibration*, **56**(3), 415–57.

Lewis, C. H. and Griffin, M. J. (1979) The effect of character size on the legibility of numeric displays during vertical whole-body vibration, *Journal of Sound and Vibration*, **67**(4), 562–5.

Loriga, G. (1911) Il lavoro con i martelli pneumatici. The use of pneumatic hammers, *Boll. Ispett. Lavoro*, **2**, 35–60.

Mansfield, N. J. and Griffin, M. J. (1996) Vehicle seat dynamics measured with an anthropodynamic dummy and human subjects. *Inter-noise '96*, Proceedings of 25th Anniversary Congress, Liverpool, Book 4, Institute of Acoustics, ISBN 1-873082 91 6, 1725–30.

Mansfield, N. J. and Griffin, M. J. (1997) Non-linearities in biodynamic responses to whole-body vertical vibration. Awaiting publication in *Journal of Biomechanics*.

Martin, B. J., Roll, J. P. and Gauthier, G. M. (1984) Spinal reflex alterations as a function of intensity and frequency of vibration applied to the feet of seated subjects, *Aviation, Space and Environmental Medicine*, **55**(1), 8–12.

McLeod, R. W. and Griffin, M. J. (1989) A review of the effects of translational whole-body vibration on continuous manual control performance, *Journal of Sound and Vibration*, **133**(1), 55–115.

McLeod, R. W. and Griffin, M. J. (1990) Effects of whole-body vibration waveform and display collimation on the performance of a complex manual control task, *Aviation, Space and Environmental Medicine*, **61**(3), 211–19.

Meddick, R. D. L. and Griffin, M. J. (1976) The effect of two-axis vibration on the legibility of reading material, *Ergonomics*, **19**(1), 21–33.

Moseley, M. J. and Griffin, M. J. (1986a) Effects of display vibration and whole-body vibration on visual performance, *Ergonomics*, **29**(8), 977–83.

Moseley, M. J. and Griffin, M. J. (1986b) A design guide for visual displays and manual tasks in vibration environments. Part I: visual displays, *Institute of Sound and Vibration Research, Technical Report* No. 133, University of Southampton.

Moseley, M. J., Lewis, C. H. and Griffin, M. J. (1982) Sinusoidal and random whole-body vibration: comparative effects on visual performance, *Aviation, Space and Environmental Medicine*, **53**(10), 1000–5.

Nielsen, S. L. and Lassen, N. A. (1977) Measurement of digital blood pressure after local cooling, *Journal of Applied Physiology, Respiratory Environment and Exercise Physiology*, **43**(5), 907–10.

O'Hanlon, J. G. and Griffin, M. J. (1971) Some effects of the vibration of reading material upon visual performance, *Institute of Sound and Vibration Research, Technical Report* No. 49.

Paddan, G. S. and Griffin, M. J. (1988a) The transmission of translational seat vibration to the head – 1. Vertical seat vibration, *Journal of Biomechanics*, **21**(3), 191–7.

Paddan, G. S. and Griffin, M. J. (1988b) The transmission of translational seat vibration to the head – II. Horizontal seat vibration, *Journal of Biomechanics*, **21**(3), 199–206.

Reason, J. T. (1970) Motion sickness: A special case of sensory rearrangement, *Advancement of Science*, **26**, June, 386–93.

Reason, J. T. (1978) Motion sickness adaptation: a neural mismatch model, *Journal of the Royal Society of Medicine*, **71**, 819–29.

Reason, J. T. and Brand, J. J. (1975) *Motion Sickness*. Academic Press, London, ISBN 0-12-584050-0.

Ribot, E., Roll, J. P. and Gauthier, G. M. (1986) Comparative effects of whole-body vibration on sensorimotor performance achieved with a mini-stick and a macro-stick in force and position control modes, *Aviation, Space and Environmental Medicine*, **57**(8), 792–9.

Sherwood, N. and Griffin, M. J. (1990) Effects of whole-body vibration on short-term memory, *Aviation, Space and Environmental Medicine*, **61**(12), 1092–7.

Sherwood, N. and Griffin, M. J. (1992) Evidence of impaired learning during whole-body vibration, *Journal of Sound and Vibration*, **152**(2), 219–25.

Taylor, W. and Pelmear, P. L. (eds) (1975) *Vibration White Finger in Industry*. Academic Press, London, ISBN 0 12 684550 6.

Wei, L. and Griffin, M. J. (1998) Mathematical models for the apparent mass of the seated human body exposed to vertical vibration. *Journal of Sound Vibration*, 212 (5), 855–74.

Wei, L. and Griffin, M. J. (1998) The prediction of seat transmissibility from measures of seat impedance. Awaiting publication in *Journal of Sound and Vibration*.

Wells, M. J. and Griffin, M. J. (1984) Benefits of helmet-mounted display image stabilisation under whole-body vibration, *Aviation, Space and Environmental Medicine*, **55**(1), 13–18.

Wilson, R. V. (1974) Display collimation under whole-body vibration, *Human Factors*, **16**(2), 186–95.

5

Fundamentals of noise and vibration control

F.J. FAHY

5.1 MOTIVATION AND CONSTRAINTS

Sound may be desired, undesired, or simply unnoticed by the recipient; all categories may be either harmless or harmful to human beings, depending upon a number of factors which are addressed in Chapter 3. Acoustic noise may therefore be defined as 'undesirable or harmful sound'. The principal harmful effect of noise on human beings is hearing damage, but it may also affect physiological and psychological health by generating stress through annoyance, and by disturbing rest and sleep. Other adverse effects include distraction from task performance, degradation of speech intelligibility and interference with the performance and enjoyment of music.

Noise can also constitute a serious threat to mechanical and electrical systems by producing excessive vibration, stress, fatigue damage or malfunction. It can have a serious negative influence on residential property prices. Finally, and very importantly, it adversely affects industrial product quality, marketability and commercial competitiveness.

There seems to be adequate motivation for developing and implementing noise control measures as effectively as possible. However, it must not be forgotten that there may exist constraints to such implementation, principally financial expense, but also many other factors, including safety, weight, volume, hygiene, ease of maintenance, overheating, restriction of access, system performance and efficiency, vulnerability to damage and aesthetic acceptability.

Any one source may transmit acoustic or vibrational disturbances via a number of parallel paths, all of which must be controlled if treatment is to be effective. Hence, the most efficacious form of noise control is that applied 'at source'. Sources of noise are exceedingly diverse in form, material and mechanism and, in order to control them by design or modification, the engineer must understand the influence of the various physical and operating parameters on their outputs. In this chapter, after an introduction to noise control target specification, the reader is provided with bases for categorizing, characterizing and quantifying noise sources in terms of sound power. The principles of *passive* noise control are then illustrated in general terms. The behaviour of sound absorbent materials and sound-reflecting partitions are presented in terms of mathematical models and descriptions of physical behaviour; practical applications follow. The chapter closes with a brief review of the principles and practice of vibration control, which should be read in conjunction with Chapter 2, 'Fundamentals of vibration'.

It should be emphasized that this chapter provides an introduction to its title subject and can in no way be considered to compete with comprehensive handbooks such as Bies and Hansen (1995) or Beranek and Vér (1992).

5.2 MEASURES OF SOUND: NOISE CONTROL TARGETS

Noise control may either be implemented at the design stage of man-made products or as a remedial measure in cases where system operation causes excessive noise. This chapter addresses technical means of passive noise control, excluding architectural and environmental planning. Noise control requires the specification of quantitative targets which are often determined by legally enforceable requirements, e.g. EU Directives. Such targets must be expressed in terms of numerical values of acoustical quantities or indices derived therefrom. This section briefly describes the principal measures of acoustical quantities and presents examples of their incorporation into derived quantities and indices which are used to express noise control targets.

5.2.1 Measures of sound

Sound may be defined as a time-varying disturbance of the density of a fluid from its equilibrium value, which is accompanied by a proportional disturbance of sound pressure, and is associated with small oscillatory movements of fluid particles. Acoustic disturbances propagate at a characteristic speed, the 'speed of sound'. The nature and characteristics of sound are described in Chapter 1. The most commonly measured acoustical quantity is sound pressure. Pressure is measured because (i) the mechanical input to the auditory system is pressure; (ii) it is the easiest acoustical quantity to measure; (iii) mechanical structures, such as walls, respond to pressure. Figures 5.1(a) and 5.1(b) illustrate the time history of pressure and a quantity termed the 'mean square pressure' which is most commonly employed to quantify a continuous acoustic disturbance. The actual pressure–time history is used to describe and quantify transient acoustic disturbances, such as sonic booms and gunshots.

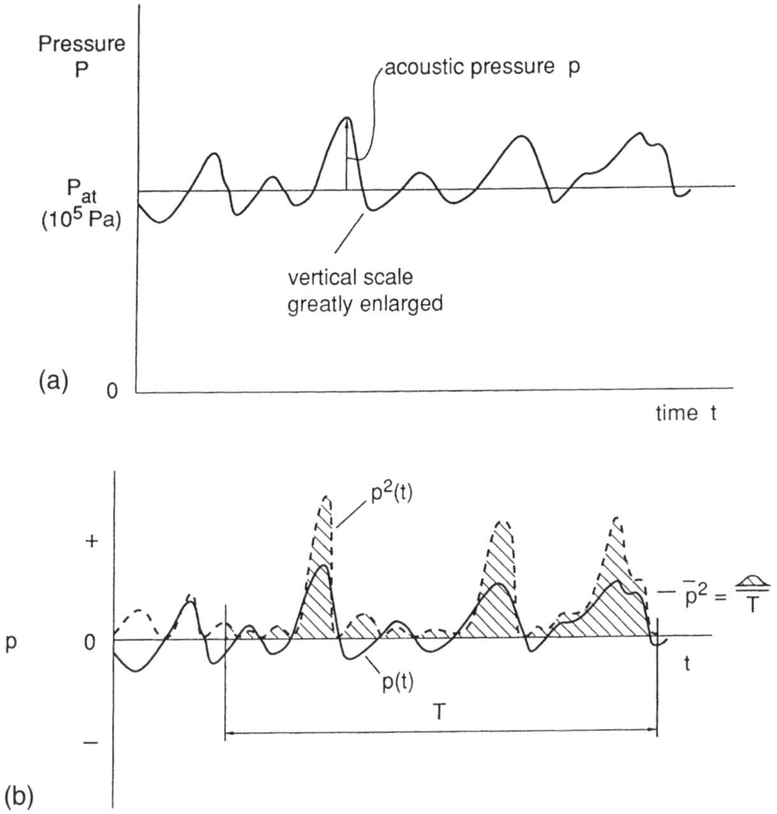

Figure 5.1 (a) Pressure–time history. (b) Illustration of evaluation of mean square pressure.

$$\text{Mean square pressure } \overline{p^2} = \frac{1}{T}\int_{t_1}^{t_1+T} p^2(t)\mathrm{d}t \qquad \mathrm{Pa}^2[(\mathrm{N/m}^2)^2] \tag{5.1}$$

Note that $\overline{p^2}$ is generally a function of the integration interval, but converges to a stable value with increase of this interval for continuous, time-stationary (steady) sound.

Sound waves transport energy in the form of kinetic energy of particle motion and potential energy of elastic strain (fractional change of fluid volume). The measure of the magnitude and direction of the local rate of energy transport is termed the *sound intensity*: its symbol is I and its unit is $\mathrm{W/m^2}$. Sound intensity is a vector quantity which should strictly be called 'sound power flux density', a more explicit description. Sound intensity is not simply related to mean square sound pressure, as many textbooks would have you believe, except under very special circumstances (Fahy, 1995).

The output of a sound source is generally quantified in terms of the radiated *sound power*; its symbol is W (or P in International Standards) and its unit is the watt. This quantity is largely independent of the environment of the source, unlike sound pressure, which also varies with distance and directional location. (Useful thermal analogies of $\overline{p^2}$ and W are the temperature in a room and the heat output of a heater; the latter varies little with the degree of thermal insulation of a room whereas the former varies greatly.) It is evident that the description of source 'strength' in terms of pressure is not appropriate, unless under closely specified conditions.

The human auditory system does not respond linearly to variations of sound pressure (fortunately!); like many other physiological and psychological perceptions it varies approximately logarithmically. Hence, logarithmic equivalents of $\overline{p^2}$, I and W are conventionally employed. Their common unit is the decibel (dB) (although the Bel is used to express sound power in computer system noise measurements). The logarithmic measures are defined as follows:

Sound pressure level

$$L_{\mathrm{p}} = 10\log_{10}\left[\overline{p^2}/p_{\mathrm{ref}^2}\right] \equiv 20\log_{10}[p_{\mathrm{rms}}/p_{\mathrm{ref}}]\mathrm{dB} \tag{5.2a}$$

Sound intensity level

$$L_{\mathrm{I}} = 10\log_{10}[I/I_{\mathrm{ref}}]\,\mathrm{dB} \tag{5.2b}$$

Sound power level

$$L_{\mathrm{W}} = 10\log_{10}[W/W_{\mathrm{ref}}]\,\mathrm{dB} \tag{5.2c}$$,

in which p_{ref}, I_{ref} and W_{ref} are 20×10^{-6} Pa, 10^{-12} W/m^2 and 10^{-12} W, respectively, and p_{rms} indicates 'root mean square'. In the case of instantaneous values of L_{p}, p_{rms} is replaced by $p(t)$.

5.2.2 Frequency decomposition of sound (frequency analysis)

Any pressure–time history (even isolated transients) may be mathematically decomposed into, or constructed from, an infinite number of sinusoids of infinite duration (Fourier theorem). The amplitudes and phases of the sinusoids are determined by the specific time history. The distributions of the amplitudes and phases as functions of frequency are termed the amplitude and phase spectra. Periodic sounds, such as compressor whines, or cyclic events, such as regularly repeated impacts, have spectra which only have non-zero components at discrete frequencies which are multiples (harmonics) of the fundamental frequency, which is the inverse of the event period; these are called 'line spectra' (Fig. 5.2(a)). Aperiodic sounds, such as subsonic jet noise or isolated impacts, have continuous frequency spectra (Fig. 5.2(b)).

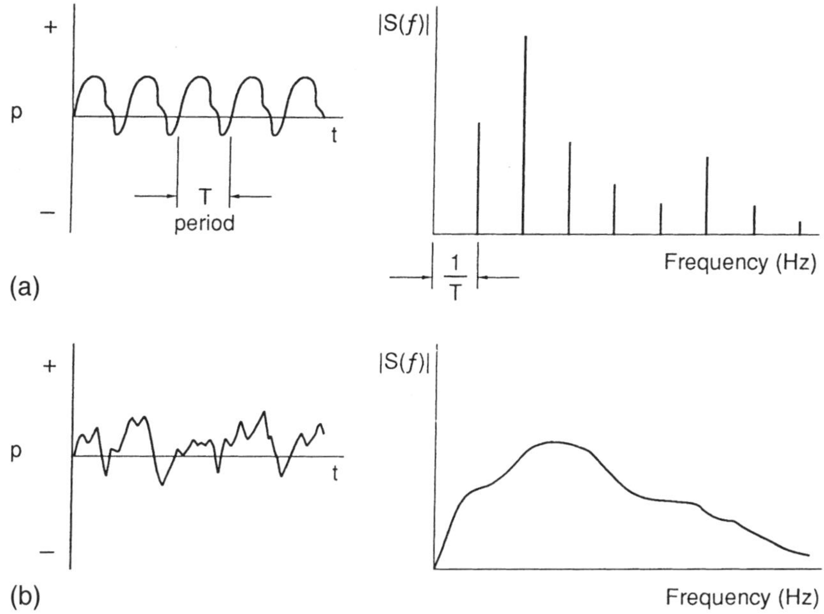

Figure 5.2 (a) Periodic and amplitude spectrum. (b) Aperiodic signal and amplitude spectrum.

Frequency decomposition (or frequency analysis) of sound is very important in acoustics and vibration for the following reasons:

1. the sensitivity of human auditory system and the vibrational response of mechanical systems to sound vary with frequency;
2. the performance of a noise control system is frequency dependent;
3. the mathematical analysis of sinusoidal (pure tone or harmonic) sound is relatively straightforward, and any sound field and its (linear) effects may be synthesized by the addition of harmonic components.

Spectral analysis may be performed either by the use of contiguous (digital or analogue) filters or by Fourier analysis via the Discrete Fourier Transform (DFT), conventionally implemented by various algorithms which are collectively described as Fast Fourier Transforms (FFT) (see Chapters 6 and 8 for details). The process of contiguous filtering is illustrated in Fig. 5.3. Note that it is not valid to add the individual p_{rms} outputs of different filters to form L_p – only $\overline{p^2}$ contributions may be added. Standard percentage bandwidth frequency bands are listed in Table 5.1.

Neither contiguous filter nor FFT analyses produce *continuous* frequency spectra, although the displays of FFT analysers would suggest otherwise. Hence the derived spectra of aperiodic sounds are approximations to the

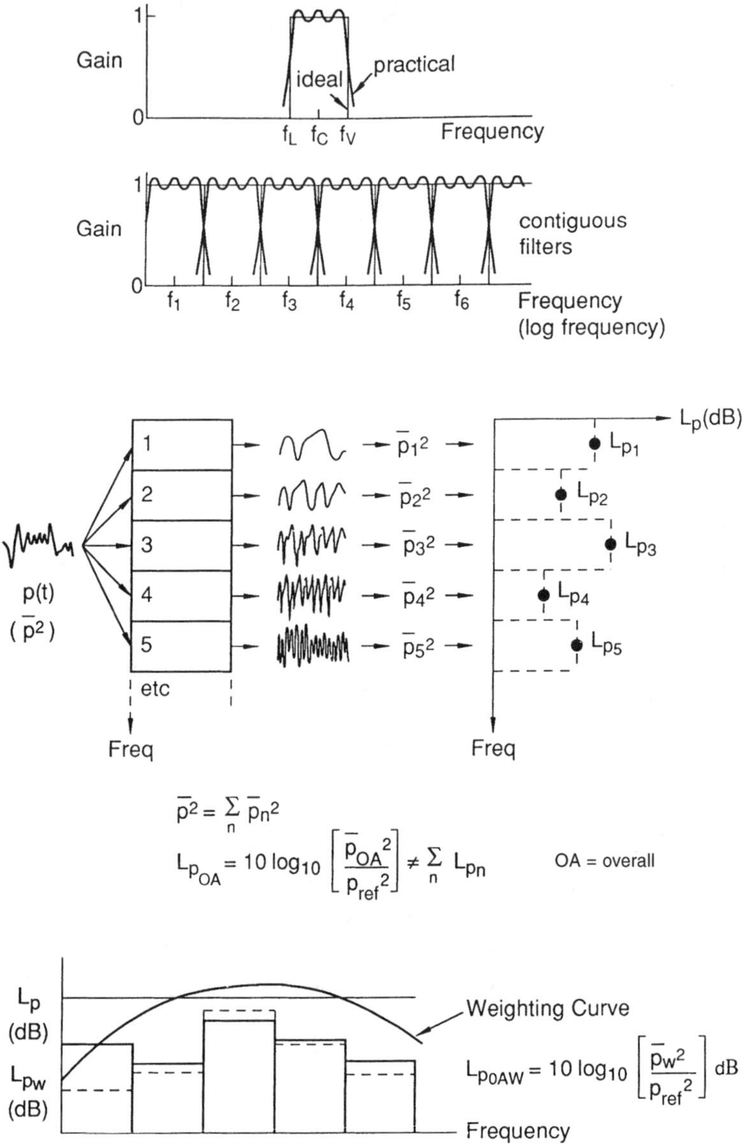

Figure 5.3 Frequency analysis by contiguous filters.

true spectra: the narrower the filter, or the finer the frequency resolution, the better the approximation (provided the analyser is given enough time to estimate each value accurately).

Table 5.1 Standard frequency bands (Hz)

Band number	*Octave band centre frequency*	*One-third octave band centre frequency*	*Band limits*	
			Lower	*Upper*
14		25	22	28
15	31.5	31.5	28	35
16		40	35	44
17		50	44	57
18	63	63	57	71
19		80	71	88
20		100	88	113
21	125	125	113	141
22		160	141	176
23		200	176	225
24	250	250	225	283
25		315	283	353
26		400	353	440
27	500	500	440	565
28		630	565	707
29		800	707	880
30	1,000	1,000	880	1,130
31		1,250	1,130	1,414
32		1,600	1,414	1,760
33	2,000	2,000	1,760	2,250
34		2,500	2,250	2,825
35		3,150	2,825	3,530
36	4,000	4,000	3,530	4,400
37		5,000	4,400	5,650
38		6,300	5,650	7,070
39	8,000	8,000	7,070	8,800
40		10,000	8,800	11,300
41		12,500	11,300	14,140
42	16,000	16,000	14,140	17,600
43		20,000	17,600	22,500

5.2.3 Weighted spectra and noise control targets

The sensitivity of the human auditory system is frequency dependent. It is therefore appropriate to 'weight' sound spectra used in estimating the subjective effect of sound to improve the correlation between objective measurements and subjective effects. The most widely used weighting

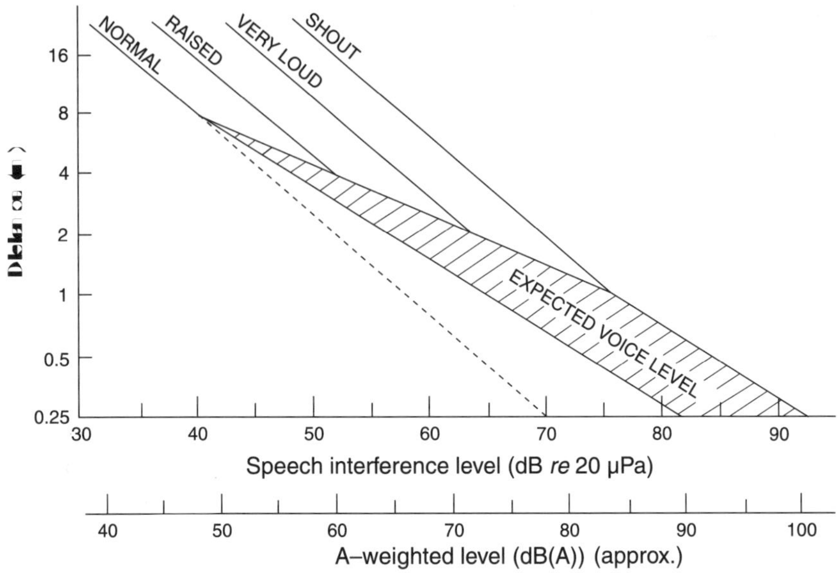

Figure 5.4 Rating noise with respect to speech interference (ANSI S3.14 – 1977).

curve is that termed the dB(A) curve, which approximates the inverse of the human equal loudness curve for sound levels typical of speech (60–70 dB). Noise control targets are often specified in terms of dB(A). Note that the performance of noise control systems, such as 'acoustic' doors and machinery enclosures, may not validly be expressed in terms of a single figure dB(A) reduction unless the noise source spectrum is also specified in detail.

Other common forms in which noise control targets may be expressed include the 'Speech Interference Level' (SIL), which is the arithmetic average of the 500, 1000 and 2000 Hz (in some standards also the 4000 Hz) octave band sound pressure levels (see Fig. 5.4), and NR, NC and RC curves (Figs. 5.5(a), 5.5(b) and 5.5(c)) (see Chapter 3 and Bies and Hansen, 1995).

5.3 SOUND SOURCES

5.3.1 What is a sound source?

This is not such a silly question as it at first seems. Take, for example, a car gear box. Vibration of the gear box casing, or its sound insulating cover, radiates sound into the car, but it is the time-varying gear meshing forces produced by the steady engine torque which generate these vibrations–which is *the* source? In this section we shall consider a source to be that

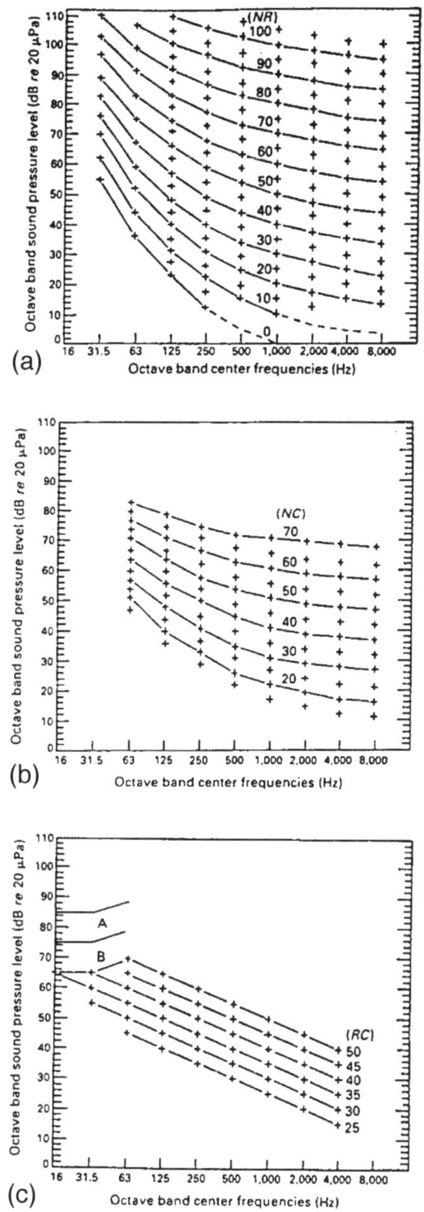

Figure 5.5 (a) Noise rating curves (NR). (b) Noise criteria curves. (NC) (c) Room criterion curves (RC).

Table 5.2 Noise generation mechanisms of a selection of machines

Classification of machine or process	*A*	*B*	*C*	*D*	*E*	*F*	*G*	*H*	*I*
Air cleaning bench for machine castings	109	√							√
Air line for cleaning down	102	√	√						√
Ball mill for powderizing materials	106				√	√	√		
Blower for heating of blanks	104	√	√						
Chain rivetting machine	111			√		√			√
Chipping hammer	118				√		√		√
Chisel, pneumatic	116				√		√		√
Concrete pipe making machine	97					√	√		
Crusher plane in quarry	110			√		√	√		
Dicing machine	100								
Diesel generator set	110	√	√			√	√		
Die casting machine	97			√		√	√	√	√
Drop hammer (peak)	111			√	√	√	√	√	√
Dust collector plane	98		√				√		
Fan (for tank cooling)	97		√				√		
Feed slide for tins	100	√			√	√			√
Fettling tool	104			√	√				√
Forge plants	107		√	√	√	√	√	√	√
Grit blasting machine	104	√				√	√		
Hammer (for plate levelling) (peak)	110			√	√	√			
Hammer (pneumatic)	122	√			√				√
Handling and transport of billets	98	√				√	√		√
Lathe (wheel)	108				√	√	√	√	
Linishing machine	104				√				
Mill-alumina	109					√	√		√
Mill-coal	107					√	√		√
Mill (flint)	101			√		√	√	√	
Mixing plant (plastics) (peak)	110						√	√	
Mound machine in foundary	101					√	√		
Nailing machines	95			√		√			√
Nut runner (air operated)	106	√	√			√			√
Peening gun	97	√		√	√	√	√		√
Perforating machine	101					√	√	√	
Plasma arc profiling machine	110	√				√	√		
Power hammer	108			√	√	√	√		√

Table 5.2 (*Continued*)

Classification of machine or process	*A*	*B*	*C*	*D*	*E*	*F*	*G*	*H*	*I*
Press-cropping	100				√	√	√	√	√
Press-laminations	98				√	√	√	√	√
Saw bench – plasterboard	106			√			√		
Saw – friction	118			√		√	√	√	
Saw – for batches of tubes	105			√	√	√	√	√	
Saw – slotting	112			√		√	√	√	
Wet riddle	106			√	√	√	√		
Woodworking circular saw	104		√	√	√	√	√	√	
Woodworking double end tenoner	100			√	√	√			
Woodworking planer	103			√	√	√	√	√	

Key
A – Noise level dB (A) (1 m)
B – Jet emission
C – Fan
D – Hammer deceleration
E – Workpiece distortion or vibration
F – Anvil or case ringing
G – Supporting structure vibration
H – Air ejection
I – Blow-off valves

component or mechanism which generates sound by acting directly on the surrounding air (or other fluid medium). Sound sources are extremely diverse in their physical forms, mechanisms and characteristics. Any single physical noise-generating system, such as a vacuum cleaner, may produce noise by means of a number of distinct mechanisms (see, for example, Table 5.2 (White and Walker, 1984)). Hence, in technical terms, it comprises a number of separate sources. In spite of this diversity, each source may be placed in one or more of three basic categories, as explained in section 5.3.3.

5.3.2 Control of sound 'at source'

From an engineering point of view, noise control is best implemented at the system design stage, with the objective of making a product inherently quiet. The European Directives on machinery noise exhort the producer to control noise 'at source' (White and Walker, 1984; EC Directives, 1986). A designer is likely to be successful in this respect only if he or she has a thorough understanding of the mechanics of sound and vibration generation; this is not easily achieved!

5.3.3 Categorization of sources of sound

In mathematical terms, physical sources may be modelled as idealized elements, or arrays thereof; these are termed 'monopoles', 'dipoles', and

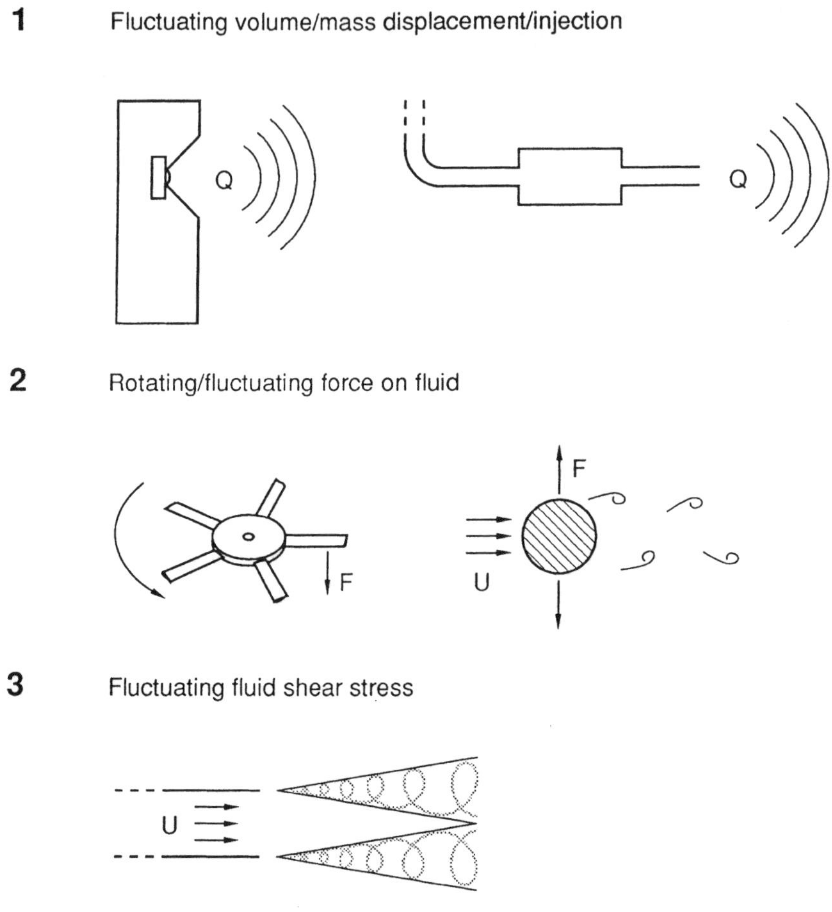

Figure 5.6 Examples of noise source categories.

'quadrupoles' (higher order sources are of no practical importance) (see Chapter 1). In physical terms, sources may be broadly categorized as follows (see Fig. 5.6).

Category 1: Sources which produce fluctuating net displacements of fluid volume (or mass) by ejecting fluid from their surfaces, displacing fluid by vibration of their surfaces, or via unsteady combustion processes. Direct radiator loudspeakers and compressed air sirens are common examples.

Category 2: Sources which apply fluctuating or moving forces to the local fluid. Rotating fan blades and singing telephone wires are common examples.

Category 3: Sources which involve no solid surfaces and in which fluctuating fluid stresses are associated with the unsteady convection of fluid momentum. The fluid jet is a common example.

Contrary to the opening statements of many elementary textbooks on acoustics, noise generation does not necessarily involve vibration of solid surfaces. (Try blowing closely on the sharp edge of a table top, whipping a rigid stick through the air, or expelling air rapidly from a bicycle pump). Vibrating solid surfaces do, however, constitute the majority of noise sources. They are difficult to categorize because their capacity for generating noise depends critically upon the spatial distributions of amplitude and phase of surface acceleration over their surfaces, which vary greatly from structure to structure.

Source categorization is important because the efficiency with which the three categories convert mechanical, electrical or chemical energy into sound decreases with increase in category number, and the variation of sound power with frequency, or speed, also varies from one to another. Since the *rate of change of excitation force* and *rate of change of volume displacement* are generally the most influential factors in determining sound power, the noise control engineer must be in a position to estimate the quantitative effect of reducing these rates on sound power. Horn loudspeakers are designed to generate sound and, with an electro-acoustical efficiency of about 40% , are among the most efficient sound generators. Most machines have far lower mechano-acoustical efficiencies.

Because most machines incorporate a number of sources falling into different categories, it is not generally possible to place them as complete units into any one category.

5.3.4 Characterization of sources of sound

In principle, sources of sound may be characterized in the following terms:

1. mechano-(or electro- or chemico-) acoustical efficiency;
2. variation of sound power (or pressure) with operational parameters, principally speed and load;
3. variation of sound power (or pressure) with time;
4. frequency spectrum of sound power (or pressure);
5. far field directivity;
6. strength and extent of the hydrodynamic near field.

5.3.4.1 Mechano-acoustical efficiency

Only mechanical and fluid dynamic sources are considered since they are the most common. The mechano-acoustical efficiency η_{ma} is defined as the ratio of radiated sound power to the mechanical power generated or absorbed by the machine. The mechano-acoustical efficiency of machines which radiate via surface vibration varies greatly with the mechanical design, the damping of the structure and the form and material of the surface structure. However

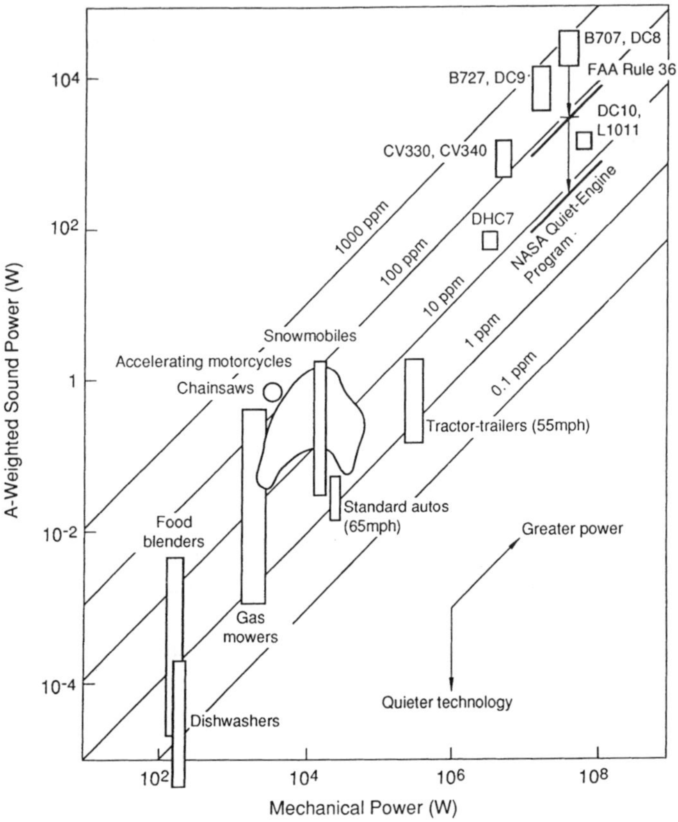

Figure 5.7 A-weighted sound power levels versus mechanical power of various sources (1997).

a measure of η_{ma} for a vibrating surface structure itself is quantifiable, as shown later. Figure 5.7 presents some examples of the mechano-acoustical efficiencies of a number of common machines, e.g. a standard car at $10^{-4}\%$. It is the very inefficiency of most sources that makes it extremely difficult to reduce noise without making significant design or operational changes.

The high efficiency of Category 1 sources is exemplified by unsilenced internal combustion (IC) engines, vane compressors, sirens and pulse jets. The outlets from IC engine exhaust pipes fall into the same category, but, fortunately, are relatively easily silenced. The efficiencies of fluid dynamic sources which act directly on the surrounding fluid are more easily related to the system parameters. Category 1 sources have an efficiency proportional to $U/c = M$, where U is the flow speed and M is the flow Mach number. The efficiency of Category 2 sources, such as unsteady flow impinging on ventilation system diffusers and airframe noise, varies as M^3; and that of

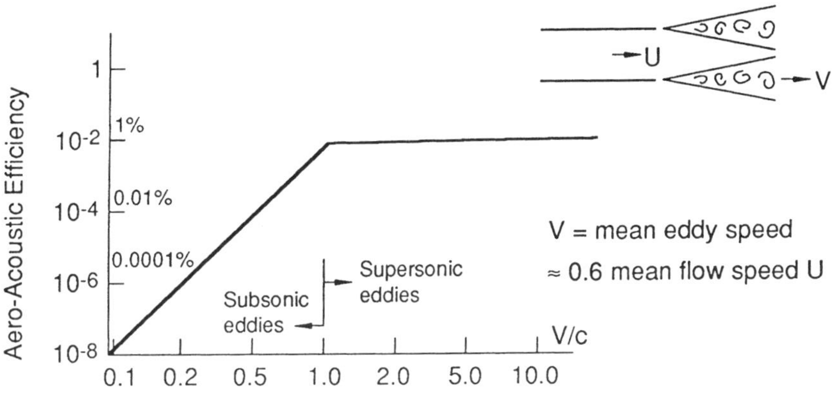

Figure 5.8 Aero-acoustic efficiency of fluid jets.

Category 3 sources varies as M^5, provided $M < 1$ (see Fig. 5.8). The extreme dependence of the latter on M underlies the success achieved by designers in reducing the exhaust jet noise of civil aircraft engines by about 25 dB over the past 25 years by introducing high by-pass ratio engines in which the exhaust Mach numbers are now about half those of Concorde. The vital importance of minimizing flow speeds in noise-sensitive environments is clear.

The audio-frequency vibration fields of most mechanical systems exhibit very complex spatial distributions of frequency-dependent amplitude and phase. It is therefore more practicable to relate sound radiation to a global measure of mean square surface vibration velocity. This is a useful measure except where highly damped, or structurally irregular, structures are excited by localized forces or displacements. We define the space-averaged mean square vibration velocity as

$$\langle\overline{v^2}\rangle = (1/S)\int_S \left[(1/T)\int_t^{t+T} v^2(\boldsymbol{x}, t)\,\mathrm{d}t\right]\mathrm{d}\boldsymbol{x} \tag{5.3}$$

in which $\boldsymbol{x}$ indicates position on a surface of area S . This may be approximately determined experimentally by measurements of the mean square velocity at a number of discrete points on a surface:

$$\langle\overline{v^2}\rangle \approx (1/N)\sum_{i=1}^{N}\overline{v_i^2}. \tag{5.4}$$

The radiation ratio (or inappropriately termed 'efficiency') is defined as the radiated sound power, normalized on radiating surface area, $<\overline{v^2}>$ and the characteristic specific acoustic impedance of the fluid thus:

$$\sigma = W_{\text{rad}}/\langle\overline{v^2}\rangle S\rho_0 c_0. \tag{5.5}$$

The time-averaged power dissipated by a modally resonant structure excited by band-limited noise is given by

$$W_{\text{diss}} = \eta_{\text{d}}\, \omega_{\text{c}}\, E = \eta_{\text{d}} \omega_{\text{c}} M_{\text{eff}} \langle \overline{v^2} \rangle \tag{5.6}$$

where η_{d} is the modal-average dissipation (or internal) loss factor, E is the time-averaged stored vibrational energy ω_{c} is the band centre frequency and M_{eff} is the effective vibrating mass. For a uniform plate, M_{eff} equals the actual mass; in this case,

$$\eta_{\text{ma}} = \frac{W_{\text{rad}}}{W_{\text{rad}} + W_{\text{diss}}} = \left[1 + \frac{\eta_{\text{d}}}{\sigma} \cdot \frac{\omega m}{\rho_0 c_0}\right]^{-1} \tag{5.7}$$

where $m = M/S$. Some typical radiation efficiency curves are shown in Fig. 5.9. Stiff, lightweight structures have radiation ratios of about unity over a wide frequency range, and are therefore good radiators and transmitters of sound, which is clearly undesirable.

5.3.4.2 Sound power vs. speed

The mechanical power transported per unit area of a fluid flow is proportional to U^3. Combining this dependence with the efficiency dependencies described above, indicates that the sound powers of Category 1, 2

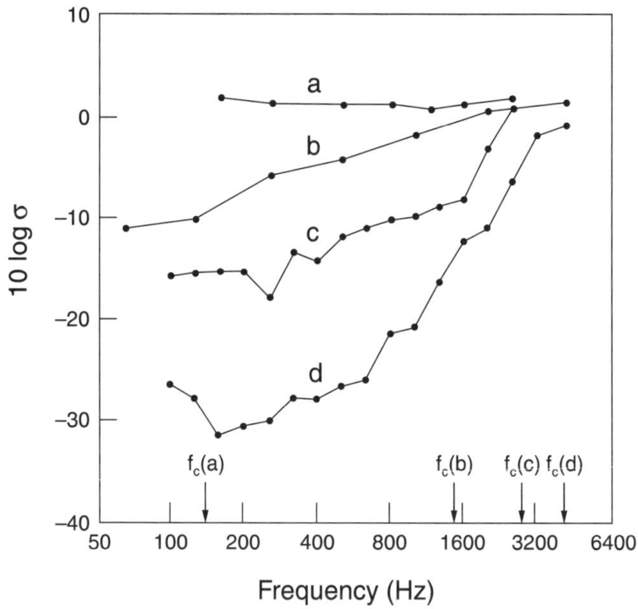

Figure 5.9 Examples of radiation index in air: (a) 14 cm thick concrete; (b) diesel engine block; (c) 13 mm thick plasterboard with lath gridwork; (d) steel pipe of 72 cm diameter and 1.3 mm wall thickness. f_{c} = critical frequency. (Cramer *et al.*, 1988).

and 3 fluid dynamic sources vary as U^4, U^6 and U^8, respectively. Interestingly, the sound powers of the least efficient sources increase most rapidly with speed. Because most mechanical systems generate noise by various different physical mechanisms, it is not always possible to attribute a unique index to the speed dependence of sound power. However, combustion noise from IC engines, which is caused by a repetitive pressure cycle in the combustion chamber varies systematically with rotational speed, as shown by Fig. 5.10; the index is between 2.5 and 7 for normally aspirated engines and between 1.6 and 2.8 for turbocharged engines. At low speeds, the index for diesel engines is lower than for petrol engines because of the rapid pressure rise which produces the characteristic diesel 'knock'. Empirical data (Bies and Hansen, 1995) suggest the following rotational speed (N) dependencies: propeller fans, N^5; electric motors, $N^{1.5}$; electrical power generators, $N^{6.6}$; and spur gears, N^3.

A generalized graphical procedure for estimating the effect of operational speed on noise radiated by vibrating structures excited by periodic forces is presented in Fig. 5.11. The cycle rate determines the fundamental frequency; the shape of each impulse determines the envelope of the spectrum and the frequency range of significant energy. The speed dependence is strongly influenced by the frequency dependence of the envelope of the force spectrum in the region where it falls most rapidly; the associated slope of the A-weighting curve also influences the 'subjective' speed-dependence. The influence of spectral slope is clearly illustrated in Fig. 5.12 which shows the speed dependence of the noise from two sources of a fuel injection cam box.

In general, speed is the enemy of the noise control engineer, but it is the friend of the production engineer – a compromise has to be found.

5.3.4.3 Sound power vs. mechanical power and load

High-power machines are not necessarily noisier than low-power machines and may be quieter. A given operational power may be achieved by a combination of high speed and low torque or low speed and high torque. The latter usually requires bulkier, heavier structures, which, if taken together with low speed, produce less noise, but is less favoured in terms of cost, efficiency and power–weight ratio. Increase of load demands increased reaction forces which, if speed is unchanged, will increase sound power approximately in proportion to their squares.

5.3.4.4 Sound power vs. time

Although frequency (Fourier) analysis is very useful and important, it may obscure the nature of noise-generating mechanisms because a characteristic time-history is smeared into hundreds of featureless, eternal sinusoids. In performing diagnostic tests, noise control engineers should always examine the time histories of radiated sound pressure or intensity for links with the

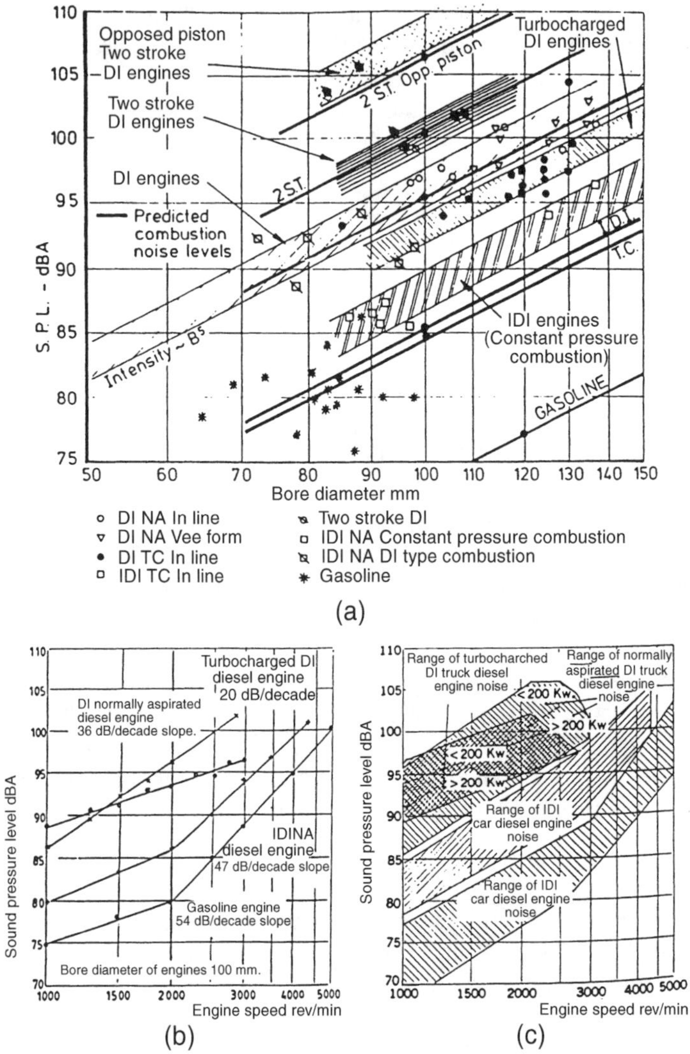

Figure 5.10 Comparison of the noise of engines with different combustion systems as a function of (a) bore diameter, (b) engine speed, and (c) engine types.

event history of the mechanical processes occurring in a source, especially when modern frequency analysers are in use. An example of a pressure–time history is shown in Fig. 5.13 and of a time-gated intensity distribution in Fig. 5.14. It is also good diagnostic practice to slow down a recording of the noise of a source so that a listener may more readily identify individual noise-generating events.

5.3.4.5 Frequency spectrum

Periodic events produce line spectra having non-zero components only at frequencies which are multiples of the inverse of the period. *Aperiodic* events produce continuous spectra. Signal processing and display procedures obscure this distinction – user beware! Clearly, a correctly resolved

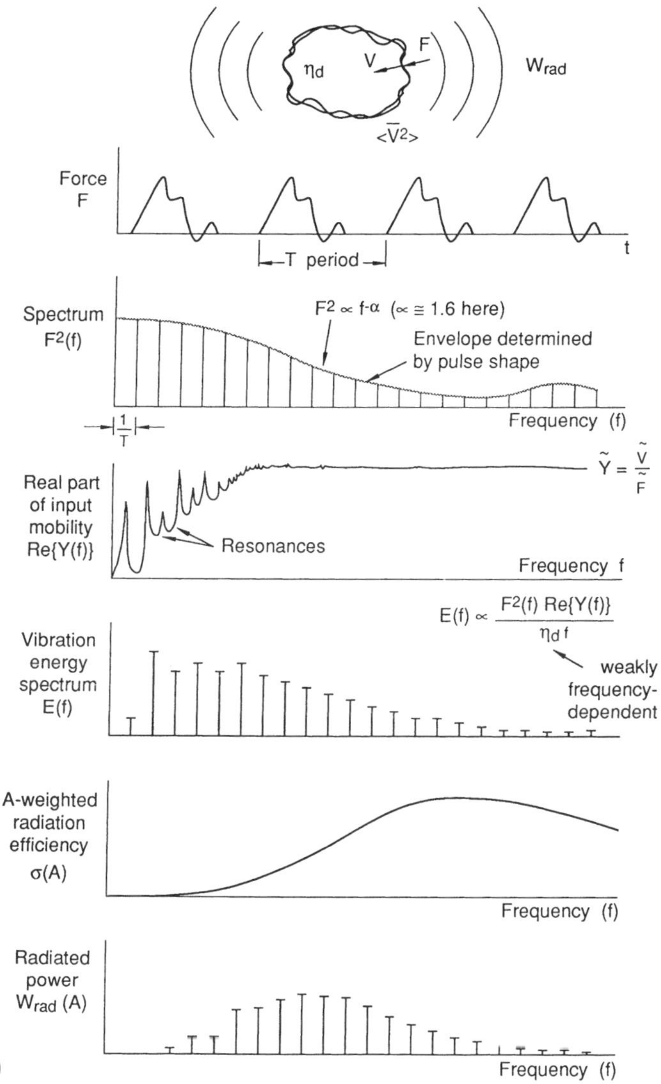

(a)

Figure 5.11 (a) Sound radiation from vibrating structures generated by periodic forces.

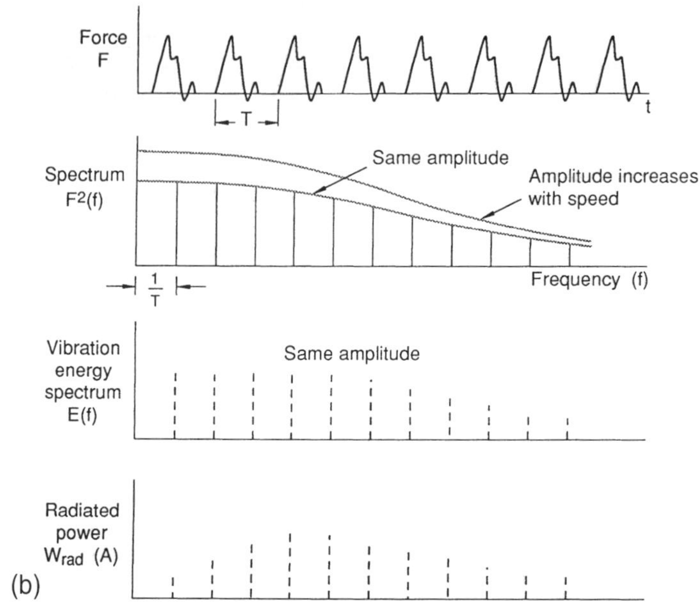

Figure 5.11 (b) Effect of doubling of operational speed on radiated power spectrum.

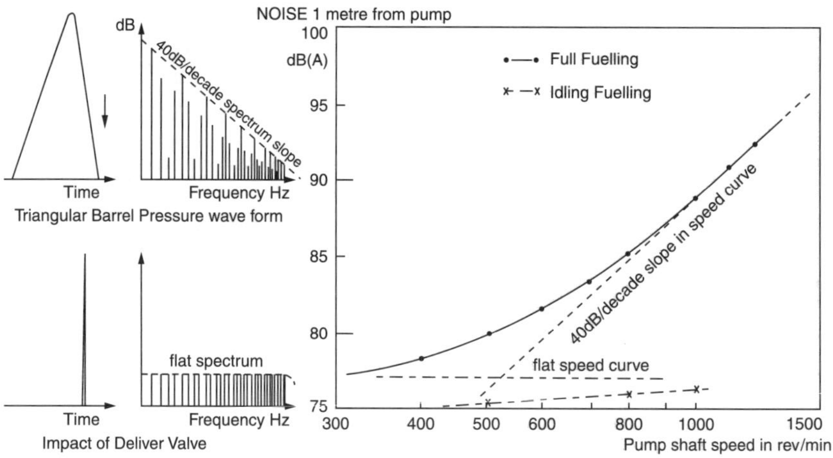

Figure 5.12 Forces on a fuel injection pump cambox and noise vs. speed curves.

frequency spectrum of sound pressure or power will reveal all the cycles inherent in the noise-generating mechanism, as illustrated by the ball bearing spectrum in Fig. 5.15 in which all the effects of imperfections are

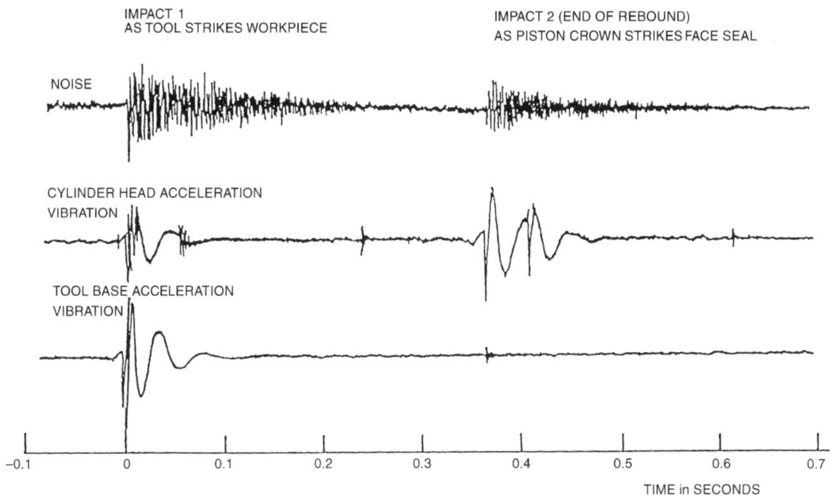

Figure 5.13 Pressure and acceleration time histories of a 'Petro forge'.

evident. Such spectra may be used for the purpose of machine condition monitoring.

Some sources generate both periodic and aperiodic noise, as exemplified by the fan noise spectrum shown in Fig. 5.16. The shape of the envelope of the line spectrum is determined by the time history of the pulse applied by each blade to the air, and the shape of the continuous spectrum, which is associated with turbulent pressure fluctuations on the blades, is determined by the flow speed and the geometric scale of the flow regime. A pure subsonic jet produces a characteristic broad-band spectrum which peaks at a frequency given by $f = 0.2\ U/d$, where U is the mean jet convection speed and d is the jet diameter at exit. The flow of a fluid over a circular cylindrical body placed with its axis normal to the flow direction generates a periodic sound of frequency given by the same formula with d equal to the cylinder diameter.

The shape of the frequency spectrum of a noise source is subjectively important because it determines the difference between the unweighted value of L_p and the value in dB(A). It is also important to the noise control engineer because it gives clues to the noise generating mechanisms and, as shown above, it determines the likely effectiveness of speed change. The selection of an appropriate form of noise control system is also intimately related to the noise spectrum.

The noise caused by impacts between solid bodies can be dramatically reduced by interposing a resilient element. The effect may be understood in terms of the change of spectrum (Fig. 5.17). By spreading out the interaction force over time, the peak force, and therefore the peak sound level, is reduced, and the spectrum falls to low levels at a frequency equal to the

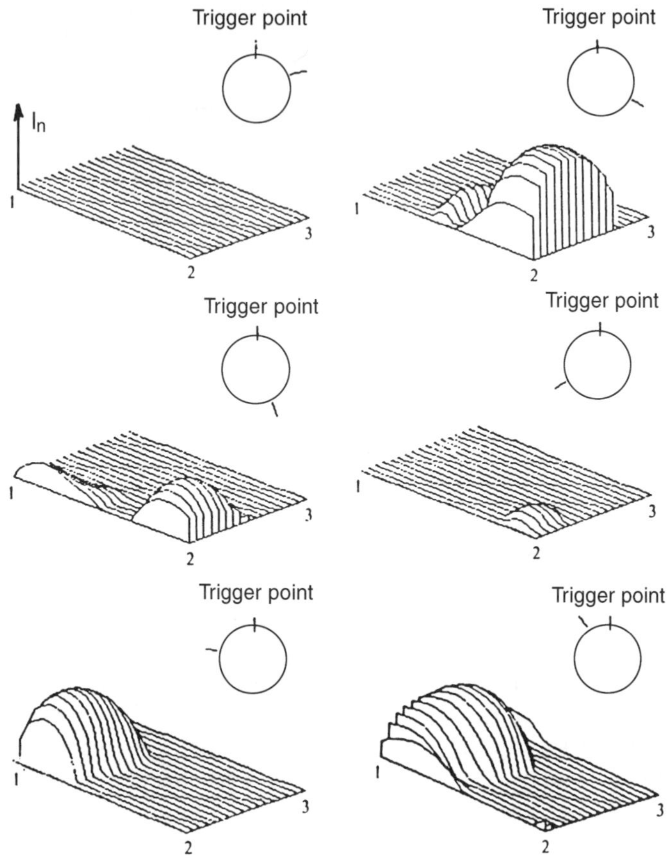

Figure 5.14 Gated intensity maps on a medium diesel engine at various points in a cycle. (Fahy, 1995).

inverse of the pulse length, thereby reducing the overall dB(A) level. Try dropping a coin (edge on) onto a bare table, and then repeat with a soft material covering – the change of momentum (impulse) is the same, but the noise is greatly reduced.

5.3.4.6 Far field directivity

Far field directivity is important in the outdoor environment because the environmental impact of a source on local residents depends upon the location of the residences relative to the source. Of particular concern are industrial exhaust stacks. Although the actual levels produced depend upon the form of source and local topography and meteorological conditions, a general method for predicting directivity of circular section stacks exists

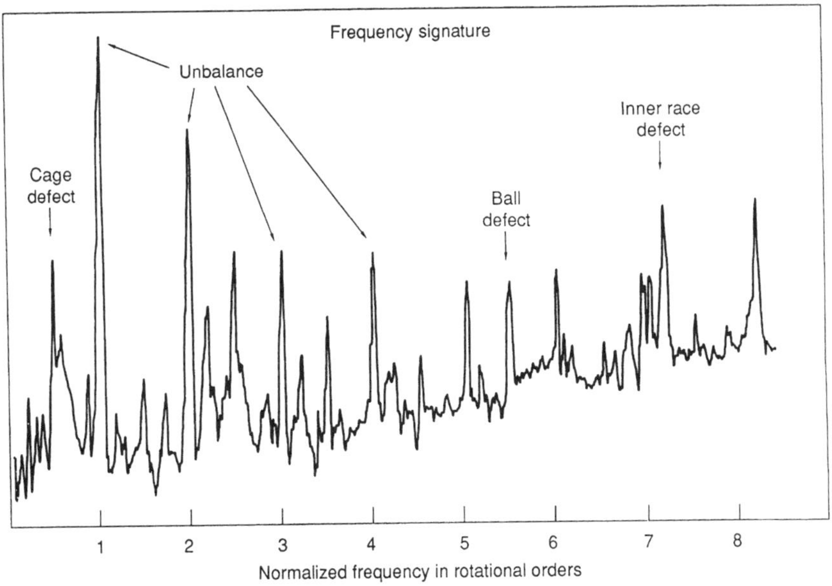

Figure 5.15 Mechanical 'signature' of a ball bearing.

(Joseph and Hewlett, 1996). Typical curves are shown in Fig. 5.18. Aircraft propeller and jet noise sources are strongly directional, as are road vehicle tyres. Noise control engineers can sometimes take advantage of directivity by optimally locating and orienting sources.

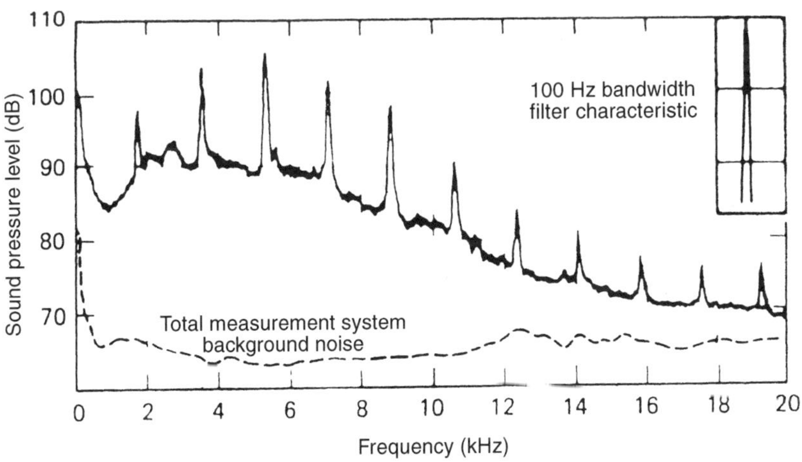

Figure 5.16 Frequency spectrum of an axial fan obtained with a 100 Hz filter.

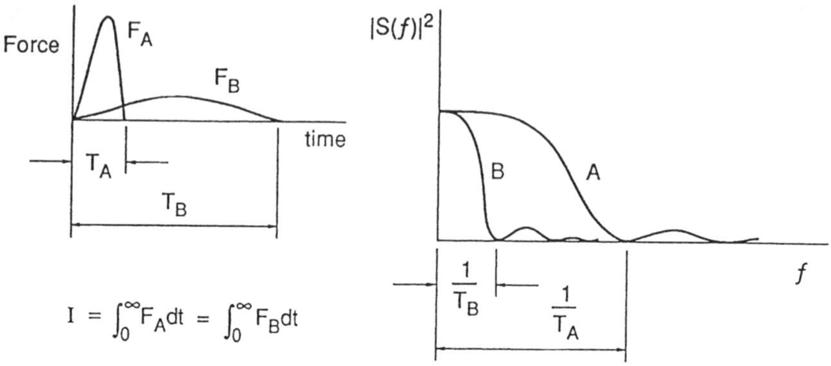

Figure 5.17 Impulsive forces and squared amplitude spectra.

5.3.4.7 Near field

Sound sources produce two near-field regions: the 'geometric' near field and the 'hydrodynamic' near field; the former is not a very useful indicator of source category. The strength and extent of the latter increases as source efficiency decreases. In the hydrodynamic near field, the particle velocities (and hence fluid momentum fluctuations) are much greater in proportion to sound pressure than in the far field. Also, energy flux (intensity) circulation is usually exhibited by the hydrodynamic near field.

5.4 NOISE SOURCE QUANTIFICATION

5.4.1 Objectives

The principal objectives of source output quantification are as follows:

1. for comparison of sound powers of machines and plant for the purpose of user selection;
2. for source labelling;
3. for predicting the sound pressure field and the associated adverse effects, e.g. hearing hazard, environmental impact;
4. to check conformance with regulatory or legal requirements;
5. to identify source mechanisms (diagnostics);
6. to understand the physics of the source in order to develop mathematical models for the purpose of making parametric studies.

5.4.2 Quantification methods

Sound power quantification methods fall into two categories: direct and indirect. In the former, the sound power is inferred directly from measure-

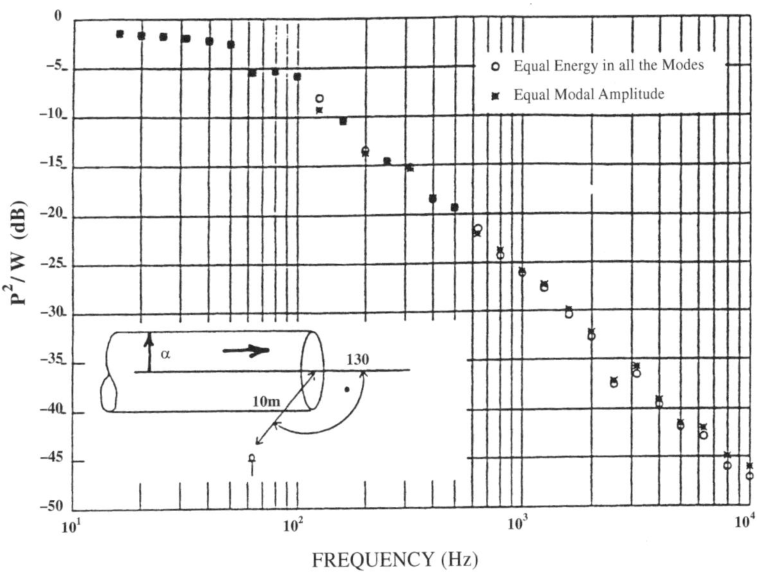

Predicted ratio of for field sound pressure level and power with a = 1m, R = 10m, θ = 130° using two different source models

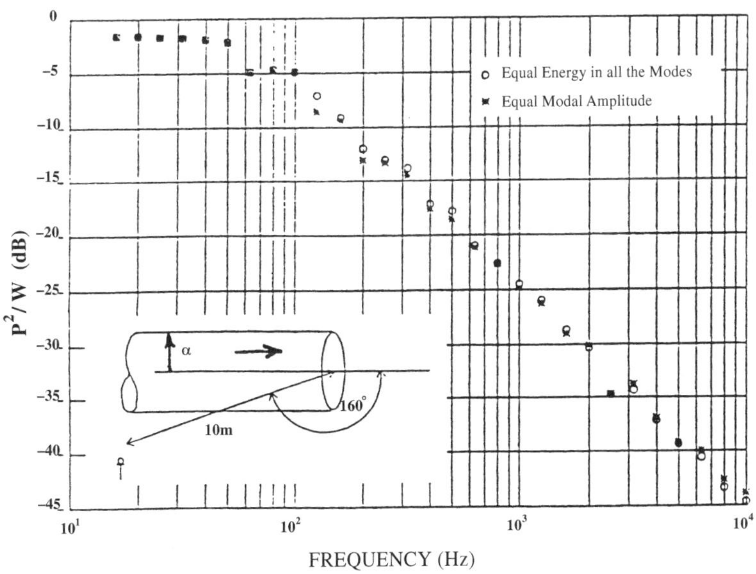

Predicted ratio of for field sound pressure level and power with a = 1m, R = 10m, θ = 160° using two different source models

Figure 5.18 Directivity of circular exhaust pipes.

ments of the radiated sound field in conjunction with an assumed field model; in the latter, it is determined either by comparison with a calibrated source (substitution method) or from measurements of the vibration velocity of a radiating surface. Sections 5.4.2.1 and 5.4.2.2 below describe direct methods and sections 5.4.2.3 and 5.4.2.4 describe indirect methods.

In order for manufacturers and users to be confident in test results, and for the purpose of satisfying legal requirements, internationally agreed standardized test methods have been developed by the International Organization for Standardization (ISO) (see Tables 5.3 and 5.4). In Europe, the CEN standards closely follow most ISO standards; in North America the ANSI and ASTM standards are somewhat different.

5.4.2.1 Pressure-based methods

Since the sound pressure level generated by a source varies with distance, direction and environmental conditions, and the presence of other (extraneous) sources adds to the sound produced by a source under test to an unpredictable degree, these methods require the isolation of the source in an acoustically controlled environment.

Anechoic, semi-anechoic test environment (ISO 3745 (3744))

A measurement surface is described around the source and divided into n segments. The principle is illustrated in Fig. 5.19(a). It is implicitly assumed that $I = \overline{p^2}/\rho_0 c_0$ or, equivalently, that $L_\mathrm{p} = L_\mathrm{I}$, and that the intensity vector lies normal to the measurement surface.

The sound pressure is sampled at one point in each segment. If we assume that

$$I_{ni} = \overline{p_i^2}/\rho_0 c_0, \tag{5.8}$$

then

$$W \approx \sum_{i=1}^{n} \left(\overline{p_i^2}/\rho_0 c_0\right) S_i.$$

Reverberant environment (ISO 3741 (3742-1/2))

The source sound power is equal to the estimated rate of energy dissipation by the walls of a reverberant room (Fig. 5.19(b)):

$$W = \langle \overline{p^2} \rangle a / 4\rho_0 c_0 \tag{5.9}$$

in which $\langle \overline{p^2} \rangle$ is the space-averaged mean square sound pressure in the room, determined either from an array of fixed microphones distributed over the room volume, or from a mechanized continuous traverse of the volume; a is the random incidence (or Sabine) absorption of the room estimated by measuring the rate of decay of $\overline{p^2}$ following impulsive excitation, or the cessation of continuous excitation, of the room (see Chapter 1). The absorption is given by

Table 5.3 Factors influencing the choice of the method

X Applicable		*ISO 3741*	*ISO 3743-1*	*ISO 3743-2*	*ISO 3744*	*ISO 3745*	*ISO 3746*	*ISO 3747*	*ISO 9614-1*	*ISO 9614-2*
Grade of accuracy	1 (Precision)	x				x			x	
	2 (Engineering)		x	x	x			x	x	x
	3 (Survey)						x		x	x
Size of source	Large source and/or not movable				x		x	x	x	x
	Small sources – movable	x	x	x	x	x	x		x	x
Test environment	Laboratory reverberation room	x								
	Special reverberation test room			x						
	Laboratory anechoic room					x (a)				
	Hemi-anechoic room				x (b)	x (a)				
	Hard-walled test room		x							
	Large room, outdoors				x (b)				x	x
	In situ, indoors						x (c)	x	x	x
	In situ, outdoors						x (c)		x	x
Background noise level	$\Delta L \geq 10$ dB	x	x	x	x	x	x	x	x	x
	$\Delta L \geq 6$ dB		x	x	x		x	x	x	x
	$\Delta L \geq 3$ dB						x		x	x
	$\Delta L < 3$ dB								x (d)	x (d)
Character of noise	Types as defined in ISO 12001				x	x	x			
	All types, except isolated bursts	x	x	x				x		
	Stationary in time								x	x
Instrumentation	Sound level meter:									
	class 1	x		x	x	x		x		
	class 2						x			

Table 5.3 (*Continued*)

X Applicable		*ISO 3741*	*ISO 3743-1*	*ISO 3743-2*	*ISO 3744*	*ISO 3745*	*ISO 3746*	*ISO 3747*	*ISO 9614-1*	*ISO 9614-2*
	Integrating sound level meter:									
	class 1	x	x	x	x	x		x		
	class 2						x			
	Frequency band filter class 1	x	x	x	x	x		x		
	Sound intensity instrument								x	x
Sound power levels obtainable	1/3-Octave-band levels	x			x	x			x	x
	Octave-band levels	x	x	x	x	x		x	x	x
	A-weighted levels	x	x	x	x	x	x	x	x (e)	x (e)
Optional information available	Other weightings	x	x	x	x	x	x	x		
	Directivity information				x	x				
	Temporal pattern				x	x	x			

(a) Environmental correction $K_2 \leq 0.5$ dB
(b) Environmental correction $K_2 \leq 2$ dB
(c) Environmental correction $K_2 \leq 7$ dB
(d) Lower limit is approximately −10 dB, but depends on measurement conditions
(e) A-weighted (63 Hz – 4000 Hz or 50 Hz – 6300 Hz)

Table 5.4 Factors influencing the choice of method

X Applicable		*ISO 3741*	*ISO 3743-1*	*ISO 3743-2*	*ISO 3744*	*ISO 3745*	*ISO 3746*	*ISO 3747*	*ISO 9614-1*	*ISO 9614-2*
Grade of accuracy	Precision					x			x	
	Engineering	x	x	x	x			x	x	x
	Survey						x		x	x
Size of source	Large sources and/or not moveable				x		x	x	x	x
	Small sources – moveable	x	x	x	x	x	x		x	x
Test environment	Laboratory reverberation room	x								
	Special reverberation test room			x						
	Laboratory anechoic room					x (a)				
	Semi-anechoic rooms				x (b)					
	Hard-walled test room		x							
	Large rooms, outdoors				x (b)				x	x
	In situ, Indoors						x (c)	x	x	x
	In situ, outdoors						x (c)		x	x
Background noise level	$L \geq 10\,\text{dB}$	see standard	x	x	x	x	x		x	x
	$L \geq 6\,\text{dB}$		x	x	x		x	x	x	x
	$L \geq 3\,\text{dB}$						x		x	x
	$L < 3\,\text{dB}$								x	x
Character of noise	All types as defined by ISO 2204				x	x	x			
	All types, except isolated bursts	x	x	x				x		
	Stationary in time								x	x

Table 5.4 (*Contd.*)

X Appplicable		*ISO 3741*	*ISO 3743-1*	*ISO 3743-2*	*ISO 3744*	*ISO 3745*	*ISO 3746*	*ISO 3747*	*ISO 9614-1*	*ISO 9614-2*
Instrumentation	Sound level meter:									
	class 1	x		x	x	x		x		
	class 2						x			
	Integrating sound level meter									
	class 1	x	x	x	x	x	x	x		
	class 2									
	Frequency band filter class 1	x	x	x	x	x		x		
	Sound intensity instrument								x	x
Sound power levels obtainable	1/3 Octave band levels	x			x	x			x	x
	Octave band levels	x	x	x	x	x		x	x	x
	A-weighted levels	x	x	x	x	x	x	x	x (d)	x (e)
Optional information available	Other weightings	x	x	x	x	x	x	x		
	Directivity information				x	x				
	Temporal pattern				x	x	x			

x = Applicable
(a) Environmental correction $K_2 \leq 0.5\,\text{dB}$
(b) Environmental correction $K_2 \leq 2\,\text{dB}$
(c) Environmental correction $K_2 \leq 7\,\text{dB}$
(d) A-weighted (50 Hz–10000 Hz)
(e) A-weighted (63 Hz–4000 Hz or 50 Hz–6300 Hz)

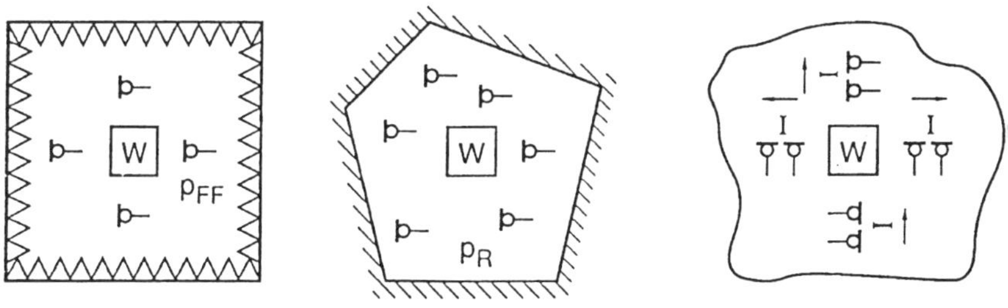

(a) Anechoic Chamber (b) Reverberation Chamber (c) Intensity:Arbitrary Environment

Figure 5.19 Schematic representation of test conditions for sound power determination.

$$a = 0.16\, V/T \tag{5.10}$$

where V is the room volume and T is the time for the sound energy to decay by a factor of 10^6 (60 dB).

The other 374-standards listed in Table 5.3 are variants on these methods with empirically derived factors correct for non-ideal conditions.

5.4.2.2 Intensity-based methods [ISO 9614-1 and 9614-2]

These methods employ the direct measurement of the component of the sound intensity vector normal to a measurement surface described around a source (Fig. 5.19(c)). An instrument complying with IEC1043 must be used (IEC 1043 1993). The Gauss integral theorem is exploited to suppress the influence of extraneous sources outside the measurement surface. Note that the suppression applies only to the sound power estimate, not to measurements of intensity at individual points – an intensity meter cannot distinguish between sound radiated by a source under test and by extraneous sources (Fahy, 1995). The measurement surface is divided up into segments and the normal intensity component is measured either at a fixed point at the centre of each segment (ISO 9614-1) or by scanning the intensity probe over each segment (ISO 9614-2). Hence

$$W \approx \sum_i I_{ni} S_i \text{ or } \sum_i \langle I_{ni} \rangle_i S_i. \tag{5.11}$$

5.4.2.3 Source substitution method [ISO 3747]

A special-purpose reference source, calibrated in terms of radiated sound power spectrum, is either placed on the inert source, or replaces a removable source, in a reasonably reverberant environment containing no extraneous sources. The space-averaged mean square sound pressure is measured with only the source under test in operation and then with only the reference source in operation. The expression for the derived sound power level is

$$L_{\mathrm{W}} = L_{\mathrm{W}'} + \langle L_{\mathrm{p}} \rangle - \langle L_{\mathrm{p}'} \rangle \qquad (5.12)$$

in which the prime indicates operation of the reference source alone.

5.4.2.4 Source surface vibration measurement

Equation (5.5) gives the radiated sound power as

$$W_{\mathrm{rad}} = \sigma \langle \overline{v^2} \rangle S \rho_0 c_0. \qquad (5.13)$$

The space-averaged mean square surface vibration velocity is estimated from measurements with an accelerometer (or laser or volume velocity transducer) and the relevant radiation efficiency is obtained from published sources (e.g. Beranek and Vér, 1993). This method of estimating sound power is immune to interference by extraneous sources.

5.4.3 Other measurements

5.4.3.1 Emission sound pressure levels

Sound pressure levels are measured at workstations (operator positions) in terms of L_{peak}, L_{eq}, etc. These values are specific to the measurement position and the acoustic environment. Recent research suggests that intensity measurement may be used instead of pressure measurement to suppress the effect of the reverberant, diffuse field of other extraneous sources present (Jonasson, 1996).

5.4.3.2 Source location techniques

There are many source location techniques, including sound intensity vector mapping, near-field acoustical holography, vibro-acoustic reciprocity and inverse methods of various types. They are too specialized for inclusion in this introductory chapter.

5.5 PRINCIPLES OF PASSIVE NOISE CONTROL

5.5.1 Summary of basic principles and implementation

5.5.1.1 Control at source

This may be effected by suppressing or eliminating the noise generating mechanisms or by reducing their effectiveness by decoupling them from their contiguous media so they generate less acoustical/vibrational power.

5.5.1.2 Control of the airborne sound transmission path(s) between source and receiver

This may be effected by means of introducing devices into the transmission path which either reflect sound waves or absorb (dissipate) their energy, or perform both functions.

5.5.1.3 Control of the structure-borne sound transmission path(s) between source and receiver

This may be effected by decoupling (vibration isolation), inserting wave-blocking devices or increasing structural damping. Modification of the impedance presented to mechanical excitation forces, or displacements, is included under heading 5.5.1.1. Reduction of radiation ratio may be included under this heading.

5.5.2 Control at source

Once sound energy is generated, it spreads by many solid and fluid paths, each of which requires the attention of the noise control engineer. It is obviously more cost-effective to suppress the noise generation at source. However, this presents the greatest technical challenge of the three approaches described above; the physics of the complete noise-generating system(s) must be understood and modifications must not adversely affect the production costs or time, the operational cost-effectiveness or the efficiency of the system. Every effort must be made to reduce the rates of change of fluid and mechanical forces involved in the source mechanism; impacts are particularly noisy. There is such a diversity of noise sources that a comprehensive review of source control procedures cannot be accommodated in this chapter. A few generic examples are illustrated in Figs 5.20(a)–(f).

5.5.3 Control of the airborne transmission path

There are two physically distinct processes employed: reflection and absorption. The latter is the more desirable because sound energy is dissipated into heat, and causes no more trouble. A few generic examples are presented in Figs 5.21(a)–(g). Both processes are exploited in combination in many noise control systems. Some generic examples are illustrated in Figs 5.22(a)–(d). Noise control enclosures, close covers and wrappings 'bottle up' the sound energy and dissipate it.

5.5.4 Control of the structure-borne sound transmission path

The various means of controlling the transmission of structure-borne sound are illustrated by generic examples illustrated in Figs 5.23(a)–(d). It should be

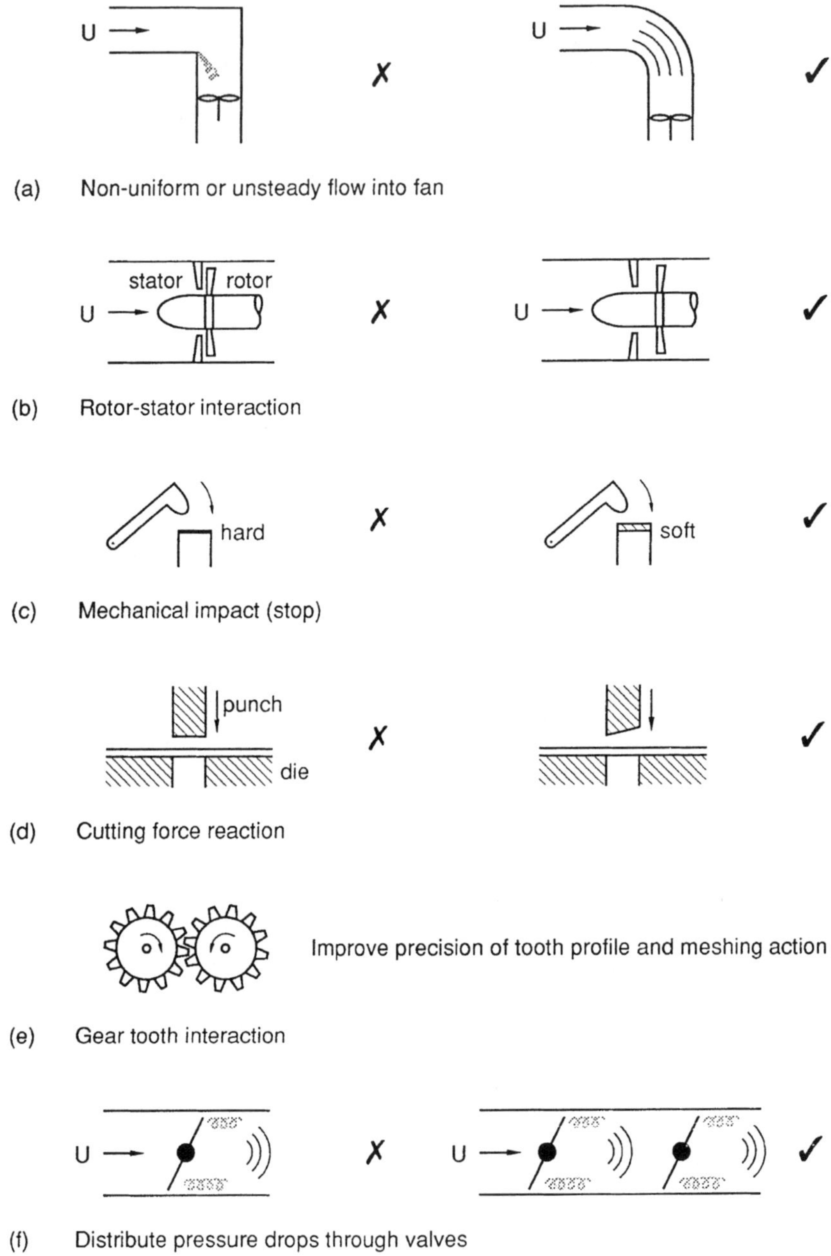

Figure 5.20 Examples of control at source.

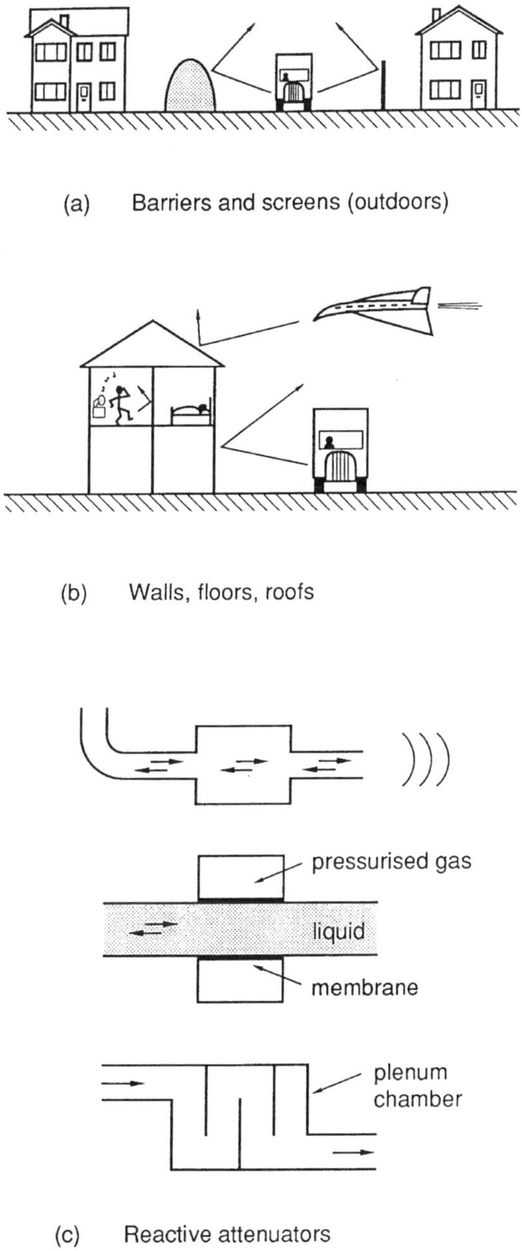

Figure 5.21 (a–c) Examples of airborne sound path control by reflection.

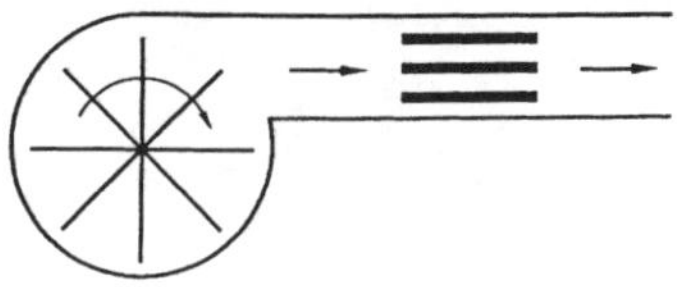

(d) Splitter silencer for ventilation system

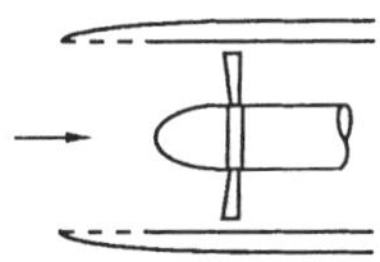

(e) Turbofan intake attenuator (non-linear)

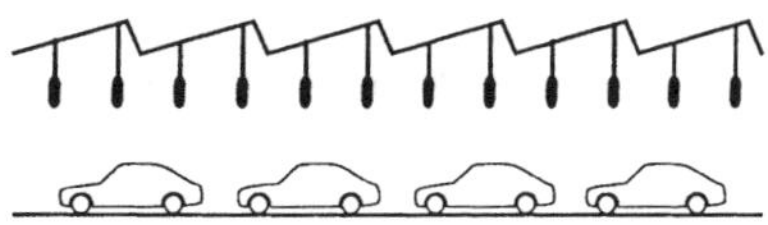

(f) Suspended absorbers over production line

(g) Ceiling absorbers (eg swimming pool, canteen)

Figure 5.21 (d–g) Examples of airborne sound path control by absorption.

emphasized that it is very difficult and costly to suppress such transmission in large structures such as buildings and ships because load-bearing structures cannot readily accommodate resilient inserts and also because of the presence of many parallel transmission paths, each of which must be treated.

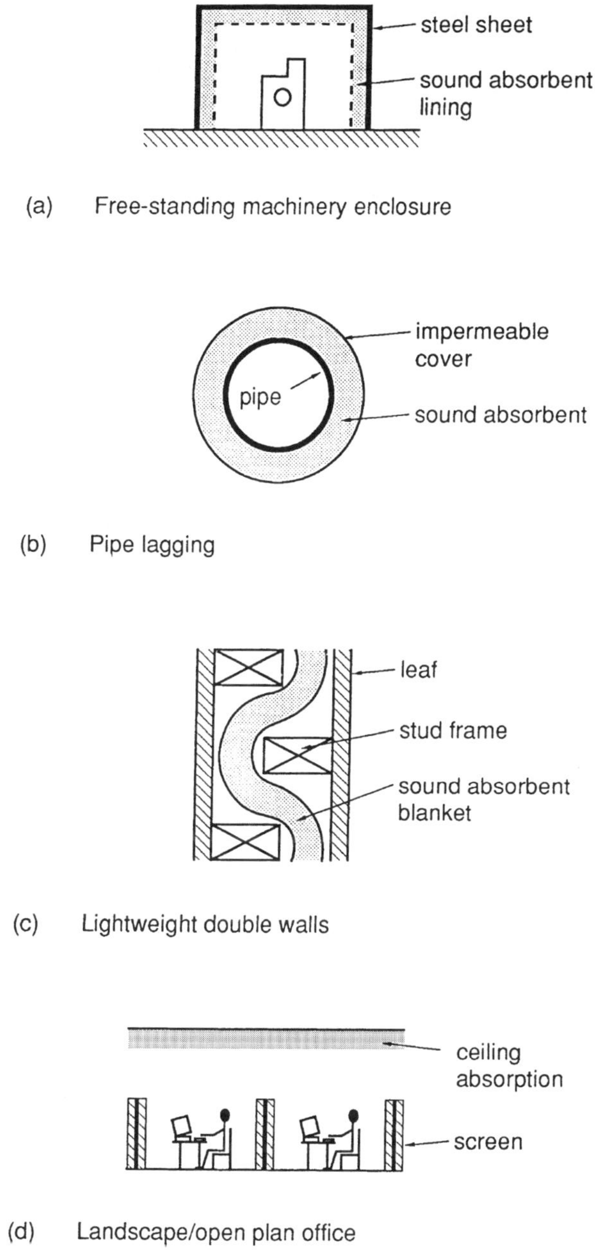

Figure 5.22 (a–d) Examples of airborne sound path control by combination of reflection and absorption.

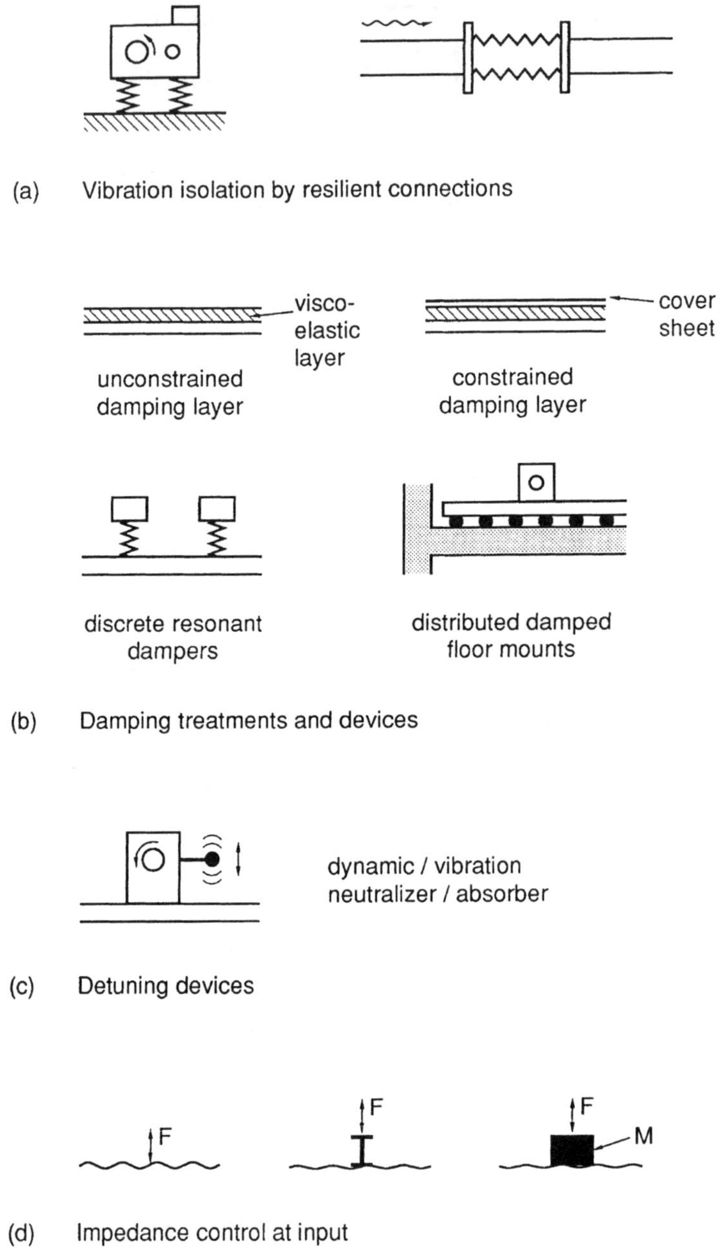

Figure 5.23 (a–d) Examples of structure-borne noise control.

Figure 5.24 A woven spillage (collection container).

It is generally impracticable to try to reduce the radiation ratio of vibrating structures, but in cases where solid surfaces can be replaced by perforated structures considerable reductions are possible (e.g. Fig. 5.24).

5.6 SOUND ABSORPTION: MECHANISMS, MATERIALS, CONSTRUCTIONS AND APPLICATIONS

5.6.1 Applications of sound absorption

The following are the principal applications of sound absorbent materials in noise control:

1. incorporation in noise control enclosures, covers and wrappings to reduce reverberant build-up of sound and hence increase insertion loss;
2. incorporation in flow ducts to attenuate sound from fans and flow control devices;
3. application to the surfaces of rooms to control reflected sound: (a) steady state SPLs in factories, offices and other buildings; (b) transient decay

(reverberation) in auditoria – considerations of intelligibility and musical appreciation; (c) reverberation in factories, canteens, swimming pools, studios etc. – comfort considerations;
4. in vehicles (walls, engine compartments, engine exhausts);
5. in lightweight walls and ceilings of buildings;
6. on traffic noise barriers to reduce reflection.

5.6.2 Mechanisms of sound absorption

The various physical mechanisms of sound absorption may be classified as follows:

1. viscous losses due to oscillatory flow in pores, or interstices, of porous materials, (e.g. mineral wool, fibreglass, *open cell* plastic foam, porous plaster, wood wool);
2. heat conduction through porous materials;
3. vibration of solid skeleton of porous materials (e.g. *closed cell* plastic foam);
4. vibration of impermeable panels by incident sound (wood, glass, metal etc.) – internal and support friction causes mechanical energy loss;
5. viscous and heat conduction losses in acoustic (Helmholtz) resonators.

5.6.3 Acoustic parameters of porous absorbing materials

Porous materials such as fibreglass and wood wool have very complex geometric forms which are not amenable to mathematical description by deterministic models (see Zwikkwer and Kosten, 1949; Allard, 1993; Ingard, 1994). Consequently quasi-continuum models, having spatially 'smeared' properties, are assumed.

There are many controlling parameters. However there are only three major parameters, flow resistivity, porosity and structure factor, which significantly influence the acoustic properties of *rigid* porous materials such as bonded fibreglass. We shall restrict our attention to those materials in which vibration of the solid skeleton is negligible.

5.6.3.1 Flow resistivity (r)

Consider a *steady* flow of air through a layer of porous material. The viscous resistance to flow gives rise to a static pressure gradient $\partial p/\partial x$. The flow resistivity r is defined as follows:

$$\partial p/\partial x = -ru' \tag{5.14}$$

where u' is defined as the volume flow rate/unit cross-sectional area. Note that u' is not the flow velocity in the pores of the material. Of course, in free air, the fluid momentum (Euler) equation relates pressure gradient and fluid acceleration,

$$\partial p/\partial x = -\rho_0(\partial u/\partial t). \tag{5.15}$$

Flow resistivity, as measured by a steady (DC) flow measurement (ISO 9053: 1991), is assumed to apply to oscillatory (AC) acoustic flow. In general, this is a valid assumption, because the oscillatory flow speeds and Reynolds numbers are very low. Various techniques for determining r from impedance tube measurements are also available (Bies and Hansen, 1995; Ren and Jacobsen, 1993).

5.6.3.2 *Porosity* (h)

Porosity is defined as the ratio of volume of voids to the total volume occupied by the porous structure; it influences the effective compressibility or bulk modulus, of the fluid. Porosity is normally between 0.90 and 0.95 for effective materials.

Consider the fluid continuity equation. In free air

$$\partial u/\partial x = -(1/k_0)\partial p/\partial t \tag{5.16a}$$

where κ_0 is the bulk modulus and $\kappa_0 = \partial p/\partial(\rho/\rho_0) = \rho_0 c_0^2$

In a porous material

$$\partial u'/\partial x = -(1/\kappa)\partial p/\partial t \tag{5.16b}$$

where $\kappa = \kappa_0/h$.

5.6.3.3 *Structure factor* (s)

The structure factor expresses the influence of the geometric form of the structure on the effective *density* of the fluid. The vibrating fluid is forced to undergo accelerations in various directions by the tortuous and irregular forms of the pores (channels) within a porous material; the effect is to increase the apparent fluid density and to decrease the effective sound speed. It is not generally possible to estimate s theoretically. Some of the physical origins of the structure factor may be identified.

1. Side pockets (Fig. 5.25). The fluid in a side pocket does not take part in the oscillatory flow in the pore, but its presence decreases the effective

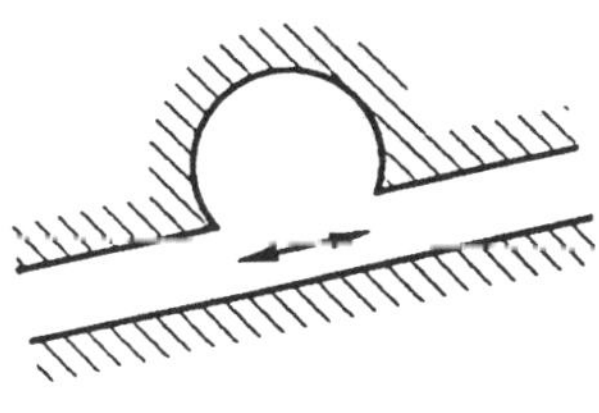

Figure 5.25 Pore and side pocket.

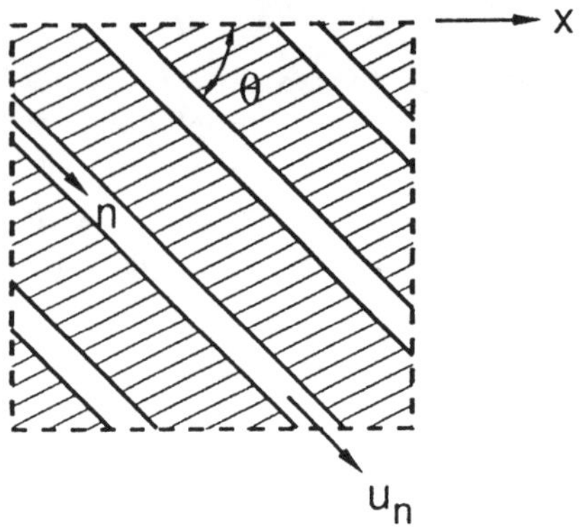

Figure 5.26 Non-axial pore orientation.

bulk modulus of the fluid by contributing to the local volume of compressible fluid. Hence the magnitude of the acoustic pressure associated with a given fluid strain gradient in the pore is smaller than in free air. In terms of the wave equation, this is equivalent to a mean fluid density larger than the free air density.

2. Non-uniformity of cross-section area of the pores. Non-axial flow in the region of a change of cross-section possess kinetic energy of non-axial motion and is equivalent to the addition of localized mass.
3. Non-axial pore orientation (Fig. 5.26). The momentum equation of inviscid fluid in a pore is

$$\partial p/\partial n = -\rho_0(\partial u_n/\partial t) \tag{5.17}$$

or

$$(\partial p/\partial x)\cos\theta = -\rho_0(\partial u'/\partial t)/h\cos\theta. \tag{5.18}$$

Hence

$$\partial p/\partial x = -(s\rho_0/h)(\partial u'/\partial t) \tag{5.19}$$

where $s \propto 1/\cos^2\theta$. The effective density is increased by a factor s. For random orientation of pores $s = 3$. Experiments show that s generally lies in the range 1.2–2.0. This contribution to structure factor may be expressed in terms of 'tortuosity' which may be measured by ultrasonic means (Allard, 1993).
4. Velocity profile in pores. The fluctuating velocity is zero at the walls of the pores and has a non-uniform distribution over the cross-section of the pore. Therefore the volume acceleration produced by a given pressure gradient is reduced; this is equivalent to an increase in fluid density, and therefore contributes to the structure factor.

5.6.4 Equation of motion of fluid in a porous material

The modified equation of motion of a fluid contained within a porous material may be written

$$\frac{\partial p}{\partial x} = -\left(\frac{s\rho_0}{h}\right)\frac{\partial u'}{\partial t} - ru'. \tag{5.20}$$

For simple harmonic motion at frequency ω,

$$\frac{\partial p}{\partial x} = -\left[\frac{s\rho_0}{h} + \frac{r}{j\omega}\right]\frac{\partial u'}{\partial t}. \tag{5.21}$$

Compare this with the free fluid Euler equation (5.15). The term in brackets is the effective density, which is complex. The effect of the parameters s, h and r is to alter the speed of propagation of acoustic disturbances from the free-air speed c (normally it is considerably lower than c), and to attenuate the disturbance as it travels through the material. Combination of equations (5.16b) and (5.21) produce a modified one-dimensional wave equation

$$\frac{1}{h}\frac{\partial^2 p}{\partial t^2} + \left(\frac{r}{s\rho_0}\right)\frac{\partial p}{\partial t} = c_1^2\frac{\partial^2 p}{\partial x^2} \tag{5.22}$$

where $c_1^2 = c_0^2/sh$

5.6.5 The propagation constant of a porous material

Solution of the modified, one-dimensional harmonic wave equation takes the form

$$p(x,t) = A\mathrm{e}^{j\omega t}\mathrm{e}^{-\gamma x} \tag{5.23}$$

where γ is known as the propagation constant; it is normally complex.

We may write

$$\gamma = \alpha + j\,\beta \tag{5.24}$$

where β is the phase constant and α is the attenuation constant.

Hence, a harmonic plane wave inside a porous material is represented by

$$p(x,t) = A\mathrm{e}^{i(\omega t - \beta x)}\mathrm{e}^{-\alpha x}. \tag{5.25}$$

α is controlled primarily by viscous losses, i.e. flow resistivity, r.

It may be shown that $\gamma = j\omega\sqrt{\rho'/\kappa}$ where $\kappa = \rho_0 c_0^2/h$ and $\rho' = s\rho_0/h + r/j\omega$. Hence the characteristic specific acoustic impedance of the medium is given by the complex quantity

$$z_\mathrm{c} = p/u' = \sqrt{\kappa\rho'}. \tag{5.26}$$

For examples, see Allard (1993). Two different layer thicknesses can be used to terminate an impedance tube to determine z_c (Bies and Hansen, 1995; Beranek and Vér, 1992; and Utsono, Tanaka and Fukikawa, 1989).

5.6.6 Impedance of a rigidly backed porous layer

Consider a layer of porous material of thickness l terminated by a rigid plane surface. If it is assumed that the wave vector in the material is directed normal to the surface we can define a specific normal surface acoustic impedance as

$$z_n = z_c \coth(\gamma l). \tag{5.27}$$

z_n can vary over a wide range of values because resonances and anti-resonances can occur due to interference between the incident and reflected waves in the material. The range of variation decreases as the attenuation constant α increases, because the wave returning to the free surface after reflection from the rigid surface is strongly attenuated. If the flow resistivity of such a porous layer is too high, waves are strongly reflected at the free surface; if it is too low, waves are not effectively absorbed in the material. A good practical compromise ('rule of thumb') is given by the following relationship:

$$r\, l \approx 3\rho_0 c_0.$$

5.6.7 Reflection and absorption of plane waves at an impedance surface

Most sound absorbing materials are not amenable to detailed mathematical modelling for the purpose of estimating the propagation constant; therefore a 'black box' representation is commonly used. Any uniform plane surface may be characterized by its surface normal specific acoustic impedance z_n which is defined, for harmonic time dependence, as the ratio of the complex amplitude of surface pressure to the complex amplitude of the component of particle velocity normal to the surface:

$$z_n = p\, e^{j\omega t}/u_n e^{j\omega t} = p/u_n. \tag{5.28}$$

In general, z_n is a function of the angle of incidence ϕ of a plane incident wave; but for many common forms of sound absorbent material it is effectively independent of ϕ because wave motion parallel to the surface is strongly attenuated by viscous dissipation. This condition is known as *local reaction*. Systems in which waves can travel freely along a surface, such as sheets of glass, metal, plywood, etc., exhibit extended reaction and z_n cannot then be uniquely defined.

The real and imaginary parts of z_n are symbolized by r_n and x_n; hence $z_n = r_n + j\, x_n$. These components are known respectively as the specific acoustic resistance and specific acoustic reactance. The time-averaged rate of work by an incident wave per unit surface area by the surface pressure, which equals the rate of sound energy absorption per unit area, is given by

$$W'_{abs} = \tfrac{1}{2}\mathrm{Re}\{p\, u_n^*\} = \tfrac{1}{2}|p|^2 \left[r_n/(r_n^2 + x_n^2)\right] \tag{5.29}$$

$$= \tfrac{1}{2}|u_n|^2 r_n \tag{5.30}$$

Pure specular reflection will occur only if the surface is plane, infinitely extended and has a uniform impedance. In practice, the surface dimensions must exceed about three wavelengths for diffraction to be negligible. The surface pressure p and the surface normal particle velocity u_n are related to the incident field pressure by the following analysis (Fig. 5.27):

$$p_i(x, y, t) = A \exp[-jk \cos \phi x] \times \exp[-jk \sin \phi y] \times e^{j\omega t} \tag{5.31}$$

$$p_r(x, y, t) = B \exp[jk \cos \phi x] \times \exp[-jk \sin \phi y] \times e^{j\omega t}. \tag{5.32}$$

At $x = 0$

$$p = (A + B) \exp[-jk \sin \phi y] \exp[j\omega t]. \tag{5.33}$$

and

$$u_n = \left\{ \frac{(A - B)}{\rho_0 c_0} \cos \phi \right\} \exp[-jk \sin \phi y] \exp[j\omega t]. \tag{5.34}$$

Therefore

$$B/A = [z_n' \cos \phi - 1]/[z_n' \cos \phi + 1] \tag{5.35}$$

where

$$z_n' = z_n / \rho_0 c_0.$$

The incident power per unit area of surface is given by

$$W_i' = I_i \cos \phi = (|A|^2 / 2\rho_0 c_0) \cos \phi. \tag{5.36}$$

The reflected power per unit area of surface is given by

$$W_r' = I_r \cos \phi = (|B|^2 / 2\rho_0 c_0) \cos \phi. \tag{5.37}$$

Hence the sound power absorption coefficient $\alpha(\phi)$ is given by

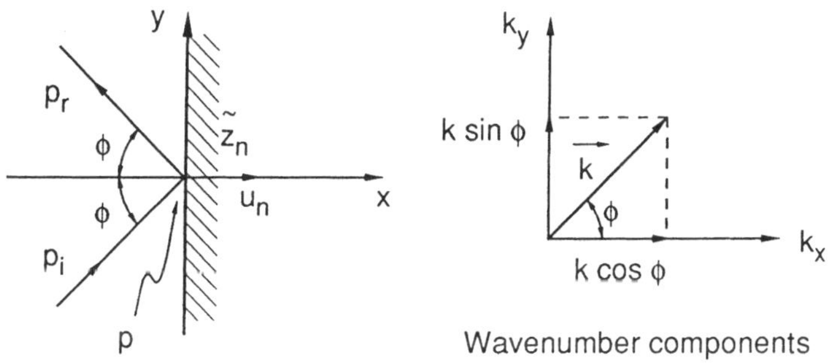

Figure 5.27 Reflection of plane sound waves by a uniform plane surface.

$$\begin{aligned}\alpha(\phi) &= 1 - |B/A|^2 \\ &= 4r'_n \cos\phi / \left[(r'_n \cos\phi + 1)^2 + (x'_n \cos\phi)^2\right]\end{aligned} \tag{5.38}$$

where $r'_n = r_n/\rho_0 c_0$ and $x'_n = x_n/\rho_0 c_0$.

$$\text{Now } \frac{\partial\alpha(\phi)}{\partial\phi} = 0 \quad \text{when } \cos\phi = |z_n|^{-1} = (r'^2_n + x'^2_n)^{-1/2} \tag{5.39}$$

$$\text{and } \alpha(\phi)_{max} = 2r'_n/(|z'_n| + r'_n). \tag{5.40}$$

For most materials in air $|z'_n| > 1$ and therefore an angle of maximum absorption exists (this is not true under water). As $|z'_n|$ increases, the angle of maximum absorption approaches 90° (grazing incidence); (see the example in Fig. 5.28). A mathematical expression for the absorption coefficient for diffuse field incidence may be derived analytically (Ingard, 1994); it may exceed or be less than that for normal incidence, depending on r'_n and x'_n. Typical experimental values for the surface impedance and absorption coefficients of a layer of porous plastic foam are presented in Fig. 5.29, in which the maximum and minimum are associated with resonance and anti-resonance frequencies at about 650–1300 Hz respectively.

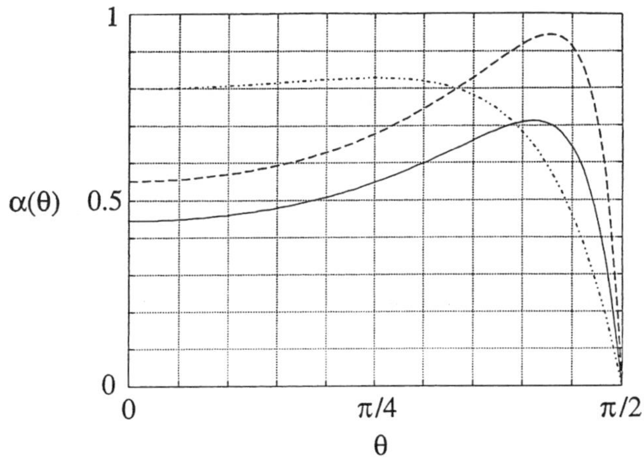

	$r_n/\rho_0 c$	$x_n/\rho_0 c$
——— (solid)	2	-3
······ (dotted)	1	-1
- - - - (dashed)	4	-2

Figure 5.28 Examples of variation of absorption coefficient with surface normal impedance.

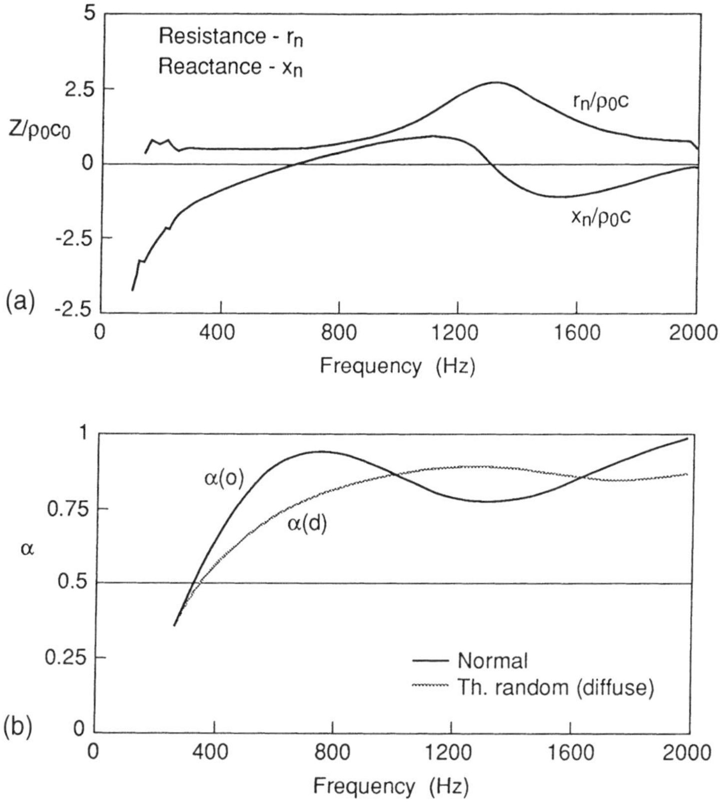

Figure 5.29 (a) Normal specific acoustic impedance. (b) Sound absorption coefficients of a 100 mm thick sheet of porous plastic foam on a rigid wall.

5.6.8 Sound absorbers

Sound absorbers take many forms (Fig. 5.30). The most common are: (i) porous/fibrous sheets (mats) covered by perforated or slotted impermeable sheets; (ii) acoustic 'tiles' constructed from gypsum or similar materials; these are permeated by a network of channels which compensate the high-flow resistance of the material; (iii) panel (membrane) absorbers which consist of thin impermeable sheets of metal, glass, plastic backed by air cavities; (iv) acoustic (Helmholtz) absorbers in which an air cavity communicates with the surrounding air via a narrow neck.

The first two forms generally provide broad band absorption over three or four octaves, whereas the resonator absorbers are effective at most over one octave. The performance of resonator absorbers in an enclosure

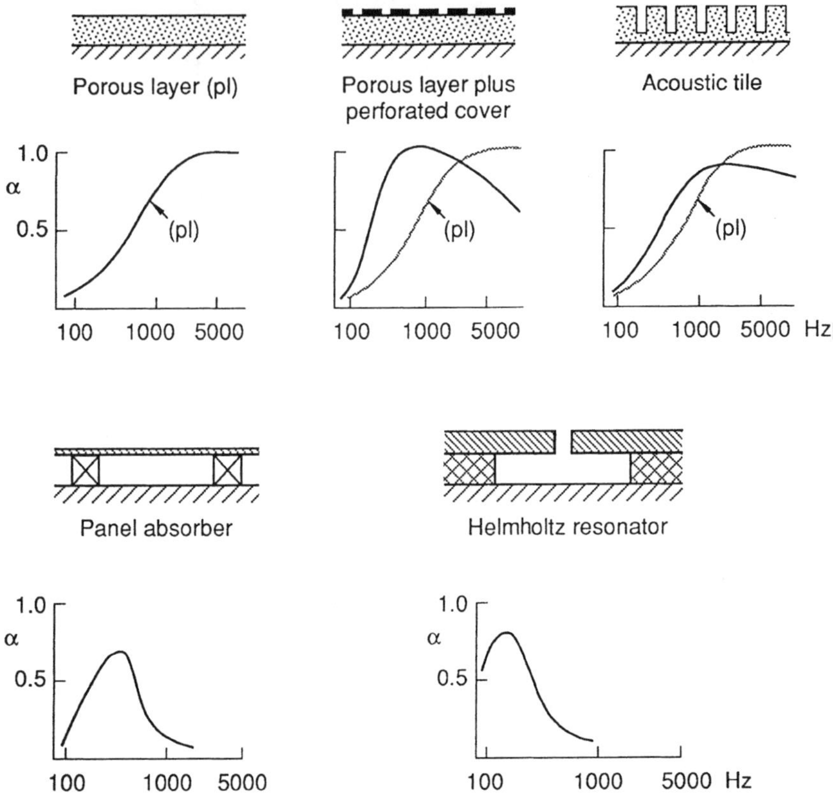

Figure 5.30 Wall absorbers with typical spectra of diffuse field sound absorption coefficients.

is very dependent upon their location in the enclosure and upon the internal (dissipation) loss factors; these must be closely matched to the radiation loss factor of the resonator for effective absorption to occur (Fahy, 1987). Helmholtz resonators are best used in arrays so that the mutual radiation impedance effects improve the possibility of impedance matching.

The diffuse field sound absorption coefficient of sound absorbers is determined by measuring the reverberation time of a reverberation chamber with and without a 10 m^2 sample placed on one surface (see equation (5.10)).

5.6.9 Some applications of sound absorbing materials

5.6.9.1 Control of reverberant sound in rooms

The simple theoretical energetic model of steady-state sound fields in rooms presented in Chapter 1 shows that surface absorbers mounted on the room surfaces are only effective in reducing reverberant sound levels; sound levels close to sources (direct fields), such as those at machine operator positions, are little affected. The contents of many rooms provide substantial sound absorption, making significant reductions in reverberant levels difficult to achieve; a doubling of sound absorption achieves a reduction in reverberant sound pressure level of only 3 dB.

The simple (Sabine) model is not applicable to large industrial halls or auditoria. The spatially uniform reverberant field indicated by this model does not exist. In industrial halls, the presence of scattering objects such as machines and stockpiles increases the sound pressure level within a few metres of a source relative to the free field value, and produces a continuous spatial decay of sound level with increasing distance from a source. The installation of hundreds of sound absorbent panels, vertically suspended in a regular array at some distance above a workspace or assembly line increases the rate of spatial decay of sound and generally lowers sound levels. Panels should placed about 1 m apart to produce optimum effect.

5.6.9.2 Improvement of the performance of noise control enclosures

Free-standing enclosures are widely used to control noise from machinery and plant. The following simple analysis is based upon the assumption of the existence of a diffuse field in an enclosure. Figure 5.31 represents an enclosure constructed from an impermeable outer sheet lined with an

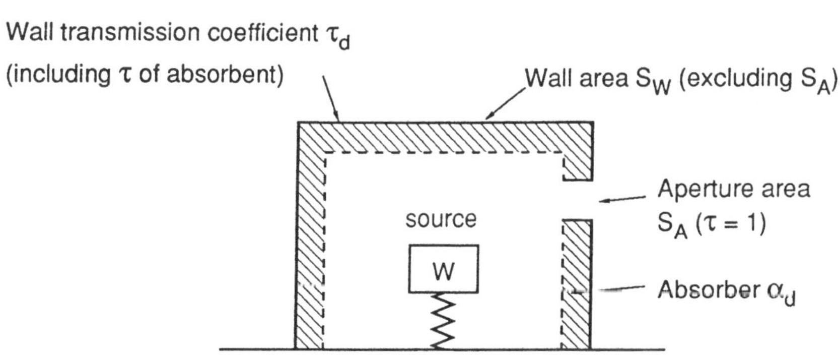

Figure 5.31 Free-standing noise control enclosure.

absorbent layer. It is assumed that the source is located so that the reverberant field dominates at the walls, which is increasingly unlikely as the absorption coefficient of the lining is increased. We assume that the damping in the outer sheet is included in the absorption coefficient and that the mass of the absorbent layer is added to the mass of this sheet.

Assuming a diffuse field

$$I = \langle\overline{p_r^2}\rangle/4\,\rho_0 c_0. \tag{5.41}$$

The power balance is given by

$$W \approx I[S_w(\alpha_d + \tau_d) + S_A] \tag{5.42}$$

assuming that $\tau_A = 1$ and that the diffuse field absorbent coefficient α_d is unaffected by wall vibration.

The power transmitted through aperture is given by

$$W_A = S_A I \approx W S_A/[(S_w(\alpha_d + \tau_d) + S_A)]. \tag{5.43}$$

The power transmitted through wall is given by

$$W_w = I\tau_d S_w = W\ S_w\tau_d/[S_w(\alpha_d + \tau_d) + S_A]. \tag{5.44}$$

The total power transmitted per unit source power is given by

$$W_T/W = (S_A + \tau_d S_w)/(S_w(\alpha_d + \tau_d) + S_A). \tag{5.45}$$

With no absorbent or damping

$$W_T/W = 1.$$

Hence the sound power insertion loss is given by

$$IL = 10\ \lg(W/W_T) = 10\ \lg\{[(S_w/S_A)(\alpha_d + \tau_d) + 1]/[(S_w/S_A)\tau_d + 1]\}\text{dB}. \tag{5.46}$$

Normally $\alpha_d \gg \tau_d$ (except at low frequencies) and $S_w \gg S_A$. Therefore

$$IL \approx 10\ \lg\{[(S_w/S_A)\alpha_d]/[(S_w/S_A)\tau_d + 1]\}\text{dB}. \tag{5.47}$$

Clearly α_d should be maximized and S_A/S_w and τ_d should be minimized. This equation is not correct for zero absorption coefficient – why?

Close-fitting covers and pipe wrapping (lagging) provide little insertion loss below about 300 Hz because of strong acoustic coupling between the underlying vibrating surface and the impermeable cover sheet. A typical example is shown in Fig. 5.32.

5.6.9.3 Sound absorption in ducts

Sound absorbing materials are widely used in attenuators installed in ducts which carry gas flow, such as ventilation ducts and gas turbine

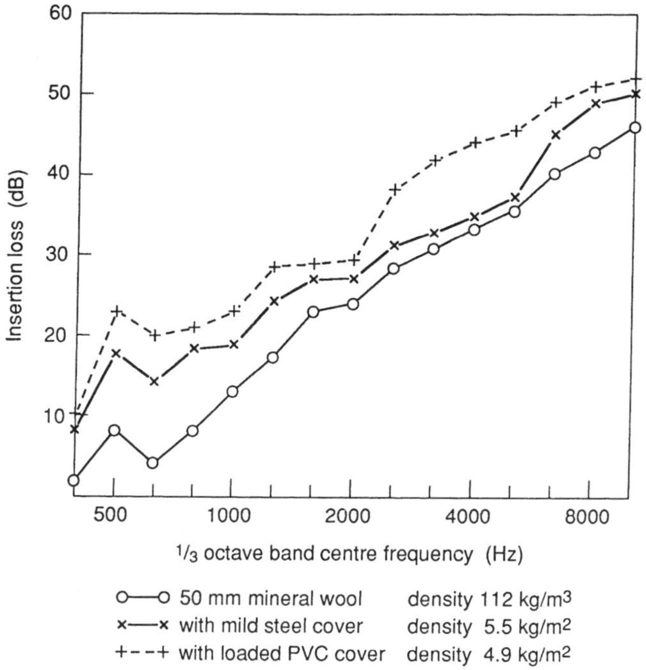

Figure 5.32 Insertion loss of 50 mm pipe lagging with and without impermeable covers.

exhaust ducts. Special-purpose materials are necessary in ducts carrying hot gas. It is generally ineffective simply to line the wall surfaces of a duct with sound absorbent (except at sharp bends) because the ratio of exposed absorbent surface per unit length of duct to the cross-sectional area through which noise passes is too small. Consequently, splitter attenuators are used, as illustrated in Fig. 5.33. The liner flow resistance must be relatively high to provide optimum attenuation. A product of flow resistivity and liner thickness of 3 to 4 times $\rho_0 c_0$ generally gives optimum performance.

Attenuation reaches a maximum in the frequency range where the acoustic wavelength is close to twice the width of the channel between the splitters; it falls considerably towards lower frequencies and moderately toward higher frequencies (see Fig. 5.34). Thicker linings provide increased low-frequency absorption but increase flow pressure losses. The necessary length of ventilation duct attenuators is usually controlled mainly by the sound power of the fan in the air handling unit in the octave frequency bands centred on 63 and 125 Hz.

Figure 5.33 Assembling a splitter attenuator. (Courtesy of Sound Attenuations Ltd).

5.7 TRANSMISSION OF AIRBORNE SOUND THROUGH PARTITIONS

5.7.1 Introduction

The transmission of sound from one volume of fluid to another through a solid partition which separates them is of central concern in noise control. The great majority of practical problems involve transmission from one volume of air to another – so-called 'airborne sound transmission'; in fact,

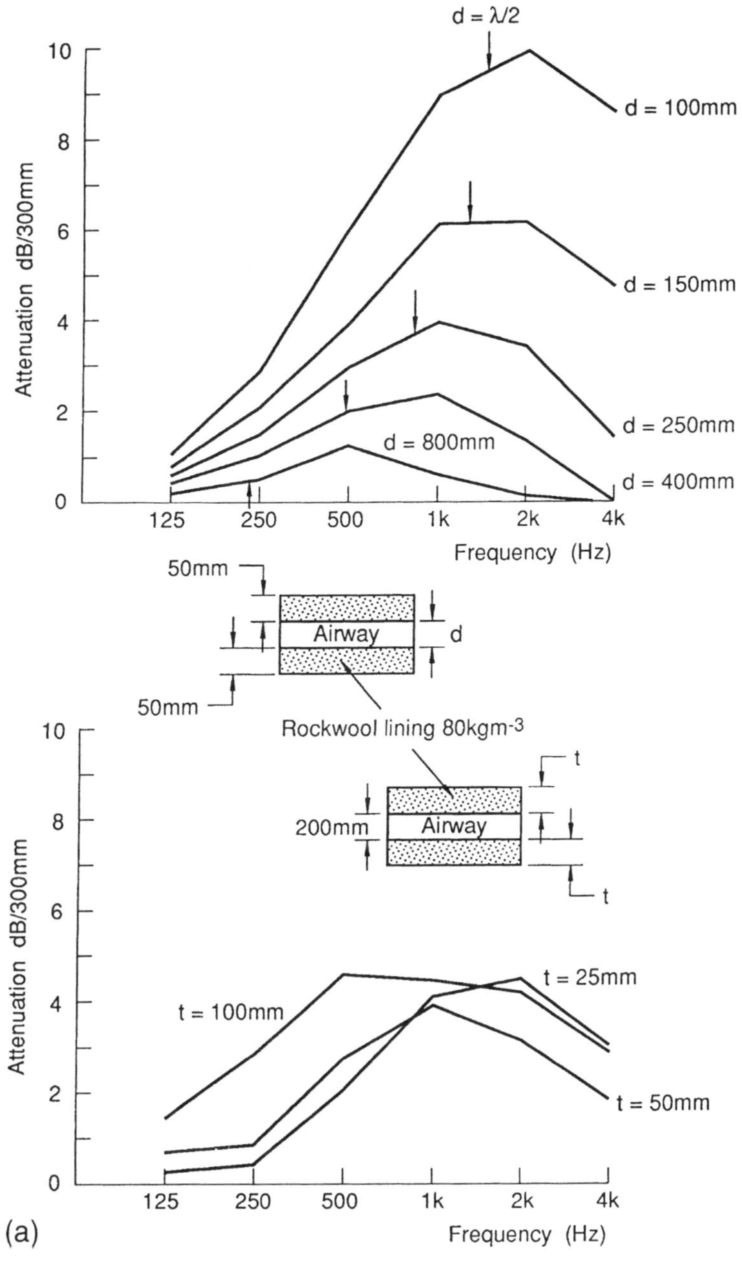

Figure 5.34 (a) Dependence of lined duct attenuation on airway width and lining thickness.

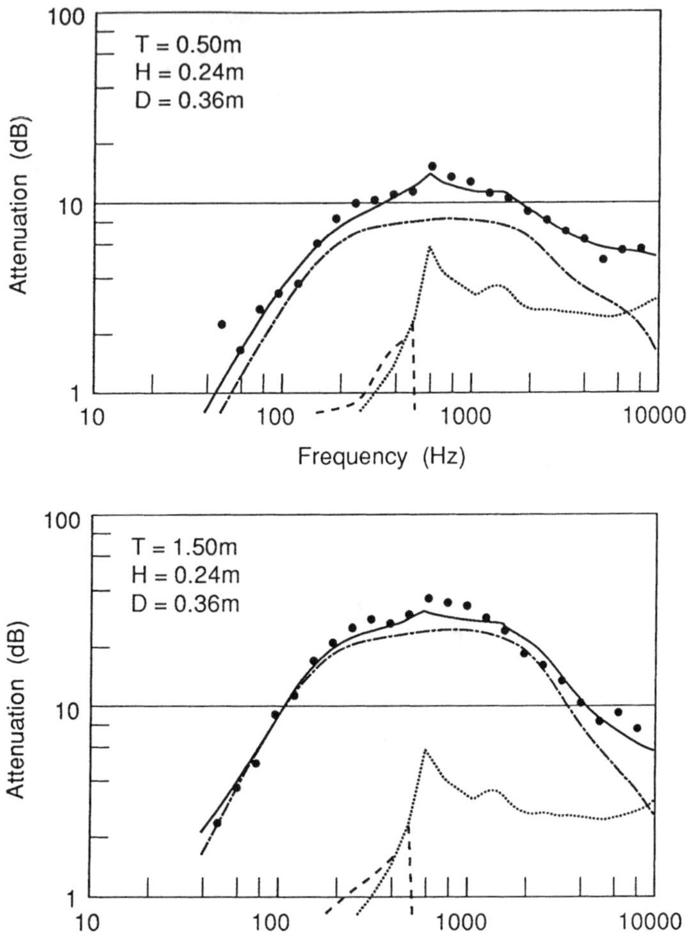

(b)

Figure 5.34 (b) Typical performance of splitter attenuators: T = baffle length; H = airway width; D = splitter thickness; • experiment; ——— theory; – – – theory, propagation loss; ······ theory, entry reflection loss; - - - theory, exit reflection loss. (Mechel, 1990).

the transmission process involves the vibrational response of the solid partition to the incident sound and the consequent radiation of sound into the air on both sides, so the process is not purely 'airborne'. The term 'structure-borne sound' is generally taken to apply to processes by which vibrational disturbances of solid structures in the audio-frequency range transport energy received from a generator over substantial distances to the location of the receiver where it generates airborne sound radiation.

The design and construction of effective and economical partitions is a central element in the practice of noise control by engineers and architects. The physics of the general problem has been most extensively studied in relation to sound transmission across partitions in buildings, e.g. walls, floor and windows; many theoretical models have been proposed and a vast amount of measured data has been gathered over the past 50 years, (e.g. Bies and Hansen, 1995). In recent years, the commercial demand to reduce noise inside passenger transport vehicles, such as cars and aircraft, together with the needs to reduce the noise levels inside factories and to minimize environmental impact of noise generated by industrial plant, such as pipes and cooling systems, have motivated theoretical and experimental studies of a wide range of partition geometries and materials. A range of examples is shown in Fig. 5.35.

In practice, partitions are often inhomogeneous, non-uniform, comprise combinations of different materials, have indeterminate boundary conditions and are subject to complex sound fields generated in enclosures such as rooms and ventilation ducts. Surprisingly, the sound transmission behaviour of many single-leaf building and noise enclosure components which have essentially plane geometry is quite well represented by a very simple mathematical model which completely ignores the boundaries of the partition and the enclosure. Partition curvature considerably complicates the vibrational behaviour and is not explicitly considered in this book (but see Fahy, 1987). Note that the terms 'partition' and 'panel' tend to be used synonymously in the literature.

The first part of this section presents the mathematical model of a single-leaf, uniform, homogeneous partition and the sound transmission characteristics predicted therefrom. The phenomenon of 'wave coincidence' and the influence of the system parameters on sound insulation performance are explained. Partitions often consist of combinations of layers of different materials – so-called multi-layer panels. Double-leaf partitions, in which two sheets of solid material are separated by an air cavity, are commonly used in buildings to provide effective airborne sound insulation at minimum weight. The second part of this section presents a largely qualitative description of the transmission behaviour of such partitions in order to indicate the influence of the various material and geometric parameters on insulation performance. The final section summarizes the standard method of measuring airborne sound insulation performance of partitions.

5.7.2 Transmission of normally incident plane sound waves through a plane partition

The model is shown in Fig. 5.36. The partition is thin, uniform, plane and infinite in extent; it is characterized by a mass per unit area m. The fluids on either side, which may be different, are specified in terms of their

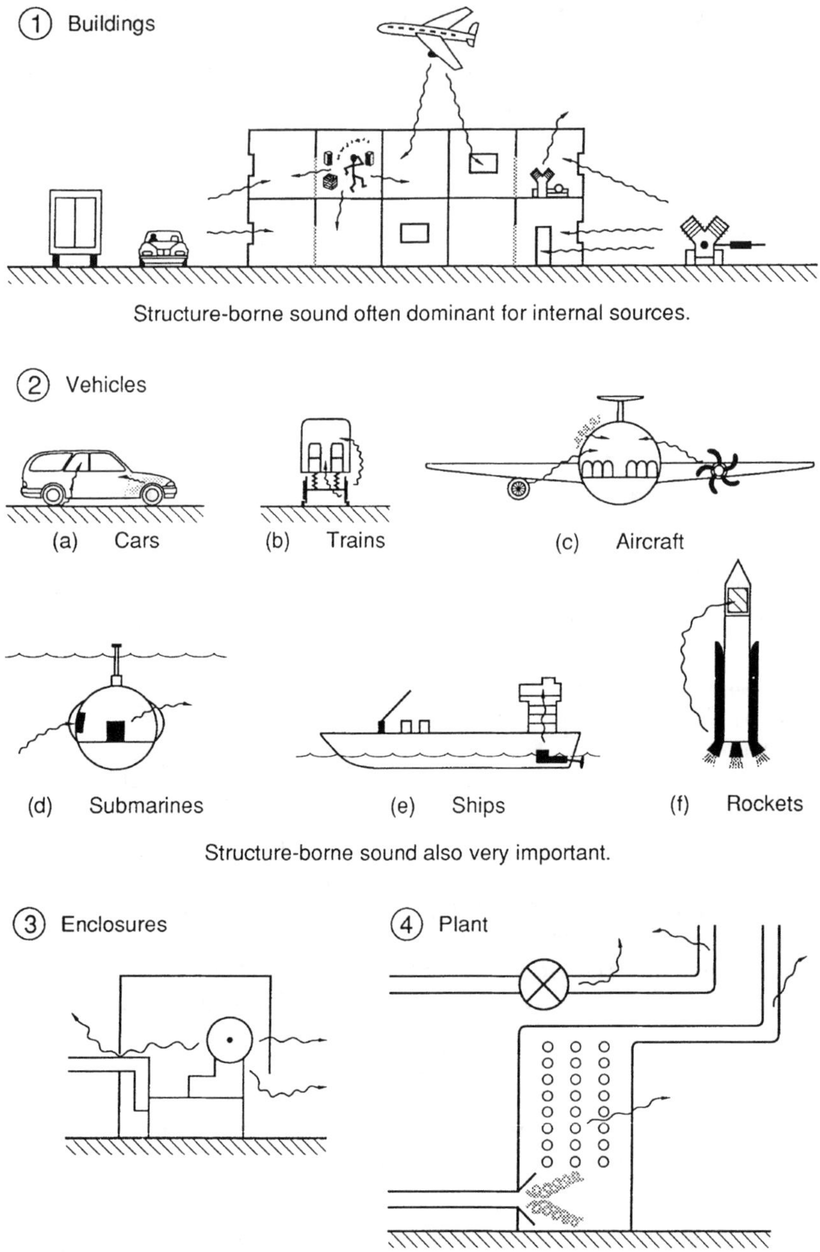

Figure 5.35 Airborne sound transmission: practical examples.

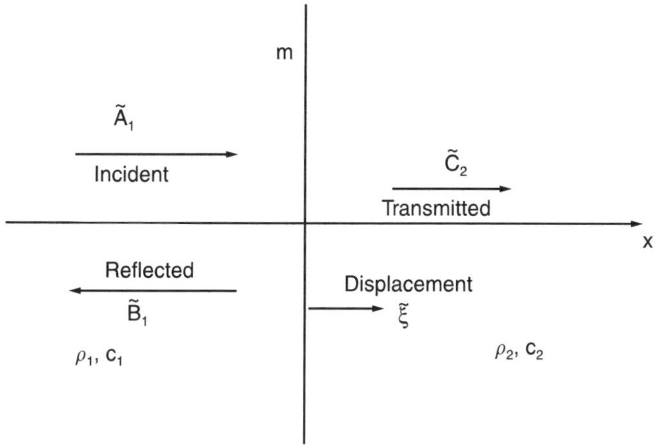

Figure 5.36 Idealized model of normal incidence transmission through a single-leaf partition.

characteristic specific acoustic impedances $\rho_1 c_1$ and $\rho_2 c_2$; their ratio $\rho_1 c_1/\rho_2 c_2$ is denoted by r. The time dependence is harmonic and denoted by $e^{j\omega t}$. The complex amplitudes of the incident, reflected and transmitted waves are denoted by A, B and C, respectively.

The equation of harmonic motion of the panel which is located in the plane $x = 0$, in which $\exp(j\omega t)$ has been dropped, is

$$-\omega^2 m\zeta = A + B - C. \tag{5.48}$$

The boundary conditions on the fluids are expressed as equalities of particle velocity and panel velocity at $x = 0$

$$\frac{A - B}{\rho_1 c_1} = j\omega\zeta = \frac{C}{\rho_2 c_2}. \qquad (5.49), (5.50)$$

B and ζ may be eliminated from equations (5.48) and (5.49) to give:

$$-\frac{\omega^2 m A_2}{j\omega\rho_2 c_2} = A + B - C. \tag{5.51}$$

From equations (5.49) and (5.50)

$$B = -rC + A \tag{5.52}$$

and from equations (5.51) and (5.52)

$$j\omega m\frac{C}{\rho_2 c_2} = 2A - C(1 + r)$$

or
$$C/A = 2\left[\frac{j\omega m}{\rho_2 c_2} + 1 + r\right]^{-1}. \tag{5.53}$$

The sound power transmission coefficient τ is defined as

$$\begin{aligned}\tau &= \text{transmitted power/incident power}\\ &= \text{transmitted intensity/incident intensity}\\ &= [|C|^2/2\rho_2 c_2]/[|A|^2/2\rho_1 c_1].\end{aligned} \tag{5.54}$$

From equation (5.53)
$$\tau = r|C/A|^2 = \frac{4r}{(1+r)^2 + (\omega m/\rho_2 c_2)^2}. \tag{5.55}$$

Note that τ is reciprocal in r, i.e. the same in both directions. The non-dimensional parameter $\omega m/\rho_2 c_2$ is a measure of the ratio of the inertial mechanical impedance per unit area of the partition to the acoustic impedance of the fluid on the transmission side; it is termed the 'fluid loading parameter'. The transmission coefficient is determined by r if the fluid loading parameter is much less than unity, i.e. the presence of the partition does not significantly affect the transmission coefficient. In special cases, such as transmission from air to water, and vice versa, r is either much greater, or much less, than unity. Since the fluid loading parameter increases linearly with frequency, whereas the fluid impedances are frequency independent, there will be a frequency at which the terms in the denominator of equation (5.55) are equal. Above this frequency, the partition significantly influences the transmission coefficient.

In the case of air-to-air transmission through solid partitions, $r = 1$, and the fluid loading parameter $(\omega m/\rho_0 c_0)$ is generally much greater than unity, so that equation (5.55) may be approximated by

$$\tau = (2\rho_0 c_0/\omega m)^2. \tag{5.56}$$

The transmission coefficient is controlled by the product of frequency and mass per unit area. Note that this simplification is not valid in cases of very low frequencies and/or very lightweight partitions such as plastic film sheets.

5.7.2.1 Transmission loss

It is conventional to express the transmission coefficient τ in logarithmic terms as a transmission loss (USA) or sound reduction index (ISO/CEN) as follows:
$$TL(R) = 10\ \log_{10}(1/\tau) \quad \text{(dB)}.$$

We shall use R in future.

Introduction of the normally assumed value of acoustic impedance for air into equation (5.56) yields the following expression for the normal incidence sound reduction index:

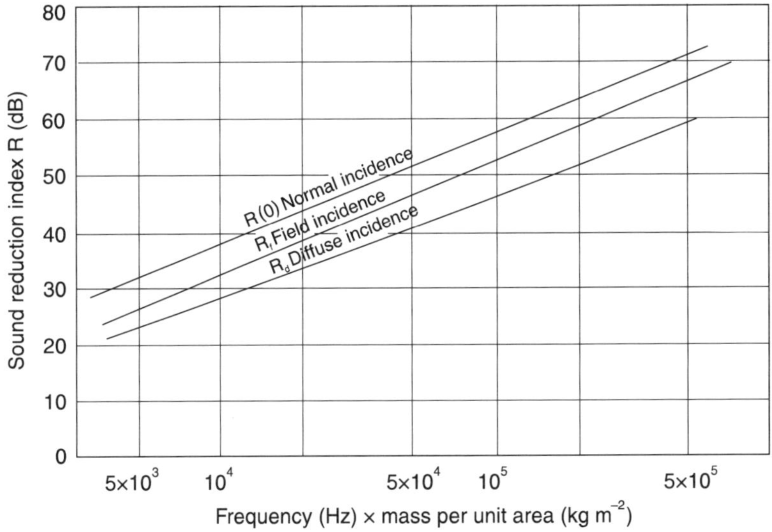

Figure 5.37 Sound reduction indices for unbounded partitions at subcritical frequencies.

$$R(0) = 20\log_{10}[fm] - 42 \quad \text{(dB)} \tag{5.57}$$

in which the (0) indicates normal incidence and f is in Hz.

This equation is said to express the *normal incidence mass law* for sound transmission. Note that $R(0)$ is determined by the product of mass per unit area and frequency (fm); thus $R(0)$ increases by 6 dB per doubling of either mass or frequency. The corresponding unique graph of $R(0)$ vs. (fm) is presented in Fig. 5.37.

5.7.3 Transmission of obliquely incident plane sound waves through a plane partition

This case is much more complicated than that of normal incidence because the incident plane acoustic wave pressure induces bending in the partition, together with the resulting elastic bending forces and associated dissipative energy losses (damping). The idealized model is shown in Fig. 5.38. The general expression for the sound power transmission coefficient is rather complicated and therefore it is reduced to the special case of the same fluids on both sides in these notes. The incident plane wave has a 'trace wavenumber' component parallel to the plane of the partition given by $k_t = k\sin\phi$. Because the partition is uniform and infinite in extent, the

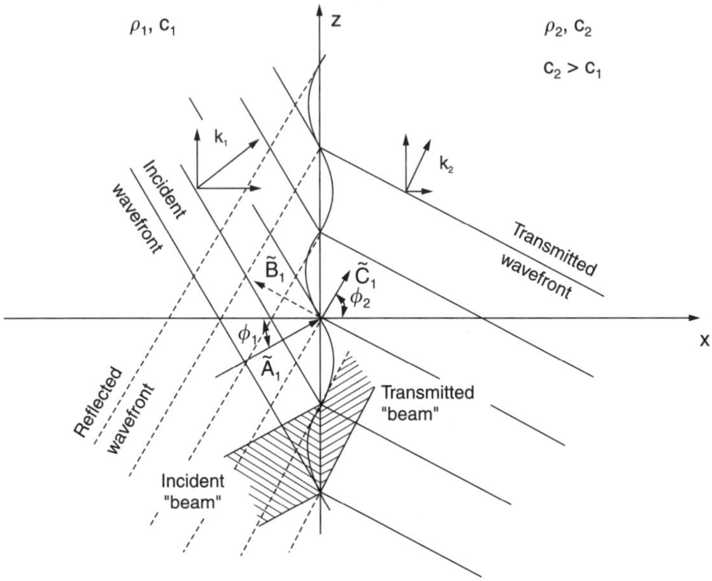

Figure 5.38 Transmission of obliquely incident sound through an unbounded flexible partition.

bending wave induced by the incident plane wave must also possess the same wavenumber.

The mathematical solution of this model is obtained by application of the thin plate bending wave equation under the action of the pressure difference across the partition, together with the kinematic boundary conditions at surface of the partition which equate the normal component of acoustic particle velocity to that of the partition (see Fahy (1987) for details). Note that refraction does not occur when the fluids are identical on each side.

The resulting expression for the sound power transmission coefficient corresponding to angle of incidence ϕ is given by

$$\tau(\phi) = \frac{(2\rho_0 c_0/\omega m)^2 \sec^2 \phi}{[(2\rho_0 c_0/\omega m)\sec\phi + \eta(k/k_b)^4 \sin^4\phi]^2 + [1 - (k/k_b)^4 \sin^4\phi]^2} \tag{5.58}$$

where k_b is the free bending wavenumber in the partition given by

$$k_b = \omega^{1/2}(m/B)^{1/4} \tag{5.59}$$

which is the wavenumber which occurs *naturally* in the partition at frequency ω. B represents the bending stiffness of the plate.

The non-dimensional parameters in equation (5.58) are the fluid loading parameter (as above), the ratio of the acoustic wavenumber to the *free* bending wavenumber in the plate at frequency ω, the dissipation loss factor η and the angle of incidence ϕ. Figure 5.39 illustrates the variation of the acoustic and free bending wavenumbers with frequency. It is seen that in all cases there exists a frequency at which the two curves intersect and at which the two wavenumbers are equal; this is termed the '*critical frequency*'. The phenomenon associated with the equality of the trace and free bending wavenumbers is termed '*coincidence*'; hence the critical frequency is sometimes referred to as the '*lowest coincidence frequency*'. For a uniform panel of bending stiffness B and mass per unit area m, the critical frequency is given by

$$k \sin\phi = (\omega/c_0)\sin\phi = k_b = \omega^{1/2}(m/B)^{1/4} \tag{5.60}$$

$$\text{or} \quad f_c = (1/2\pi)(c_0^2/\sin^2\phi)(m/B)^{1/2}. \tag{5.61}$$

In the special case of a uniform, homogeneous partition of material density ρ_m, Young's modulus E and thickness h, $B = Eh^3/12(1-\nu^2)$, and this expression becomes

$$f_c = c^2/1.8\,h\,c_l \tag{5.62}$$

where $c_l = (E/\rho_m)^{1/2}$ is the speed of compressional waves (sound) in a bar of the material. Note that the product of thickness and critical frequency hf_c is uniquely determined by the material parameters. Hence a table of values of hf_c may be drawn up for a range of common materials (Table 5.5). Examples: 6 mm glass, $f_c = 2100$ Hz; 100 mm dense concrete, $f_c = 190$ Hz.

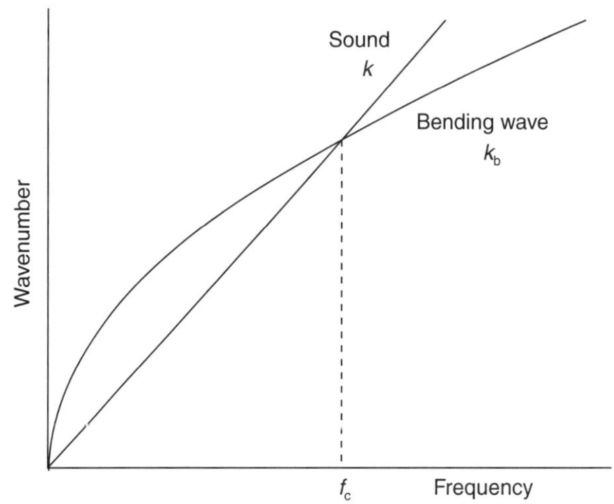

Figure 5.39 Dispersion curves for sound and bending waves.

Table 5.5 Product of plate thickness and critical frequency in air (20 °C)[a]

Material	hf_c (m s^{-1})
Steel	12.4
Aluminium	12.0
Brass	17.8
Copper	16.3
Glass	12.7
Perspex	27.7
Clipboard	23[b]
Plywood	20[b]
Asbestos cement	17[b]
Concrete	
dense	19[b]
porous	33[b]
light	34[b]

[a] To obtain values in water, multiply by 18.9
[b] Variations of up to ±10% possible

Lightweight stiff partitions have low critical frequencies and heavy limp ones have high critical frequencies.

5.7.3.1 *Frequencies well below the critical frequency*

At frequencies well below the critical frequency (sub-critical frequencies), we may assume that $k \sin\phi / k_b \ll 1$. If, in addition, $(\omega m/\rho_0 c_0) \gg 1$, the expression for τ reduces to

$$\tau(\phi) \approx (2\rho_0 c_0/\omega m \cos\phi)^2. \tag{5.63}$$

This has the same form as equation (5.56) for normal incidence with ωm replaced by $\omega m \cos\phi$. This indicates that the effective mass $m \cos\phi$ reduces as the angle of incidence increases, and that the *partition mass* per unit area is the only controlling partition parameter. The product of frequency and effective mass uniquely determines $R(\phi)$ as in the normal incidence mass law; we can therefore refer to the *oblique incidence mass law*. The sound reduction index for waves incident at angle ϕ is given by

$$R(\phi) = R(0) + 20 \log_{10}(\cos\phi) \quad \text{(dB)} \tag{5.64}$$

which is less than $R(0)$ for all non-zero angles. This expression is not valid for $\phi = 90°(\pi/2)$.

In many practical cases in buildings, the sound field is formed by the reflection of sound waves from many surfaces and approximates to the ideal model termed a 'diffuse field' in which plane waves are assumed to propagate in all directions with equal probability (Chapter 1). It is therefore of interest to derive an expression for the sound reduction index corresponding to diffuse field incidence which is given by

$$R_{\mathrm{d}} = R(0) - 10\ \log_{10}[0.23R(0)] \quad (\mathrm{dB}). \tag{5.65}$$

Note that $R_{\mathrm{d}} < R(0)$. This may be termed the *diffuse incidence mass law*.

Many measurements of sound reduction index of real (finite) partitions suggest that the R_{d} gives results which are rather too low. An *empirical* formula derived from these measurements is

$$R_{\mathrm{f}} = R(0) - 5 \quad (\mathrm{dB}) \tag{5.66}$$

where R_{f} is termed the '*field incidence sound reduction index*'.

The various forms of the mass-controlled sound reduction index are plotted against the frequency–mass product in Fig. 5.37.

5.7.3.2 Frequencies around and above the critical frequency

As shown by Fig. 5.40, the acoustic trace wavenumber curve corresponding to any single angle of incidence intersects the partition free bending wavenumber curve, the frequency of intersection decreasing with angle. Hence there is a coincidence frequency corresponding to *each angle of incidence*, and there is a coincidence angle corresponding to every frequency *above the critical frequency*. This phenomenon may be demonstrated graphically by means of a wavefront diagram (Fig. 5.41).

Coincidence corresponds to the disappearance of the second term in square brackets in the denominator of equation (5.58), i.e $(k/k_{\mathrm{b}}) \sin\phi = 1$, in which case the damping term controls the transmission if $\eta > (2\rho_0 c_0/\omega m)\sec\phi$; this is made clear by considering the result of putting η to zero. The form of variation of $R(\phi)$ with frequency for individual angles of incidence is illustrated by Fig. 5.42 in which the damping-controlled coincidence 'dip' associated with each angle is clearly seen. The depth of the dip decreases as damping is increased.

It is not easy to interpret equation (5.58) in terms of parametric control of $R(\phi)$. A helpful graphical construction is presented in Fig. 5.43. At any one super-critical frequency, waves incident at angles less than the coincidence angle at that frequency are *mass* controlled, those incident at higher angles are *stiffness* controlled, and those incident at angles close to the coincidence angle are *damping* controlled. The general form of the diffuse incidence sound reduction index is shown in Fig. 5.44.

A general expression for the diffuse-field sound reduction index at frequencies well above the critical frequency was published by Cremer in 1942; the influence of damping is clearly seen:

$$R_{\mathrm{d}} = R(0) + 10\ \log_{10}(f/f_{\mathrm{c}} - 1) + 10\ \log_{10}\eta - 2 \quad (\mathrm{dB}). \tag{5.67}$$

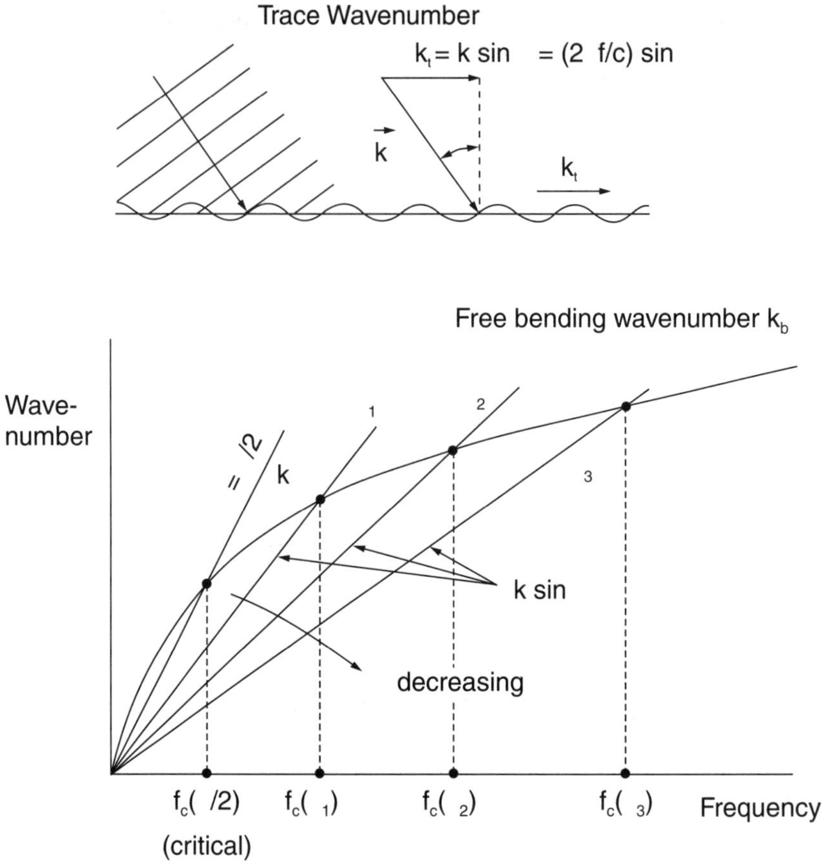

Figure 5.40 Illustration of the variation of coincidence frequency with angle of incidence.

5.7.3.3 Bounded partitions

The effect of partition boundaries on sound transmission has been analysed by Sewell (see Fahy, 1987). A comparison of the result of applying his expression to a glass panel is compared with measured results in Fig. 5.45. The principal effect of 'finiteness' is an increase in low frequency R_d to values above $R(0)$.

Note that increasing the *damping* of a partition will in most cases not increase R below the critical frequency, although it will reduce the resonant response of the partition. This is because sound is transmitted below the critical frequency primarily by non-resonant vibration, except in a small frequency range close to the fundamental resonance frequency of a

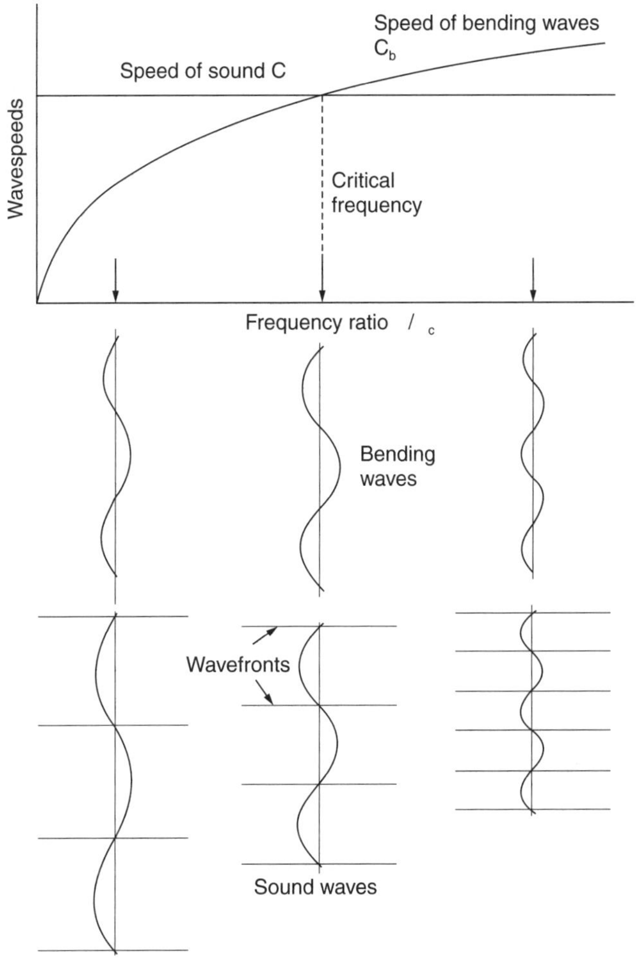

Figure 5.41 Coincidence diagram. (Trace the sound wavefronts on to a transparent sheet and superimpose them on the corresponding bending waves at various angles to illustrate coincidence).

partition, and at higher frequencies in cases of very lightly damped and rather small panels (Fahy, 1987).

5.7.4 Sound transmission through double-leaf partitions

Theoretical and experimental analyses of sound transmission through single-leaf partitions show that the sound reduction index at a given frequency generally increases by 5 to 6 dB per doubling of mass, provided that no significant flanking or coincidence-controlled transmission occurs.

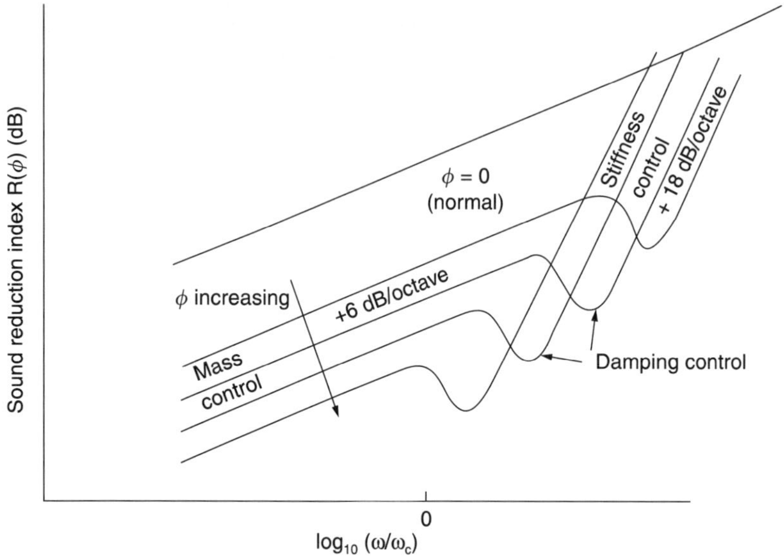

Figure 5.42 Variation of *R* with frequency for single angle of incidence.

In practice, it is often not only necessary for structures to have low weight, but also to provide high transmission loss; examples include the walls of aircraft fuselages, partition walls in tall buildings, and movable walls between television studios. This requirement can clearly not be met by single-leaf partitions.

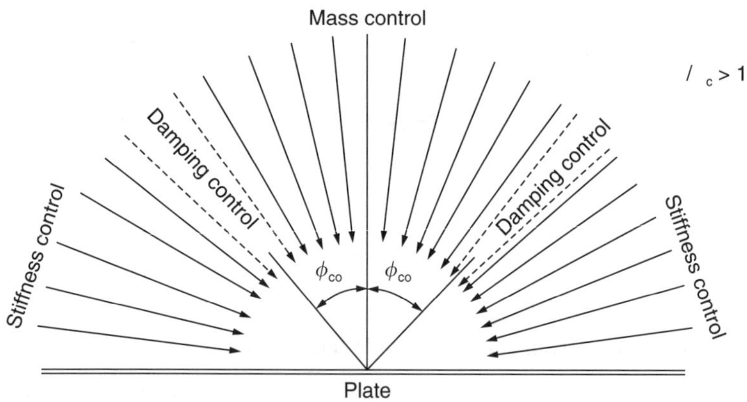

Figure 5.43 Regions of incidence angle associated with mass, stiffness, and damping control of transmission at a single supercritical frequency.

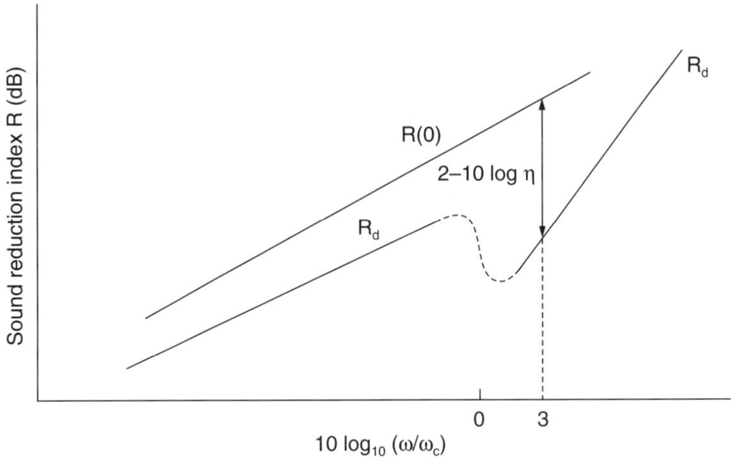

Figure 5.44 General form of the theoretical diffuse field sound reduction index curve.

The most common solution to this problem is to employ constructions comprising two separate leaves separated by an air space or cavity. It would be most convenient if the sound reduction index of the combination were to equal the sum of those of the two leaves when used as single-leaf partitions. Unfortunately the air in the cavity dynamically couples the two leaves, with

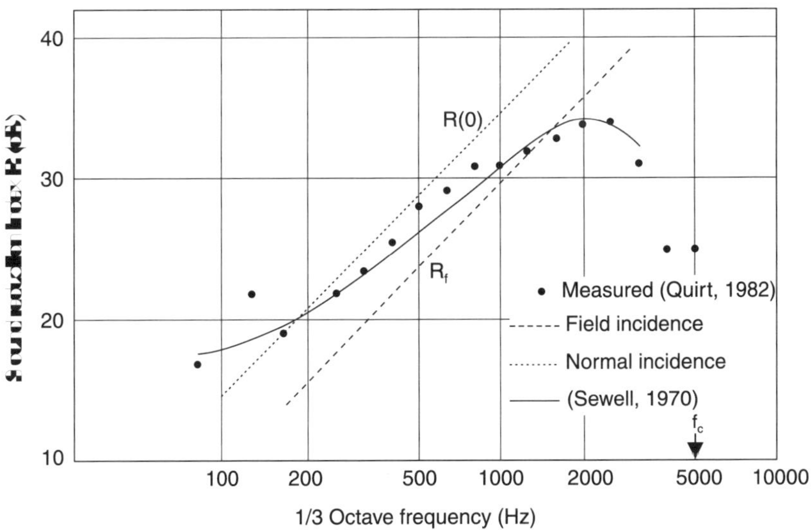

Figure 5.45 Sound reduction index of 3 mm thick glazing.

the result that the sound reduction index of the combination may fall below this ideal value, sometimes by a large amount.

It is clear from a superficial review of the available literature that theoretical analysis of the sound transmission behaviour of double-leaf partitions is far less well developed than that of single-leaf partitions, and that consequently greater reliance must be placed upon empirical information. The reason is not hard to find; the complexity of construction and the correspondingly larger number of parameters, some of which are difficult to evaluate, militate against the refinement of theoretical treatments. In particular, it is difficult to include the effects of mechanical connections between leaves, and of non-uniformly distributed mechanical damping mechanisms, in mathematical models.

5.7.4.1 Theoretical performance

A double-leaf partition consisting of two plane partitions separated by a layer of air (Fig. 5.46) behaves in a somewhat similar manner to a vibration isolation system consisting of a mass (leaf 2) mounted on a damped spring (air layer) to isolate it from the vibrational motion of a supporting base structure (leaf 1) (see Chapter 2). The vibration isolator is ineffective below the resonance frequency of the mounted mass–spring system, and actually amplifies the base motion in the region of the resonance frequency, to a degree controlled by the damper.

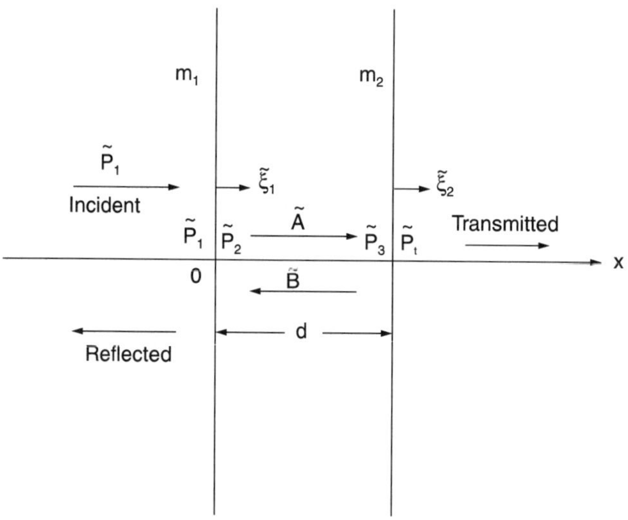

Figure 5.46 Idealized model of normal incidence sound transmission through a double-leaf partition.

A double-leaf system has a 'mass-air-mass' resonance frequency given by

$$f_0 = (1/2\pi)(\rho_0 c_0^2/d)^{1/2}[(m_1 + m_2)/m_1 m_2]^{1/2} \quad (5.68)$$

in which d is the thickness of the air layer and m_1 and m_2 are the masses per unit area of leaves 1 and 2 respectively. Below f_0, the stiffness of the air layer causes the panels to move virtually as a single partition and $R(0)$ given by equation (5.57) with $m = m_t = m_1 + m_2$.

In the frequency region close to f_0, the sound reduction index is *less* than that of a single partition of mass m_t. Above f_0, the vibration isolation produced by the air spring is effective and the $R(0)$ rises at 18 dB octave (6 dB/octave for m_t plus 12 dB/octave due to vibration isolation).

In the mathematical model which leads to equation (5.68), it is assumed that the thickness d of the air layer is much less than an acoustic wavelength, i.e. $k_d \ll 1$. The 18 dB octave rise in $R(0)$ continues until $kd \approx \pi$ which corresponds to an acoustic resonance frequency of the air cavity between rigid walls. At this frequency, and its integer multiples ($kd = n\pi$), the air cavity 'locks' the two leaves together and $R(0)$ falls $R(0, m_t)$. At the intermediate frequencies, given by $kd = (2n - 1)\pi/2$, $R(0)$ reaches a maximum given by $R(0) = R(0, m_1) + R(0, m_2) + 6$ (dB). The general forms of the $R(0)$ curves for various mass ratios m_1/m_2 are presented in Fig. 5.47.

In practice, the detailed behaviour shown in Fig. 5.47 is not observed in experimentally measured values of R because measurements are usually

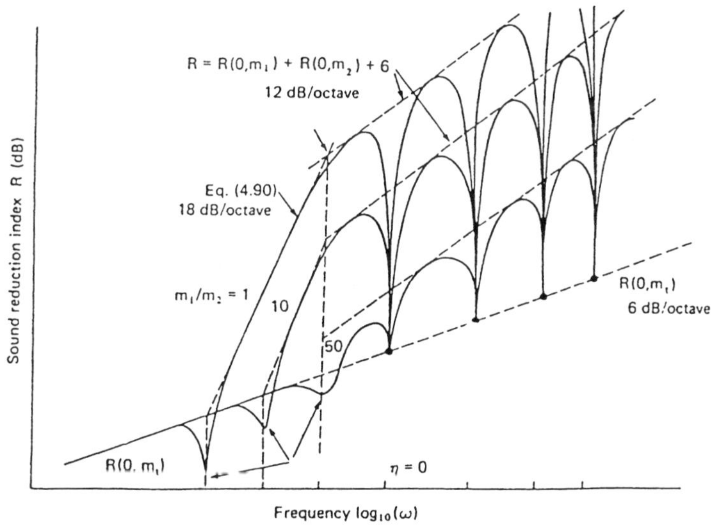

Figure 5.47 Illustration of the theoretical effect on normal incidence sound reduction index of varying leaf mass ratios, while keeping the total mass constant.

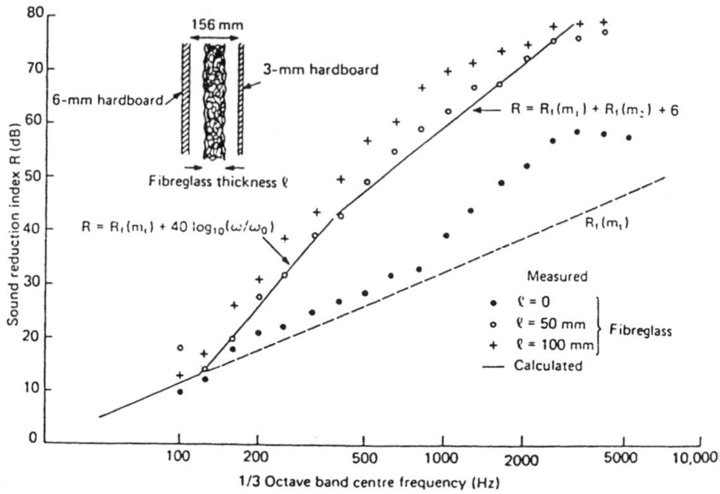

Figure 5.48 Performance of a mechanically isolated double-leaf partition.

made in one-third octave bands, the incident field is diffuse (not normal), and the air cavity usually contains sound absorbent material, which together obscure the individual maxima and minima. However, the general forms of the envelopes of the curves are observed.

5.7.4.2 Practical considerations

In order to achieve good low-frequency sound insulation f_0 must be kept as low as possible which requires a combination of heavy leaves separated by large distance. The importance of incorporating sound absorption material in the air cavity is illustrated by Fig. 5.48.

The individual leaves are normally mounted on a frame, usually of timber, but more effective if thin sheet metal. The mechanical connections 'short-circuit' the air spring and degrade the performance of the partition unless care is taken to minimize this path by the use of separate frames for each leaf on very flexible metal connections. The general effects are seen in Fig. 5.49.

5.7.5 Measurement of sound reduction index

A partition to be tested is placed between two mechanically and acoustically isolated reverberation rooms (Fig. 5.50). A broadband noise source generates an approximation to a diffuse field in the source room (1) and estimates of the space-average mean square sound pressures in room 1 and in receiver room (2) are made either by means of an array of fixed microphones or by

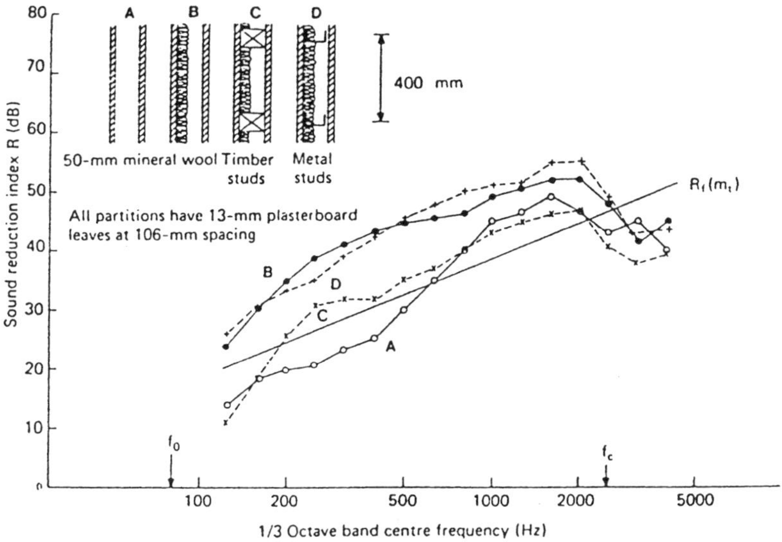

Figure 5.49 Influence of stud type on performance.

means of a continuous spatial microphone sweep. Measurements are made on one-third octave bands, as specified in ISO 140-3.

The sound power incident on a partition of area S is given by $W_i = \left(\langle \overline{p_1^2} \rangle / 4\rho_0 c_0\right) S$ of which a proportion $W_t = \tau W_i$ is transmitted into room 2 where it produces a space-average mean square pressure given by

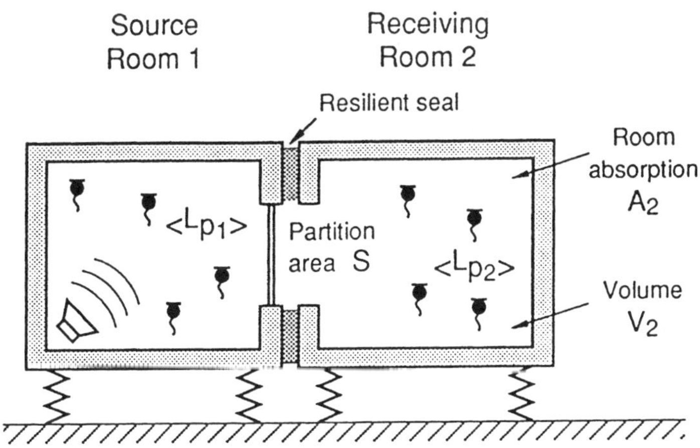

Figure 5.50 Schematic of airborne sound transmission test suite.

$\langle \overline{p_2^2} \rangle = 4\rho_0 c_0 W_t / a_2$, where a_2 is the absorption of room 2 (see Chapter 1). Consequently $\langle \overline{p_2^2} \rangle / \langle \overline{p_1^2} \rangle = S\tau / a_2$ from which

$$R = 10\log_{10}(1/\tau) = L_{p1} - L_{p2} + 10\log_{10}(S/a_2) \quad \text{(dB)}. \tag{5.69}$$

The absorption a_2 of room 2 is determined from measurements of the reverberation time T in room 2 via the Sabine formula $T = 0.16\,V_2/a_2$, where V_2 is the volume of room 2.

5.8 NOISE BARRIERS (SCREENS)

Noise barriers or screens are employed outdoors to reduce the levels of noise caused by road or rail traffic, or industrial operations, at the facades of nearby buildings. They are also used in offices and industrial workspaces to attenuate the propagation of noise or speech sounds. This section is restricted to a brief introduction to the performance characteristics of outdoor barriers. An appreciation of the phenomenon of diffraction is essential to the understanding of the acoustic performance of barriers. It is clear from everyday experience that you do not have to be able to see someone on the other side of a garden wall, or round the corner of a building, to be able to hear them. Sound waves can 'bend around' barriers to a degree essentially governed by a non-dimensional parameter called the Fresnel number, defined by Fig. 5.51 in relation to a point monopole source and a semi-infinite, thin, rigid, infinitely long barrier. The insertion loss of such a barrier is shown in Fig. 5.52 as a function of the Fresnel number (Maekawa, 1974). It is seen that the insertion loss is positive even when the

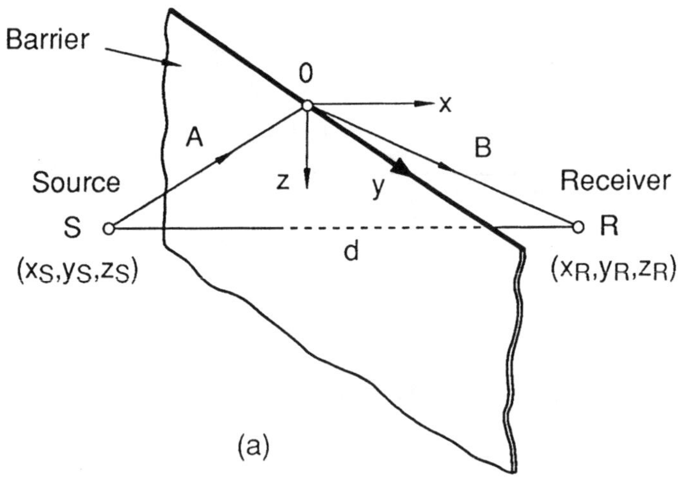

Figure 5.51 $N = \frac{2}{\lambda}(A + B - d)$ Fresnel number.

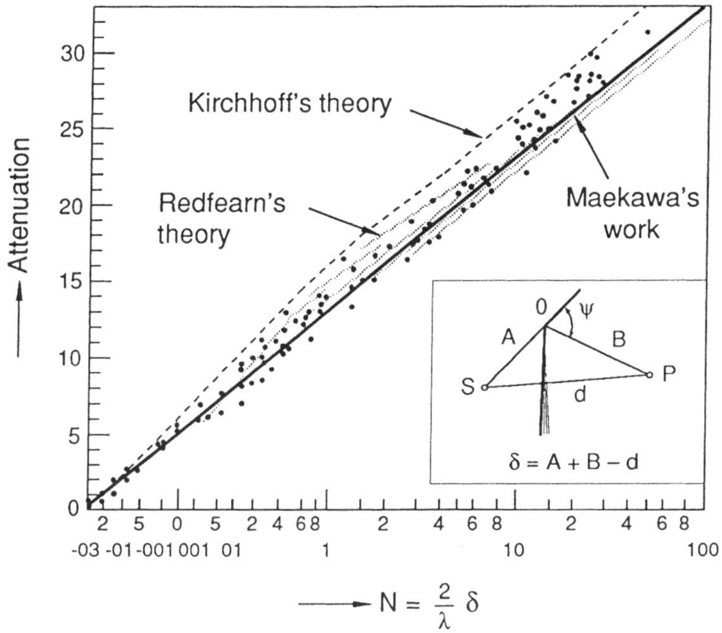

Figure 5.52 Sound attenuation by a semi-infinite screen in free space. Horizontal scale, logarithmic scale in the region of $N > 1$, adjusted so that the experimental curve becomes a straight solid line in the region of $N < 1$. Depending on whether $N > 0$ or $N < 0$, the receiving point P lies in the illuminated region or in the geometrical shadow, respectively. Attenuation is relative to propagation in free space. •, experimental value measured by pulsed tone ($\psi = 0 - 90°$).

line joining source and receiver passes above the top of the barrier. The equivalent performance for an infinitely long, incoherent line source (often used to approximate to a stream of cars) is shown in Fig. 5.53 (Maekawa 1974); it is inferior to that for a point source.

In practice, the performance of a barrier depends upon the following principal factors: barrier geometry (height, length and cross-sectional form); presence of sound-absorbing materials on the barrier; any apertures in the barrier; form and frequency spectrum of the source; locations of source and receiver relative to the barrier; topography and acoustic impedance of the ground between source and receiver positions; micro-meteorological conditions, particularly wind speed and direction; presence of nearby reflecting/scattering objects.

It is clearly impracticable in this volume to present a comprehensive survey of the influences of these many factors on barrier performance: for this the reader is referred to Beranek and Vér (1992). However, the following general points should be appreciated. It is generally not practicable to exceed an

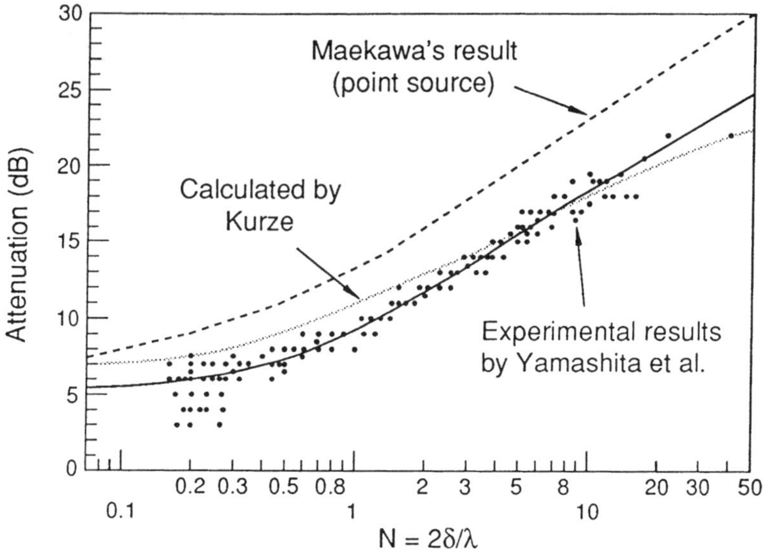

Figure 5.53 Sound attenuation by a thin screen versus Fresnel number N for an incoherent line source.

insertion loss of 15 dB(A) without prohibitively high cost. The reduction of noise from individual, well-separated trucks in low-speed urban traffic is different than that from a dense line of passenger cars partly because trucks act more like individual 'point' sources and partly because the exhaust outlet, engine and cooling fan are more elevated. The presence of a porous ground surface attenuates the propagation of sound travelling close to the surface, an effect readily observed by climbing a rising field beside a busy road. A barrier may destroy this effect, thereby producing a negligible, or even negative, insertion loss. The main effect of wind is to cause turbulence to be generated by the top of the barrier which reduces downwind barrier performance and increases upwind performance; it also creates large time-dependent fluctuations in insertion loss. The presence of absorbent materials on the road/rail side of a barrier increases performance by suppressing reflections between the side of the vehicle and the barrier. The presence of absorbent on and around the top of a barrier increases performance.

5.9 GENERAL VIBRATION CONTROL STRATEGIES

A large proportion of noise radiators take the form of vibrating solid surfaces. Among the most common are industrial machinery and plant, building components, domestic equipment and vehicle engines and body structures. In most cases, reduction of the surface vibration will produce proportional changes in radiated sound. It is, of course, preferable to reduce the vibrational excitation forces (or displacements) at source, but this is often impractical for technical or economic reasons. There are four principal alternative passive vibration control strategies:

1. *Vibration isolation*: This involves the insertion of resilient (flexible) elements between the structure which is directly subject to the source of vibration and other structures to which it is connected.
2. *Damping*: This generally involves the application of highly dissipative materials to the radiating structure to convert a large proportion of the vibrational energy into heat and therefore to reduce the vibration level. Damping may also be effected by connecting ancillary structures to a structure so as to produce enhanced dry friction losses or by attaching distributed damped resonators.
3. *Impedance mismatching*: This involves the design or modification of the structures where they are attached to the directly excited structure so as to reduce the transfer of vibrational energy from one to the other.
4. *Dynamic neutralization*: This involves the connection of ancillary mechanical resonators either to the directly excited structure, or to connected structures. In a narrow frequency range around resonance, they apply strong dynamic reaction forces which suppress the vibrational response of the structures to which they are attached.

5.10 VIBRATION CONTROL FOR NOISE REDUCTION

The principles and practice of impedance mismatching are outside the scope of this introductory chapter. The principle of dynamic neutralization is explained in Chapter 2. In application to noise control, this technique is generally only applicable to harmonic excitation at a fixed frequency, such as that of the fundamental tone of an aircraft propeller.

5.10.1 Damping treatments

The application of damping treatment to reduce sound radiation by a vibrating structure is successful only in cases where the radiation is caused by *resonant response* of the structure in one or more of its natural modes, or where modes *vibrate freely* (the structure 'rings') following the application of an impulsive input such as a mechanical impact. To be effective, it is also necessary that the additional damping greatly exceeds that naturally existing in the structure. Damping treatment is generally only effective when applied to rather thin structural components, such as metal panels of a few millimetres in thickness.

Surface damping treatments in the form of layers of viscoelastic materials take two generic forms. An 'unconstrained' damping layer is applied in the

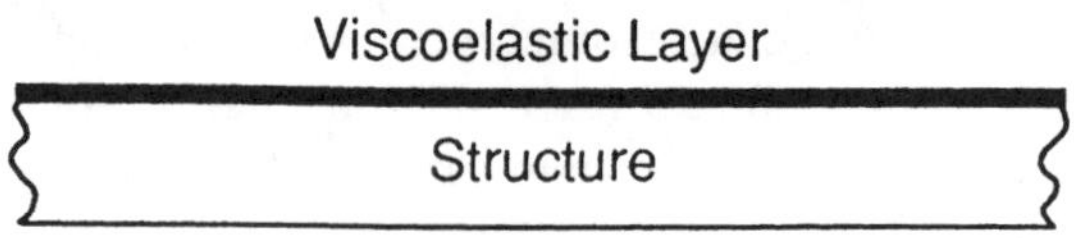

Figure 5.54 Unconstrained layer damping treatment.

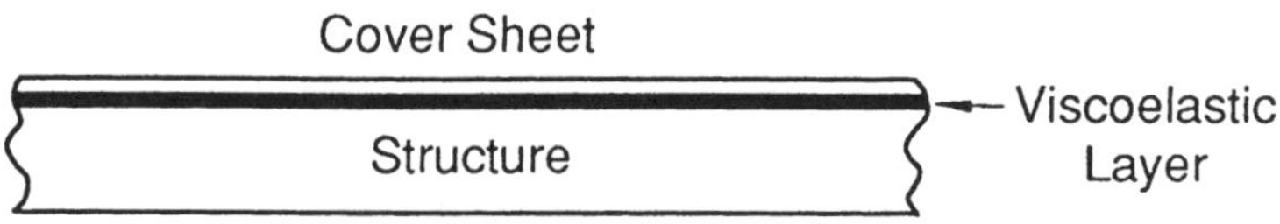

Figure 5.55 Constrained layer damping treatment.

form of a single homogeneous layer either sprayed, painted or bonded onto the surface of the structure (Fig. 5.54). A 'constrained' damping layer consists of a layer of viscoelastic material onto the surface of which is bonded a thin sheet of high elastic modulus (Fig. 5.55). This cover sheet constrains the viscoelastic layer to deform in shear so that the energy dissipation is enhanced by comparison with the unconstrained layer (Cremer, Heckl and Ungar, 1988). Single viscoelastic layers may also be sandwiched between two or more structural sheets of comparable thickness, but they reduce the static stiffness of the composite structure compared with that of a homogeneous structure of the same thickness. Multiple constrained layers (sandwiches) may also employed. The selection of layer thicknesses and loss factors may be optimized to produce maximum effectiveness when applied to a given structure.

As an alternative to viscoelastic layers, the damping of a sheet structure may be significantly increased by attaching another similar sheet to it by means of 'spot' connections (screws, rivets, welds). The combination of dry friction and air-pumping which results from vibrational motion can dissipate energy very effectively, but this type of structure is prone to corrosion and wear. Various forms of dissipative material may be also be attached to steel structures by means of magnetic forces; this is very useful for temporarily damping components which are subject to impact during a manufacturing process. The attachment to a structure of many small damped mechanical resonators tuned to a range of different frequencies can also produce significant damping, but the added weight is often unacceptable in practice.

In case where the radiating structure is excited by localized vibrational forces or displacements (e.g. a sump cover on an automotive engine, there is a limit to the effectiveness of surface damping treatment. It does not significantly reduce the structural vibration level in the immediate vicinity of

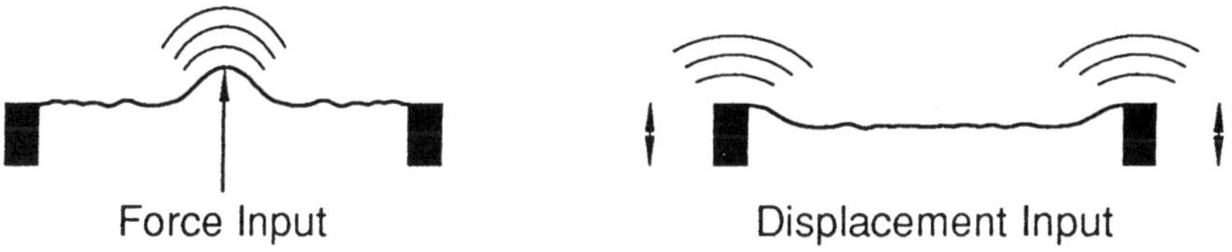

Figure 5.56 Residual radiation from heavily damped structure.

the region of excitation, and, however effective the damping, the sound radiated directly from these regions cannot be reduced below a residual level (Fig. 5.56) (Fahy, 1987).

5.10.2 Vibration isolation (low-frequency models)

The simplest model of vibration isolation consists of a concentrated mass mounted on a linear elastic spring supported by a *rigid* foundation (Fig. 5.57). For forced excitation of the mass, the effectiveness of the vibration isolator (spring) is expressed in terms of the magnitude of the ratio of the force transmitted to the foundation to the harmonic excitation force. For base excitation, the effectiveness is expressed in terms of the ratio of mass displacement to base displacement. The mathematical analyses are presented in Chapter 2.

The force transmissibility is significantly less than unity for excitation frequencies in excess of about three times the resonance frequency of the mass–spring system. Although this model is widely applied in practice to select vibration isolator stiffness, it can be highly misleading and can lead to poor and even disastrous results, principally because very few foundations,

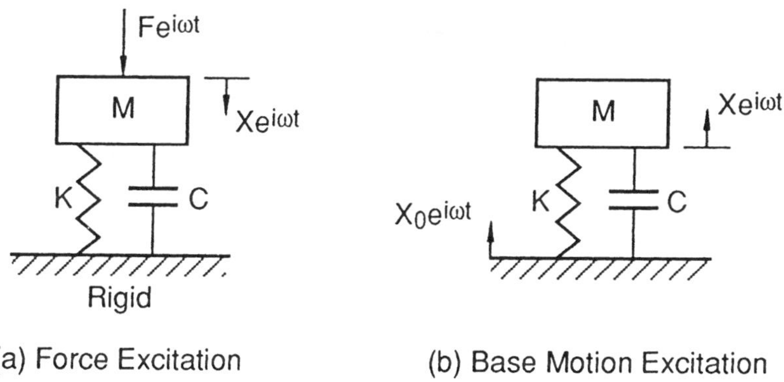

Figure 5.57 Single-degree-of-freedom isolator model.

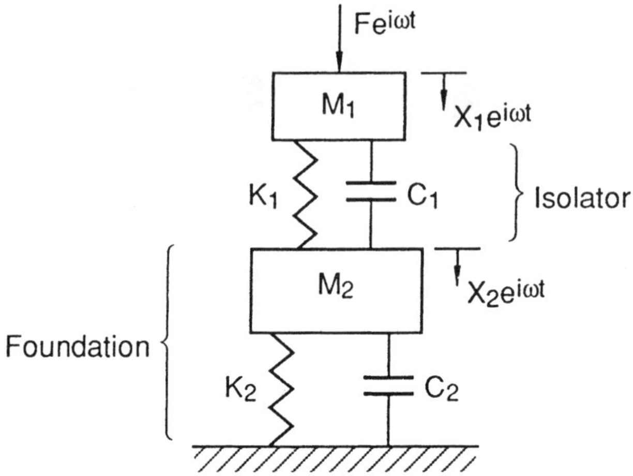

Figure 5.58 Two-degree-of-freedom isolator model.

except heavy concrete building foundations at ground level, are effectively rigid.

A somewhat more realistic model is provided by a two-degree-of-freedom system in which the foundation is represented by a second mass–spring system (Fig. 5.58). This system is analysed in Chapter 2. It has two undamped natural frequencies given by

$$\omega^2 = \{[\omega_1^2 + \omega_{2c}^2] \pm [(\omega_1^2 + \omega_{2c}^2)^2 - 4\omega_1^2\omega_2^2]^{1/2}\}/2$$

where ω_1, ω_2 and ω_{2c} are equal to $(K_1/M_1)^{1/2}$, $(K_2/M_2)^{1/2}$ and $[(K_1 + K_2)/M_2]^{1/2}$, respectively. In many cases, ω_2 is considerably higher than ω_1, in which case the two natural frequencies of the combined system correspond closely to ω_1 and ω_2.

The effectiveness of the isolator is greatly reduced at frequencies close to the foundation resonance. The foundation response is proportional to the stiffness of the isolator spring which must therefore be made sufficiently low to compensate for the effect of the foundation resonance. Hence, in cases of single-frequency excitation (e.g. out-of-balance excitation of a rotating machine), any foundation must be designed to ensure that no resonance frequency lies close to the excitation frequency. Even this model is not generally applicable to low-frequency isolator design because any rigid body has six degrees of freedom, and therefore six natural modes and frequencies.

5.10.3 Vibration isolation (high-frequency model)

None of the above models is appropriate to the selection of vibration isolation systems for the audio-frequency range because directly excited structures, such as machines, and structures connected to them, exhibit many resonances within this range. In principle, one could evaluate all the modes and natural frequencies of these structures (for example, by means of finite-element analysis) and design the isolation system accordingly. In practice, such deterministic models are highly unreliable for predicting all except a few of the lowest-order modes of complex structures because the dynamic properties of the structures are imprecisely known; the large modelling and computational effort required is therefore unjustified. An alternative, generalized model is required.

5.10.3.1 Mobility

In order to make an appropriate selection of resilient connections between complex structures, it is necessary to characterize dynamically both the isolator and the structures at the connection interface. For the purpose of illustration in this introductory course, it will be assumed that the connection may be considered as a point and that motion is only possible in one direction. The relevant descriptor is termed the '*driving point mobility*'. It is defined as the ratio of the complex amplitude of velocity at a point of excitation to the complex amplitude of an exciting force. In these notes it is symbolized by Y, but M is also commonly used. Note that for a single-degree-of-freedom at the driving point, the mobility is the inverse of the impedance. An example of a mobility spectrum of a long span concrete floor of a hospital is shown in Fig. 5.59. The many resonances are clearly

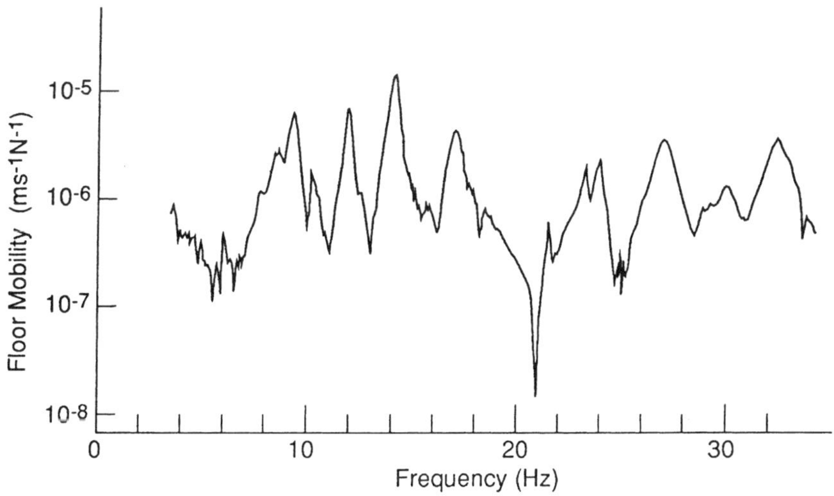

Figure 5.59 Driving point mobility of a long-span concrete floor.

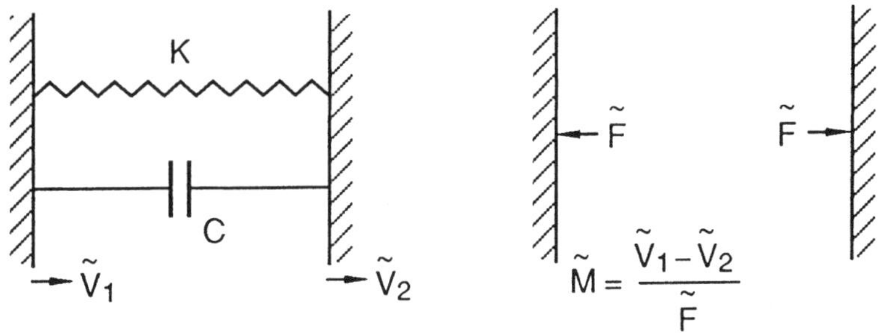

Figure 5.60 Differential mobility of a massless isolator.

observed. A large range of load-bearing engineering structures exhibit maximum (resonance) driving point mobilities of about 10^{-5} m/sN.

5.10.3.2 Mobility of an isolator

Resilient elements exhibit modes and resonances like every other structure, but well below their fundamental resonance frequencies it is reasonable to model them as lumped, massless elements consisting of a combination of an elastic spring and a viscous dashpot (Fig. 5.60).

The dynamic behaviour of this model may be represented in terms of its *differential* mobility (i.e. *relative* velocity of the two ends divided by the applied force). Simple analysis shows the differential mobility to be given by

$$Y = 1/[K/j\omega + C] = [C + jK/\omega]/[(K/\omega)^2 + C^2]. \tag{5.70}$$

If $K/\omega \gg C$, the isolator is stiffness controlled (low frequencies).

If $K/\omega \ll C$, the isolator is damping controlled (high frequencies).

Note that this simple model indicates that low damping is beneficial. However, in practice, sufficient damping is necessary at high frequencies to suppress the adverse effects of standing wave resonances within the isolator element; a compromise is necessary.

5.10.3.3 A model of high-frequency vibration isolation

The model is shown in Fig. 5.61. A generalized source structure S is directly excited by some applied force(s). As a result, vibrational (incident) waves arrive at the point of connection to the isolator I. In the *absence* of the isolator the resulting *free* normal velocity of the surface is denoted by V_f. The isolator is then attached to the source and the resulting reaction force at the connection point generates a wave in S having normal surface velocity V_r in the opposite direction to that of the incident wave. The isolator generates a wave of normal surface velocity V_t in the receiving structure R.

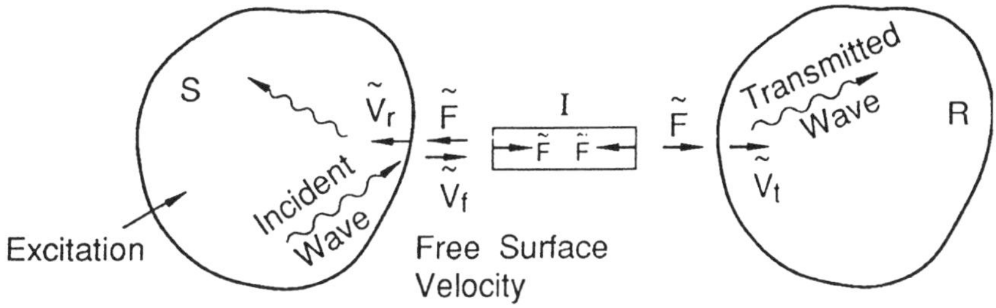

Figure 5.61 High frequency mobility isolator model.

The forces on each end of the isolator are only equal if it is assumed to be *massless*.

Let V_f = the velocity amplitude of the *free* surface (without isolator), V_r = the velocity amplitude due to the isolator force F, and V_t = the velocity amplitude due to the isolator force F,

$$V_f - V_r = V_f - F\ Y_s \quad (Y_s \text{ at isolator connection}) \tag{5.71}$$

$$V_t = F\ Y_R \tag{5.72}$$

$$F = [(V_f - V_r) - V_t/Y_I \quad (\text{definition of } Y_I). \tag{5.73}$$

Rearranging equation (5.71) $\quad F = (V_f - V_t)/Y_I - V_r/Y_I$.

From equation (5.72) $\quad F = (V_f - F\ Y_R)/M_I - F\ Y_s/Y_I$.

Therefore $F[1 + Y_R/Y_I + Y_s/Y_I] = V_f/Y_I$. (5.74)

From equations (5.72) and (5.74) $V_t/V_f = \dfrac{Y_R}{Y_I + Y_R + Y_s}$. (5.75)

If $Y_I = 0$ (rigid connection) $V_t/V_f = \dfrac{Y_R}{Y_R + Y_s}$. (5.76)

$$\text{Isolator effectiveness } E = \left|\frac{(V_t/V_f)_{Y_I=0}}{(V_t/V_f)_{\text{isolator}}}\right|$$

$$= \left|1 + \frac{Y_I}{Y_R + Y_s}\right|. \tag{5.77}$$

It is therefore necessary to make $|Y_I| \gg |Y_R + Y_s|$ for high isolator effectiveness.

The result of this analysis demonstrates that, to be effective, the differential mobility of the isolator must greatly exceed the mobilities of *both the source and the receiver*. The mobility of a structure reaches

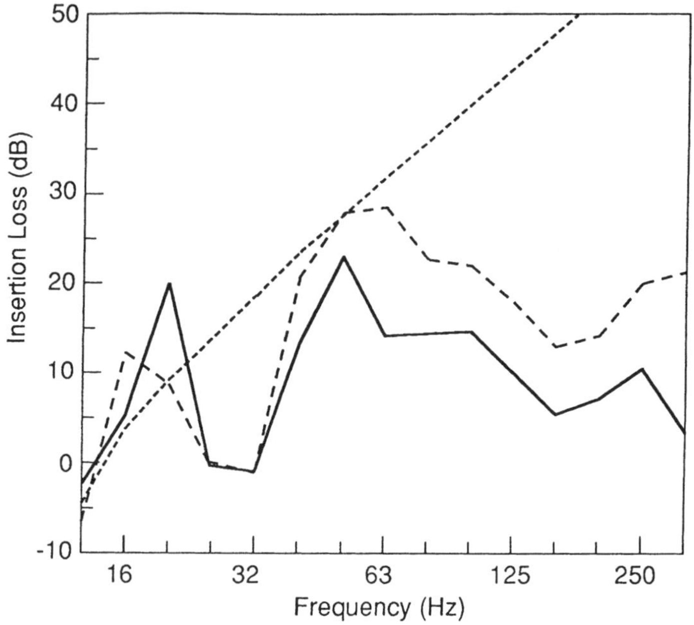

Figure 5.62 Foundation force insertion loss of an isolator under a fan unit: - - - - - - single mass–spring–damper theory; – – – – prediction based upon measured fan mobility; —— measured.

maximum values at resonance, and therefore it is very difficult to provide sufficient dynamic isolation between multi-resonant structures without compromising static stability, especially if one or both are light in weight. A comparison between the performance of a spring isolator under fan unit and that predicted by equation (5.77) is shown in Fig. 5.62, together with the prediction from the simple single-degree-of-freedom model (Granhäll and Kihlman, 1980). The finite mobility of the unit at the mounting point accounts for the difference.

It is common to mount machinery and plant components on heavy blocks in order to improve vibration isolation. It can be seen from the above analysis that the mechanism of improvement at high frequencies is the decrease of source mobility.

5.10.3.4 Frequency-average mobility

Although the detailed modal behaviour and frequency response of complex structures is completely unpredictable over most of the audio-frequency range, it is possible to characterize their frequency-average dynamic behaviour on the basis of rather simple geometric and material data. An

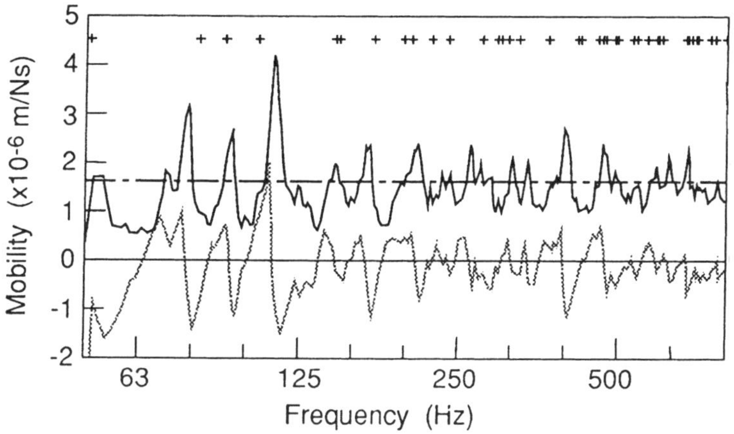

Figure 5.63 Measured mobility of a 6 × 4 m^2× 200 mm thick concrete floor: —— real part; ······ imaginary part; – - – theoretical infinite plate mobility (real); + predicted modal resonance frequencies.

example of the frequency-average mobility of a concrete floor is shown in Fig. 5.63 (Craik, 1996). A range of examples for plates and beams is presented in Cremer, Heckl and Ungar (1988), and White and Walker (1984). Consequently, high-frequency vibration isolation design is based upon the frequency-average mobility characteristics of connected structures.

5.11 REFERENCES

Allard, J. F. (1993) *Propagation of Sound in Porous Media*, Elsevier Science Publishers, London.

Beranek, L. L. and Vér, I. L. (eds) (1992) *Noise and Vibration Control Engineering*, John Wiley and Sons Inc., New York.

Bies, D. A. and Hansen, C. H. (1995) *Engineering Noise Control*, 2nd edn, E&FN Spon, London.

Craik, R. J. M. (1996) *Sound Transmission through Buildings using Statistical Energy Analysis*, Gower Publishing, Aldershot.

Cremer, L., Heckl, M. and Ungar, E. (1988) *Structure-borne Sound* (2nd edn), Springer Verlag, Berlin.

EC Directives (1986) 86/188/EEC 'Protection of workers from risks related to exposure to noise at work' and 89/392/EEC 'Essential health and safety requirements relating to the design and construction of machinery'.

Fahy, F. J. (1987) *Sound and Structural Vibration*, Academic Press, London.

Fahy, F. J. (1995) *Sound Intensity*, 2nd edn, E&FN Spon, London.

Granhäll, A. and Kihlman, T. (1980) The use of mechanical impedance data in predicting isolation efficiency. *Noise Control Engineering*, **14** (2), 88–93.

Ingard, U. (1994) *Sound Absorption Technology*, Noise Control Foundation, Poughkeepsie, New York.

International Electrotechnical Commission (1993) IEC 1043: Instruments for the measurement of sound intensity.

Jonasson, H. (1996) Determination of emission sound pressure levels. Comparisons between a new intensity method and traditional methods. *Proceedings of Internoise 96, Liverpool, UK*, 2667–672.

Joseph, P. F. and Hewlett, D. A. K. (1996) *The Determination of Exhaust Stack Sound Power from Ground Based Sound Pressure Measurements*. ISVR Contract Report No. 96/18, University of Southampton.

Maekawa, Z. (1974) Environmental sound propagation. *Proceedings of the Eighth International Congress on Acoustics*, London.

Mechal, F. (1990) Numerical results to the theory of baffle-type silencers. *Acustica*, **72**, 7–20.

Ren, M. and Jacobsen, F. (1993) A method of measuring the dynamic flow resistance and reactance of porous materials. *Applied Acoustics*, **39**, 265–76. (See also *Journal of Sound and Vibration*, **103**, 567–72).

Utsuno, H., Tanaka, T. and Fujikawa, T. (1989) Transfer function method for measuring the characteristic impedance and propagation constant of porous materials. *Journal of the Acoustical Society of America*, **86**, 637–43.

White, R. G. and Walker, J. G. (eds) (1984) *Noise and Vibration*, Ellis Horwood, Chichester, England.

Wu'Qunli (1988) Empirical relations between acoustical properties and flow resistance of porous plastic open-cell foam. *Applied Acoustics*, **25**, 141–8.

Zwikker, C. and Kosten, C. W. (1949) *Sound Absorbing Materials*, Elsevier Publishers, Amsterdam.

5.12 QUESTIONS

1. A manufacturer advertises an 'acoustic' door as providing X dB(A) noise reduction. Explain why this statement is scientifically unsatisfactory, using diagrams as appropriate. Present an alternative, more satisfactory, form of statement of acoustic performance of the door.
2. The surface normal specific acoustic impedance of a surface is given by $\tilde{z}_n/\rho_0 c = 1.8 - 4\,j$. Determine by analysis the angle of incidence for maximum absorption coefficient and the value of this maximum.
3. It is necessary, for hygienic purposes, to cover a sound absorber consisting of a thick sheet of mineral wool, with a thin plastic sheet of mass 1 kg m^{-2}. By treating the plastic sheet as a *simple partition in free space* (i.e. no interaction with the mineral wool), estimate the approximate reduction in absorption coefficient of the mineral wool at frequencies of 100, 1000 and 10,000 Hz. The normal incidence absorption coefficient of the uncovered mineral wool sheet at these frequencies is 0.3, 0.8 and 1.0 respectively.
   Show the derivation of all expressions used.
   [Hint: consider the transmission of incident sound, and of sound reflected from the absorber, to be independent processes.]

4 (a) A noise control enclosure placed around a floor-mounted machine consists of a rectangular box constructed of thin steel sheet, fully

lined with a sound absorbent sheet. The dimensions of the box are 3 m × 3 m × 2 m (height). It incorporates an aperture of dimensions 600 mm × 300 mm. The diffuse field sound reduction index of the enclosure in the 1000 Hz one-third octave band is 20 dB and the diffuse incidence sound absorption coefficient of the lining, when placed against a rigid surface, averages 0.80 in this band.

Derive an expression for the sound power insertion loss of this enclosure and determine its value in the 1000 Hz one-third octave band, carefully explaining the assumptions made in your analysis.

(b) Explain why your analysis would probably give inaccurate results at 100 Hz, and at all frequencies in the absence of an absorbent lining.

5. It is required to vibration isolate a water pipe from the wall of a building on which it is mounted because it is transmitting excessive flow noise. The pipe is made of copper and has an outer diameter of 20 mm and a wall thickness of 2.5 mm. The building wall is made of dense concrete and has a thickness of 100 mm. The mobilities of the pipe and the wall are given by the following expressions.

$$\mathrm{M_p} = (1 - j)/[4B^{1/4}\omega^{1/2}m^{3/4}] \quad \text{and} \quad M_{\mathrm{w}} = 1/[8(B\rho h)^{1/2}]$$

where ω denotes radian frequency, ρ denotes material density, m denotes mass per unit length and B denotes bending stiffness. The elastic modulus and density of copper are 12.5×10^{10} Nm^{-2} and 8.9×10^{3} kgm^{-3} and the corresponding properties of dense concrete are 2.6×10^{10} Nm^{-2} and 2.3×10^{3} kgm^{-3}. Determine the mobility of an isolator which gives an effectiveness of 100 at a frequency of 500 Hz.

ACKNOWLEDGMENT

Figure 5.36–5.39, 5.41–5.49 and Table 5.5 are reproduced by permission of the publishers from *Sound and Structural Vibration* (1985) by F.J. Fahy, Academic Press, London.

6

Fundamentals of signal processing

J.K. HAMMOND

6.1 INTRODUCTION

Signal processing is the science of applying transformations to measurements, to facilitate their use by an observer or a computer, and Digital Signal Processing (DSP) is the enabling technology for applications across all disciplines and sectors.

The study of sound and vibration is increasingly dependent on the use of special-purpose signal analysers or software packages which are easily installed on a variety of computing platforms. The accessibility and convenience of DSP analysis modules and procedures can sometimes create a deceptive air of simplicity in often complicated phenomena. It is important that practitioners, while availing themselves of the full range of DSP capabilities, should have a clear understanding of the fundamentals of the science of signal processing and so be fully aware of the assumptions, implications and limitations inherent in their analysis methods.

The aim of this chapter is to provide an introduction to signal processing, emphasizing the basis of the methodology.

Data analysis involves the three phases of data acquisition, processing and interpretation (of the results of the processing) and, of course, all three are linked in any application. The last phase is naturally very much related to the subject under investigation, but the first two may be discussed independently of specific applications. A vast body of theory and methodology has been built up as a consequence of the problems raised by the need for data analysis; this is often referred to as 'signal analysis' or 'time series analysis' depending on the context. This chapter concerns methods of analysis of time histories which have usually been obtained as measurements in the course of conducting an experiment, or when recording the behaviour of a physical system. We shall limit our considerations to 'basic' procedures; inevitably the treatment will be superficial owing to the restricted scope afforded by a single chapter.

6.1.1 Data classification, generation and processing

Figure 6.1 illustrates a broad categorization of time histories (Bendat and Piersol, 1971). A basic distinction is the designation of a signal as 'random' or 'deterministic' where by 'random' we mean one that is not exactly predictable. Very often, processes are mixed and the demarcations of Fig. 6.1 are not easily applied.

Additionally, it is often convenient to regard a time history as arising from some 'data generation model' which may be physical or conceptual (e.g. appropriately filtered white noise), as in Fig. 6.2.

We shall neglect chaotic signals which may be generated by deterministic non-linear processes but nevertheless have unpredictable behaviour. The processing strategy we employ may be significantly influenced by prior conceptions regarding the 'type' of signal and its mode of generation, leading to either *parametric* or *non-parametric* analysis.

These categorizations are a rather simplistic view of modern signal processing which employs a variety of noise reduction, identification and inversion techniques based within optimization theory, matrix theory and

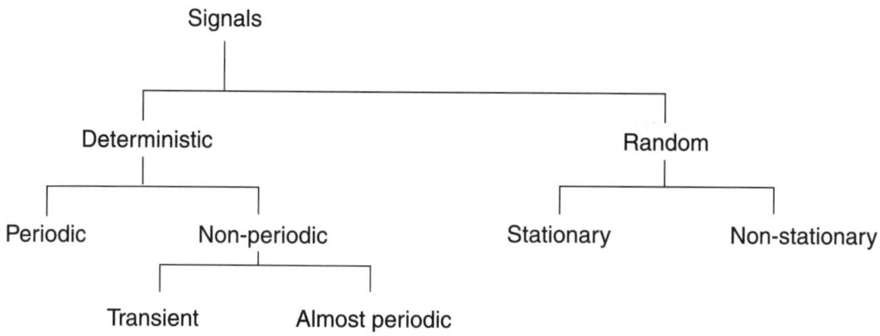

Figure 6.1 Types of signals.

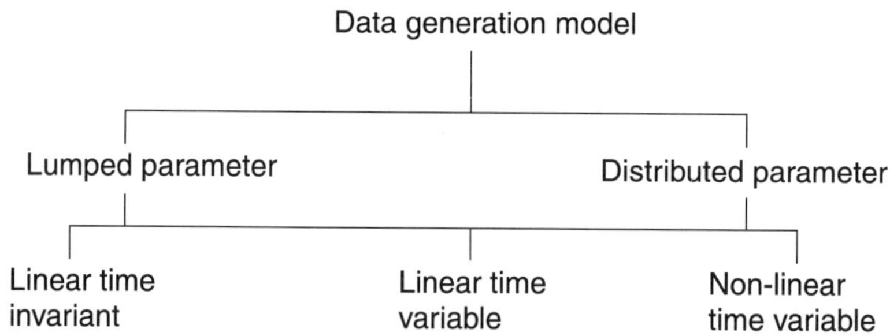

Figure 6.2 Data generation models.

probabilistic methods. However, it is important to be aware in any analysis that procedures based on prior knowledge (or often merely assumptions) may bias the outcome. We should strive always to keep in mind the First Principle of Data Reduction (Ables, 1974): 'The result of any transformation imposed on the experimental data shall incorporate and be consistent with all relevant data and be maximally non-commital with regard to unavailable data.'

6.1.1.1 Objectives of data analysis

Generally speaking, the objective of data analysis is to highlight/extract information contained in a signal that direct observation may not reveal. Sometimes only 'gross' characteristics of a signal may be required (e.g. mean value, mean square level, total energy) in which case some simple processing will produce the required quantities whilst ignoring other information contained in the signal which might only unnecessarily confuse matters. At other times, it may be necessary to probe more deeply in an attempt to extract features of interest.

6.1.1.2 Methods of data analysis

The signals that we shall consider are time histories (often electrical outputs from transducers), or equivalent (e.g. rough ground height profiles where a space variable is the equivalent of time) and useful information extraction procedures may be performed while remaining in the 'time domain'. However, very often transform methods are employed since they offer simplifications both in theory and physical interpretation, and 'frequency domain methods' in particular are commonly used. The signals recorded from physical phenomena are generally 'continuous time' in nature (sometimes called analogue data), but developments in technology have now ensured that most signal processing is achieved digitally. This means the data is sampled, and the analysis of discrete (in time) data requires that the analyst should have a good grasp of the fundamentals of digital signal processing, which is now a vast area of study. The next two sections deal with elements of Fourier analysis of continuous time signals.

6.2 FOURIER ANALYSIS OF CONTINUOUS TIME SIGNALS

This section is concerned with the representation of signals as the sum of simpler components, namely sinusoids and cosinusoids. As a preliminary, we briefly discuss the (Dirac) delta function and the concept of convolution.

6.2.1 Preliminary – the delta function

The delta function (sometimes called the unit impulse function) is written $\delta(t)$ and is defined by

$$\delta(t) = 0 \quad t \neq 0$$
$$\int_{-\infty}^{\infty} \delta(t)\mathrm{d}t = 1. \tag{6.1}$$

Clearly this is not a function in the ordinary sense. It belongs to the class of 'generalized' functions. We can intuitively think of it as a very narrow, tall spike at $t = 0$, i.e. consider a function (Figure 6.3)

$$\delta_\varepsilon(t) = \frac{1}{\varepsilon} \quad -\frac{\varepsilon}{2} < t < \frac{\varepsilon}{2}$$
$$= 0 \quad \text{elsewhere}$$

and so $\int_{-\infty}^{\infty} \delta_\varepsilon(t)\mathrm{d}t = 1$. This defines a unit 'impulse'.

Now visualize $\delta(t) = \lim_{\varepsilon\to 0} \delta_\varepsilon(t)$. We represent it as an arrow whose height indicates the 'strength' of the impulse (Fig. 6.4).

Another interpretation may be given in terms of the unit step function

$$u(t) = 1 \quad t > 0$$
$$= 0 \quad t < 0$$

so that

$$\delta(t) = \lim_{\varepsilon\to 0} \delta_\varepsilon(t) = \frac{\mathrm{d}}{\mathrm{d}t}u(t) \tag{6.2}$$

i.e. the delta function is the derivative of the unit step function. The introduction of this concept permits us to differentiate functions having discontinuities.

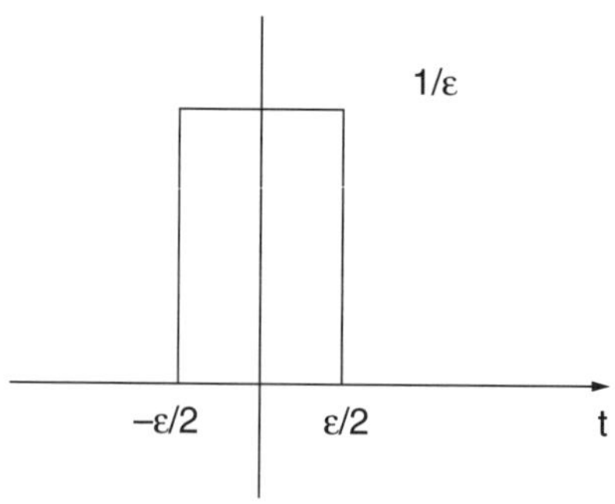

Figure 6.3 An impulse of unit magnitude.

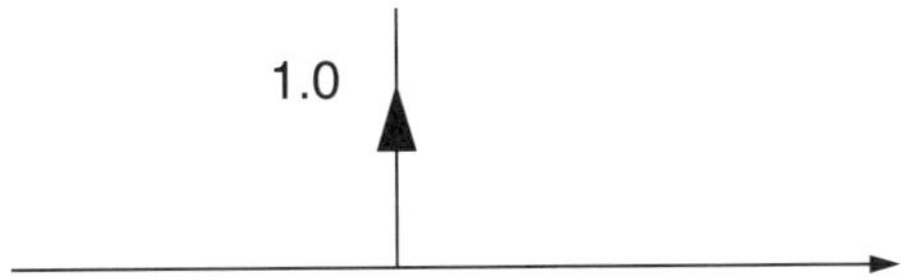

Figure 6.4 Representation of the delta function.

A delta function located at $t = a$ is written $\delta(t - a)$ and if $f(t)$ is an 'ordinary' function, then

$$\int_{-\infty}^{\infty} f(t)\delta(t-a)\mathrm{d}t = f(a). \tag{6.3}$$

This 'sifting property' is very useful and we also note that

$$\int_{-\infty}^{\infty} \mathrm{e}^{\pm jat}\mathrm{d}t = 2\pi\delta(a); \qquad \int_{-\infty}^{\infty} \mathrm{e}^{\pm 2\pi jat}\mathrm{d}t = \delta(a) \ . \tag{6.4}$$

(N.B. We emphasize that our heuristic exposition may be made more formal using 'Distribution Theory'; see, for example, Zadeh and Desoer, 1963).

6.2.2 The Fourier series and Fourier integral

In Fig. 6.1 we defined three types of deterministic signal, namely, periodic, transient and almost periodic. Periodic data is analysed using the Fourier series, transient data uses the Fourier integral, and some classes of almost periodic signals (and also damped signals) may be studied using the Prony series (Spitznogle and Quazi, 1970; Davies, 1983). The basic difference between the Prony series and the Fourier series is that in the former case the frequencies are not assumed known, and so both (complex) amplitude and frequency are computed, while only complex amplitudes are calculated at particular frequencies using the Fourier method. The Prony method is computationally more 'difficult' but can be very attractive under certain circumstances (e.g. where data lengths are short and where the data matches the model of a sum of damped oscillations).

If a signal repeats itself exactly every T_{p} seconds, then $x(t)$ may be represented as a sum of sinusoids and cosinusoids whose frequencies are integer multiples of the fundamental frequency $1/T$ Hz, together with a constant representing the 'offset' of the signal. Each harmonic component is characterized by its amplitude and phase. It is very convenient to express these oscillations in complex form which leads to expressing the real valued signal $x(t)$ as

$$x(t) = \sum_{n=-\infty}^{\infty} C_n e^{j2\pi nt/T_p} \; . \tag{6.5}$$

The amplitude C_n is now complex, representing both amplitude and phase of each component, and the concept of a negative frequency has been introduced by the term $n \cdot \frac{1}{T}$ when n is negative. If $x(t)$ is a transient then the discrete set of sinusoids and cosinusoids becomes a continuum and we write

$$x(t) = \int_{-\infty}^{\infty} X(f) e^{j2\pi ft} \mathrm{d}f \; . \tag{6.6}$$

$X(f)$ is given by

$$X(f) = \int_{-\infty}^{\infty} x(t) e^{-j2\pi ft} \mathrm{d}t \; . \tag{6.7}$$

The specification of C_n or $X(f)$ computed from $x(t)$ is equivalent to the original time history, since we can reconstruct the signal from these. The frequency f is in Hz and $\omega = 2\pi f$ is the frequency in rad/s. We note that the properties of the delta function (particularly (6.3)) and (6.4) may be used to 'invert' (6.6) to give (6.7).

6.2.2.1 Random signals and non-stationarity

It is a logical step to ask if the representations (6.5) and (6.6) may be generalized to allow the representation of random signals. 'Randomness' and the nature of the randomness (stationary or non-stationary) introduce some significant conceptual aspects (Priestley, 1967). In summary, a stationary random signal may be expressed as

$$x(t) = \int_{-\infty}^{\infty} e^{j\omega t} \mathrm{d}Z_x(\omega) \tag{6.8}$$

which is a generalization of (6.6). (This is a Fourier–Stieltjes form – only reducing to the Riemann form (6.6) when $Z_x(\omega)$ is differentiable as for 'transients' above.) We will sidestep this mathematical correctness in section 6.6 in our treatment of stationary random signals.

If the process is *non-stationary*, then while this form *may* still be appropriate there are significant changes that occur to the corresponding spectra (due to *non-orthogonality* of $\mathrm{d}Z_x(\omega)$) and so concepts of two-dimensional (frequency-frequency or time-frequency) spectra have been introduced. The whole subject of time–frequency analysis has attracted considerable interest but is beyond the scope of this introductory chapter and the reader is referred to Priestley (1967), Cohen (1989) and Hammond and White (1996) for further details.

6.3 SOME RESULTS IN SIGNAL AND SYSTEM ANALYSIS

In this section we establish some basic results in signal and system analysis that are needed in data analysis. Concepts that are described include convolution (in both time and frequency domains) and the Hilbert transform.

6.3.1 The effect of delay

If a signal $x(t)$ has Fourier transform $X(f)$, then a delayed signal $x(t - t_0)$ has Fourier transform $e^{-j2\pi f t_0} X(f)$. The effect of the delay in the frequency domain is the multiplier $e^{-j2\pi f t_0}$, which has unit magnitude and phase $\phi(f) = -2\pi f t_0$. The derivative of the phase with respect to f is $\frac{d\phi(f)}{df} = -2\pi t_0$, i.e., proportional to delay t_0 for all frequencies. Written in terms of ω we can write $-\frac{d\phi}{d\omega} = t_0$.

The quantity $-\frac{d\phi}{d\omega}$ is called the group delay of the signal. In this case, the delay is the same for all frequencies because this is a pure delay (there is no dispersion). If $-\frac{d\phi}{d\omega}$ is a *nonlinear* function of ω then dispersion is present and phase distortion of the signal occurs.

6.3.1.1 Echoes

If a pure echo is present in a signal, $y(t)$, this may be modelled as

$$y(t) = x(t) + ax(t - t_0)$$

where $x(t)$ is the 'original' signal, a is the echo amplitude and t_0 is the 'epoch' of the echo. The Fourier transform of $y(t)$ is

$$Y(f) = (1 + ae^{-j2\pi f t_0})X(f) \ .$$

The effect of the echo is contained in the term $(1 + ae^{-j2\pi f t_0})$ which has an oscillatory form as a function of frequency in both amplitude and phase. The squared amplitude is $(1 + a^2 + 2a \cos 2\pi f t_0)$ which imposes a harmonic form on $|X(f)|^2$. Accordingly, the appearance of such a feature in an energy (or power) spectrum is an indicator of an echo. Additional echoes and dispersion obviously complicate the picture. The analysis of such signals leads to the introduction of cepstral analysis (Bogert, Healey and Tukey, 1963) which in turn was generalized to homomorphic deconvolution (Oppenheim and Schafer, 1975).

6.3.2 Convolution

The *convolution* of two time histories is introduced via the concept of linear filtering. In the figure below, $x(t)$ is the excitation of a linear system with impulse response function $h(t)$ producing response $y(t)$.

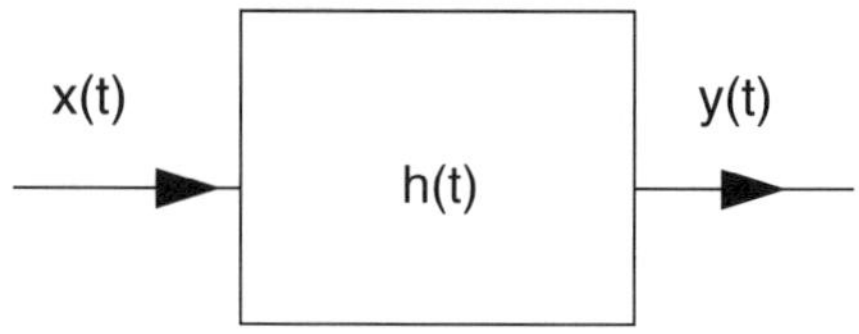

Figure 6.5 A single-input single-output system.

With reference to Fig. 6.5, since $h(t)$ is the response to a unit impulse $\delta(t)$, then for a time invariant system, $h(t - t_1)$ is the response to $\delta(t - t_1)$. Splitting up $x(t)$ into elemental impulses $x(t_1)\Delta t_1$ then the response at time t due to such an impulse is $h(t - t_1)x(t_1)\Delta t_1$ if the system is linear (see Fig. 6.6). Adding up all such responses and letting $\Delta t_1 \to 0$ yields

$$y(t) = \int_{-\infty}^{t} h(t - t_1)x(t_1)\mathrm{d}t_1. \tag{6.9}$$

The upper limit is t because we assume the system is causal. Using the substitution $t - t_1 = \tau$, this may be expressed as

$$y(t) = \int_{0}^{\infty} h(\tau)x(t - \tau)\mathrm{d}\tau \tag{6.10}$$

when $h(\tau)$ now has the role of a weighting (or memory) function as depicted in Fig. 6.7.

The convolution of the functions h and x is often written $y = h * x$. This expression of the effect of a linear filter in the time domain is considerably simplified by taking the Fourier transform, which yields

$$Y(f) = H(f)X(f), \tag{6.11}$$

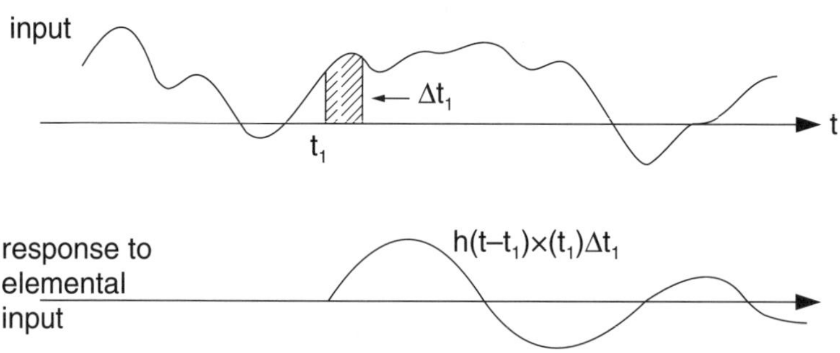

Figure 6.6 The response of a system to an 'elemental' impulse.

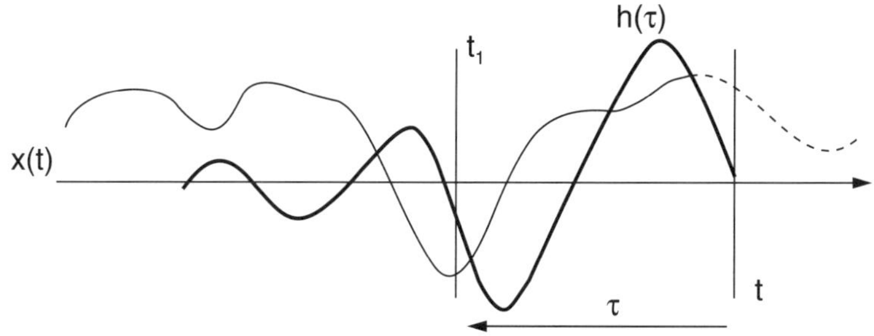

Figure 6.7 The impulse response function considered as a weighting (memory) function.

i.e. convolutions transform to products. $H(f)$ is the frequency response of the linear filter and is the Fourier transform of the impulse response function.

6.3.3 Windowing

Suppose $x(t)$ is known only for $-T/2 < t < T/2$, or we choose to 'truncate' the signal to that length.

It is as though we are looking at the data through a 'window' $w(t)$ where, in this case,

$$\begin{aligned} w(t) &= 1 \quad |t| < T/2 \\ &= 0 \quad |t| > T/2 \end{aligned}$$

giving the truncated signal $x_T(t) = w(t)x(t)$.

Fourier transforming $x_T(t)$ we obtain (for a general $w(t)$)

$$X_T(f) = \int_{-\infty}^{\infty} X(g)W(f-g)\mathrm{d}g \tag{6.12}$$

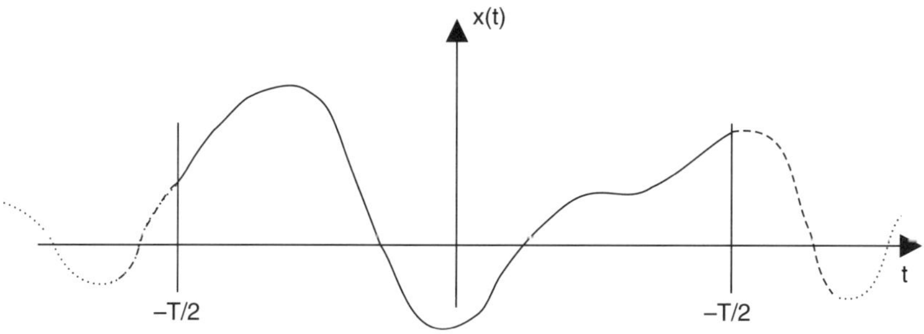

Figure 6.8 Data truncation with a rectangular window.

i.e. the Fourier transform of the *product* of two time functions is the convolution of their transforms (compare this with equations (6.10) and (6.11)).

For the particular case of the 'rectangular' data window above,

$$W(f) = \frac{T\sin(\pi fT)}{(\pi fT)} \qquad \text{(Fig. 6.9)}$$

and through the convolution operation widens the 'true' Fourier transform $X(f)$, so distorting it.

This distortion is sometimes called smearing (due to the main lobe) and 'leakage' since components of $X(f)$ at values other than $g = f$ 'leak' through the side lobes to contribute to the value of X_T at f. For example, suppose $x(t) = 2\pi pt$, then

$$X(f) = \tfrac{1}{2}[\delta(f+p) + \delta(f-p)]$$

and so

$$X_T(f) = \tfrac{1}{2}[W(f+p) + W(f-p)]$$

showing that the shape of the spectral window (in the frequency domain) replaces the delta function.

Comparison of theoretical and achieved spectra are depicted in Fig. 6.10.

If two (or more) closely spaced harmonic components occur in a signal, then the distortion may mean that the components cannot (easily) be resolved (see Fig. 6.11).

To separate two peaks at frequencies f_1, f_2, it is necessary to use a record length T of order $T \geq 2/(f_2 - f_1)$ for a rectangular window.

We note the rectangular window is a particularly 'poor' window as far as the side lobes are concerned (the highest side lobe is 13 dB below the main peak with asymptotic roll off of 6 dB/octave). The sharp corners of the data window result in this rippling effect. However, the main lobe is narrower than other windows.

Numerous alternatives have been suggested to reduce leakage in Fourier transform calculations. By tapering the windows to zero, the side lobes are

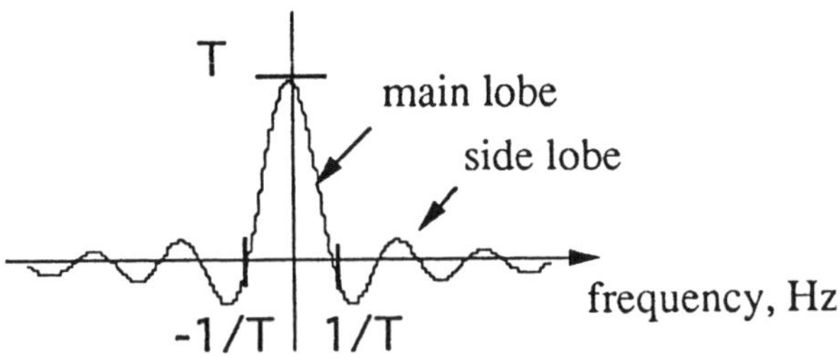

Figure 6.9 Fourier transform of the rectangular window.

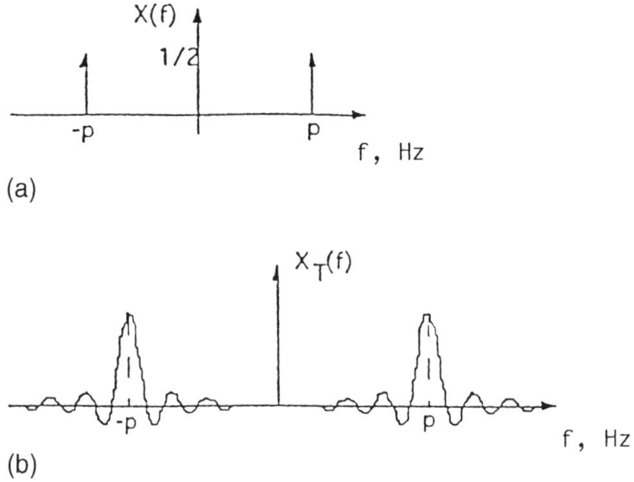

Figure 6.10 Spectra of a cosine wave: (a) theoretical; (b) windowed.

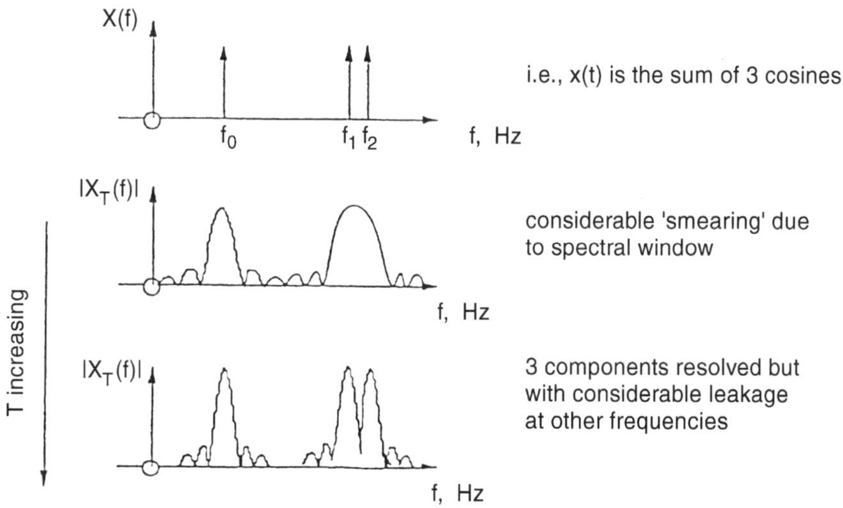

Figure 6.11 Effect of windowing on the modulus of the Fourier transform of three cosines.

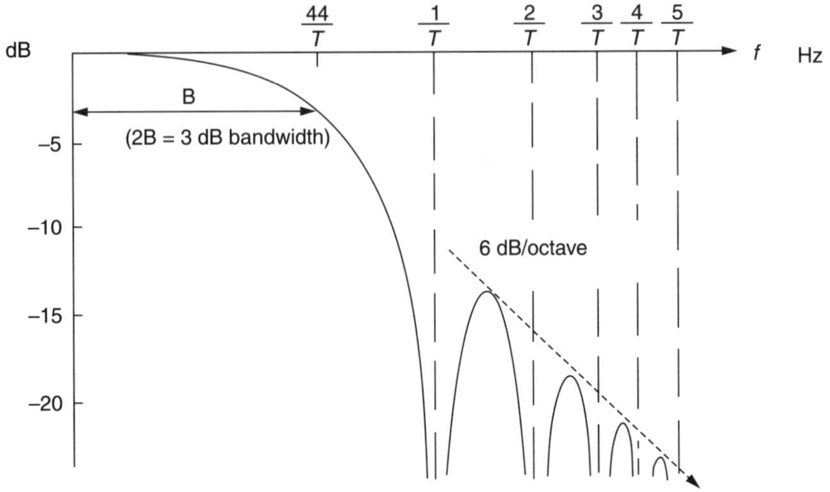

Figure 6.12 Frequency domain characterization of the rectangular window.

reduced but at the expense of widening the main lobe (increasing smearing); see section 6.3.4. The frequency domain behaviour of windows is often given in decibels (dB) where the characteristics have been normalized to unity gain (0 dB) at zero frequency, and some terms are defined in Fig. 6.12, which relates to the rectangular window.

The properties of a few commonly used windows are summarized in Table 6.1 (see, for example, Childers and Durling (1975) for the appropriate formulae).

The paper by Harris (1978) contains a comprehensive treatment of windows and their effects.

6.3.4 The uncertainty principle

A property of the Fourier transform of a signal is that the narrower a signal characterization is in one domain the wider is its characterization in the

Table 6.1 Properties of a few commonly used windows

Window (length T)	*Highest side lobe* (dB)	*3 dB bandwidth*	*Asymptotic roll off (dB octave)*	*First zero crossing*
Rectangle	−13	0.88 $1/T$	6	$1/T$
Bartlett	−26	1.25 $1/T$	12	$2/T$
Hann(ing) or Tukey	−32	1.4 $1/T$	18	$2/T$
Hamming	−42	1.3 $1/T$	6	$3/T$

other. An extreme example is a delta function $\delta(t)$ whose Fourier transform is a constant. Another is an oscillation $e^{j2\pi pt}$ whose Fourier transform is $\delta(f-p)$. This fundamental property of signals is summarized in the so-called 'Uncertainty Principle' as follows (Hsu, 1970).

If $x(t)$ is a signal with finite energy, i.e., $||x||^2 = \int_{-\infty}^{\infty} x^2(t)\mathrm{d}t < \infty$, and has Fourier transform $X(\omega)$, then defining the following

$$\bar{t} = \frac{1}{||x||^2} \int_{-\infty}^{\infty} t x^2(t)\mathrm{d}t \qquad \Delta t^2 = \frac{1}{||x||^2} \int_{-\infty}^{\infty} (t-\bar{t})^2 x^2(t)\mathrm{d}t$$

and

$$||X||^2 = \int_{-\infty}^{\infty} |X(\omega)|^2 \mathrm{d}\omega \qquad \bar{\omega} = \frac{1}{||X||^2} \int_{-\infty}^{\infty} \omega |X(\omega)|^2 \mathrm{d}\omega$$

$$(\Delta\omega)^2 = \frac{1}{||X||^2} \int_{-\infty}^{\infty} (\omega - \bar{\omega})^2 |X(\omega)|^2 \mathrm{d}\omega$$

($\bar{t}, \bar{\omega}$ are measures of location and $\Delta t, \Delta\omega$ are measures of spread), then it can be shown that

$$\Delta\omega\Delta t \geq \tfrac{1}{2} \ . \tag{6.13}$$

We say that the Bandwidth-Time (BT) product of a signal has a lower bound of $\frac{1}{2}$ where bandwidth and time here refer to the 'extent' of the signal in these two domains. Clearly, as $\Delta\omega$ gets narrower, Δt must get wider (i.e., the inverse spreading property). It is this inequality that is at the heart of 'difficulties' in time–frequency analysis methods which are Fourier based, i.e. if one wants a 'local' Fourier transform then increasing the 'localization' leads to poorer resolution in frequency and vice-versa, i.e. one cannot simultaneously obtain arbitrarily fine 'resolution' in *both* frequency and time domains.

It is noted that the Gaussian pulse e^{-at^2} (whose Fourier transform is another Gaussian pulse) has the minimum BT product of $\frac{1}{2}$.

6.3.5 The Hilbert transform

The Hilbert transform of a signal $x(t)$ is defined as the convolution of $x(t)$ with $1/\pi t$ and is denoted $\hat{x}(t)$, i.e.

$$\hat{x}(t) = \frac{1}{\pi t} * x(t). \tag{6.14}$$

This concept may be arrived at via analytic function theory. Despite the singular nature of the non-causal filter $1/\pi t$, many useful analytic results can be obtained. For example, the Hilbert transform of $\cos \Omega t$ is $\sin \Omega t$ (and that of sin Ωt is $-\cos\Omega t$) and hence a Hilbert transform is sometimes referred to as a 90° phase shifter.

Transformation to the frequency domain is helpful as the Fourier transform of the convolution operation can be expressed

$$\widehat{X}(\omega) = H(\omega)X(\omega) \tag{6.15}$$

where

$$\begin{aligned} H(\omega) &= j \qquad \omega < 0 \\ &= -j \quad\ \omega > 0 \\ &= 0 \qquad \omega = 0\ . \end{aligned} \tag{6.16}$$

This result may be used to obtain simple analytic results as above (rather than the time domain convolution).

The Hilbert transform has considerable significance in signal processing analysis and processing. For our purpose here, we note that it is used in the definition of the 'instantaneous frequency' of a signal in the following way.

Combining the original signal $x(t)$ with its Hilbert transform $\hat{x}(t)$ in the form

$$\sigma_x(t) = x(t) + j\hat{x}(t)$$

creates what is referred to as the 'analytic' signal or 'pre-envelope' signal $\sigma_x(t)$, which is complex and hence expressible as

$$\sigma_x(t) = A_x(t)e^{j\phi_x(t)} \tag{6.17}$$

where $A_x(t) = \sqrt{x^2 + \hat{x}^2}$ is called the instantaneous amplitude,
$\phi_x(t) = \tan^{-1}(\hat{x}(t)/x(t))$ is the instantaneous phase, and $\dot{\phi}_x$ is the instantaneous frequency.

For the trivial case $x(t) = \cos\Omega t$, the analytic signal is $e^{j\Omega t}$ and the instantaneous frequency is Ω, as one would expect. These definitions of instantaneous amplitude, phase and frequency have come to be accepted for a wide class of signals, and are particularly relevant to amplitude-modulated and frequency-modulated signals, e.g. for a chirp, when tracking a 'variable' frequency may be significant.

6.3.6 Instantaneous frequency and group delay

In section 6.3.1, the notion of group delay was introduced and in section 6.3.5 the idea of instantaneous frequency was noted. Both these notions are important as starting points for time–frequency analysis. One can conceive of a plot of group delay vs ω, i.e. time vs frequency and, quite separately, a plot of $\dot{\phi}$ vs t, i.e. (instantaneous) frequency vs time, yielding apparently two *different* ways of leading to time–frequency analysis.

Both appear to offer the possibility of combining time and frequency in a plane to provide the basis of time–frequency analysis. The question arises as to whether the two time–frequency laws are the same or not – specifically, whether they are inverse functions of each other. In general they are not, and Boashash (1987) describes conditions that must be met by a signal $x(t)$ for these two time–frequency laws to be equivalent. These conditions are that the signal should be 'asymptotic', i.e. that the BT product (referred to in section 6.3.4) should be 'large' and that the signal should be 'monocomponent', i.e. $\omega_i = \dot{\phi}$ should be invertible. The concept of monocomponent has not been rigorously defined but is intuitively clear, for example two chirps, each with different instantaneous frequency added together, lead to a signal that is not monocomponent. Computation of an overall $\dot{\phi}$ would result in some form of average. So, if a signal is monocomponent and asymptotic, then the instantaneous frequency and group delay are equivalent. The reader is referred to two review papers for time–frequency methods (Cohen, 1989; Hammond and White, 1996).

6.4 THE EFFECTS OF SAMPLING

This section introduces the concepts associated with converting a continuous time signal $x(t)$ into a sequence of numbers obtained by observing the signal every Δ seconds.

6.4.1 Analogue-to-digital conversion

An analogue-to-digital converter (ADC) is a device that operates on a continuous time history (input) and produces a sequence of numbers (output) that are sample values of the input. It is convenient to regard the process as consisting of two stages, namely 'sampling' and 'quantization'. In Fig. 6.13, $x(n\Delta)$ is the exact value attained by time history $x(t)$ at time $t = n\Delta$ (Δ is the sample interval for uniform rate sampling). $\tilde{x}(n\Delta)$ is the representation of $x(n\Delta)$ in a computer and differs from $x(n\Delta)$ since a finite number of bits are used to represent each number. Actual ADCs do not consist of two separate components (as in the 'conceptual' (Fig. 6.13)), and a variety of types is available.

The problem of quantization error is treated by Oppenheim and Schafer (1975), Rabiner and Gold (1975), Childers and Durling (1975) and Otnes and Enochson (1978). The output $\tilde{x}(n\Delta)$ can be written as

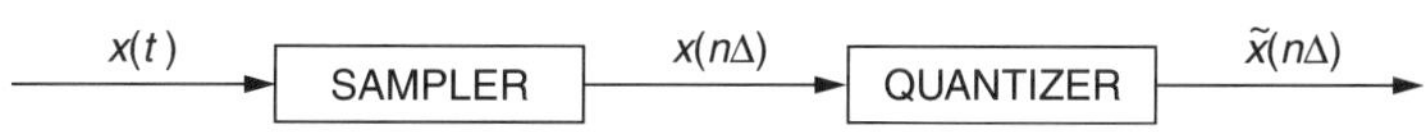

Figure 6.13 Conceptual model of analogue-to-digital conversion.

$\tilde{x}(n\Delta) = x(n\Delta) + e(n\Delta)$, and finite word length effects are described by treating the error $e(n\Delta)$ as random 'noise'. The probability distributions describing the error depend on the particular way in which the quantization is done. Often the error is ascribed a uniform distribution (with zero mean) over one quantization step and furthermore is assumed to be stationary and 'white'. For a b-bit word length (excluding sign), and if the full range of the ADC corresponds to X volts, say, then the variance of the error is $\sigma_e^2 = \left(\frac{X}{2^b}\right)^2/12$. Further, if the data x is assumed to be a random signal and σ_x^2 denotes the variance of $x(n\Delta)$, then a measure of signal-to-noise ratio is $S/N = 10\log_{10}\left(\frac{\sigma_x^2}{\sigma_e^2}\right)$. Choosing $\sigma_x = X/4$ (to 'minimize' clipping) then $S/N = 6b - 1.24$ dB, giving a value of about 70 dB for a 12-bit ADC. This is reduced further by practical considerations (see Otnes and Enochson, 1978) relating to the quality of the acquisition system.

It is vital that the rate at which $x(t)$ is sampled should be high enough to avoid aliasing (see section 6.4.3). This usually means that $x(t)$ should be low-pass filtered using analogue filters preceding the ADC. The cut-off frequency and cut-off rate of these 'anti-alias' filters should be chosen with particular applications in mind, but very roughly speaking if the 3 dB point is a quarter of the sampling frequency and the cut-off rate better than 48 dB/octave, then this at least ensures a 40–50 dB reduction at the folding frequency (half the sampling frequency; see section 6.4.2), and so probably an acceptable level of aliasing (though we must emphasize that this may not be adequate for some applications). Selection of sample rate is very important. While aliasing is to be avoided, unnecessarily high sample rates are wasteful. The specific application (the bandwidth of interest) and the characteristics of the antialias filter should dictate the selection of an 'optimal' rate.

6.4.2 The Fourier transform of a sequence

Since the calculations of Fourier coefficients and transforms are usually carried out digitally, it is necessary to re-examine the Fourier methods we described for data, but now in discrete form. We might treat a discrete version of (6.6), for example, as an approximation to the 'correct' result. However, completely rigorous, exact theories can be developed for discrete data which are, of course, closely related to the continuous time counterparts, but there are fundamental differences, and it is often more convenient to consider the problems of discrete data as self-contained and relate the results to those of the continuous domain only as and when they are required, accounting at that stage for any errors or approximations that might have been incurred in using discrete methods in place of the continuous operations. Having said this, we will nevertheless attempt to relate the Fourier analysis of continuous and discrete data.

6.4.2.1 Impulse train modulation

One way of introducing the Fourier transform of a sequence involves the use of the mathematical notion of 'ideal sampling' of a continuous wave. If an analogue signal $x(t)$ is to be sampled every Δ seconds, it is convenient to model the sampled signal as the product of the continuous signal with a 'train' of delta functions $i(t)$ where

$$i(t) = \sum_{n=-\infty}^{\infty} \delta(t - n\Delta), \tag{6.18}$$

so that the sampled signal, $x_s(t)$, is modelled as $x_s(t) = x(t)i(t)$. The Fourier transform of $x_s(t)$ is $X_s(f)$, and using the properties of the delta function (6.3 and 6.4), is written

$$X_s(f) = \sum_{n=-\infty}^{\infty} x(n\Delta)e^{-j2\pi fn\Delta} \ . \tag{6.19}$$

Multiplying both sides by $e^{j2\pi fn\Delta}$ and integrating with respect to f from $-1/2\Delta$ to $1/2\Delta$ gives

$$x(n\Delta) = \Delta \int_{-1/2\Delta}^{1/2\Delta} X_s(f)e^{j2\pi fn\Delta}df \ . \tag{6.20}$$

Equations (6.19) and (6.20) relate the *sequence of numbers* $x(n\Delta)$ to the quantity $X_s(f)$ which is termed the Fourier transform of the sequence.

A natural question to ask at this stage is 'How is $X_s(f)$ (or as it is often written, $X(e^{j2\pi f\Delta})$) related to $X(f)$?' The answer to this can be obtained as follows. Since $i(t)$ is periodic, we represent it as a Fourier series, i.e.

$$i(t) = \frac{1}{\Delta} \sum_{n=-\infty}^{\infty} e^{(2\pi jnt)/\Delta} \ . \tag{6.21}$$

Now substituting this into $x_s(t) = x(t)i(t)$ and Fourier transforming gives an alternative right-hand side to (6.19), namely

$$X(e^{j2\pi/\Delta}) = \frac{1}{\Delta} \sum_{n=-\infty}^{\infty} X(f - (n/\Delta)) \ . \tag{6.22}$$

This important equation (depicted graphically in Fig. 6.14) relates the Fourier transform of a continuous signal and the Fourier transform of the sequence formed by sampling the signal at equispaced intervals. It will be discussed further in the explanation of 'aliasing' in section 6.4.3.

6.4.2.2 The z-transform

An alternative route to equation (6.19) is to use the notion of the *z*-transform of a sequence (used in the solution of difference equations). The

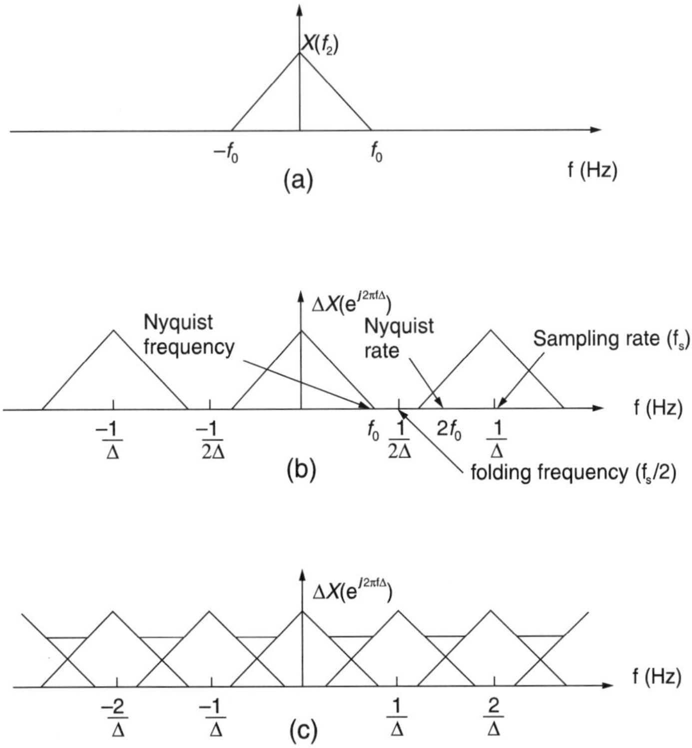

Figure 6.14 The effect of sampling on the Fourier transform.

z-transform of a sequence of numbers $x(n)$, say (where the notion of time is suppressed by setting the sampling interval to unity) is

$$X(z) = \sum_{n=-\infty}^{\infty} x(n) z^{-n} \ . \tag{6.23}$$

z is a complex number and $X(z)$ is a function of a complex variable. If z is allowed to take values on the unit circle in the z-plane, i.e., $z = e^{j\omega} = e^{j2\pi f}$, then (6.23) is

$$X(e^{j2\pi f}) = \sum_{n=-\infty}^{\infty} x(n) e^{-j2\pi f n} \tag{6.24}$$

which is seen to correspond identically with (6.19), with $\Delta = 1$.

6.4.3 Aliasing

Equation (6.22) describes how the frequency components of the sampled waveform relate to that of the continuous waveform, and Fig. 6.14 explains

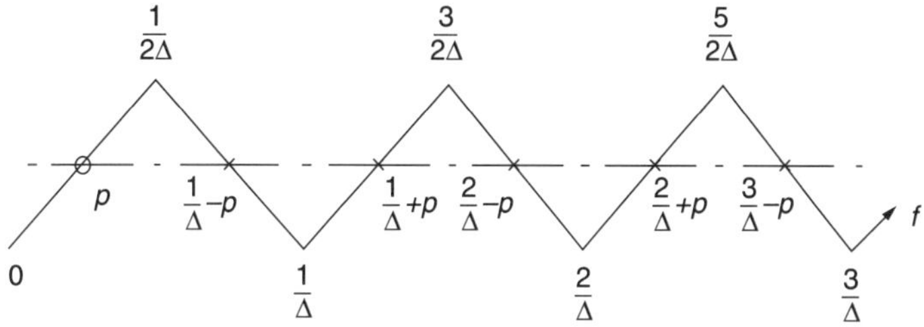

Figure 6.15 Possible aliases of frequency p Hz.

this pictorially. In Fig. 6.14(a) it is assumed that $X(f) = 0$ for $|f| > f_0$ (say), and Fig. 6.14(b) is a plot of $\Delta X(e^{j2\pi/\Delta})$ assuming that the sampling rate $f_s = 1/\Delta$ is such that $f_0 < 1/2\Delta$. Some commonly used terms are defined on the diagram.

If $f_0 > f_s/2$, there is an overlapping of the shifted versions of $X(f)$ resulting in a distortion of the frequency description for $|f| < 1/2\Delta$ as in Fig. 6.14(c). This 'distortion' is due to the fact that high frequencies in the data are indistinguishable from lower frequencies owing to the sampling rate not being fast enough. Specifically, consider $x(t) = \cos 2\pi pt$ sampled at $f_s = 1/\Delta$, i.e., at $t = n\Delta$ (with, say, $p < 1/2\Delta$). This may be written

$$x(n\Delta) = \tfrac{1}{2}\left(e^{j2\pi n\Delta p} + e^{-j2\pi n\Delta p}\right) .$$

Now letting frequency p above be replaced by $\pm p + (k/\Delta)$, $k = 1, 2, \ldots$, results in an identical sequence. So, if a frequency component were detected at p Hz, any one of these higher frequencies might have been responsible for this rather than a 'true' component at p Hz. This phenomenon is called *aliasing*. The values $\pm p + (k/\Delta)$ can be seen graphically for some p between 0 and $1/2\Delta$ by 'pleating' the frequency axis as in Fig. 6.15.

To avoid aliasing, the sample rate must be chosen to be greater than twice the highest frequency contained in the signal (see the discussion in section 6.4.1 regarding the 'anti-alias' filters).

6.4.4 The discrete Fourier transform

The equation (6.19) is still not a practical proposition, and a finite sum would be more appropriate which, for an N-point sequence, could be written

$$\sum_{n=0}^{N-1} x(n\Delta)e^{-j2\pi fn\Delta} . \tag{6.25}$$

Obviously, if this involved truncating the data there would be some distortion explained earlier by the discussion on windowing. Assuming for the moment that we can ignore this, we can think of simplifying our task by only evaluating f at specific points, say at $k/N\Delta$ Hz, $k = 0, \ldots, N-1$. Then

$$X(e^{j2\pi k/N}) = \sum_{n=0}^{N-1} x(n\Delta)e^{-j(2\pi/N)nk} \quad . \tag{6.26}$$

Or, adopting the more usual notation (and setting Δ to unity for convenience),

$$X(k) = \sum_{n=0}^{N-1} x(n)e^{-j(2\pi/N)nk} \quad . \tag{6.27}$$

Multiplying both sides by $e^{j(2\pi/N)mk}$ and summing over k gives

$$x(n) = \frac{1}{N}\sum_{k=0}^{N-1} X(k)e^{j(2\pi/N)nk} \quad .$$

The pair (6.26) and (6.27) constitute the discrete Fourier transform (DFT) and are the form that are suitable for machine computation.

It is important to realize that even though the original sequence $x(n\Delta)$ may have been zero for n outside the range $0 \to N-1$, the act of 'sampling in frequency' has imposed a *periodic structure* to the sequence, i.e. it follows from (6.27) that $x(n+N) = x(n)$.

We are now in a position to summarize the various Fourier transforms we have defined and their differences, and this is done in Figs 6.16 and 6.17, where we follow the method of display given by Randall (1977).

It is instructive to follow (diagramatically) the interpretation of a DFT of an aperiodic function that is truncated, as in Fig. 6.17 (see also Brigham (1974)).

6.4.5 The fast Fourier transform

The evaluation of the DFT can be accomplished efficiently by a set of algorithms known as the fast Fourier transform (FFT). This section briefly outlines the basic 'decimation in time' method for a radix 2 FFT and follows Oppenheim and Schafer (1975) and Rabiner and Gold (1975).

6.4.5.1 The radix 2 FFT

Writing $e^{-j(2\pi/N)}$ as W_N then the DFT is

$$X(k) = \sum_{n=0}^{N-1} x(n)W_N^{nk} \quad . \tag{6.28}$$

The FFT algorithms essentially exploit the periodicity and symmetry properties of W_N^{nk}, and so reduce the number of multiply and add operations

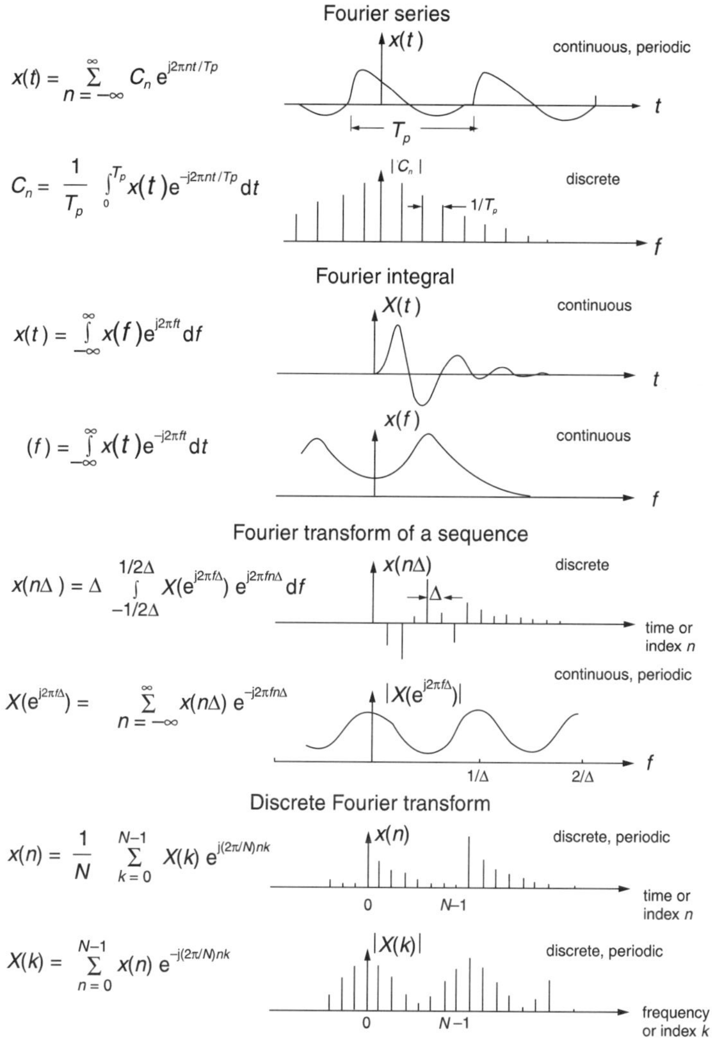

Figure 6.16 Summary of Fourier transforms.

needed to calculate $X(k)$ from about N^2 to approximately $N \log_2 N$ which for $N = 1024$ is a reduction by a factor of about 100.

We shall take N to be a power of two $(= 2^\nu)$ and so obtain a base-2 or radix-2 algorithm. As the name 'decimation in time' suggests, the sequence $x(n)$ is successively decomposed into smaller subsequences. In fact, $N/2$ point sequences $x_1(n), x_2(n)$ can be defined as the even and odd members of $x(n)$, so that

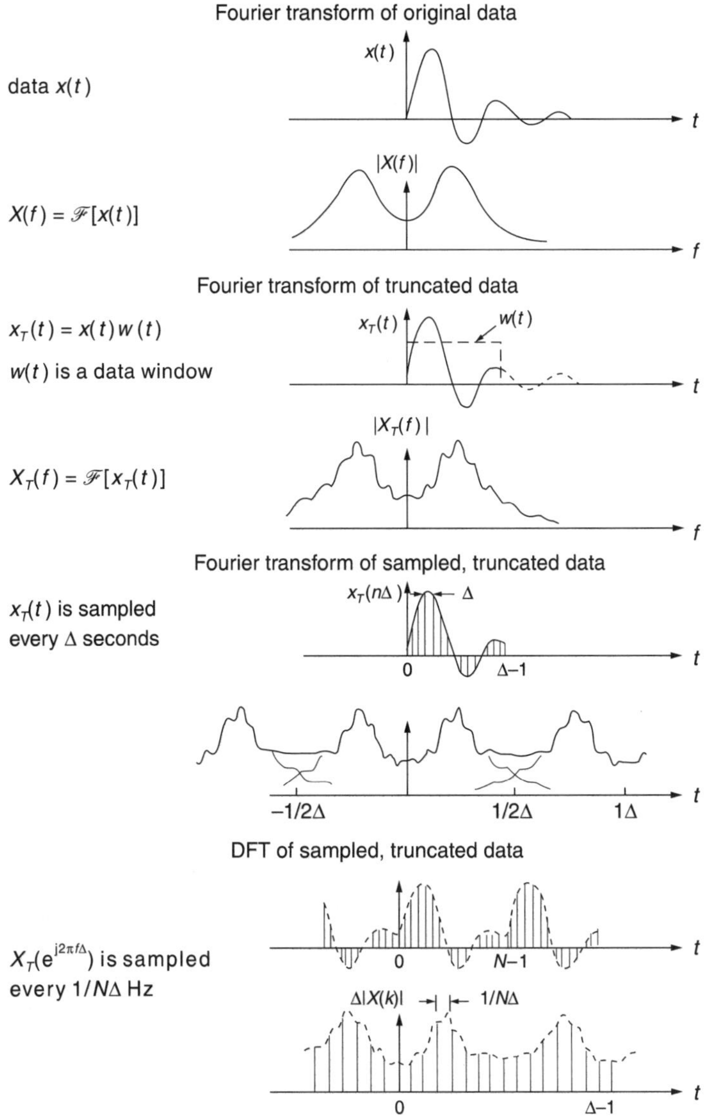

Figure 6.17 Relationship of a DFT to the Fourier integral.

$$X(k) = \sum_{n=0}^{\frac{N}{2}-1} x(2n) W_N^{2nk} + \sum_{n=0}^{\frac{N}{2}-1} x(2n+1) W_N^{(2n+1)k} \quad .$$

Then, noting that $W_N^2 = W_{N/2}$ this can be written

$$X(k) = \sum_{n=0}^{\frac{N}{2}-1} x_1(n) W_{N/2}^{nk} + W_N^k \sum_{n=0}^{\frac{N}{2}-1} x_2(n) W_{N/2}^{nk}$$

or

$$X(k) = X_1(k) + W_N^k X_2(k) \ , \tag{6.29}$$

where $X_1(k), X_2(k)$ are $N/2$ point DFTs of $x_1(n), x_2(n)$ whose 'usual' properties are used to calculate $X(k)$ for $0 \le k \le N-1$.

Some signal flow graph terminology is defined in Fig. 6.18 and Fig. 6.19 shows the first decomposition for the case $N = 8$, showing the two 4-point transforms making up the full 8-point transform.

The same is now done to the 4-point transforms by dividing each of $x_1(n), x_2(n)$ into two sequences of even and odd members, resulting in four 2-point transforms. Following Rabiner and Gold (1975), the overall scheme is

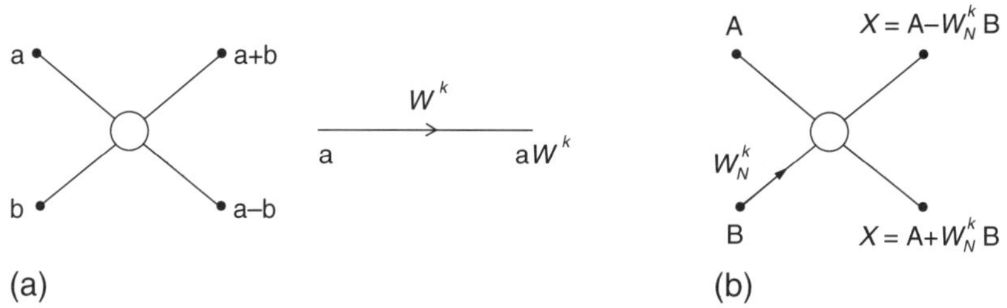

Figure 6.18 Signal flow terminology.

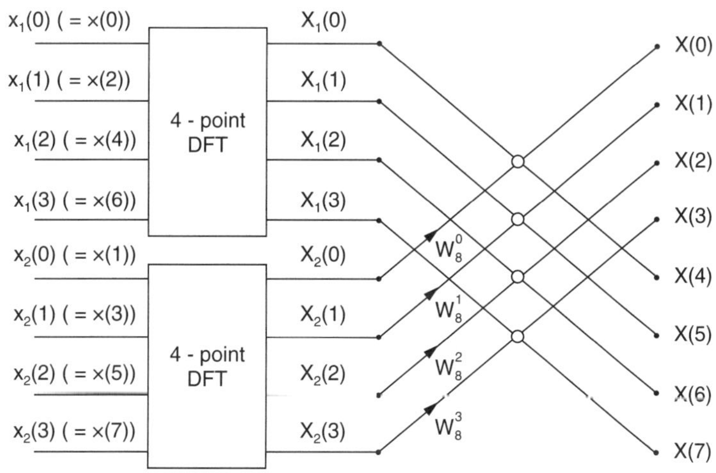

Figure 6.19 An 8-point DFT in terms of two 4 point DFTs.

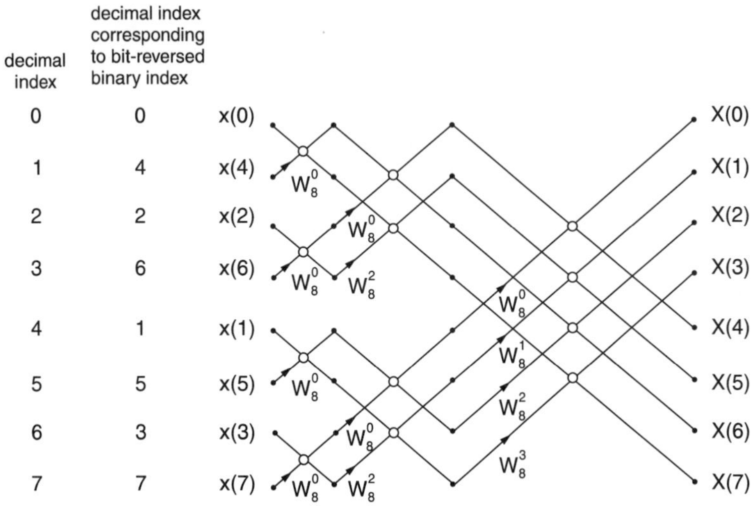

Figure 6.20 The decimation in time FFT algorithm.

shown in Fig. 6.20, where the basic operation is the 'decimation in time butterfly' (Fig. 6.20).

There are $N/2$ butterflies at each stage, and if $N = 2v$ there are $v = \log_2 N$ stages, and with approximately two multiply/add operations 'per butterfly' we have of the order of $N \log_2 N$ operations in all. (Note that the output array $X(k)$ is in the correct order, but the input array is shuffled. In fact it is in 'bit reversed' order.)

6.5 RANDOM PROCESSES

6.5.1 Introduction

In section 6.2.2, we discussed the analysis of deterministic signals using Fourier methods. This section is devoted to the treatment of non-deterministic signals. We may visualize a sample of such a signal as shown in Fig. 6.21.

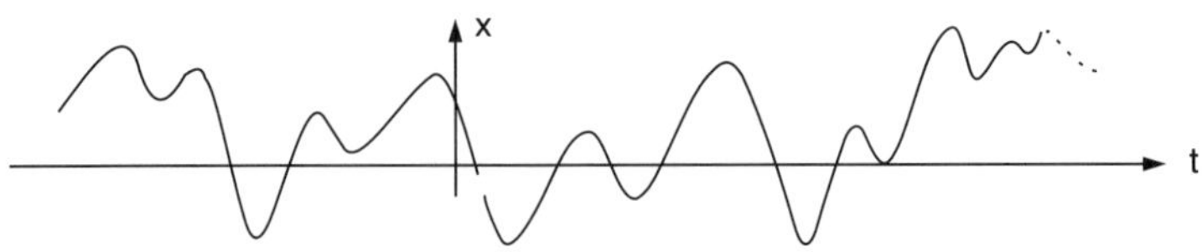

Figure 6.21 Sample time history.

An example might be the record of wave height at a point at sea. The essential feature of such a signal is that its value cannot be predicted *exactly* in advance. We shall briefly summarize the way in which such random processes are described.

6.5.2 Probability theory

The mathematical theory underlying models describing *uncertain* situations is probability theory. It is often explained by considering examples taken from games of chance involving coins, dice, etc. We need to define a few terms.

(a) An experiment of chance

An experiment whose outcome is not predictable is called an experiment of chance.

(b) Sample space

The collection (set) of all possible outcomes of an experiment is called its sample space (denoted Ω), with individual elementary outcomes as ω (not to be confused in this context with frequency).

(c) Event

Any collection (subset) of points in the sample space is an event. Individual events in the sample space are called *elementary events*. We see that *events* are collections of elementary events.

To each event E in Ω, we now wish to assign a number which measures our belief that E will occur. This is the *probability* of event E and it is written $\text{Prob}(E)$ or $P(E)$ and satisfies $0 \leq P(E) \leq 1$.

6.5.3 Random variables and probability distributions

In many cases we may not wish to define the outcome of an experiment in terms of elements of the sample space but in terms of a convenient set of numbers. This leads to the definition of a random variable. A *random variable* is a function defined on a sample space. We write $X(\omega) = x_i$ to denote that the random variable X takes the value x_i for elements ω that comprise the event. If the x_i are countable, the random variable is *discrete* and if not it is *continuous*. Some processes may be mixed.

6.5.3.1 *Probability distributions for discrete random variables*

Definition: If random variable X takes on only a discrete set of values $x_1, x_2, \ldots,$ then its probability distribution is characterized by specifying the probabilities with which X assumes these values, i.e.

$$P(X = x) \text{ for } x = x_1, \ldots,$$

$P(X = x)$ is called the *probability function* of random variable X.

6.5.3.2 *The probability distribution function*

Definition: The distribution function $F(x)$ of a random variable X is defined as

$$F(x) = P[X \leq x] = \text{Prob}[X \text{ taking on a value up to and including } x].$$

For a discrete process there are jumps in the function.

6.5.3.3 *Continuous distributions*

Now it is no longer possible to enumerate all possible outcomes, i.e. the number of elements in the sample space is infinite and uncountable and so the probability that $X = x$, say, is 0. A more useful approach is to think of the probability of X lying within *intervals* on the x axis, i.e.

$$P[X > a], \quad P[a \leq X \leq b] \ .$$

Starting from the distribution function $F(x) = P[X \leq x]$ we are led to the *probability density* function as follows.

6.5.3.4 *Probability density functions*

Consider the probability of X lying in an interval x to $x + \delta x$, then

$$P[x < X \leq x + \delta x] = F(x + \delta x) - F(x) \ .$$

Then $\frac{P[x<X\leq x+\delta x]}{\delta x}$ is a measure of the probability/unit length near x and if $\lim\limits_{\delta x \to 0} \frac{P[x<X\leq x+\delta x]}{\delta x}$ exists it is called the *probability density function* (p.d.f.) and is denoted $p(x)$ as

$$p(x) = \frac{\mathrm{d}F(x)}{\mathrm{d}x} \tag{6.30}$$

so that

$$F(x) = \int_{-\infty}^{x} p(u)\mathrm{d}u \ . \tag{6.31}$$

It follows that

1. $p(x) \geq 0$
2. $\int_{-\infty}^{\infty} p(x)\mathrm{d}x = 1$
3. $P[a < X < b] = \int_a^b p(x)\mathrm{d}x$.

Note: we can define the probability density function for discrete random variables if we include delta functions.

6.5.3.5 Joint distributions

Now extend our considerations to two or more random variables (say X, Y, i.e. a bivariate process). Then the probability of $X < x$ *jointly occurring* with $Y \leq y$ is

$$P[X \leq x, Y \leq y] = F(x, y) \ .$$

Just as for the univariate case we can define a joint density function

$$\begin{aligned} p(x, y) &= \lim_{\substack{\Delta x \to 0 \\ \Delta y \to 0}} \frac{P[x \leq X \leq x + \Delta x,\ y \leq Y \leq y + \Delta y]}{\Delta x \Delta y} \\ &= \frac{\partial^2 F(x, y)}{\partial x \partial y} \ . \end{aligned} \tag{6.32}$$

6.5.4 Expectations of functions of a random variable

Let X be a random variable, the *expected value* of X is defined

$$E[X] = \int_{-\infty}^{\infty} xp(x)\mathrm{d}x \ . \tag{6.33}$$

We use the notation $\mu_x = \int_{-\infty}^{\infty} xp(x)\mathrm{d}x$ for this 'mean value' or 'first moment' of X.

More generally, if Y is a function of X, i.e. if $Y = g(X)$, then the average value of Y is

$$E[Y] = E[g(X)] = \int_{-\infty}^{\infty} g(x)p(x)\mathrm{d}x \ . \tag{6.34}$$

This may be generalized to functions of several random variables. In a bivariate case (X, Y) with random variables (X, Y) and if W is a function of X and Y, i.e. $W = g(X, Y)$ then the average value of W is

$$E[W] = \iint_{-\infty}^{\infty} g(x, y)p(x, y)\mathrm{d}x\mathrm{d}y \ . \tag{6.35}$$

6.5.4.1 Moments of a random variable

The moments of a random variable are often taken as relatively simple indicators of the behaviour of a process. The first few moments are noted below.

$$\text{First moment (mean) } E[X] = \mu_x = \int_{-\infty}^{\infty} xp(x)\mathrm{d}x \ . \tag{6.36}$$

This is a measure of location.

$$\text{Second moment (mean square)} = E[X^2] = \int_{-\infty}^{\infty} x^2 p(x)\mathrm{d}x \ . \tag{6.37}$$

This is a measure of spread relative to the origin.

We usually take 'central moments', i.e. moments about the mean. The second moment about the mean is the *variance*

$$V(X) = \mathrm{Var}(X) = \sigma_x^2 = E[(X - \mu_x)^2] = \int_{-\infty}^{\infty} (x - \mu_x)^2 p(x)\mathrm{d}x \ . \tag{6.38}$$

This is a measure of spread relative to the mean level.

The standard deviation is $\sigma_x = \sqrt{V(X)}$.

6.5.4.2 The normal distribution (Gaussian distribution)

A random variable is normally distributed (or has a Gaussian distribution) if it has a probability density function given by

$$p(x) = \frac{1}{\sqrt{2\pi}\sigma} \mathrm{e}^{-(x-\mu)^2/2\sigma^2} \tag{6.39}$$

where

$$E[X] = \mu, \qquad \sigma^2 = \int_{-\infty}^{\infty} (x - \mu)^2 p(x)\mathrm{d}x. \tag{6.40}$$

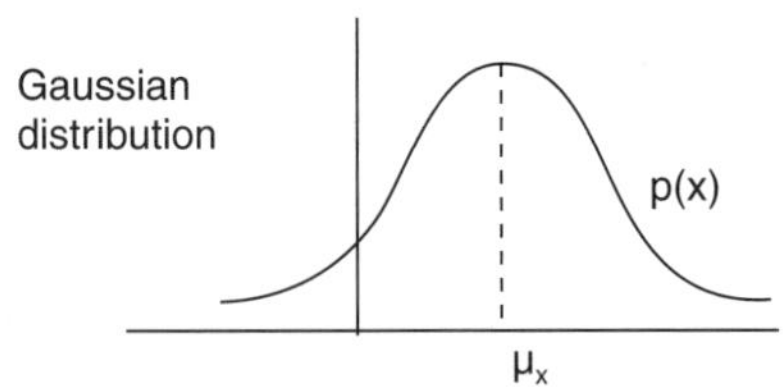

Figure 6.22 A Gaussian probability density function.

This distribution is completely characterized by the two parameters μ and σ. The importance of this distribution is stressed by the *Central Limit Theorem* which (roughly speaking) says that the sum of independent random variables $S_n = \sum_{k=1}^{n} X_k$ tends to a Gaussian random variable as n gets large, regardless of the individual distributions of the X_k.

6.5.4.3 Higher moments

Higher moments of random variables give useful information about the probability density function of the process. For example, a measure of the departure from symmetry or 'skewness' of a distribution is the third moment about the mean $E[(X-\mu)^3]$. This is sometimes measured in terms of the coefficient of skewness

$$\gamma_1 = \frac{E[(X-\mu_x)^3]}{\sigma_x^3} \tag{6.41}$$

$\gamma_1 = 0$ for a Gaussian distribution.

The degree of *flattening* of a p.d.f. near its mean is sometimes measured by the coefficient of *kurtosis*

$$\gamma_2 = \frac{E[(X-\mu_x)^4]}{\sigma_x^4} - 3\,\text{(fourth moment about the mean)} \tag{6.42}$$

$$\gamma_2 = 0 \text{ for a Gaussian distribution .}$$

If $\gamma_2 > 0$, the distribution is more peaky than Gaussian (leptokurtic).

If $\gamma_2 < 0$, the distribution is more flattened than Gaussian (platykurtic).

Sometimes only $\frac{E[(X-\mu_x)^4]}{\sigma_x^4}$ is computed and the value relative to 3 is considered.

If a process is Gaussian, $\gamma_1 = 0$ and $\gamma_2 = 0$ (or $\frac{E[(X-\mu_x)^4]}{\sigma_x^4} = 3$). Consequently, monitoring the third and fourth moments allows detection of non-Gaussianity. Since many *nonlinear* systems exhibit *non-Gaussian* responses, then γ_1 and γ_2 can be used to detect (and perhaps characterize) *nonlinearity*. The fourth moment (kurtosis) is used as an indicator of damage in machinery condition monitoring. Early damage in rotating machinery rolling elements often results in accelerometer signals whose kurtosis increases significantly beyond 3 owing to the impacting arising from faults.

6.5.4.4 Bivariate processes

These moments may be generalized to jointly distributed random variables, e.g. the second joint moment is

$$E[(X-\mu_x)(Y-\mu_y)] = \iint_{-\infty}^{\infty} (x-\mu_x)(y-\mu_y)p(x,y)\mathrm{d}x\mathrm{d}y \tag{6.43}$$

and is called the *covariance* function relating X and Y.

Some definitions relating to bivariate processes follow.

X, Y are:

1. *uncorrelated* if $E[XY] = E[X]E[Y]$;
2. *orthogonal* if $E[XY] = 0$;
3. *independent* if $p(x, y) = p(x)p(y)$.

A joint distribution for Gaussian (bivariate) process (X, Y) is expressed as follows.

If $E[X] = \mu_x; E[Y] = \mu_y$

and $E[(X - \mu_x)^2] = \sigma_x^2; \quad E[(Y - \mu_y)^2] = \sigma_y^2; \quad E[(X - \mu_x)(Y - \mu_y)] = \sigma_{xy}$.

Then

$$p(x, y) = \frac{1}{2\pi} \frac{1}{|S|^{1/2}} \exp\left\{ -\frac{1}{2} (\boldsymbol{v} - \underline{\mu}_v)^{\mathrm{T}} S^{-1} (\boldsymbol{v} - \underline{\mu}_v) \right\} \tag{6.44}$$

where $S = \begin{bmatrix} \sigma_x^2 & \sigma_{xy} \\ \sigma_{xy} & \sigma_y^2 \end{bmatrix}$ and $\boldsymbol{v} = \begin{bmatrix} x \\ y \end{bmatrix}, \underline{\mu}_v = \begin{bmatrix} \mu_x \\ \mu_y \end{bmatrix}$.

6.5.5 Stochastic processes

Time has played no part so far in our description of random processes. Now consider a random variable (defined on a sample space) which is time-dependent. A possible notation is $X_t(\omega)$. (Note again that ω denotes an element in the sample space and *not* frequency.) In fact this is abbreviated $X(t)$ and implies that for each point in time t in the range $-\infty < t < \infty$ a random variable $X(t)$ exists which has a sample space $-\infty < X(t) < \infty$ and an associated probability density function $p_t(x)$ (or $p(x(t))$). We shall see that it is necessary to specify the joint probability density functions associated with an arbitrary set of times $t_1, t_2, \ldots, t_n$. Thus a time series can be described as an ordered set of random variables $X(t)(-\infty < t < \infty)$ for continuous time series. An ordered set of random variables is a *stochastic process*.

6.5.5.1 The ensemble

An observed time series $x(t)$ may be regarded as one realization of an infinite number of functions which might have been observed, i.e. this set may be visualized as being generated by a large number of independent repetitions of the same experiment under conditions which are essentially identical, i.e. for any fixed t, $X(t)$ is a random variable with p.d.f. $p(x(t))$. This is sometimes written $\{X(t)\}$, see Fig. 6.23.

Comment on notation: strictly $X(t)$ denotes the random variable X for a particular value of t and $x(t)$ denotes a particular realization. However, notation varies and we, too, will sometimes use $x(t)$ when we should really write $X(t)$.

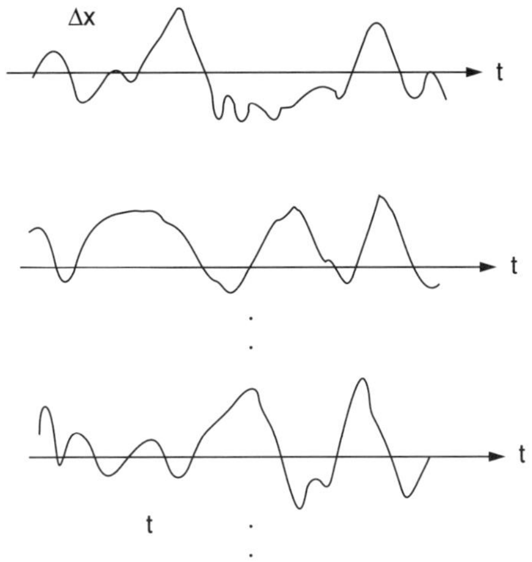

Figure 6.23 An ensemble of time histories.

6.5.5.2 Probability distributions associated with a stochastic process

We conceive of a probability density function for a stochastic process as follows. Let the distribution function be $F_1(x,t)$ where

$$F_1(x,t) = \text{Prob}[X(t) \le x] \ .$$

So $\text{Prob}[x < X(t) \le x + \Delta x] = F_1(x + \Delta x, t) - F_1(x,t)$.
Therefore

$$\lim_{\Delta x \to 0} \frac{\text{Prob}[x < X(t) \le x + \Delta x]}{\Delta x} = \frac{\text{d}F_1(x,t)}{\text{d}x}$$

or

$$p_1(x,t) = \frac{\text{d}F_1(x,t)}{\text{d}x} \tag{6.45}$$

i.e. $p_1(x,t)$ is a probability density function which evolves with time (Fig. 6.29).

Similarly, we may define joint probability density functions, relating $x(t)$ at times t_1 and t_2.

If $F_2(x_1, t_1; x_2, t_2) = \text{Prob}[X(t_1) \le x_1 \text{ jointly with } X(t_2) \le x_2]$ then the joint probability density function is

$$p_2(x_1, t_1; x_2, t_2) = \frac{\partial^2 F_2(x_1, t_1; x_2, t_2)}{\partial x_1 \partial x_2} \ . \tag{6.46}$$

In general, a univariate stochastic process is described by the n-th order joint probability density function

$$p_n(x_1, t_1; x_2, t_2; \ldots; x_n, t_n) \ .$$

For our purposes we shall only consider p_1, p_2.

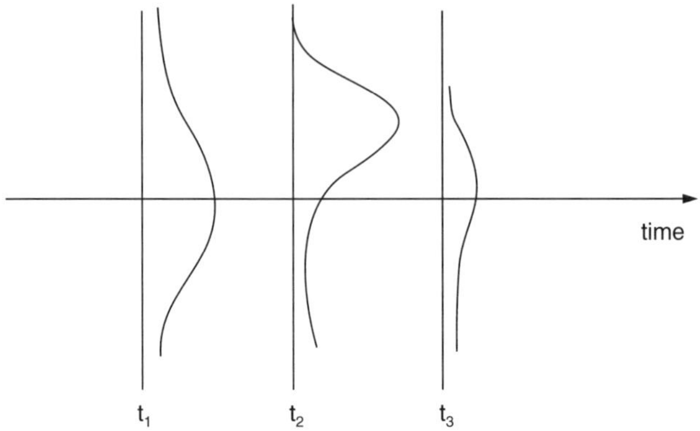

Figure 6.24 Evolving probability density of a non-stationary process.

6.5.5.3 Moments of a stochastic process

We define moments for stochastic processes just as we did for random variables, noting now that time is involved. Each moment is now time-dependent, for example,

$$\text{Mean } \mu_x(t) = E[X(t)] = \int_{-\infty}^{\infty} xp(x, t)\mathrm{d}x \quad \text{(1st moment)}$$

$$\text{Mean square } E[X^2(t)] = \int_{-\infty}^{\infty} x^2 p(x, t)\mathrm{d}x \quad \text{(2nd moment)}$$

$$\text{Variance } \sigma_x^2(t) = E[X - \mu_x(t)^2] = \int_{-\infty}^{\infty} (x - \mu_x)^2 p(x, t)\mathrm{d}x.$$

The expected value operator E may be interpreted as an average using probabilistic concepts or as an 'ensemble average'.

6.5.5.4 Stationarity

In general, the properties of a stochastic process are time-dependent. To simplify matters we often assume a sort of 'steady state' has been reached in the sense that the statistical properties are unchanged under a shift in time, i.e. (a) $p_1(x, t) = p(x)$. This means that μ_x, σ_x^2 are constant. (b) $p_2(x_1, t_1; x_2, t_2)$ is a function of $(t_2 - t_1)$ only and not t_1 and t_2 explicitly.

If conditions (a) and (b) hold, the process is said to be *weakly stationary* or simply *stationary*. If $p_n(x_1, t_1; x_2, t_2; \ldots; x_n, t_n) = p_n(x_1, t_1 + T; \; x_2, t_2 + T; \ldots)$ for all n, i.e. p_n is invariant under a time shift, then we say the process is *completely* stationary. We shall be satisfied with conditions (a), (b). Since all practical processes change as time evolves, the assumption of stationarity is an approximation. However, we shall see that it has important consequences. For example, in many situations we may only have a single data record available for analysis (rather than a set of records) and so rather than averaging across the *ensemble* we will average along *time*, i.e. we will use *time averages* in place of *ensemble averages*. Clearly, *stationarity* is a necessary condition if this is to be useful.

6.5.5.5 Covariance (correlation) functions

For a random process $\{x(t)\}$ a widely used average is the *autocovariance function* which is defined as

$$E[(X(t_1) - \mu_x(t_1))(X(t_2) - \mu_x(t_2))] = R_{xx}(t_1, t_2). \tag{6.47}$$

This is a measure of the 'degree of association' of the signal at time t_1 with *itself* at time t_2 ($t_2 \geq t_1$). If the mean value is not subtracted, $E[X(t_1)X(t_2)]$ is often referred to in the engineering literature as the *autocorrelation function*. (We note that sometimes a *normalized* autocovariance function is referred to as the autocorrelation function – and this is sometimes called an autocorrelation coefficient – so, beware of terminology!)

Since we shall deal mainly with stationary processes, we shall simplify the above by noting that the statistical properties are unchanged under a time shift, so

$$E[(X(t_1) - \mu_x)(X(t_2) - \mu_x)] = R_{xx}(t_2 - t_1) \tag{6.48}$$

i.e. a function of the difference $(t_2 - t_1)$ only. We usually write $t_1 = t$ and $t_2 = t + \tau$, so

$$E[X(t) - \mu_x)(X(t + \tau) - \mu_x)] = R_{xx}(\tau). \tag{6.49}$$

τ is called the lag.

Note: when $\tau = 0$, $R_{xx}(0) =$ Variance$[X(t)]$; also, as τ gets large, it seems reasonable that for random processes the average $E[X(t)X(t + \tau)]$ should approach zero, since the values of $X(t)$ for large time separations are 'less associated', i.e. $R_{xx}(\tau)$ may have the shape as in Fig. 6.25.

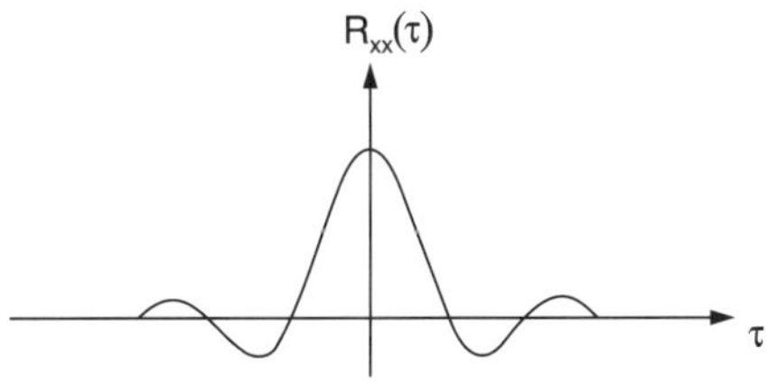

Figure 6.25 An autocorrelation function.

Note: $R_{xx}(\tau)$ is an *even* function of τ since

$$E[x(t)x(t+\tau)] = E[x(t-\tau)x(t)] \ .$$

6.5.5.6 *The cross covariance (cross correlation) function*

When two random processes are involved $\{X(t), Y(t)\}$, e.g. input/output processes, then we generalize the above and define the cross covariance function

$$R_{xy}(t_1, t_2) = E[(X(t_1) - \mu_x(t_1))(Y(t_2) - \mu_y(t_2))]. \tag{6.50}$$

This is a measure of the degree of association of signal X at t_1 with signal Y at t_2. If we assume stationarity of both signals, then R_{xy} is a function of $t_2 - t_1$.

As above, let $t_1 = t$ and $t_2 = t + \tau$ and set mean values to zero to write

$$R_{xy}(\tau) = E[X(t)Y(t+\tau)] \ . \tag{6.51}$$

Note: $R_{xy}(\tau)$ is *not* an even function. In fact,

$$R_{xy}(\tau) = R_{yx}(-\tau) \ .$$

We list some properties of covariance functions.

Properties of covariance functions

The autocovariance function (ACVF)

Define $\rho_{xx}(\tau) = \frac{R_{xx}(\tau)}{R_{xx}(0)}$; $\rho_{xx}(\tau)$ is the normalized (non-dimensional) form of the ACVF, then

1. $R_{xx}(\tau) = R_{xx}(-\tau)$ (the ACVF is 'even');
2. $\rho_{xx}(0) = 1; R_{xx}(0) = \sigma_x^2$;
3. $|R_{xx}(\tau)| \le R_{xx}(0) \quad \text{so} -1 \le \rho_{xx}(\tau) \le 1$;
4. $R_{xx}(\tau)$ is a continuous function of τ for a continuous stochastic process.

The cross-covariance function (CCVF)

Define $\rho_{xy}(\tau) = \frac{R_{xy}(\tau)}{\sqrt{R_{xx}(0)R_{yy}(0)}}$ (the cross-correlation coefficient), then

1. $R_{xy}(-\tau) = R_{yx}(\tau)$ N.B. $R_{xy}(\tau)$ is neither odd nor even in general.
2. $|R_{xy}(\tau)|^2 \le R_{xx}(0)R_{yy}(0)$.
3. $-1 \le \rho_{xy}(\tau) \le 1$.

Note: The correlation coefficient is useful when the two quantities involved are different in scale and dimension.

6.5.5.7 Ergodic processes and time averages

The definitions of the moments above have all involved the underlying theoretical probability distributions of the stochastic processes or been interpreted equivalently as ensemble averages. Often we find that we have only a single realization (record) of limited length from which to make estimates of these averages we have defined. This motivates the question as to whether averages *along the record*, i.e. a *time average* might be used in place of an ensemble average. We have noted this already in section 6.5.5.4 and decided that *stationarity* is a prerequisite if we are to follow this up. In fact it turns out that many processes have the desirable property that *some ensemble averages* are equivalent to *time averages* over one record. Such an average is said to be *ergodic*. Sometimes processes are termed *weakly ergodic* if the mean value and covariance functions may be calculated by performing time averages in place of ensemble averages.

We add that while the concept of ergodicity is given prominence in many texts it can only be approached correctly through estimation theory (section 6.7) and it must be related directly to the particular average in question, e.g. mean value, ACVF etc. We *cannot* simply refer to a *process* as ergodic.

Nevertheless, we shall anticipate that many processes *are* ergodic with respect to the mean and ACVF and CCVF implying that we can write the following definitions for these averages.

$$\text{The mean value } \mu_x = \lim_{T\to\infty} \frac{1}{T}\int_0^T x(t)\mathrm{d}t \left[\text{or } \lim_{T\to\infty} \frac{1}{T}\int_{-T/2}^{T/2} x(t)\mathrm{d}t\right] \tag{6.52}$$

i.e. we get the same value for the time average over any single member of the ensemble as $E[X(t)]$.

Similar arguments hold for the CCVF

$$R_{xy}(\tau) = \lim_{T\to\infty} \int_0^T (x(t) - \mu_x)(y(t+\tau) - \mu_y)\mathrm{d}t \ . \tag{6.53}$$

6.5.5.8 Examples

The autocovariance function

A sinusoid

Let $x(t) = X\sin(\omega t + \theta)$ where θ is a random variable with uniform probability density function

$$\begin{aligned} p(\theta) &= \frac{1}{2\pi} \quad 0 \le \theta \le 2\pi \\ &= 0 \quad \text{elsewhere} \ . \end{aligned}$$

From the results of earlier sections,

$$\mu_x = E[x(t)] = \int_{-\infty}^{\infty} X \sin(\omega t + \theta) p(\theta) d\theta = 0$$

$$R_{xx}(\tau) = E[x(t)x(t+\tau)] = \int_{-\infty}^{\infty} X^2 \sin(\omega t + \theta) \sin(\omega t + \tau + \theta) p(\theta) d\theta$$

$$= \frac{X^2}{2} \cos \omega \tau \ .$$

Note that the autocovariance function is 'phase blind', i.e. it is independent of θ. Note also that the same result can be obtained using a time average over a single record of the process (and allowing the record length T to get large).

Asynchronous random telegraph signal

Consider a time function that switches between two voltage levels $+a, -a$ (see Fig. 6.26).

The crossing times t_i are random and we shall assume the times t_i constitute a Poisson process with parameter λ, i.e. probability of k crossings in time τ is

$p_k = \frac{e^{-\lambda|\tau|}(\lambda|\tau|)^k}{k!}$, where λ is the number of crossings/unit time.

The mean value, $\mu_x = 0$, and

$$R_{xx}(\tau) = E[x(t)x(t+\tau)] = a^2 \text{ or } - a^2 \ .$$

The result is a^2 if an even number of crossings occurs in time τ and $-a^2$ if an odd number of crossings occurs in time τ.

So

$$R_{xx}(\tau) = \sum_{k=0}^{\infty} [a^2 p_{2k} - a^2 p_{2k+1}]$$

$$= a^2 e^{-\lambda|\tau|} \left[\sum_{k=0}^{\infty} \frac{(\lambda|\tau|)^{2k}}{2k!} - \frac{(\lambda|\tau|)^{2k+1}}{(2k+1)!} \right]$$

$$R_{xx}(\tau) = a^2 e^{-2\lambda|\tau|}$$

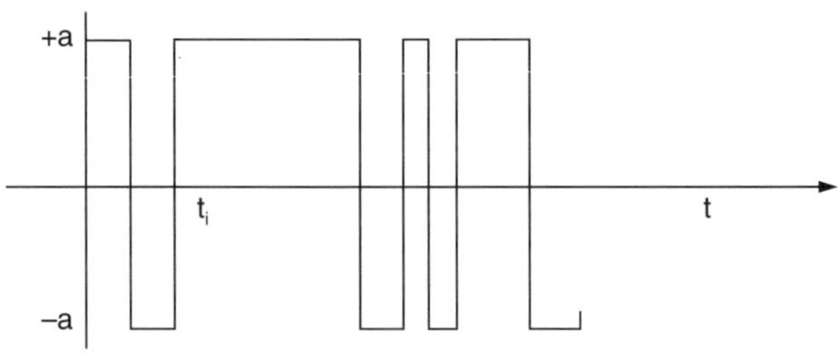

Figure 6.26 Asynchronous random telegraph signal.

i.e. the autocovariance function is an exponentially decaying function where λ controls the decay rate.

White noise

This is a theoretical concept – but which is, nevertheless, very useful. We can lead up to it by considering a limiting form of the above. As $\lambda \to \infty$ above $R_{xx}(\tau)$ gets very narrow, i.e. the process is very erratic. In order that $R_{xx}(\tau)$ does not 'disappear' completely we intuitively let the parameter a get large to compensate. We see that $R_{xx}(\tau) \to k\delta(\tau)$. This leads to the idea that a 'completely erratic' random process will have a delta function-like autocovariance function, i.e. $R_{xx}(\tau) = k\delta(\tau)$. Processes with this property are called 'white noise'. We shall only be able to approximate such processes in practice and often refer to 'band limited white noise', which anticipates the spectral description of such processes.

Note: Processes with a variety of probability distributions can have a delta function autocorrelation, so simply saying a process is 'white' does not convey its probabilistic structure.

The cross covariance (correlation) function

Consider two functions

$$x(t) = A\sin(\omega t + a) + B$$

$$y(t) = C\sin(\omega t + c) + D\sin(n\omega t + d) \ .$$

Using the time averaged form, the cross correlation is

$$R_{xy}(\tau) = \tfrac{1}{2}AC\cos(\omega\tau + (a - c)) \ .$$

Note: the cross correlation function seeks out the part of $y(t)$ that 'matches' or fits $x(t)$ *and it retains the relative phase* $(a - c)$, i.e. the cross correlation detects the delay relating the similar parts of x and y.

A signal buried in noise

Assume the noise and signal are uncorrelated, e.g. $y(t) = s(t) + n(t)$. If $s(t)$ is a sine wave and $n(t)$ is broad band noise (uncorrelated with s), then

$$\begin{aligned} R_{yy}(\tau) &= E[(s(t) + n(t))(s(t+\tau) + n(t+\tau))] \\ &= R_{ss}(\tau) + R_{nn}(\tau) \ . \end{aligned}$$

Since $R_{nn}(\tau)$ decays quickly, then $R_{ss}(\tau)$ dominates the result for larger values of τ, and so this provides a way of identifying the sine component embedded in the noise.

6.5.5.9 Spectra

So far our considerations of stochastic processes have been restricted to the time domain. We now introduce the frequency domain. We have seen that Fourier methods may be applied to 'periodic' and 'transient' phenomena – we shall now see that Fourier methods may be applied to random (stationary) processes. We shall draw on the work in sections 6.2 and 6.3.

Consider a sample function of a random process $x(t)$ and let us look at a truncated version $x_{\mathrm{T}}(t)$ (Fig. 6.27)

$$\begin{aligned} x_{\mathrm{T}}(t) &= x(t), \quad |t| < \tfrac{T}{2} \\ &= 0 \qquad \text{elsewhere} . \end{aligned}$$

We shall study the frequency decomposition of the power of this signal. Since $x_{\mathrm{T}}(t)$ is 'pulse-like' (i.e. transient), it has a Fourier representation

$$x_{\mathrm{T}}(t) = \int_{-\infty}^{\infty} X_{\mathrm{T}}(f) e^{2\pi jft} df .$$

The total energy of the signal is $\int_{-\infty}^{\infty} x_{\mathrm{T}}^2(t)dt$ which tends to infinity as T gets large and so we consider the average

$$\frac{1}{T}\int_{-\infty}^{\infty} x_{\mathrm{T}}^2(t)dt \text{ which is the average power.}$$

By Parseval's theorem

$$\frac{1}{T}\int_{-\infty}^{\infty} x_{\mathrm{T}}^2(t)dt = \frac{1}{T}\int_{-T/2}^{T/2} x^2(t)dt = \frac{1}{T}\int_{-\infty}^{\infty} |X_{\mathrm{T}}(f)|^2 df \qquad (6.54)$$

Now as $T \to \infty$,

$$\lim_{T\to\infty} \frac{1}{T}\int_{-T/2}^{T/2} x^2(t)dt = \int_{-\infty}^{\infty} \lim_{T\to\infty} \frac{|X_{\mathrm{T}}(f)|^2}{T} df. \qquad (6.55)$$

The quantity $\frac{|X_{\mathrm{T}}(f)|^2}{T}$ is called the sample (raw; see below) spectral density and is given the notation $\hat{S}_{xx}(f)$ and since the left-hand side of the above

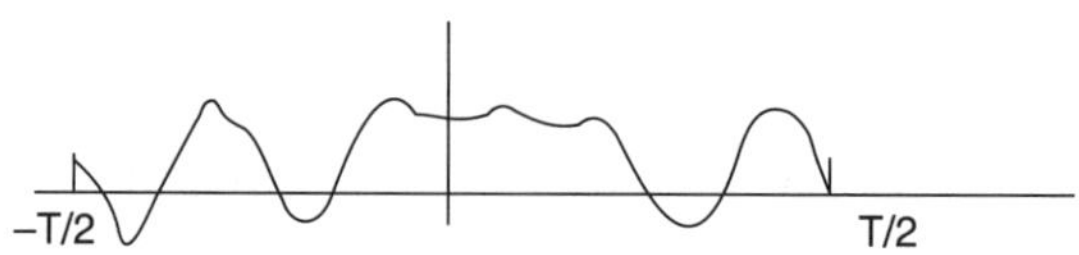

Figure 6.27 A truncated sample function.

equation is the average power of the sample function it is tempting to define $\lim_{T\to\infty} \frac{|X_T(f)|^2}{T}$ as the power spectral density. Unfortunately, this is not useful since this quantity does not converge in a statistical sense. When we come to estimation theory (section 6.7), we shall see that if this quantity is evaluated from a data length $T' > T$ the answers are just as erratic as for the shorter data length, i.e. the use of 'more data' does not 'improve' our estimate. In fact, we have now come across an estimate for which the ergodic property does not hold. However, we shall see that a very useful theoretical function may be introduced by averaging this quantity to 'remove' the erratic behaviour, i.e. consider the average

$$E\left\{\lim_{T\to\infty}\frac{1}{T}\int_{-T/2}^{T/2} x^2(t)\mathrm{d}t\right\} = E\left\{\int_{-\infty}^{\infty}\lim_{T\to\infty}\frac{|X_T(f)|^2}{T}\mathrm{d}f\right\} . \tag{6.56}$$

This means that the variance of the process is

$$\sigma_x^2 = \int_{-\infty}^{\infty} S_{xx}(f)\mathrm{d}f , \tag{6.57}$$

where

$$S_{xx}(f) = \lim_{T\to\infty}\frac{E|X_T(f)|^2}{T} . \tag{6.58}$$

This states that the average power of the process (the variance) is decomposed over frequency in terms of the function $S_{xx}(f)$. This function is the *power spectral density function* of the process. This is a useful function which has a direct physical interpretation. It can also be related directly to the autocovariance function in the following way.

$$S_{xx}(f) = \int_{-\infty}^{\infty} R_{xx}(\tau)e^{-2\pi jf\tau}\mathrm{d}\tau \tag{6.59}$$

and

$$R_{xx}(\tau) = \int_{-\infty}^{\infty} S_{xx}(f)e^{2\pi jf\tau}\mathrm{d}f . \tag{6.60}$$

These relations are sometimes called the Wiener–Khinchin theorem.

Note: If ω is used, the equivalent result is

$$S_{xx}(\omega) = \int_{-\infty}^{\infty} R_{xx}(\tau)e^{-j\omega\tau}\mathrm{d}\tau$$

and

$$R_{xx}(\tau) = \frac{1}{2\pi}\int_{-\infty}^{\infty} S_{xx}(\omega)e^{j\omega\tau}d\omega \ .$$

The location of the factor 2π may be interchanged or indeed replaced with $1/\sqrt{2\pi}$ for symmetrical form.

Comments on the power spectral density function

1. If $x(t)$ is in volts, $S_{xx}(f)$ has units of volt^2/Hz.
2. $S_{xx}(f)$ is an even function of frequency.
3. A one-sided power spectral density can be defined called

$$\begin{aligned} G_{xx}(f) &= 2S_{xx}(f) \quad & f > 0 \\ &= S_{xx}(f) \quad & f = 0 \\ &= 0 \quad & f < 0 \ . \end{aligned}$$

Examples of power spectra

(a) If $R_{xx}(\tau) = \sigma^2 e^{-\lambda|\tau|}$, then

$$S_{xx}(f) = \int_{-\infty}^{\infty} R_{xx}(\tau)e^{-2\pi jf\tau}d\tau = \frac{2\lambda\sigma^2}{\lambda^2 + 4\pi^2 f^2}$$

i.e. an exponentially decaying correlation function results in a 'low frequency' power spectral density function.

(b) If $R_{xx}(\tau) = k\delta(\tau)$ then $S_{xx}(f) = k$

i.e. a 'narrow' autocorrelation function results in a broad spectrum.

(c) If $R_{xx} = \frac{X^2}{2}\cos 2\pi f_0\tau$, then

$$S_{xx}(f) = \frac{X^2}{4}\delta(f + f_0) + \frac{X^2}{4}\delta(f - f_0)$$

i.e. an oscillatory correlation corresponds to spikes in the spectral density.

The cross-spectral density function

This may be defined (generalizing the Wiener–Khinchin theorem) as

$$S_{xy}(f) = \int_{-\infty}^{\infty} R_{xy}(\tau)e^{-j2\pi f\tau}d\tau \tag{6.61}$$

with the inverse

$$R_{xy}(\tau) = \int_{-\infty}^{\infty} S_{xy}(f)e^{j2\pi f\tau}df \ . \tag{6.62}$$

Alternatively, $S_{xy}(f)$ may be defined as

$$S_{xy}(f) = \lim_{T\to\infty} E\frac{\{X_T^*(f)Y_T(f)\}}{T} \tag{6.63}$$

where $X_T(f)$, $Y_T(f)$ are Fourier transforms of $x(t), y(t)$ for $\frac{-T}{2} \leq t \leq \frac{T}{2}$.

Note: The cross-spectral density is in general complex and has the interpretation of being the frequency domain equivalent of the cross-covariance function, i.e. $S_{xy}(f) = |S_{xy}(f)|e^{j\arg S_{xy}(f)}$, where $|S_{xy}(f)|$ is the *cross-amplitude spectrum* and it shows whether frequency components in one time series are associated with large or small amplitudes at the same frequency in the other series. Arg $S_{xy}(f)$ is the *phase spectrum* and this shows whether frequency components in one series lag or lead the components at the same frequency in the other series.

Properties and associated definitions

1.
$$S_{xy}(f) = S_{yx}^*(f). \tag{6.64}$$

   This may be proved using the fact that $R_{xy}(\tau) = R_{yx}(-\tau)$

2. A one-sided cross-spectral density $G_{xy}(f)$ is defined

$$\begin{aligned} G_{xy}(f) &= 2S_{xy}(f) && f > 0 \\ &= S_{xy}(f) && f = 0 \\ &= 0 && f < 0. \end{aligned} \tag{6.65}$$

3. The coincident spectral density (co-spectrum) and quadrature spectral density (quad-spectrum) are defined as

$$G_{xy}(f) = C_{xy}(f) - jQ_{xy}(f), \quad f \geq 0 \ .$$

   C_{xy}, Q_{xy} are the so-called co- and quad-spectra, respectively.

4. An inequality satisfied by the cross-spectral density is

$$|G_{xy}(f)|^2 \leqslant G_{xx}(f)G_{yy}(f) \tag{6.66}$$

   (or $|S_{xy}(f)|^2 \leqslant S_{xx}(f)S_{yy}(f)$).

   We shall (later) define a 'coherence function' as $\frac{|G_{xx}|^2}{G_{xx}G_{yy}}$, i.e. a normalized cross-spectral density function.

6.6 INPUT–OUTPUT RELATIONSHIPS AND SYSTEM IDENTIFICATION

We shall now develop input–output relationships for linear systems operating with random inputs. We shall consider time invariant systems and stationary excitations.

6.6.1 Single input single-output systems

Figure 6.5 depicts a linear, time invariant system with input $x(t)$ and output $y(t)$.

The response due to an input starting at t_0 is

$$y(t) = \int_{t_0}^{t} h(t - t_1)x(t_1)\mathrm{d}t_1 \ . \tag{6.67}$$

If (i) the system is stable and (ii) the response to a stationary $x(t)$ has reached a 'steady state', i.e. $y(t)$ is also stationary, then

$$y(t) = \int_{-\infty}^{t} h(t - t_1)x(t_1)\mathrm{d}t_1$$

or

$$y(t) = \int_{0}^{\infty} h(\tau)x(t - \tau)\mathrm{d}\tau \ . \tag{6.68}$$

Since the input is stationary and random, it is characterized by μ_x and $R_{xx}(\tau)$ (for convenience set $\mu_x = 0$). The output autocovariance function $R_{yy}(\tau)$ is

$$R_{yy}(\tau) = E[y(t)y(t+\tau)] = E\left[\int_0^\infty \int_0^\infty h(\tau_1)x(t-\tau_1)h(\tau_2)x(t+\tau-\tau_1)\mathrm{d}\tau_1\mathrm{d}\tau_2\right]$$

i.e.

$$R_{yy}(\tau) = \int_0^\infty \int_0^\infty h(\tau_1)h(\tau_2)R_{xx}(\tau+\tau_1-\tau_2)\mathrm{d}\tau_1\mathrm{d}\tau_2. \tag{6.69}$$

The frequency domain equivalent is more useful. Fourier transforming this equation gives

$$S_{yy}(f) = |H(f)|^2 S_{xx}(f) \tag{6.70}$$

where $H(f) = \int_0^\infty h(\tau)\mathrm{e}^{-2\pi jf\tau}\mathrm{d}t$ is the system frequency response function.

The frequency domain form is clearly simpler than the time domain version and indicates how the power spectral density of the input is 'shaped' by the frequency response characteristic of the system, and the output variance is the area under the $S_{yy}(f)$ curve. The expression above is purely real (phase plays no part). The following expression introduces phase.

The input–output cross covariance function $E[x(t)y(t+\tau)]$ is

$$R_{xy}(\tau) = E[x(t)y(t+\tau)] = E\left[\int_0^\infty h(\tau_1)x(t)x(t+\tau-\tau_1)\mathrm{d}\tau_1\right]$$

$$R_{xy}(\tau) = \int_0^\infty h(\tau_1) R_{xx}(\tau - \tau_1) \mathrm{d}\tau_1. \tag{6.71}$$

The frequency domain equivalent of this is obtained by transforming this to give

$$S_{xy}(f) = H(f) S_{xx}(f). \tag{6.72}$$

Note that this expression contains the phase of the frequency response function, i.e. $\arg S_{xy}(f) = \arg H(f)$. The expression (6.72) is often the basis of system identification schemes because of this property.

6.6.2 The ordinary coherence function

The coherence between input $x(t)$ and output $y(t)$ is defined

$$\gamma_{xy}^2(f) = \frac{|G_{xy}(f)|^2}{G_{xx}(f) G_{yy}(f)} = \frac{|S_{xy}(f)|^2}{S_{xx}(f) S_{yy}(f)} \tag{6.73}$$

and from the inequality (6.66),

$$0 \le \gamma_{xy}^2(f) \le 1 \ . \tag{6.74}$$

For the single input-single output system in Fig. 6.5, the coherence function may be shown (using the basic input–output relationships) to be

$$\gamma_{xy}^2 = \frac{|H|^2 G_{xx}^2}{G_{xx} |H|^2 G_{xx}} = 1 \ .$$

So, if two processes are *linearly* related, the coherence is unity, and if $x(t)$ and $y(t)$ are completely unrelated (linearly), the coherence function is zero. If the coherence function is greater than zero but less than one, then x and y are in part linearly related but, in addition, at least one of the following holds:

1. the measurements are 'noisy'
2. x and y are not *only* linearly related
3. y is an output due to input x *and other inputs*.

6.6.2.1 Example

Consider the effect of measurement noise (see Fig. 6.28). $y_m(t)$ is a measurement. The coherence between x and $y_m(t)$ is

$$\gamma_{xy_m}^2(f) = \frac{|G_{xy_m}(f)|^2}{G_{xx}(f) G_{y_m y_m}(f)} \ .$$

Assuming x and n are uncorrelated, with power spectra G_{xx} and G_{nn}, and using the standard input–output relationships, then

$$\gamma_{xy_m}^2(f) = \frac{1}{1 + \frac{G_{nn}(f)}{G_{yy}(f)}} = \frac{G_{yy}}{G_{yy} + G_{nn}}$$

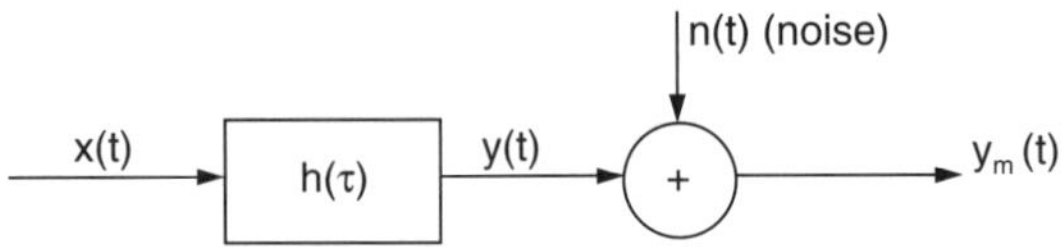

Figure 6.28 The effect of measurement noise on the output.

i.e. $\gamma^2_{xy_{\hat{m}}}$ is fractional portion of the power of the output $y_m(t)$ which is contributed by input $x(t)$, and $1 - \gamma^2_{xy_{\hat{m}}}$ is the fractional portion of the power of the output *not* accounted for by input $x(t)$.

The term *coherent output power* describes the power of that part of $y_m(t)$ that is due to x, i.e. $G_{yy}(f) = \gamma^2_{xy_m}(f)G_{y_my_m}(f)$. Noise power (or incoherent output power) is that part of $y_m(t)$ that is due to noise n, i.e. $G_{nn}(f) = [1 - \gamma^2_{xy_m}(f)]G_{y_my_m}(f)$.

6.6.3 System identification

This section is concerned with the identification of linear time invariant systems using random input–output data. As a preliminary, we consider a problem in least squares estimation.

Assume that we make measurements of two quantities x and y and that we intend to find a linear relationship between them of the form $y = ax$. If, for each x_i, the associated value of y is y_i, then we might plot the results as a scatter diagram (Fig. 6.29).

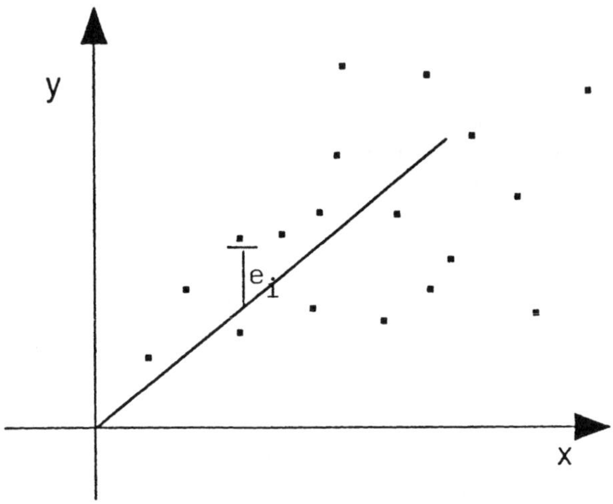

Figure 6.29 Scatter diagram relating variables x (input) and y (output); errors on output.

Now consider three situations:

1. Variable x_i is known exactly but y_i may be 'noisy'.
In this case, we attempt to find the linear relation by forming an error $e_i = y_i - \hat{a}x_i$ and find the $\hat{a}$ that minimizes $\sum e_i^2$. This is

$$\hat{a} = \frac{\sum x_i y_i}{\sum x_i^2} \tag{6.75}$$

(see Fig. 6.29).

2. Variable y_i is known exactly but x_i may be 'noisy'.
Now, we form an error, $e_i = \left(x_i - \frac{y_i}{\hat{a}}\right)$
and minimizing $\sum e_i^2$ yields

$$\hat{a} = \frac{\sum y_i^2}{\sum x_i y_i}. \tag{6.76}$$

See Fig. 6.30
3. Both variables x_i and y_i are noisy. In this case the perpendicular distance to the line $y = \hat{a}x$ is minimized (see Fig. 6.31). This is referred to as the Total Least Squares approach and the error is

$$e_i = \frac{(y_i - \hat{a}x_i)}{\sqrt{1 + \hat{a}^2}}$$

and $\hat{a}$ is found by minimizing $\sum e_i^2$ with respect to $\hat{a}$.

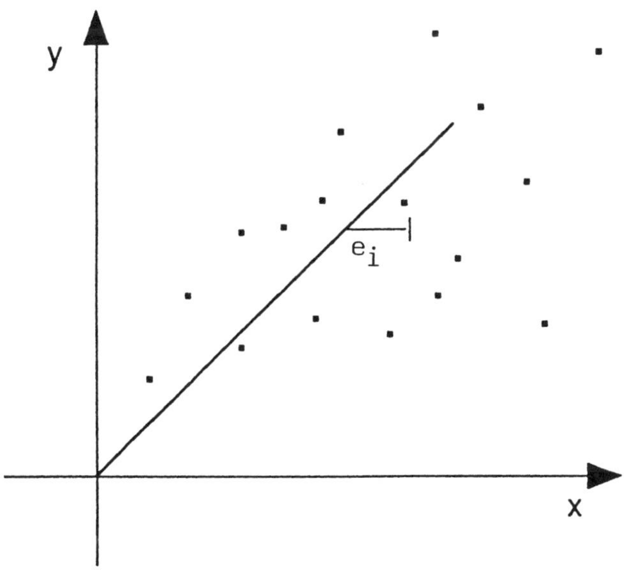

Figure 6.30 Scatter diagram; errors on input.

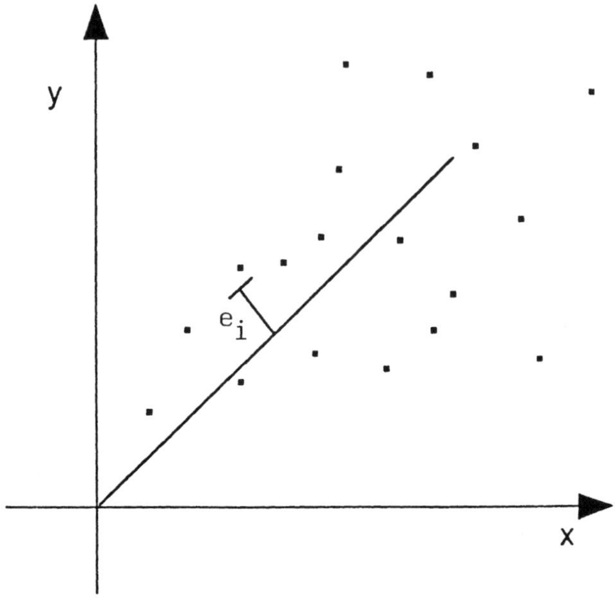

Figure 6.31 Scatter diagram; errors on input and output.

Following these preliminary observations, we return to linear system identification.

6.6.3.1 *Frequency response identification*

For a linear system we know that

$$Y(f) = H(f)X(f) \ .$$

Let us suppose that for each f we have a series of results $X_i(f)$ and $Y_i(f)$. (We shall assume appropriate use of estimation theory, as in section 6.7). Then by using the section above we seek $H(f)$ (analogous to parameter a) relating X and Y for each f. The only difference here is that the quantities are complex, and so the analysis requires appropriate modification.

For case (1), namely that of *output noise* only, we obtain

$$\hat{H}(f) = \frac{\sum X_i^*(f)Y_i(f)}{\sum |X_i(f)|^2} \tag{6.77}$$

which, with suitable averaging and limiting operations, is $\frac{S_{xy}(f)}{S_{xx}(f)}$. This is sometimes referred to as estimator $H_1(f)$, which is unbiased with respect to the presence of output noise.

For case (2), namely that of input noise only, we see that

$$\hat{H}(f) = \frac{\sum |Y_i|^2}{\sum X_i Y_i^*} \tag{6.78}$$

which leads to $\frac{S_{yy}}{S_{xy}^*(f)}$. This is sometimes referred to as estimator $H_2(f)$, which is unbiased with respect to input noise.

For case (3), the results are somewhat more complicated and lead to the so-called H_v estimator (Leuridan *et al.*, 1985) and sometimes called the H_s estimator (Wicks and Vold, 1986).

The effects of feedback

It is often the case that feedback is present in a dynamical system (see, for example, Fig. 6.32). Here signal r is the (controlled) input and n is an extraneous input. H is the system to be determined.

Forming $H_1 = \frac{S_{xy}}{S_{xx}}$ leads to $\frac{HS_{rr}+G^*S_{nn}}{S_{rr}+|G|^2S_{nn}}$ which is certainly not the required H. Indeed, if S_{nn}/S_{rr} is large then the ratio S_{xy}/S_{xx} is $1/G$. It can be easily verified, however, that the use of the three signals r, x and y and forming the ratio of two cross-spectral densities S_{ry}/S_{rx} (sometimes referred to as estimator H_3) yields the required $H(f)$.

6.7 RANDOM PROCESSES AND ESTIMATION

In section 6.5, random processes were described in terms of probability functions, correlation functions, and spectral densities. The definition of these 'theoretical' concepts serves as the basis upon which measured data may be interpreted. In this section we introduce and discuss estimation methods for random signals, based on a single recording of the process.

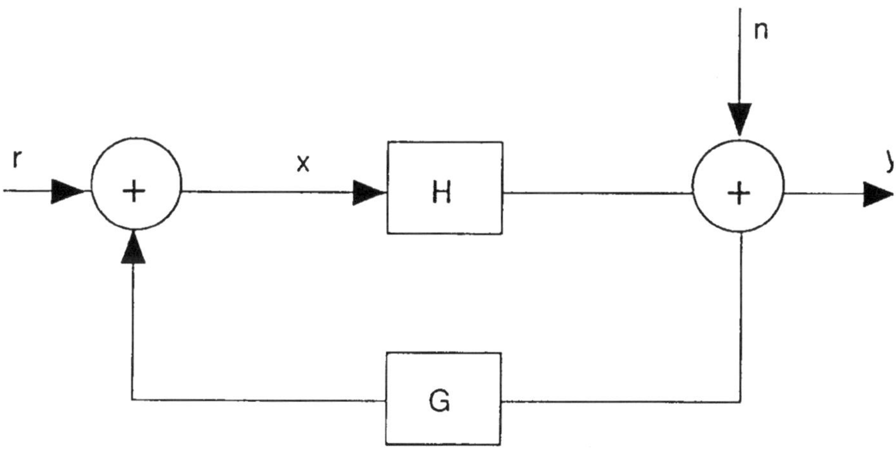

Figure 6.32 A system with feeback.

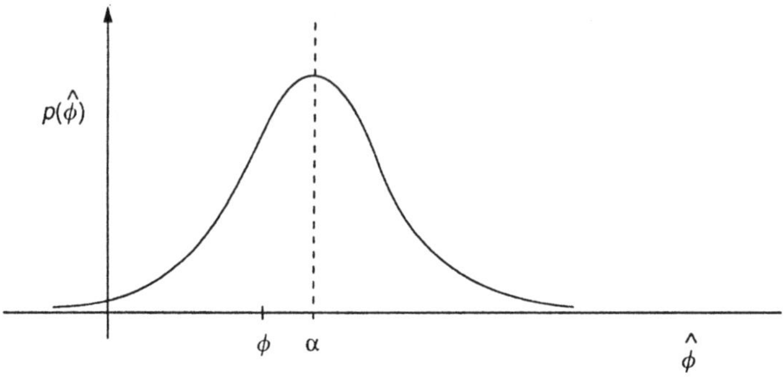

Figure 6.33 A two-input system with feedback.

6.7.1 Estimator errors and accuracy

Suppose the mean value of a random process is μ_x (unknown to us) and we only have a record length T of the signal. A reasonable estimate $\bar{x}$ of μ_x is

$$\bar{x} = \frac{1}{T}\int_0^T x(t)\mathrm{d}t \ . \tag{6.79}$$

The value obtained is a sample value of a random variable $\overline{X}$, having its own probability distribution (called a sample distribution), and $\bar{x}$ is a single realization. Each $\bar{x}$ computed from a (different) record of length T will, in general, be different. If the estimation procedure is to be useful we would hope that (i) the scatter of values of $\bar{x}$ is not 'too great' and should lie 'close' to μ_x; (ii) the more data we take (the larger T is) the 'better' the estimate. We can formalize these ideas as follows.

Let ϕ be a parameter we wish to estimate (in the introductory example above ϕ is μ_x) and let $\hat{\Phi}$ be an estimator for ϕ, then $\hat{\Phi}$ is a random variable with the probability distribution, for example, as in Fig. 6.33, where it can be seen that estimates $\hat{\phi}$ of random variable $\hat{\Phi}$ would predominantly take values near α.

The sampling distribution $p(\hat{\phi})$ is often difficult to obtain, and so we shall settle for a few summarizing properties.

6.7.1.1 Bias error

The bias of an estimator is defined

$$b(\hat{\Phi}) = E[\hat{\Phi}] - \phi \ . \tag{6.80}$$

This is the difference between the average of the estimator and the true value. If $b(\hat{\Phi}) = 0$, then $\hat{\Phi}$ is 'unbiased'. While it might seem desirable to use an unbiased estimator, it is sometimes prudent to tolerate some bias if it means that the variability of the estimate is reduced (relative to that of an unbiased estimator).

6.7.1.2 Variance

The variance of an estimator is a measure of the spread of $\hat{\Phi}$ about its own mean value

$$\text{Variance}[\hat{\Phi}] = E[(\hat{\Phi} - E(\hat{\Phi}))^2] \ . \tag{6.81}$$

The square root of the variance is the standard deviation ($\sigma(\hat{\Phi})$) of the estimator. We would hope that this would be 'small', i.e. the probability density function should be 'peaky', but this requirement often increases the bias. A measure of the relative importance of bias and variance is the mean square error.

6.7.1.3 Mean square error (mse.)

The mse of an estimator is a measure of the spread of values of $\hat{\Phi}$ about the value ϕ, i.e.

$$\text{m.s.e.}[\hat{\Phi}] = E[(\hat{\Phi} - \phi)^2] \ . \tag{6.82}$$

The right-hand side of (6.82) can be rewritten as

$$\text{m.s.e.}[\hat{\Phi}] = \text{Var}[\hat{\Phi}] + b^2[\hat{\Phi}] \tag{6.83}$$

showing that m.s.e. reflects both variance and bias.

If the m.s.e. reduces as the sample size (amount of data used) increases, then the estimator is *consistent*.

6.7.1.4 Confidence intervals

The estimates $\hat{\phi}$ referred to so far are *point estimates*, i.e. single values. It is frequently desirable to know that a parameter is likely to fall within a certain *interval of values*. For example, if we estimate a mean value $\bar{x}$ as 5, perhaps it is 'likely' that μ_x lies in the interval 4.5–5.5. These estimates are *interval estimates*, and are called *confidence intervals* when we attach a number describing the likelihood of the parameter falling within the interval.

If, for example, we state 'a 95% confidence interval for μ_x is (4.5, 5.5)', then this means we are 95% confident that μ_x lies in the range (4.5, 5.5). Note that this does *not* mean that the probability that μ_x lies in the interval (4.5, 5.5) is 0.95 (μ_x is not a random variable and so we do not assign probabilities to it). We mean that if we could realize a large number of samples and find a confidence interval of μ_x for *each* sample, then approximately 95% of these intervals would contain μ_x.

In order to calculate confidence intervals, we need to know the sampling distribution of the estimator. We shall return to this problem for a specific estimator.

6.7.2 Estimators for stochastic processes

The problem of the study of estimators for stochastic processes is a wide one, and some idea of the extent of the field can be obtained from Jenkins and Watts (1968). Several other texts are not as detailed and give useful summaries (Beauchamp (1973); Bendat and Piersol (1971, 1980); Otnes and Enochson (1978)).

We shall pick out the autocorrelation function and power spectral density as examples, and try to indicate some of the features that should be considered when analysing a sample time history. Following this we briefly discuss cross spectra and system identification. Unless otherwise stated, we shall assume we are dealing with realizations of a continuous, stationary random process. In practice, data is sampled so that the computation may be done and is therefore subject to the attendant 'aliasing' problems. It is assumed throughout that the sampling rates are 'sufficiently high'.

6.7.3 The autocorrelation function

6.7.3.1 Definition

If we have a signal defined for $0 \leq t \leq T$, then there are two (Jenkins and Watts, 1968) commonly used estimators of the function $R_{xx}(\tau)$. Following Jenkins and Watts, we write them as $\hat{R}_{XX}(\tau)$ and $\hat{R}'_{XX}(\tau)$ where (assuming a zero mean process)

$$\begin{aligned} \hat{R}_{XX}(\tau) &= \frac{1}{T}\int_0^{T-|\tau|} X(t)X(t+|\tau|)\mathrm{d}t \qquad 0 \leq |\tau| \leq T \\ &= 0 \qquad\qquad |\tau| > T \end{aligned} \tag{6.84}$$

and

$$\begin{aligned} \hat{R}'_{XX}(\tau) &= \frac{1}{T-|\tau|}\int_0^{T-|\tau|} X(t)X(t+|\tau|)\mathrm{d}t \qquad 0 \leq |\tau| \leq T \\ &= 0 \qquad\qquad |\tau| > T. \end{aligned} \tag{6.85}$$

Often the expression (6.85) is used since it is unbiased, but it is the m.s.e. that should be the criterion for a choice between the two.

6.7.3.2 Statistical considerations

For τ in the range $0 \leq |\tau| \leq T, \hat{R}'_{XX}(\tau)$ is unbiased, and $E[\hat{R}_{XX}(\tau)] = R_{xx}(\tau)\left[1 - \frac{|\tau|}{T}\right]$ showing (6.86) to be 'asymptotically unbiased'.

An approximate expression for the variance $\hat{R}_{XX}(\tau)$ is

$$\mathrm{Var}[\hat{R}_{XX}(\tau)] \approx \frac{1}{T} \int_{-\infty}^{\infty} [R_{xx}^2(r) + R_{xx}(r+\tau)R_{xx}(r-\tau)]\mathrm{d}r$$

showing this to be a consistent estimator. When $\tau \ll T$ there is little to choose between the two estimators, but as τ approaches T the variance of the unbiased estimator, $\hat{R}'_{XX,}$ diverges (hardly surprising considering the divisor in (6.85)).

Another important feature of the estimators is that values computed for neighbouring values of τ are samples of (in general) strongly correlated random variables, meaning that $\hat{R}_{XX}(\tau)$ is more strongly correlated than the original time series $x(t)$, so the estimate may not decay as rapidly as one might expect (see section 5.3 of Jenkins and Watts, 1968).

6.7.4 The power spectral density

Estimation methods for spectra considered here will relate to the 'traditional' methods rather than the 'parametric' methods. The standard methods are: (i) the Fourier transform of the autocorrelation function (indirect method); (ii) the direct method of averaging periodograms; (iii) filtering (either bandpass or complex demodulation (heterodyning) and low-pass filtering). We shall outline (i) and (ii). An interesting paper (Bingham, Godfrey and Tukey, 1967) considers various aspects (computational and statistical) of spectral estimation.

6.7.4.1 The Fourier transform of the autocorrelation function

Definition

An estimate of the power spectral density from a length of data T (here assumed to be defined for $|t| < T/2$) is

$$\hat{S}_{XX}(f) = \frac{|X_{\mathrm{T}}(f)|^2}{T} = \int_{-T}^{T} \hat{R}_{XX}(\tau)\mathrm{e}^{-j2\pi f\tau}\mathrm{d}\tau \ . \tag{6.86}$$

This is termed the 'raw' spectral density since it turns out (see the section) which follows) that the variability of this estimator is independent of the data length T.

To reduce the sampling fluctuations, the estimate must be smoothed in the frequency domain, and this can be achieved by multiplying the autocorrelation function estimate by a 'lag window' $w(\tau)$ which has a Fourier transform $W(f)$ (a 'spectral window'), and Fourier transforming the product. This defines the estimator $\tilde{S}_{XX}(f)$) where

$$\tilde{S}_{XX}(f) = \int_{-T}^{T} \hat{R}_{XX}(\tau)w(\tau)e^{-j2\pi f\tau}d\tau \ , \tag{6.87}$$

which by section 6.3.3 is the convolution of the 'raw' spectral density with $W(f)$, i.e. $\hat{S}_{XX}(f)^{*}W(f)$. This estimation procedure may be viewed as a smoothing operation in the frequency domain. In the time domain, the lag window can be regarded as reducing the 'importance' of values of $\hat{R}_{XX}$ as τ increases (since the variability of the estimator increases as τ increases).

Statistical properties

In order to discuss the properties of estimators $\hat{S}_{XX}$ and $\tilde{S}_{XX}$ it is necessary to first outline a few preliminary ideas in statistics (Jenkins and Watts, 1968).

The chi-squared distribution

If $X_1, X_2, \ldots, X_n$ are n independent random variables each having a normal distribution with zero mean and unit standard deviation, and if

$$\chi_n^2 = X_1^2 + X_2^2 + \ldots + X_n^2 \tag{6.88}$$

then the distribution of χ_n^2 is called the chi-squared distribution with 'n degrees of freedom' (the number of degrees of freedom representing the number of independent random variables X_i entering the expression). The first two moments of χ_n^2 are

$$E[\chi_n^2] = n \quad \text{and} \quad \mathrm{Var}[\chi_n^2] = 2n \ . \tag{6.89}$$

If a set of independent χ^2 variables are added together, i.e.

$$\chi_v^2 = \chi_{n_1}^2 + \chi_{n_2}^2 + \cdots + \chi_{n_3}^2 \tag{6.90}$$

then the sum is χ_v^2 (as indicated), and $v = n_1 + n_2 + \cdots + n_k$.

These properties of the chi-squared distribution are used in the development of some of the following results (see Jenkins and Watts).

Bias

$\hat{S}_{XX}(f)$ is an asymptotically unbiased estimator, i.e. $E[\hat{S}_{XX}(f)] = S(f)$ as $T \to \infty$.

Variance

Since $\tilde{S}_{XX}(f)$ is a squared quantity (and if we assume $X(t)$ is Gaussian) then it can be argued that $2\hat{S}_{XX}(f)/S_{xx}(f)$ is distributed as a chi-squared random variable with two degrees of freedom, i.e. χ_2^2 (for all values of data length T). From (6. 89) it follows therefore that the variance of $\hat{S}_{XX}(f)$ is $S_{XX}^2(f)$. So $\hat{S}_{XX}(f)$ is an inconsistent estimator of $S_{xx}(f)$, and the standard deviation of the estimate is as large as the quantity being estimated.

Smoothed spectral estimators

Since $\tilde{S}_{XX}(f)$ is a convolution of $\hat{S}_{XX}(f)$ with $W(f)$ it turns out that $\tilde{S}_{XX}(f)$ is biased. In fact (for large T)

$$E[\tilde{S}_{XX}(f)] \approx \int_{-\infty}^{\infty} S_{xx}(g)W(f-g)\mathrm{d}g \tag{6.91}$$

and

$$\mathrm{Bias}[\tilde{S}_{XX}(f)] \approx \int_{-\infty}^{\infty} [w(\tau)-1]R_{xx}(\tau)\mathrm{e}^{-j2\pi f\tau}\mathrm{d}\tau \ . \tag{6.92}$$

Equation (6.91) indicates how the estimate is distorted by smearing and 'leakage' of the spectral density owing to the spectral window width and shape (side lobes), and the window discussions earlier relate directly to this situation. Owing to the complicated shape of the $W(f)$ functions, we can only comment broadly that the general effect is to reduce the dynamic range of the spectra, i.e. peaks are underestimated and troughs are overestimated, and the effect is reduced as the spectral window gets narrower, i.e. as the data window T gets larger. However, the windows cannot be made too narrow since then there is little smoothing and the random errors increase. Consequently there is always the problem of trading bias errors against random errors. (The bias problem is sometimes referred to under 'bandwidth considerations'.)

Since $\tilde{S}_{XX}$ is a weighted sum of values of $\hat{S}_{XX}$, it is argued (Jenkins and Watts) that $\frac{n\tilde{S}_{XX}(f)}{S_{xx}(f)}$ is approximately distributed as χ_n^2 where n (the number of degrees of freedom) $= 2BT$ and B is the resolution bandwidth. It follows that

$$\mathrm{Var}[\tilde{S}_{XX}(f)] = \frac{S_{xx}^2(f)}{n/2} = \frac{S_{xx}^2(f)}{BT} \ , \tag{6.93}$$

and in Jenkins and Watts (1968) estimator accuracies are tabulated for various lag windows.

The choice of window should depend on whether the concern is for statistical *stability* (low variance) or small bias. If the bias is small for all f, then we have high *fidelity*. Of course in a general situation one must be careful with both. For example, if the spectral function has narrow peaks of importance we may willingly tolerate some loss of stability to resolve the peaks properly, while if the spectral function is smooth then bias errors are not likely to be so important.

Confidence intervals

We now give 'interval estimates' based on the 'point estimates' for the smoothed spectral density. Since $n\tilde{S}_{XX}(f)/S_{xx}(f)$ is distributed as a χ_n^2 random variable then the probability density function has the appearance as in Fig. 6.34.

If we choose a number α $(0 < \alpha < 1)$ such that sections of area $\alpha/2$ are marked off as shown by the points $x_{n,\alpha/2}$ and $x_{n,1-\alpha/2}$ then

$$\mathrm{Prob}[x_{n,\alpha/2} \leq \frac{n\tilde{S}_{XX}(f)}{S_{xx}(f)} \leq x_{n,1-\alpha/2}] = 1 - \alpha \ .$$

The points $x_{n,\alpha/2}$ and $x_{n,1-\alpha/2}$ can be obtained from tables of χ_n^2 for different α and the inequality 'solved' for the true spectral density $S_{xx}(f)$ to yield the equivalent inequality

$$\frac{n\tilde{S}_{xx}}{x_{n,1-\alpha/2}} \leq S_{xx}(f) \leq \frac{n\tilde{S}_{xx}}{x_{n,\alpha/2}} \ .$$

So for a particular point estimate $\tilde{S}_{xx}$ the $100(1-\alpha)\%$ confidence limits for S_{xx} are

$$\frac{n}{x_{n,1-\alpha/2}}\tilde{S}_{xx}, \quad \frac{n}{x_{n,\alpha/2}}\tilde{S}_{xx} \ . \tag{6.94}$$

The confidence *interval* is the difference between them.

Note that on a linear scale the confidence interval depends on the estimate, but on a log scale the interval is $\log(n/x_{n,\alpha/2}) - \log(n/x_{n,1-\alpha/2})$ which is a constant independent of $\tilde{S}_{xx}$.

6.7.4.2 Segment averaging

A method of spectral estimation that has become very popular (owing mainly to its speed of computation) is that of segment averaging discussed in Jenkins and Watts' book (section 6.3.4) as Bartlett's procedure and in some detail by Welch (1967) and Yuen (1979). The books by Bendat and Piersol (1971), Otnes and Enochson (1978) and Beauchamp (1973) also discuss it in detail. The basic procedure is outlined with reference to Fig. 6.35.

A data length T is split up into q segments of length T_r (in this case non-overlapping) and the raw periodogram formed for each segment, i.e.

$$\hat{S}_{xx_i}(f) = \frac{1}{T_r}|X_{T_{ri}}(f)|^2 \quad \text{for} \quad i = 1, \ldots, q \tag{6.95}$$

and to reduce the fluctuations the average

$$\tilde{S}_{xx}(f) = \frac{1}{q}\sum_{i=1}^{q} \hat{S}_{xx_i}(f) \tag{6.96}$$

is formed and it can be argued (very approximately) that since each $2\hat{S}_{XX_i}/S_{xx}$ is a χ_2^2 random variable, then $2\tilde{S}_{XX}q/S_{xx}$ is $\chi_{2_q}^2$ from which it follows that

$$\text{Var}[\tilde{S}_{XX}] \approx \frac{S_{xx}^2(f)}{q} \tag{6.97}$$

and since the resolution bandwidth B is approximately $1/T_r$ then

$$\text{Var}[\tilde{S}_{XX}] \approx \frac{S_{xx}^2(f)}{BT} \quad \text{(as before)} . \tag{6.98}$$

While this summarizes the essential features of the method, Welch (1967) and Bingham, Godfrey and Tukey (1967) give more elaborate procedures and insight. An important aspect is the use of data windows (or linear tapering as it is sometimes called) that are assigned to each segment (it is important to realize that the word linear here does not refer to the window shape but to the fact that it operates on the data directly and not on the autocorrelation function where it is sometimes called quadratic tapering). Furthermore, the segments may be chosen to overlap each other. Linear windowing in effect ignores some data because of the tapered shape of the window. Intuitively the overlapping of segments compensates for this in some way, and Welch (1967) has results for this case though it is not easy to relate these results to the indirect method of Fourier transforming the autocorrelation function.

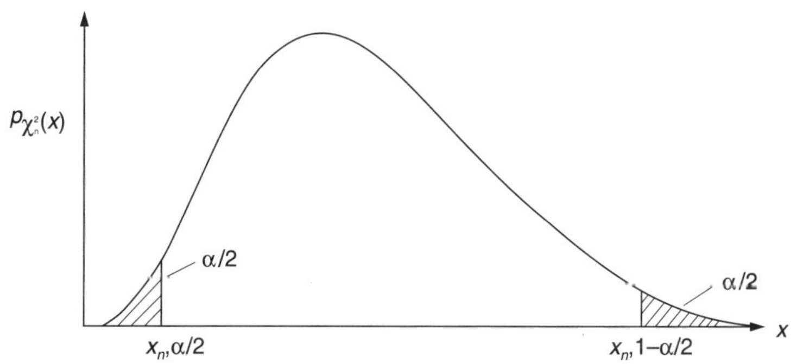

Figure 6.34 The creation of confidence intervals from the sample distribution.

The use of a data window on a segment before transformation reduces leakage but has the disadvantage of also reducing the total energy of that segment, and this must be compensated. This results in the calculation of modified periodograms for each segment of the form

$$\hat{S}_{xx_i}(f) = \frac{\dfrac{1}{T_r}\left|\displaystyle\int_{i\text{-th interval}} x(t)w(t)e^{-j2\pi ft}\,dt\right|^2}{\dfrac{1}{T_r}\displaystyle\int_{-T_r/2}^{T_r/2} w^2(t)\,dt} \tag{6.99}$$

where the denominator compensates for the 'energy reduction'.

6.7.5 Cross-spectra, coherence and transfer functions

6.7.5.1 Cross-spectra

The basic considerations above also relate to cross-spectral estimation together with some additional features. Detailed results can be found in Chapter 9 of Jenkins and Watts (1968). If the smoothed cross-spectral density $\tilde{S}_{xy}(f)$ is written as $|\tilde{S}_{xy}(f)|\exp(j\arg\tilde{S}_{xy}(f))$, then the variances of the amplitude ($|\tilde{S}_{xy}(f)|$) and phase ($\arg\tilde{S}_{xy}$) estimators are proportional to $1/BT$ but are also strongly dependent on the coherence $\gamma^2_{xy}(f)$. As Jenkins and Watts emphasize, the sampling properties of the cross-amplitude and phase estimators may be dominated by the (uncontrollable) influence of the coherency spectrum rather than by the (controllable) influence of the smoothing factor $1/BT$.

6.7.5.2 The coherence function

The estimate for the coherence function is made up from estimates of smoothed power and cross-spectra, i.e.

$$\tilde{\gamma}^2_{xy} = \frac{|\tilde{S}_{xy}(f)|^2}{\tilde{S}_{xx}(f)\tilde{S}_{yy}(f)}\,.$$

It should be noted that if 'raw' spectra are used in this estimate it is a simple matter to verify that the sample coherence function is unity for all frequencies for any signals x and y. Statistical errors and confidence limits are given in Bendat and Piersol (1971), Otnes and Enochson (1978), Bendat and Piersol (1980), and Carter (1980) discusses the bias of the estimator. Results derived by Jenkins and Watts show that the bias is proportional to the square of the derivative of the phase spectrum between x and y. This means that the estimator is sensitive to delays between x and y and can also be expected to be inaccurate when, for example, x and y are input and

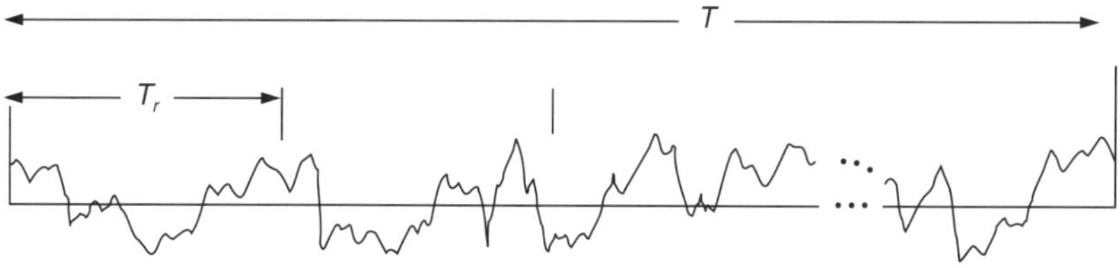

Figure 6.35 The basis of segment averaging.

output for a lightly damped oscillator. The bias error can be reduced by improving the resolution of the estimate. In Fig. 6.36 the coherence function is estimated for simulated input/output results for an oscillator with centre frequency $f_0 = 1$ Hz. The theoretical value of coherence is unity and the resolutions used are shown on the figure.

6.7.5.3 Frequency response functions

These are estimated using smoothed spectra from the expression $\tilde{H}_1(f) = \tilde{S}_{xy}(f)/\tilde{S}_{xx}(f)$, and results for errors and confidence limits can be found in Bendat and Piersol (1971), Otnes and Enochson (1978), and

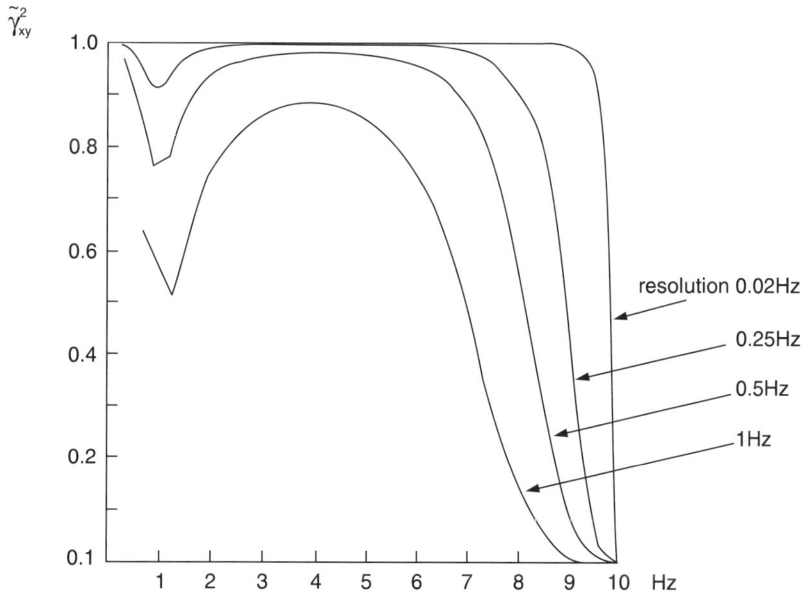

Figure 6.36 The effect of bias on the estimated coherence function

Bendat and Piersol (1980). A few results from Bendat and Piersol are quoted.

The normalized random error is defined as the standard deviation of the error divided by the quantity being estimated, i.e. $\varepsilon_r = \frac{\sigma(\hat{\phi})}{\phi}$ assuming $\phi \neq 0$. The factor n_d refers to the number of averages used and assumes the data segments used are mutually uncorrelated.

Estimator	*Random error*
$\tilde{S}_{xx}$	$\frac{1}{\sqrt{n_d}}$
$\lvert\tilde{S}_{xy}\rvert$	$\frac{1}{\lvert\gamma_{xy}\rvert/\sqrt{n_d}}$
$\arg \tilde{S}_{xy}(f)$	$\frac{(1-\gamma_{xy}^2)^{1/2}}{\lvert\gamma_{xy}\rvert\sqrt{2n_d}}$
$\tilde{\gamma}_{xy}^2(f)$	$\frac{\sqrt{2}[1-\gamma_{xy}^2]}{\lvert\gamma_{xy}\rvert\sqrt{n_d}}$
$\lvert\tilde{H}_1(f)\rvert$	$\frac{(1-\gamma_{xy}^2)^{1/2}}{\lvert\gamma_{xy}\rvert\sqrt{2n_d}}$

6.8 CONCLUDING REMARKS

This chapter has attempted to highlight and summarize some of the essential aspects of the fundamentals of signal processing that are commonly used for data analysis. Practitioners should have a good understanding of the underlying assumptions and limitations of the validity of the many procedures that are now so accessible on special purpose and PC-based analysers. Inevitably, 'real' data does not conform neatly to the various categories described and so the analyst must draw on a variety of methods to elicit the information contained in a signal.

6.9 REFERENCES

Ables, J.G.(1974) Maximum entropy spectral analysis. *Astronomy and Astrophysics Supplement Series*, **15**, 383–93.

Bendat, J.S. and Piersol, A.G. (1971) *Random Data: Analysis and Measurement Procedures*. Wiley Interscience, New York.

Bendat, J.S. and Piersol, A.G. (1980) *Engineering Applications of Correlation and Spectral Analysis*. Wiley Interscience, New York.

Beauchamp, K.G. (1973) *Signal Processing*. George Allen &Unwin.

Bingham, C., Godfrey, M. and Tukey, J.W. (1967) Modern techniques of power spectrum estimation. *IEEE Trans. Audio and Electroacoustics* **AU-15** (2).

Boashash, B. (1987) Nonstationary signals and the Wigner-Ville distribution. *Proceedings of the ASTED International Symposium on Signal Processing and its Applications*, Brisbane, Australia, 261–268.

Bogert, B., Healey, M. and Tukey J. (1963) The frequency analysis of time series for echoes. In M. Rosenblatt (eds.), *Proceedings of Symposium on Time Series Analysis*, New York: John Wiley and Sons, 209–243.

Brigham, E.O. (1974) *The Fast Fourier Transform*. Prentice Hall, Englewood Cliffs, New Jersey.

Carter, G.C. (1980) Bias in magnitude-squared coherence estimation due to misalignment. *IEEE Trans. on Acoustics, Speech and Signal Processing*. **ASSP-28** (1).

Childers, D. and Durling, A. (1975) *Digital Filtering and Signal Processing*. West Publishing Co.

Cohen, L. (1989) Time frequency distributions – a review. *Proceedings of the IEEE* **77** (7), 941–981.

Davies, P. (1983) A recursive approach to Prony parameter estimation. *Journal of Sound and Vibration* **89** (4), 571–583.

Hammond, J.K. and White, P.R. (1996) The analysis of nonstationary signals using time-frequency methods. *Journal of Sound and Vibration* **190** (3), 419–447.

Harris, F.J. (1978) On the use of windows for harmonic analysis with the discrete Fourier transform. *Proc. IEEE* **66** (1).

Hsu, H.P. (1970) *Fourier Analysis*. Simon & Schuster.

Jenkins, G.M. and Watts, D.G. (1968) *Spectral Analysis and its Applications*. Holden Day, San Francisco.

Leuridan, J., De Vis, D., Van der Auweraer, H. and Lembregts, F. (1985) *Proceedings of 4th International Modal Analysis Conference, Los Angeles*, 908–918.

Oppenheim, A.V. and Schafer, R.W. (1975) *Digital Signal Processing*. Prentice Hall, Englewood Cliffs, New Jersey.

Otnes, R.K. and Enochson, L. (1978) *Applied Time Series Analysis* Vol. 1. *Basic Techniques*. John Wiley and Sons.

Priestley, M.B. (1967) Power spectral analysis of nonstationary processes. *Journal of Sound and Vibration* **6**, 86–97.

Rabiner, L.R. and Gold, B. (1975) *Theory and Application of Digital Signal Processing*. Prentice Hall, Englewood Cliffs, New Jersey.

Rader, C.M. (1970) An improved algorithm for high-speed autocorrelation with application to spectral estimation. *IEEE Trans Audio and Electroacoustics* **AU-18**.

Randall, R.B. (1977) *Frequency Analysis*. Brüel and Kjaer Publications, Naerum, Denmark.

Spitznogle, F.R. and Quazi, A.H. (1970) Representation and analysis of time-limited signals using a complex exponential algorithm. *Journal of the Acoustical Society of America* **47**, No.5(1), 1150–1155.

Welch, P.D. (1967) The use of the fast Fourier transform for the estimation of power spectra: a method based on time averaging over short, modified periodograms. *IEEE Trans. Audio and Electroacoustics* **AU-15**.

Wellstead, P.E. (1979) Nonparametric methods of system identification. In R. Isermann (ed.), *Proceeding of 5th IFAC Symposium on Identification and System Parameter Estimation, held at Darmstadt*. Pergamon Press.

Wicks, A.L. and Vold, H. (1986) The H_s frequency response function estimator. *Proceedings of 4th International Modal Analysis Conference, Los Angeles*, 897-9.
Yuen, C.K. (1979) Comments on modern methods for spectrum estimators. *IEEE Trans. Acoustics, Speech and Signal Processing* **ASSP** -27(3).
Zadeh, L. and Desoer, C. A. (1963) *Linear System Theory*. McGraw Hill Book Co.

6.10 QUESTIONS

1. With reference to the $x(t)$ and $h(t)$ shown in the figure below, explain what is meant by the convolution of two signals, $y(t) = h(t)^*x(t)$. Calculate and sketch $y(t)$.

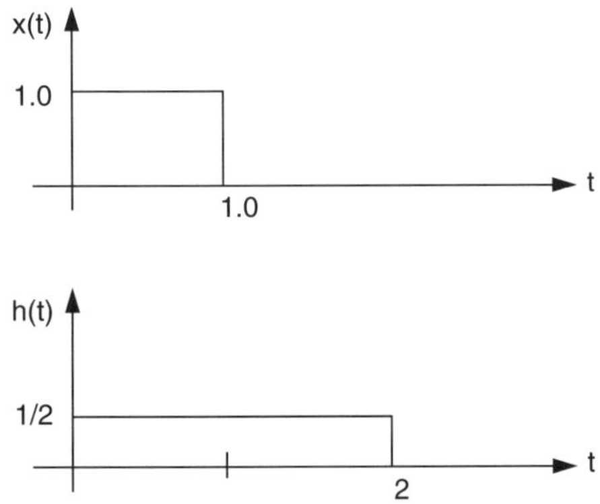

2. For the system shown below, the processes $x_1(t), y_1(t), n_1(t)$ and $n_2(t)$ are stationary random processes, and n_1, n_2 and x_1 are mutually uncorrelated.

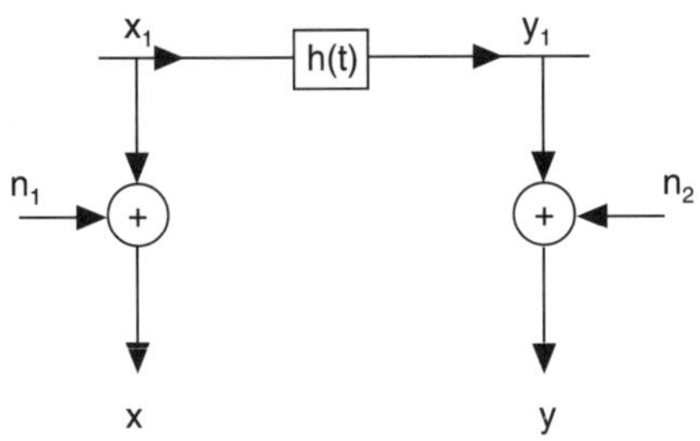

(i) Obtain expressions for

$$H_1(f) = \frac{S_{xy}(f)}{S_{xx}(f)} \quad \text{and} \quad H_2(f) = \frac{S_{yy}(f)}{S_{yx}(f)}$$

(ii) If $\frac{S_{n_1 n_1}}{S_{x_1 x_1}} = \frac{S_{n_2 n_2}}{S_{x_1 x_1}} = \rho$, explain how you would use expressions H_1 and H_2 as estimators for $H(f)$ if $H(f)$ has both resonances and anti-resonances.

3. Two signals $x(t)$ and $h(t)$ are defined below:

$$\begin{aligned} x(t) &= \mathrm{e}^{-at} \sin bt \quad & t \geq 0 \\ &= 0 & t < 0 \end{aligned}$$

and $h(t) = \delta(t) + 0.1\delta(t - t_0)$ ($\delta(t)$denotes a delta function).
With reference to these signals, illustrate carefully what is meant by the convolution of two signals, i.e. $y(t) = h(t)^* x(t)$, where * denotes convolution. Sketch and compare the Fourier transforms of $x(t)$ and $y(t)$.

4. If the aim is to 'identify' a system using a random test signal, and it is expected that the system behaves like a second-order single degree-of-freedom system with resonant frequency about 100 Hz and damping ratio less than 0.05, then select, with reasons, the following:

   (a) the bandwidth of the excitation signal;
   (b) the sample rate for digital analysis of the data;
   (c) the amount of data to be captured (assume that FFT sizes are limited to 1024 points).

   For the parameters you have selected, describe the errors that may be expected in the estimates of $S_{yy}(f), S_{xy}(f)$ and the coherence function, $\gamma_{xy}^2(f)$.

5. (a) A signal is shown below.

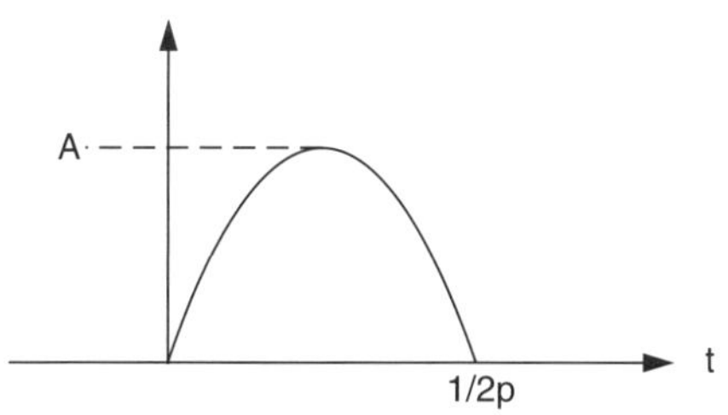

It is described by

$$\begin{aligned} x(t) &= A \sin 2\pi pt \quad & \text{for } 0 < t < \frac{1}{2p} \\ &= 0 & \text{elsewhere.} \end{aligned}$$

Calculate the Fourier transform $X(f)$ of $x(t)$ and sketch the modulus and phase of this transform.

(b) A second signal, $x_1(t)$, is formed from $x(t)$ and is described by

$$x_1(t) = x(t) + 0.5x(t - \Delta).$$

How does the Fourier transform of x_1(t) differ from that of $x(t)$?
(c) A third signal, $x_2(t)$, is formed from $x(t)$ and is described by

$$x_2(t) = \sum_{n=-\infty}^{\infty} x\left(t - \frac{n}{2p}\right).$$

What is the Fourier representation of $x_2(t)$?

7

Fundamentals of underwater acoustics

T.G. LEIGHTON

7.1 INTRODUCTION

The topic of underwater acoustics has traditionally concentrated on ocean acoustics, with the emphasis on sonar. This field is covered by a chapter in the companion volume to this book, and by a number of texts (Tolstoy and Clay, 1966; Albers, 1972; Ross, 1976; Clay and Medwin, 1977; Kinsler *et al.*, 1982; Urick, 1983; Burdick, 1984; Coates, 1990). Therefore this chapter, rather than reiterating this material, will cover the principles underlying sound transmission through water by relating main themes common to diverse applications of underwater sound. The progression of ideas is outlined in Fig. 7.1.

Following this introduction, the chapter is divided into three sections, as shown in the figure. Section 7.2 examines the basics of sound propagation in water, as predicted by linear acoustics, recalling the dependence of the sound speed on the density and bulk modulus of the liquid. Specifically, the section opens with an examination of how this dependence affects ocean acoustics, through the variation of the sound speed with hydrostatic pressure, temperature, salinity and bubble presence. This leads to a discussion of sound speed variation with depth, and how this so-called 'sound speed profile' varies with latitude, season and meteorological conditions. There follow descriptions of how the characteristics of such profiles can lead to the formation of oceanic sound channels, and how propagation over long ranges in such channels is influenced by attenuation from both geometrical spreading and acoustic absorption. Since past exploitation of such channels has been extensively covered in the above references, the text is then restricted to a more recent application of such channels, namely acoustic thermometry of the ocean climate.

Key to the long-range propagation exploited in acoustic thermometry is the signal-to-noise ratio and, through examination of ambient noise, other

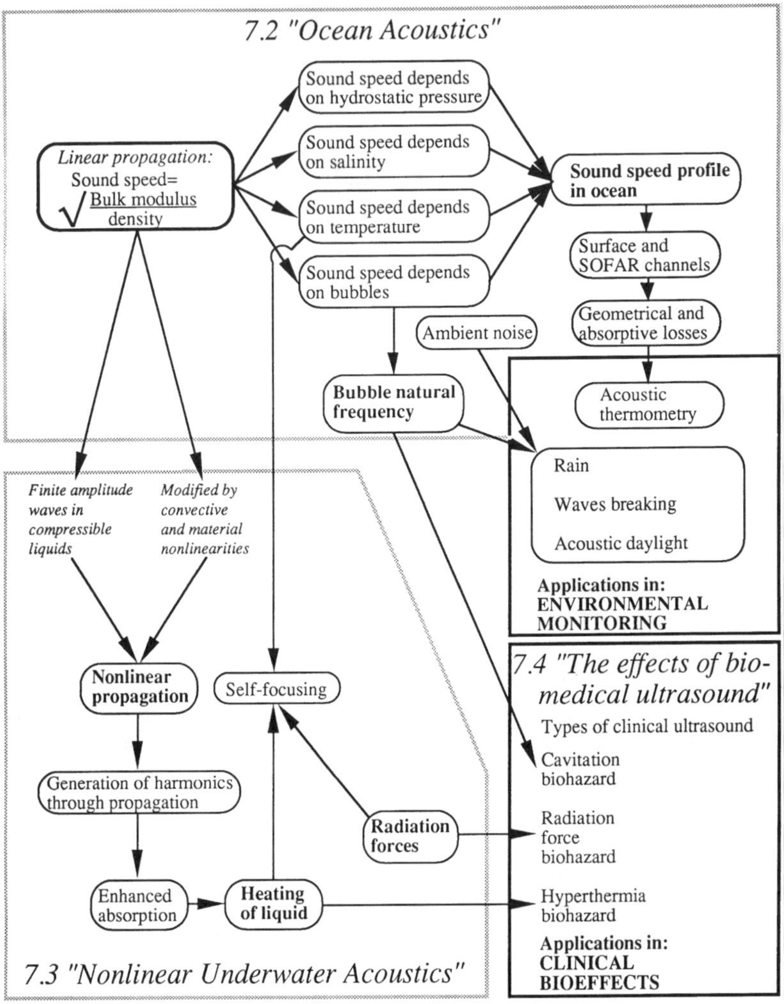

Figure 7.1 Schematic of the progression of ideas through the chapter, showing conceptual links. The titles of the boxes for sections 7.2, 7.3, and 7.4 are given. The course begins in the top left-hand of the figure.

techniques for environmental monitoring using underwater acoustics are then described. This is done through expansion of the topic of bubble acoustics, introduced earlier through their effect on the sound speed. The text outlines how the sound emissions generated by bubbles on entrainment give information about rainfall over the ocean, and about wave breaking (which, in turn, relate to the fluxes of mass, momentum and energy between the atmosphere and the ocean). Other sources of ambient noise, and their roles in the exploration of the oceans, are then described.

Section 7.3 of the text describes the nonlinear propagation of finite amplitude sound. Nonlinear propagation is compared with the linear model which was discussed thoroughly in Chapter 1, and there follows a description of key consequences (the generation of harmonics and shocks, self-focusing, enhanced heating of the liquid and radiation force effects). These effects, and the potential for cavitation, are combined in section 7.4 in a discussion of the possibility of ultrasonic bioeffects, which may be required in therapy but unwanted when diagnostic ultrasound is used.

By examining the subject through the underlying physics, diverse applications (and their similarities) are illustrated. The range encompasses clinical effects on the sub-millimetre scale and, at the other extreme, propagation paths of order 1000 km in studies of ocean climate. The latter application illustrates how today the use of sound to probe the oceans can involve the dedication of huge resources (staff, finance, equipment, etc.). Indeed in the years following the Second World War, the investment in underwater acoustics has increased possibly more than in any other branch of acoustics. It is perhaps ironic that in advancing from simple sound speed measurements in 1827 using an underwater sound source (a bell) and a reference in-air electromagnetic wave (light), through the use of underwater acoustics in navigation, detection and exploration, the ocean basin-scale experiments conducted today bear strong similarities with that early experiment, and the same physics links both with applications of medical ultrasound.

7.2 OCEAN ACOUSTICS

7.2.1 History

The history of underwater acoustics is presented in detail in the references cited at the start of this chapter, and therefore only a brief outline is given here. In 1490 Leonardo da Vinci wrote: 'If you cause your ship to stop, and place the head of a long tube in the water and place the outer extremity in your ear, you will hear ships at a great distance from you.' This early statement contained features still important today in the use of passive sonar, da Vinci drawing attention to the following requirements: low self-noise, acoustic coupling to overcome impedance mismatch, a receiver transducer, and a recognition system. This system was used as late as the First World War, when directivity was gained by the employment of a second tube.

As will be seen shortly, of key importance to ocean acoustics is the sound speed in the water. The first measurements of this were made in 1827 in Lake Geneva/Leman by Daniel Colladon, a Swiss physicist, and Charles François

Sturm, a French mathematician. From the time interval between the observation of a flash of light and the sound from an underwater bell, both signals having been generated simultaneously, they calculated the speed of sound in the lake to be 1435 m/s. The water temperature was 8 °C. By the end of that century, knowledge of the sound speed in water allowed the exploitation of such signals in navigation. Figure 7.2 shows schematically how, from the time-interval t_1 between hearing a foghorn and an underwater bell (both sounded simultaneously on a lightship at time $t = 0$), the distance X of the observer to the lightship could be calculated. The speeds of sound in air, c_a, and water, c_w, are known, and the interval t_1 is measured. If the underwater signal takes time t_0 to reach the observer, then the air signal takes time $t_0 + t_1$, and

$$X = c_w t_0 = c_a t_0 + c_a t_1 \tag{7.1}$$

implying

$$t_0 = \frac{c_a t_1}{(c_w - c_a)} \tag{7.2}$$

and

$$X = \frac{c_w c_a t_1}{(c_w - c_a)} \; . \tag{7.3}$$

Significant advances followed once transducers were introduced to interconvert electrical and mechanical signals. The piezoelectric effect was discovered in 1880 by Pierre and Jacques Curie, who noted that electric charges appear on faces of certain stressed crystals. The effect is reversible, and therefore can be exploited in both transmitters and receivers, ceramics and polymers often being used nowadays in preference to crystals. Also reversible is the magnetostrictive effect, discovered by James Joule in around 1840, who observed that certain materials change length in a magnetic field.

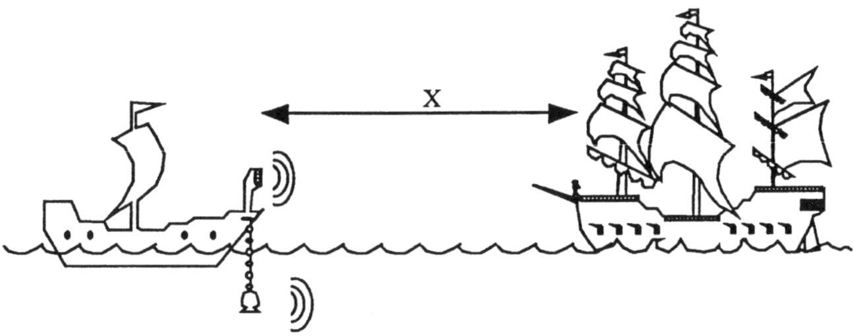

Figure 7.2 Schematic of use of airborne and underwater sound to determine distance to a lightship.

The use of alternating electric and magnetic fields can therefore be exploited to generate acoustic waves of a specific spectral content. Significant advances took place in the first half of this century, due in considerable part to the world wars, such that acoustic techniques, including sonar (retrospectively an acronym for 'sound navigation and ranging') represent probably the most powerful way of exploring the oceans. An example of passive sonar was described above, in da Vinci's description of listening for ships. An example of active sonar, where the scattering of an acoustic pulse previously transmitted by the detecting agent gives the range (from the travel time) and direction (from the directionality of the transmitter and/or receiver), is shown schematically in Fig. 7.3.

The transmission of an acoustic signal and the subsequent location of a target through detection of the acoustic reflections from it (echo-location) is a technique used by many mammals. Bats employ frequencies f in the range 30 kHz to 80 kHz, while toothed whales use frequencies up to 200 kHz (Leighton, 1994). One reason for the difference is that the speed of sound in water is roughly five times that in air, and so to obtain the same degree of spatial resolution higher frequencies must be employed in water. In echo-location, the limit of spatial resolution is roughly given by the size of the wavelength used, as illustrated by a simple estimation from the phase speed, c_0, of the wavelength, λ. The speed of sound in air at 20 °C and under one atmosphere of pressure is 343 m/s. Since $c_0 = f\lambda$ (see section 1.1 of Chapter 1), the wavelength at 80 kHz is 4.3 mm, indicating that bats can echo-locate objects down to a few mm size, corresponding to their diet of insects. Interestingly, those bats which eat fruit rather than insects tend not to use echo-location but instead find their larger, stationary food by sight. The speed of sound in water is roughly 1480 m/s, and so at 80 kHz the wavelength is nearly 2 cm. To achieve a similar degree of spatial resolution as bats, toothed whales must exploit higher frequencies; at 200 kHz, the

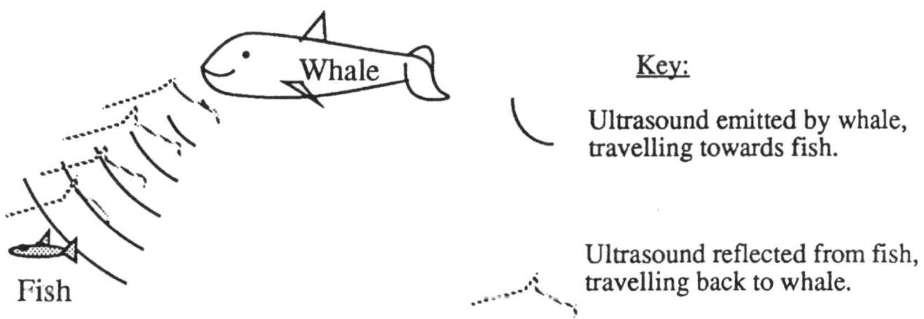

Figure 7.3 Schematic illustration of active sonar. The example illustrates monostatic active sonar, where the source and receiver are located at the same position; with bistatic sonar they are not.

maximum frequency toothed whales exploit, the wavelength in water is around 7 mm.

7.2.2 The speed of sound

7.2.2.1 The factors affecting the oceanic sound speed

From section 1.3 of Chapter 1, it is clear that a model of linear acoustic propagation demonstrates that the sound speed, c_0, depends on the equilibrium density, ρ_0 of a liquid, and on the bulk modulus, K. For a fixed mass of fluid, the bulk modulus equals $K = \mathrm{d}P/(\mathrm{d}\rho/\rho) = -\mathrm{d}P/(\mathrm{d}V/V)$ (where P is the total fluid pressure and $\mathrm{d}V/V$ the proportional change in volume of a fluid element). Specifically,

$$c_0 = \sqrt{\frac{K}{\rho_0}} \ . \tag{7.4}$$

In seawater, this dependence is not simple. The temperature-dependent nature of even pure water is, for example, anomalous in that it achieves maximum density at 4 °C. It is, however, anomalous in other ways. There is a strong dependency of the sound speed in water on temperature, as shown in Fig. 7.4 for pure water under one atmosphere of pressure. In many liquids, the sound speed decreases with increasing temperature, but water is anomalous in that it reaches a maximum (at 74 °C for pure water under one atmosphere), and in the temperature range of the ocean the sound speed increases with increasing temperature, T. The magnitude of the effect in the ocean is

$$\frac{\Delta c_0}{\Delta T} \approx 5 \ (\mathrm{m/s})/^{\circ}\mathrm{C} \ . \tag{7.5}$$

The sound speed is also dependent on salinity, S, and varies as

$$\frac{\Delta c_0}{\Delta S} \approx 1 \ (\mathrm{m/s})/(\text{part per thousand}) \tag{7.6}$$

and increases with hydrostatic pressure P_{h}, as

$$\frac{\Delta c_0}{\Delta P_{\mathrm{h}}} \approx 1.6 \times 10^{-6} \ (\mathrm{m/s})/\mathrm{Pa} \tag{7.7}$$

(Urick, 1983; Del Grosso, 1974). (These relations are given in the references in non-S.I. units. Differentiation of equation (7.8) will yield such results.) The sound speed in the ocean depends on a range of other phenomena, notably the subsurface bubble layer if one exists (section 7.2.2.5(b)).

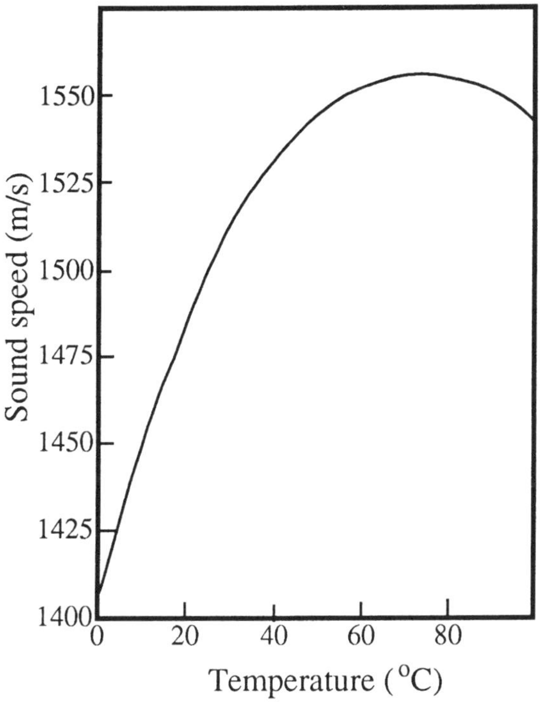

Figure 7.4 The temperature variation of the speed of sound in pure water, for linear propagation under one atmosphere of pressure. (Reprinted Leighton (1994) by permission of the publishers, Academic Press, London.)

At first sight equations (7.5) to (7.7) may give the impression that the hydrostatic pressure has only a minor effect on the sound speed compared to the other two main factors. However this is not the case because, whilst the water temperature may vary from a surface value of, say, 10 °C to 0 °C at greater depths, the pressure increases by an atmosphere for approximately every 10 m increase in depth, and achieving around 600 atmospheres at a depth of 6000 m. In open deep water, salinity usually has only a marginal effect on the sound speed.

7.2.2.2 The sound speed profile

The factors introduced above combine in the ocean to produce a variation of sound speed with depth known as the sound speed profile. A typical example is shown in Fig. 7.5. The profile allows division of such an oceanic section into characteristic regions. Consider that area of the figure which portrays the ocean at depths of greater than 200 m. This area comprises two

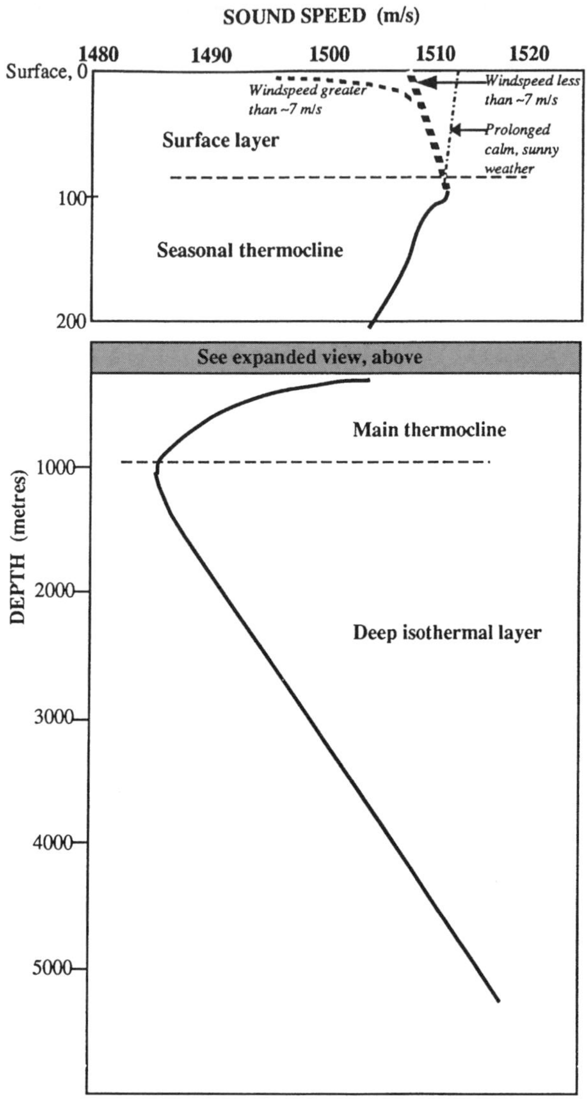

Figure 7.5 A typical sound speed profile. The upper 200 m are shown in magnification. In the surface layer, three possible sound speed profiles are shown, corresponding to the following wind conditions: prolonged calm, sunny weather (dot-dash line); a windspeed less than about 7 m/s (thick dashed line); windspeed in excess of about 7 m/s (thin dashed line).

main regions. Extending from the minimum in the sound speed profile is the Deep Isothermal Layer, where the temperature is a few degrees centigrade above 0 °C. (Strictly, while the approximation of isothermal may be applied to the deepest layer, its characteristics depend on the water density, which is in turn a function of both temperature and salinity. Therefore any stated temperature, e.g. 4 °C, would be valid only for a particular salinity, and in fact there may be spatial variation in temperature at a given depth.) The effect of hydrostatic pressure on the sound speed dominates. At high latitudes, this layer extends to the surface. This is because the temperature near the surface at the poles is close to 0 °C, and since sound speed in this temperature range decreases with decreasing temperature, both the pressure and temperature effects cause the sound speed to decrease with decreasing depth. However, at medium and low latitudes, the temperature tends to increase with decreasing depth for regions within 1 km or so of the surface.

This gives rise to the Main Thermocline, in the region above the minimum in the sound speed axis. If one were moving towards the surface from the very cold water found at the greatest depths, it would be in this region that the major increase in water temperature occurs. In the Main Thermocline, there is little seasonal variation, the water at these depths being relatively insulated from the surface by the layers above. The main thermocline extends in the figure to within 200 m of the surface. The layers above this are shown in expanded view. These, being nearer the surface, are more prone to surface influence (seasonal, and, in the shallower waters, meteorological, effects).

In the Seasonal Thermocline, the temperature decreases with depth. This region is well-defined in summer and autumn. During winter and spring, and in the Arctic, it merges with surface layer. The reason is that a well-defined thermocline needs a strong temperature gradient, which requires high temperatures near the surface since the depths are thermally more stable and cold. In the Arctic, the surface is always cold, so the thermocline is weak. Elsewhere, this situation holds in winter and spring because the sea is slow to warm up and to cool down.

Nearest the surface (0–100 m depth in the figure) is the Surface Layer, which is subject to daily and local changes, wind, and temperature. If it has been calm and sunny for long periods, the temperature decreases with depth in this region, and the thermocline extends to the surface (see the dot-dash line in the figure). However this layer may be mixed up by wind and circulations so that it becomes isothermal (hence the alternative name of this region is the Mixed Layer). If windspeed is > 7 m/s, bubbles dominate in the upper 10 m or so and reduce the sound speed drastically. Two dashed lines are shown in the figure. The thick one (on the right) shows a steady decrease in sound speed resulting from a pressure effect in an isothermal surface layer that has been mixed up by low windspeeds (below 7 m/s, say 3 m/s). The thin dashed line (on the left) shows a marked decrease in sound speed, relevant to wind speeds in excess of 7 m/s (say, 10 m/s), resulting from bubble entrainment. It should however be noted that the influence of the bubbles on the sound speed will be dependent on the acoustic frequency – see section 7.2.2.5(b).

The sound speed profile shown in Fig. 7.5 reflects the body of information which has been gathered over the years through submerged sensor measurements. However, such surveying is costly and the profile will vary both in time and in the horizontal plane since, for example, temperature and salinity may do so. Figure 7.6 shows concurrent sections from cruise 59 of the research vessel *Charles Darwin* showing, against a common latitude scale, measurements of the temperature and of the acoustic backscatter (from zooplankton) obtained from a 150 kHz Acoustic Doppler Current Profiler (ADCP). The transect of this region of the NE Atlantic took four days, and so shows two normal daily migrations of zooplankton at around 44° and 46° latitude, a behaviour which is interrupted by the Polar Front at around 49° (visible on the thermal record).

The figure shows how the ocean contains many features on the mesoscale (approximately 100 km) order. These can be thought of as being the equivalent of features (e.g. depressions and anticyclones, fronts etc.) which can be observed on weather maps. However the size scale of such features tends to be an order of magnitude smaller in the ocean (e.g. eddies have 100 km width, as opposed to 1000 km width), although the inertia and heat capacities associated with such features in the ocean are vastly greater than those observed in the atmosphere. Eddy features have been reviewed by Robinson (1983). Currents representing the motion of millions of tonnes of water exist, and these have significance to climate and ocean transport (Wunsch, 1981). The interaction between currents and land masses (e.g. Cape Horn) represent one source of eddies, and large-scale flows such as the Gulf Stream can be a significant source. Examination of such behaviour has been facilitated by the development of high-resolution model simulations of ocean circulation (Webb *et al.*, 1991). Timescale variability is in terms of days to months. In fact, inhomogeneities in ocean parameters occur right across the dimension scale, from basin-scale (of a dimension similar to, for example, the Atlantic or Pacific basins) to micro-scale. Small-scale circulations and turbulence can have sub-metre and sub-second scales. One source of sound speed variation on such a scale, wave-entrained bubbles, is discussed in section 7.2.2.5(c).

Figure 7.6 Concurrent sections from a cruise showing, as a function of latitude and depth, the temperature and the absolute acoustic backscatter obtained from a 150 kHz Acoustic Doppler Current Profiler (ADCP). The backscatter is referenced to the scattering cross-section of 1 m^2 (square) of target per 1 m^3 (cube) of water. Supplied by kind permission of H. Roe.

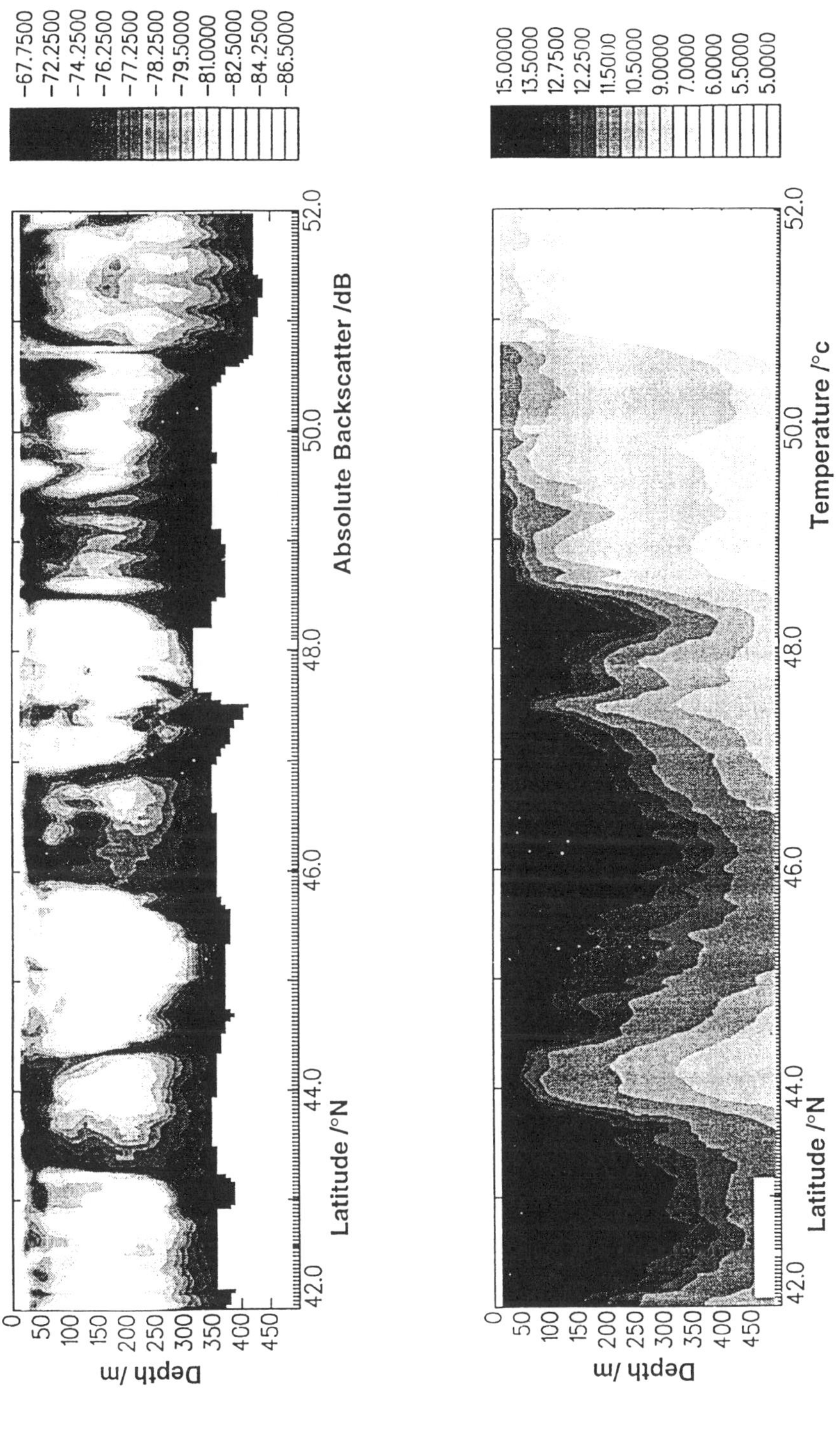
-67.7500
-72.2500
-74.2500
-76.2500
-77.2500
-78.2500
-79.5000
-81.0000
-82.5000
-84.2500
-86.5000
Depth /m
Latitude /°N
Absolute Backscatter /dB
15.0000
13.5000
12.7500
12.2500
11.5000
10.5000
9.0000
7.0000
6.0000
5.5000
5.0000
Depth /m
Latitude /°N
Temperature /°c

7.2.2.3 Ray bending

(a) Qualitative ray bending

Huygens principle allows the graphical construction of the new position of a wavefront given its previous position (a short time earlier) and knowledge of the local sound speed. As a medium is traversed by waves, each point on a wavefront can be treated as a point source of secondary wavelets, and so the position of the wavefront a small time later can be found as the envelope of these wavelets (adequate for all manner of simple, linear waves). A near-surface channel is formed when the sound speed increases with depth. This might occur in polar regions, because the sea surface is colder than the depths; or elsewhere because in the near-surface region of the ocean (say to depth 10 m), there might be a bubble population which decreases in concentration with increasing depth, and causes the sound speed to increase with depth (see section 7.2.2.5(b) and (c)). The resulting ray path is shown in Fig. 7.7. From successive wavefronts, wavelets are drawn. The radius each achieves in a given time interval is proportional to the local sound speed (i.e. which here increases with depth). As such, successive wavefronts are angled upwards, indicating that the region is upwardly refracting. Reflection occurs at the sea surface, such that waves propagate in a 'duct' or 'channel' (Fig. 7.7). The result is that much of the acoustic energy is confined within a layer, rather than spreading in all directions equally.

A similar effect is found when a minimum in the sound speed profile occurs, as discussed in the preceding section. This causes the wavefronts to angle towards the depth where the sound speed is a minimum, as shown in Fig. 7.8 using a Huygens wavelet construction. Of the two rays shown in the figure, the one which makes the greater angle with the sound channel axis on intersection is the one along which a sound pulse would travel to large

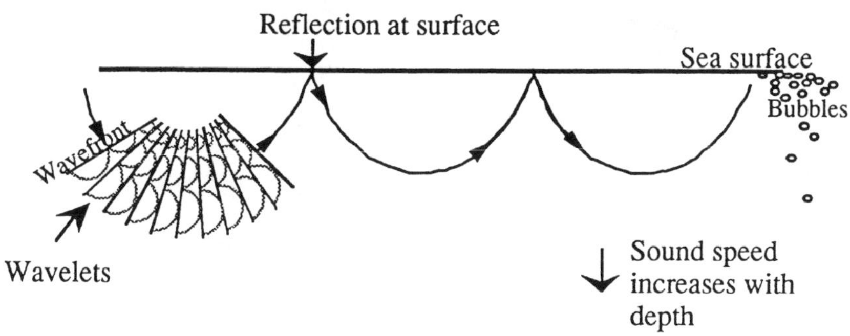

Figure 7.7 The use of a Huygens' construction to illustrate ray paths in a surface channel.

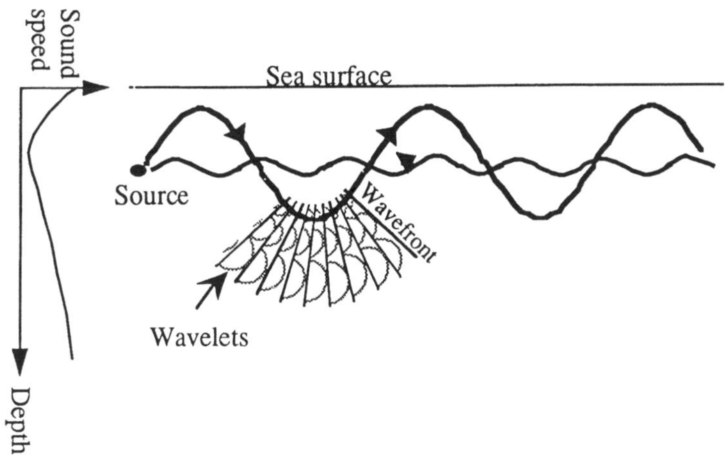

Figure 7.8 The use of a Huygens' construction to illustrate ray paths in the deep sound channel, the axis of which is located at the minimum in the sound speed profile.

distance with the least propagation time. This is because it travels through regions having a greater sound speed. The channel which forms at the minimum of the sound speed profile is called the SOFAR (sound fixing and ranging) channel. This is because acoustic signals within it can be detected at long ranges (section 7.2.2.4) making it suitable for use in locating downed aviators (who might drop explosives designed to detonate at the depth of the channel axis).

(b) Quantitative ray bending

The dependence of the sound speed on temperature, salinity, and hydrostatic pressure has been empirically determined through experimentation and a number of approximate formulations exist (Leroy, 1969; Del Grosso, 1974; Medwin, 1975; Chen and Millero, 1976; Chen *et al.*, 1977; Mackenzie, 1981) and have been compared (Spiesberger and Metzger, 1991).

Adapting Medwin's 1975 formulation to separate specifically the hydrostatic pressure terms from the depth dependence gives the following equation for the sound speed (c) in the ocean as a function of temperature ($T/^{\circ}$C), salinity (S, in parts per thousand, ppt) and hydrostatic pressure (P_h/P_a, a term which excludes atmospheric pressure and so is zero at the ocean surface):

$$\begin{aligned} c_0 = {} & 1449.2 + 4.6T - 0.055T^2 + 0.0003T^3 \\ & + (1.39 - 0.012T)(S - 35) + 1.74 \times 10^{-6} P_h \ \text{m/s} \end{aligned} \tag{7.8}$$

where c_0 is in metres per second. Consider a region of ocean 6000 m deep, where the salinity is uniform at 35 parts per thousand. This formulation will now be employed to determine the path of a ray projected horizontally at the surface in a model region of the ocean. In this region, the temperature falls linearly from a surface value of 10 °C to a temperature of 0 °C at a depth of 1000 m. The temperature is constant at 0 °C from 1000 m depth to the bottom. This calculation will be performed by first proving that, in regions where the sound speed varies linearly with depth, the ray paths form arcs of circles.

To prove that if the sound speed varies linearly with depth, the rays follow arcs of circles

To begin, Snell's law allows statement of the relation between the sound speed at a particular depth and the angle, θ, which a ray at that depth makes locally with the horizontal. If at depth d_n the sound speed is c_n and the ray path has an angle θ_n to the horizontal, then

$$\cos\theta_1/c_1 = \cos\theta_2/c_2 = \cos\theta_i/c_i \tag{7.9}$$

where the terms are defined through the scheme shown in Fig. 7.9. A sound ray is initially travelling horizontally, passing through point X_i at a depth d_i with speed c_i. It then passes through point X_1 at a depth d_1 with speed c_1,

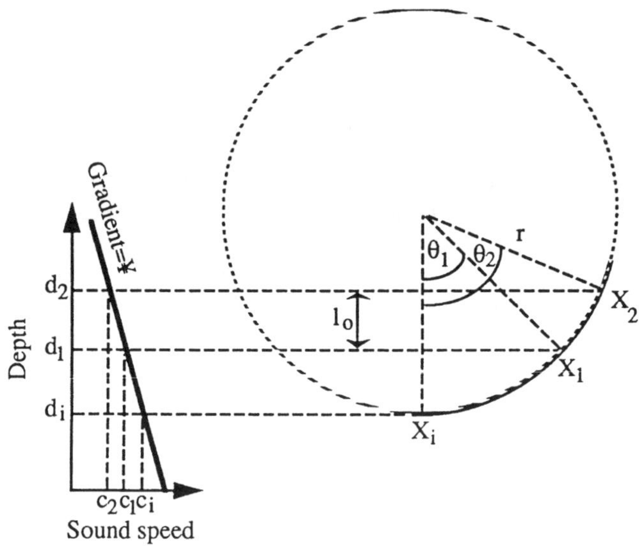

Figure 7.9 Ray paths in a region of ocean where the sound speed increases linearly with depth.

and then passes point X_2 at a depth d_2 with speed c_2. The ray is an arc of a circle if it satisfies two conditions:

The first condition is that, if the ray follows a circle arc, then the circle has radius r such that the distance l_0 on the figure is:

$$l_0 = r\cos\theta_1 - r\cos\theta_2 = d_1 - d_2\ . \tag{7.10}$$

Since the variation is linear, with gradient ¥,

$c_1 = c_i + ¥(d_1 - d_i)$ and $c_2 = c_i + ¥(d_2 - d_i)$ which implies that

$$d_1 - d_2 = (c_1 - c_2)/¥\ . \tag{7.11}$$

The second condition is that, if the circle arc is a ray, Snell's Law must be satisfied, i.e. from equation (7.9):

$$c_1 = c_i\cos\theta_1 \text{ and } c_2 = c_i\cos\theta_2 \text{ (since } \cos\theta_i = 1)\ . \tag{7.12}$$

Substituting equation (7.11) into (7.10) gives:

$$(c_1 - c_2)/r¥ = \cos\theta_1 - \cos\theta_2\ . \tag{7.13}$$

Substituting equation (7.12) into (7.13) gives:

$$(c_1 - c_2)/r¥ = c_1/c_i - c_2/c_i, \text{ i.e. } \frac{(c_1 - c_2)}{r¥} = \frac{(c_1 - c_2)}{c_i}$$

$$\Rightarrow r = \frac{c_i}{¥} \tag{7.14}$$

If both conditions are satisfied, and the ray is a circle arc. Similarly, if the sound speed decreases with depth,

$$r = -\frac{c_i}{¥} \tag{7.15}$$

the negative sign in the context of the geometry of Fig. 7.9 indicating that, in such circumstances, the ray curves downwards (in agreement with the discussions of section 7.2.2.3a). In general therefore, if h is the depth below the surface, then

$$\left|\frac{dc_0}{dh}\right| = \frac{c_i}{\text{radius of curvature}}\ . \tag{7.16}$$

With this formulation, it is possible to use equation (7.8) to approximate for the sound speed profile by two regimes in which the sound speed varies linearly with depth. The first regime is at depths above the sound speed axis, and the second is below it.

To evaluate the linear approximations to the sound speed profile

The values of the sound speeds, and the sound speed gradient, for the example section of the ocean chosen can be found from equation (7.8). For $S = 35$ parts per thousand, the sound speed varies as

$$c_0 = 1449.2 + 4.6\,T - 0.055\,T^2 + 0.0003\,T^3 + 1.74 \times 10^{-6}\,P_h \text{ m/s} \; . \quad (7.17)$$

The temperature is given by

$$T/^\circ\text{C} = 10 - 10\,h/1000 = 10 - 0.01\,h \text{ (for } 0 \le h/\text{m} \le 1000) \quad (7.18)$$

$$T/^\circ\text{C} = 0 \quad (\text{for } 1000 \le h/\text{m} \le 6000) \quad (7.19)$$

and the hydrostatic pressure is related to the depth through

$$1.74 \times 10^{-6}\,P_h = 1.74 \times 10^{-6} \rho_0 g h \approx 0.017h \quad (7.20)$$

where g is the acceleration due to gravity. Substituting equations (7.18) to (7.20) into equation (7.17) gives the sound speed as a function of depth only in the two regions above and below 1000 m depth:

$$\begin{aligned} \frac{c_0(h \le 1000\text{ m})}{\text{m/s}} &\approx 1449.2 + 4.6(10 - 0.01h) \\ &- 0.055(10 - 0.01h)^2 + 0.0003(10 - 0.01h)^3 + 0.017h \\ &= 1490 - 0.0189h - 4.6 \times 10^{-6}h^2 - 3 \times 10^{-10}h^3 \\ &\text{(for } 0 \le h/\text{m} \le 1000) \end{aligned} \quad (7.21)$$

$$c_0(h \ge 1000\text{ m}) \approx 1449.2 + 0.017h \text{ m/s} \quad (\text{for } 1000 \le h/\text{m} \le 6000). \quad (7.22)$$

At $h = 1000$ m, equations (7.21) and (7.22) give sound speeds of 1466.3 m/s and 1466.2 m/s respectively. While equation (7.22) predicts that below 1000 m depth the sound speed varies linearly with depth, above $h = 1000$ m the following relation (derived from equation (7.22)) is only approximately true:

$$c_0(h \le 1000\text{ m}) \approx 1490 - 0.0189h \text{ m/s} \quad (\text{for } 0 \le h/\text{m} \le 1000) \quad (7.23)$$

giving, at $h = 1000$ m, a speed of sound of 1471.1 m/s, in disagreement with equations (7.21) and (7.22), indicating the error in the approximation. The linear sound speed gradient assumed for the two regimes is therefore, from equations (7.22) and (7.23):

$$\mathrm{d}c_0/\mathrm{d}h \approx -0.0189 \text{ s}^{-1} \quad (\text{for } 0 \le h/\text{m} \le 1000) \quad (7.24)$$

$$\mathrm{d}c_0/\mathrm{d}h = +0.017 \text{ s}^{-1} \quad (\text{for } 1000 \le h/\text{m} \le 6000) \; . \quad (7.25)$$

To find the approximate path followed by a ray projected horizontally at the surface

It is now possible to consider a ray which is projected at the surface in a horizontal direction, and estimate the greatest depth achieved by the ray, and the horizontal distance between the source and the point where the ray returns to the surface. Figure 7.10 schematically shows the ray in question,

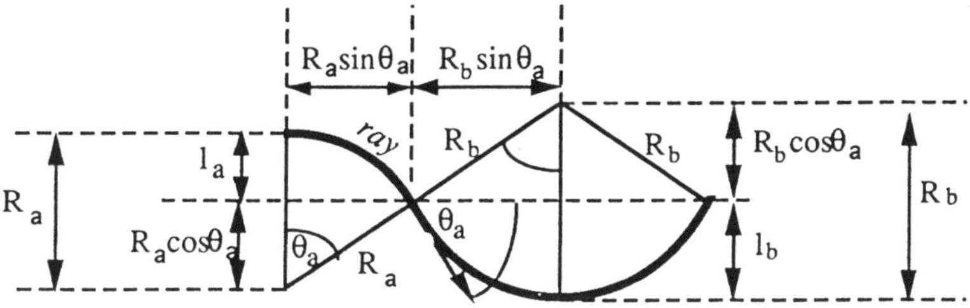

Figure 7.10 A ray, travelling from the surface to the position of its deepest descent, and then back to the 1000 m depth, within the model ocean profile (the radii of curvature are shown much reduced for clarity).

travelling from the surface to the position of its deepest descent, and then back to the 1000 m depth.

Consider when the ray is at depths less than 1000 m. It moves in an arc of a circle of radius R_a, which can be calculated by substituting the sound speed gradient (= −0.0189) from equation (7.24) and sound speed at the surface (the depth at which the wave travels horizontally) into, say, equation (7.15). The sound speed at the surface is found from equation (7.21) for the conditions $h = 0$:

$$c_0(h=0) = 1490 - 0.0189h - 4.6 \times 10^{-6}h^2 - 3 \times 10^{-10}h^3 = 1490 \text{ m/s} \ . \tag{7.26}$$

Therefore from equation (7.16) the radius of curvature at depths of less than 1000 m is:

$$|R_a| = |c_0(h=0)/(\mathrm{d}c/\mathrm{d}h)| = 1490/0.0189 \approx 78.836 \text{ km} \ . \tag{7.27}$$

After descending a depth $l_a = 1000$ m, it has travelled through an arc angle of θ_a, where from Fig. 7.10, the geometry indicates

$$l_a = R_a(1 - \cos\theta_a) \text{ implying}$$

$$\theta_a = \cos^{-1}(1 - l_a/R_a) = \cos^{-1}(1 - 1000/78\,836) = 9.14° \ . \tag{7.28}$$

By the time it reaches a depth of 1000 m (for the first time) it is a distance x_a from the source, which can be evaluated by substituting values from equations (7.27) and (7.28) into geometrical consideration of Fig. 7.10:

$$x_a = R_a \sin\theta_a = 12\,517 \text{ m} \ . \tag{7.29}$$

Now consider the ray when it is at depths of greater than 1000 m. It travels in an arc of radius R_b, which can be calculated using knowledge of the sound speed gradient (= 0.017 from equation (7.25)) and sound speed

c_b at the depth $h_b = (1000\ \mathrm{m} + l_b)$ at which the wave travels horizontally. The sound speed c_c can be found by using Snell's law to compare the ray at depths of 1000 m and h_b. Substituting values for θ_a from equation (7.28), and for $c(h = 1000)$ from equation (7.22), into equation (7.9) gives:

$$\cos\theta_a/c(1000) = 1/c_b$$
$$\Rightarrow c_b = c(1000)/\cos\theta_a = 1485.0\ \mathrm{m/s}\ . \tag{7.30}$$

Therefore from equation (7.16) the radius of curvature at depths of greater than 1000 m is:

$$|R_b| = |c_b/(\mathrm{d}c/\mathrm{d}h)| = 1484.8/0.017 = 87\,355\ \mathrm{m}\ . \tag{7.31}$$

Knowledge of this radius allows, from geometrical consideration of Fig. 7.10, estimation of the maximum depth attained by the ray, h_b, which is given by

$$h_b = (1000\ \mathrm{m} + l_b) = (1000\ \mathrm{m} + R_b(1 - \cos\theta_a)) = 2108\ \mathrm{m}\ . \tag{7.32}$$

Applying symmetry considerations to Fig. 7.10, the horizontal distance travelled before the ray reaches the surface is given by

$$2(R_a \sin\theta_a + R_b \sin\theta_a) = 2(12\,517\ \mathrm{m} + 13\,868\ \mathrm{m}) = 52.770\ \mathrm{km}\ . \tag{7.33}$$

In this model ocean profile, the ray, which was projected horizontally at the surface, returns to the surface nearly 53 km later, then to repeat the motion (so long as the sound speed profile persists). (To visualize the scale, imagine a pulse, descending from the surface into the dark depths at speeds in excess of a kilometre a second. The direction followed dips at a shallow, but ever-increasing, angle to the horizontal, reaching a maximum of $\sim 9°$ at 1 km depth. It has taken the pulse about 8.5 seconds to travel the path length of $\sim$ 12.6 km thus far. The angle now begins to decrease as the pulse descends further, to depths even sperm whales cannot attain, the sound speed increasing until at a depth of $\sim$ 2.1 km the pulse is travelling horizontally. Here there is no light, little dissolved oxygen, and few fish. Since starting, the pulse has travelled a path length of $\sim$ 26.5 km, taking around 18 seconds. The pulse begins to rise now, the angle to the horizontal first increasing then decreasing until, at the surface 53 km from the start position, it is horizontal once more. The 36 s taken is very close to the time required were a horizontal direct path to exist, illustrating how, because of the shallow angles involved, when viewed on the large scale the path is not dissimilar to the straight line. This will be re-examined in Fig. 7.11). Such distances (53 km) are characteristic of 'convergence zones'. These zones will be discussed in the companion volume to this book, as will the related phenomena of 'shadow zones' and 'caustics'. These are regions devoid of rays, and regions of partial focusing respectively.

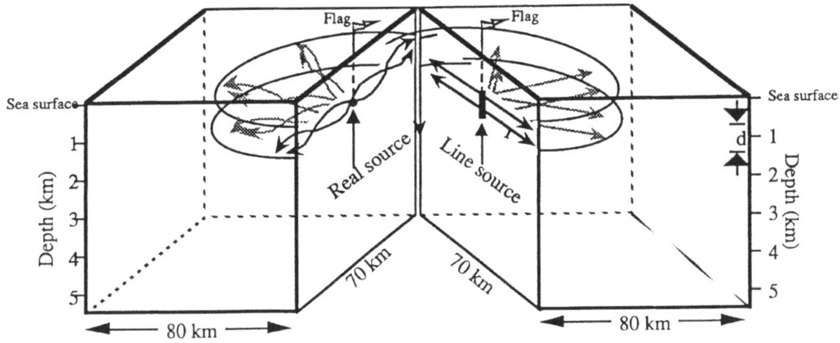

Figure 7.11 Schematic illustration showing how, to first-order approximation, ray propagation in the SOFAR channel may be likened to cylindrical spreading. A section of ocean of over 5 km depth, having a sound speed minimum at 1 km depth, is cut into two halves, which are folded away from one another for clarity. A sound speed profile similar to that shown in Fig. 7.5 is assumed. The horizontal distance scale is much less than the vertical, such that the half-spaces are each 80 km × 70 km in the horizontal plane (distances estimated from equation 7.33). The vertical plane of the cut passes through the acoustic source, which is placed at 1 km depth (and marked at the surface by a flagged buoy). Two of the ray paths from the buoy are illustrated, propagating into the half-space of a disc which roughly indicates the vertical extent of the SOFAR channel. If the sound speed profile is uniform throughout the section of ocean illustrated, the ray propagation will be identical in all horizontal radial directions out from the source (i.e. there is rotational symmetry about the vertical axis which passes through the source). However the rays in only five directions per half-space are shown for clarity. In the right half-space the ray paths for cylindrical spreading from a line source are shown, indicating the degree of similarity with the oceanic rays paths, which are sketched in the left-hand half. The similarity, when viewed on the scale of tens of kilometres, between the curved ray path and the straight-line horizontal path, was discussed in the paragraph below equation (7.33).

7.2.2.4 Long-range propagation in the SOFAR channel

(a) Geometrical losses

Having indicated the nature of ray propagation in acoustic channels, in this section the reduction in signal-to-noise amplitude due to geometrical spreading will be discussed. Absorptive losses which convert acoustic energy into heat will not be included until section 7.2.2.4(b).

In section 1.13, it was shown that when acoustic waves diverge with spherical symmetry from a point source, the intensity $I(r)$ decreases with distance from the source r as an inverse-square law, e.g. for a source of constant power W:

$$I(r) = \frac{\text{Source power}}{\text{Area perpendicular to rays}} = \frac{W}{4\pi r^2} \ . \tag{7.34}$$

If it is assumed that in the far field the intensity is proportional to p^2 (this is true for plane and spherical waves; see sections 1.8 and 1.13), then the acoustic pressure amplitude in waves diverging with spherical symmetry from a point source decreases as r^{-1}.

In the SOFAR channel however, the rays do not diverge spherically. To a first approximation, they resemble rays which spread with cylindrical symmetry from a line source, as shown in Fig. 7.11. Let the power per unit length emitted by a line source be W/d where, as shown in Fig. 7.11, d is the width of the cylinder, and also the length of the line source (which emits rays only perpendicular to its length). The intensity $I(r)$ at radius r (measured on a radius perpendicular to the line source, as shown in Fig. 7.11) is therefore:

$$I(r) = \frac{\text{Source power}}{\text{Area perpendicular to rays}} = \frac{W}{2\pi rd} \ . \tag{7.35}$$

Therefore, if there are no absorptive losses, and if the approximation is made that a cylindrically diverging system can model the SOFAR channel, then geometrical spreading causes the acoustic intensity to decrease with distance as $1/r$. If it is further assumed that, within the limits of the approximation, intensity is proportional to p^2 (see above), the model predicts that acoustic pressure amplitude will decrease with distance as $1/\sqrt{r}$. The geometrical losses are therefore less than those expected for spherical divergence.

It is not necessary to limit geometrical considerations to making the approximation that the rays follow radial paths perpendicular to a given central axis. Urick (1983) compares the intensities close to, and distant from, the source. From their ratio the concept of 'transmission loss' will now be introduced. Consider two rays which leave the source, where the sound speed is c_1, at angles $\theta_1 + \Delta\theta/2$ and $\theta_1 - \Delta\theta/2$. The power radiated by the source into $\Delta\theta$ is ΔW. The rays, travelling with rotational symmetry about the vertical axis which passes through the centre of the source, subtend an area ΔA_1 at a radius r_1 from source, such that the power flux through ΔA_1 (the shaded area on the sphere of radius r_1 in Fig. 7.12) is:

$$I(r_1) = \frac{\Delta W}{\Delta A_1} = \frac{\Delta W}{2\pi r_1 \cos\theta_1 (r_1 \Delta\theta)} \ . \tag{7.36}$$

No matter how complicated the ray paths, the two rays travel to a point X where the sound speed is c_2, a horizontal distance r from the source (see Figs 7.11 and 7.12), where they both subtend an angle of approximately θ_2 to the horizontal, and have a vertical separation Δh. Maintaining the assumption of no absorptive losses, and assuming that there has been no leaking out of energy contained between the two rays, then between them the rays still

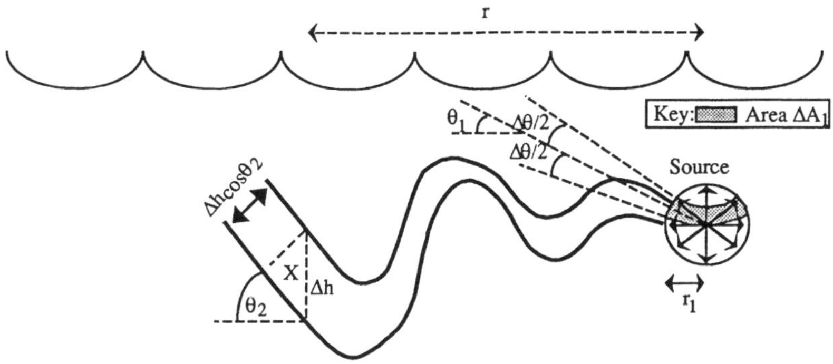

Figure 7.12 Schematic of two rays used for calculation of geometric spreading losses (dimensions and angles not to scale). After Urick (1983).

carry power ΔW. Therefore the intensity at X corresponds to the distribution of this power over the area element ΔA_2 between the rays and perpendicular to them. Recalling the rotational axis of symmetry, the width of this area strip corresponds to the distance between the two rays at X, measured normal to them, which from Fig. 7.12 is $\Delta h \cos \theta_2$. Because of the cylindrical symmetry, the area ΔA_2 is

$$\Delta A_2 = 2\pi r \Delta h \cos \theta_2 \ . \tag{7.37}$$

Therefore the intensity at X under these conditions is

$$I(r_2) = \frac{\Delta W}{\Delta A_2} = \frac{\Delta W}{2\pi r \Delta h \cos \theta_2} \ . \tag{7.38}$$

Several parameters, which may be unknown, are eliminated when the ratio of I_1/I_2 is taken. Indeed, this quantity relates not to the source, but to the characteristics of the propagation paths and therefore to the ocean environment. It is therefore convenient to discuss the proportion by which intensity is reduced through this ratio, in terms of the Transmission Loss, *TL*, which is here given by

$$TL = 10 \log_{10}\left(\frac{I_1}{I_2}\right) = 10 \log_{10}\left(\frac{\Delta A_2}{\Delta A_1}\right) \tag{7.39}$$

giving

$$TL = 10 \log_{10}\left(\frac{2\pi r \Delta h \cos \theta_2}{2\pi r_1^2 \Delta\theta \cos \theta_1}\right) \ . \tag{7.40}$$

Through application of Snell's Law (equation (7.9)), the angles which the rays subtend to the horizontal can be eliminated, allowing the transmission loss to be expressed in terms of the angular ray separation, sound speeds, and accessible distances:

$$TL = 10\log_{10}\left(\frac{r\Delta h c_2}{r_1^2\Delta\theta c_1}\right) \ . \tag{7.41}$$

Absorptive losses will also cause a proportional loss in intensity which (in the linear regime) reflects the properties of the ocean. Absorption can therefore be incorporated into the transmission loss as additive logarithmic components. Examples of this are given in the companion volume to this book. A brief discussion of the sources of absorption follows.

(b) Attenuation losses through absorption

In the previous section, the contribution to the decrease in signal amplitude which results from the geometrical spreading was discussed. In that scenario, the acoustic energy was conserved during propagation. However in practice this will not be the case, and during propagation the acoustic energy will be converted, ultimately, to heat, through a range of mechanisms. In this section, these mechanisms will be discussed. This phenomenon, as will be seen in section 7.4.3.1, is also an important consideration in the use of medical ultrasound, because of the loss in signal amplitude and also for bioeffect considerations.

From Chapter 1, a plane progressive pressure wave of single frequency, travelling in the x-direction in an absorbing medium, can be expressed as

$$p = p_0 e^{j(\omega t - kx)} e^{-\alpha x} \ . \tag{7.42}$$

Given that in this wave intensity is proportional to the square of the pressure, then it will decay as $e^{-2\alpha x}$. The units of the attenuation coefficient, α, which is defined through equation (7.42) and therefore has dimensions of $[\text{length}]^{-1}$, are termed nepers per metre (Np/m). The phenomenon may also be quantified by the *logarithmic absorption coefficient*, α_{ab}, which can be illustrated using plane waves (in which the geometric losses are zero), through consideration of the intensity $I(x)$ as a function of x, the distance from source. Since, as shown earlier, the intensity is proportional to p^2, equation (7.42) allows expression of the proportional intensity change during propagation as:

$$\frac{dI}{I} = -2\alpha\, dx \tag{7.43}$$

integration of which gives

$$I_2 = I_1 \exp(-2\alpha(x_2 - x_1)) \ . \tag{7.44}$$

The logarithmic absorption coefficient, α_{ab}, relates to the intensity reduction over a given distance (between, say, ranges x_1 and x_2, where the intensities are I_1 and I_2 respectively):

$$\alpha_{ab} = \frac{10\log_{10}(I_1/I_2)}{x_2 - x_1} \tag{7.45}$$

which, if $x_2 - x_1$ equals one metre, is numerically equal to $20\alpha\log_{10}e$. The logarithmic absorption coefficient is measured in dB per km in ocean acoustics, and dB/cm in biomedical ultrasonics, because absorption tends to increase with frequency. It should be noted that at a given frequency the penetration of sound in water is much greater than that in air. However since ocean acoustics often employs much greater propagation distances than aero-acoustics, and since biomedical ultrasonics employs much higher frequencies, in many circumstances the effects of absorption can be very significant in applications of underwater acoustics. Values of α, the amplitude attenuation constant, are given for various materials in Table 7.1 for biomedically relevant frequency of 1 MHz. (*Derating* is required when using test results obtained with medical ultrasound fields in water to determine the appropriate field properties (e.g. peak negative pressure) that would be attained *in vivo* in body fluids, to account for attenuation and nonlinear propagation.) Values pertinent to the ocean will be indicated after a consideration of the relevant mechanisms.

The acoustic absorption of seawater is much higher than that of pure water (30 times greater in the frequency range 5–50 kHz). The causes of this are mechanisms associated with shear and volume viscosity, and, below 100 kHz, a dominant loss mechanism is the ionic relaxation of magnesium sulphate. The cumulative contributions of these effects to the *logarithmic absorption coefficient*, α_{ab}, are shown as a function of frequency in Fig. 7.13. Below 5 kHz, the attenuation is higher than would be expected through the ionic relaxation of magnesium sulphate alone, and the extra mechanisms proposed include: high amplitude shock wave effects; diffraction and scattering from the SOFAR channel (not strictly a process by which acoustic

Table 7.1 Specific acoustic impedance and attenuation

	Density (kg/m^3)	*Speed of sound in the bulk* (m/s)	*Specific acoustic impedance* (kg/m^{2}s)	*Amplitude attenuation coefficient* (Neper/m) *at 1 MHz*
Air at STP	1.2	330	400	138
Water	1000	1480	1.5×10^6	0.0253
Castor oil	950	1500	1.4×10^6	10.9
PMMA (polymethyl-methacrylate)	1190	2680	3.2×10^6	23.0
Aluminium	2700	6400	1.7×10^7	0.0207
Brass	8500	4490	3.8×10^7	0.230

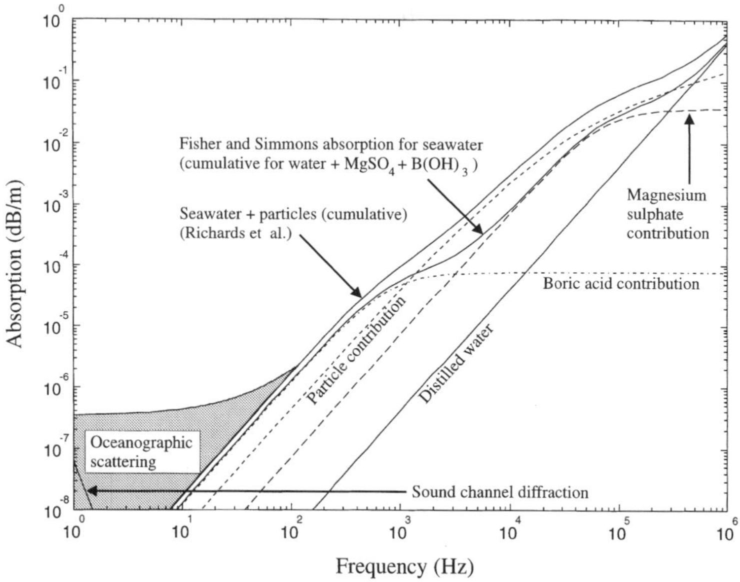

Figure 7.13 The frequency dependence of contributions to the logarithmic absorption coefficient by: pure water; the contribution from magnesium sulphate ionic relaxation; the contribution from boron–borate relaxation; the Fisher and Simmons calculation of the cumulative effects of absorption in seawater, including the contributions from magnesium sulphate ionic relaxation and from boron-borate relaxation; and the contributions from oceanographic scattering and, at low frequency, diffraction from the SOFAR channel. Seawater contributions were computed using the Fisher and Simmons methods, employing a model for seawater of Lyman and Fleming (1940) for $T = 4\,°C$, $S = 35$ ppt, and an ambient pressure of 1 bar. The contribution of a particulate suspension to the absorption was computed using the model of Richards *et al.* (1996), for a concentration of 0.2 kg/m^3 of spherical quartz particles of 3 μm radius; both this contribution, and the effect on the cumulative absorption, when this is added to the Fisher and Simmons seawater model, are shown. (T.G. Leighton, R.C. Evans).

energy is converted into heat, but included here since the energy in such ray paths does not fit the cylindrical spreading model of the previous section); eddy viscosity; and the influence of solid particulates (Richards *et al.*, 1996). Current theory attributes a contribution to the excess absorption at these lower frequencies to boron-borate relaxation, a process which is dependent on the pH of the ocean. Ionic processes involving magnesium carbonate contribute to a lesser extent. Below about 100 Hz scattering by large inhomogeneities in the SOFAR sound channel contribute to the absorption. These processes are detailed by Urick (1983).

(c) Ocean acoustic thermometry

Because of the geometrical effects outlined in section 7.2.2.4(a), acoustic signals will propagate to long range in the SOFAR channel. This facility has been exploited over the years in many ways, which are detailed in the references given at the start of the chapter, and as such will not be considered here. However in recent years a new technique has arisen which exploits this effect, Acoustic Thermometry of Ocean Climate (ATOC), and this will be discussed briefly. For further details see, for example, Munk and Baggeroer (1994), which is an introductory article to a further seventeen papers in the same volume.

Consider Fig. 7.8. As discussed in section 7.2.2.3(a), of the two rays shown in the figure, the one which makes the greater angle with the sound channel axis on intersection is the one along which a sound pulse would travel to distance with the least propagation time. This is a general trend, which occurs since the longer path travels through regions of greater sound speed. Now consider an experiment in which a single impulse is generated from an acoustic source placed on the axis of the SOFAR channel. The receiver will detect a series of impulses corresponding to the different travel times along the various ray paths. If a propagation model using an assumed temperature profile predicts arrivals at a distant receiver hydrophone of a form sketched in Fig. 7.14(a), but instead the experiment records arrivals of

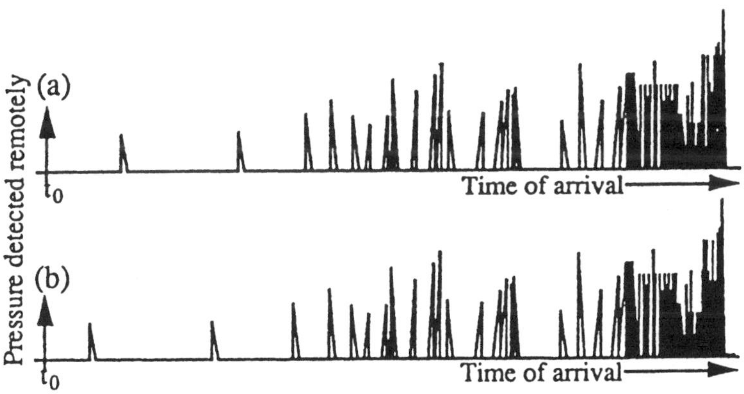

Figure 7.14 A schematic illustration of arrival times of acoustic pulses detected remotely when a source delivers a single impulse into the SOFAR channel. The earliest arrivals (to the left of the trace) correspond to ray paths which travel far away from the sound speed axis. To the right, the closely-spaced arrivals are from ray paths which keep nearer to the axis. In this imagined experiment, (a) shows the predicted time-history from an assumed sound speed profile, and (b) illustrates a time-history the detection of which would indicate that off-axis temperatures are greater than those assumed in the model profile employed in (a).

the form sketched in Fig. 7.14(b), the indications are that the on-axis temperatures assumed in the model are correct, but that off-axis the sea is warmer (since earlier arrivals, as discussed above, correspond to the rays that diverge further off-axis).

This is a very simple example but illustrates the principle which, it was proposed, might be exploited to monitor global warming. Global warming measurements are often complicated by spatial variation. Since the speed of sound is temperature dependent in water, the spatially averaged temperature can be inferred using propagation over large distance, up to thousands of kilometres.

The sound speed increases by 4.6 m/s per degree centigrade. It is estimated that, at around 1 km depth in the ocean, the average temperature might be expected to increase by 0.004 K per year (an exponential extrapolation of a 20 mK/year increase at the ocean surface, which would predict a 1.8 mm/year sea level rise resulting from thermal expansion alone – Baggeroer and Munk (1992)). Therefore over an 18 000 km travel path (half way around the world), over a year the travel times would be expected to decrease by around 0.15 s, which is clearly measurable. While explosive sources had generated sound detectable at global ranges, to detect such changes a well-characterized, non-explosive source was required.

The source site was identified at Heard Island in the Southern Indian Ocean – from here both the Atlantic and Pacific Basins could be insonified. Being at high latitudes, it was in a region where the minimum in the sound speed occurred at relatively shallow depths, around 300 m depth because of the colder surface temperatures (see section 7.2.2.2). Such a locally shallow SOFAR channel was necessary because of the limitations of the sources (which were supplied by the US Navy).

There were ten sources, deployed five at a time, when the experiment was carried out in January 1991. The plan was to send coded signals, which could be match-filtered to measure travel times to better than 0.1 s. The acoustic signals were high amplitude (209 dB re 1 μPa at 1 m) to enable detection at long range. The choice of frequencies can be illustrated by reference to Fig. 7.13. Clearly the argument is for the use of low frequencies, because absorption tends to increase with frequency. However the lowest frequencies are not ideal because of source and signal considerations (e.g. the noise from shipping becomes significant – see section 7.2.3.2), and below 10 Hz there is no great benefit in reducing frequency because oceanographic scattering generates losses which do not decrease with frequency significantly (see Fig. 7.13).

Frequency limitations of the propagation models to be used are also a consideration. For the Heard Island tests, a carrier frequency of 57 Hz was chosen to avoid the electricity mains frequencies of 50 Hz (e.g. UK) and 60 Hz (USA). Subsequent experiments included, in April 1994, propagation of 19.6 Hz signals over 2600 km paths in ice-covered water under the Arctic

(Mikhalevsky and Baggeroer, 1995), where the sound speed profile is usually upwardly refracting from the bottom to the surface because the surface temperatures are colder than those at depth.

7.2.2.5 Sea surfaces and the sound speed

The parameters which control the sound speed, and the consequences and pertinent phenomenon in the deep ocean (e.g. long-range propagation in the SOFAR channel and the role in this of attenuation) have been discussed in section 7.2.2. The sound speed in the sea boundaries, the bottom and the atmosphere/ocean interface will now be introduced.

(a) The sea bottom

The acoustics of the sea bottoms are detailed in the references cited at the outset of this chapter and will not be discussed in depth here. Suffice to say that bottoms may have a range of compositions from hard rock to sediments; that though the density of the bottom is usually greater than that of water, the sound speed can be either greater than or less than water (mud bottoms alone might have a sound speed greater than or less than water); that waves other than compressional may propagate (shear waves, surface waves etc.); that the surface may have varying degrees of roughness; and that a given bottom may contain inhomogeneities and even be stratified, giving rise to multiple reflections. The bottom can play a particularly important role in the acoustics of shallow water regions.

(b) The effect on the sound speed of the bubble population

When a bubble population is added to previously bubble-free water, the bulk average density of the resultant 'bubbly water' (ρ_b) is less than that of bubble-free water (ρ_w). However, as seen from equation (7.4), the sound speed depends not only on the density but also on the bulk modulus, and it is the ratio of these which determines how the sound speed in bubbly water (c_b) differs from the sound speed in bubble-free water (c_w). How the bulk modulus of bubbly water (K_b) differs from that of bubble-free water (K_w) is not simple. The bulk modulus is defined through the ratio of the proportion change in volume (dV/V) to the change in total fluid pressure (dP) which brings about the volume change:

$$K = -V\frac{dP}{dV} \quad . \tag{7.46}$$

The negative sign is employed because, in the quasi-static case, an increase in pressure leads to a reduction in volume, so that dP and dV normally have opposite sign, making K a positive quantity. Question 7.5.2(b) illustrates from the quasi-static case a simple calculation showing how the sound speed might be reduced by the introduction of bubbles. It is important to realize, however, that the effect on the sound speed of pulsating bubbles is a dynamic one. Here the relative phases of the oscillating pressure in the liquid and the bubble volume change are not so simply related as in the quasi-static case. This is because the bubble acts as an oscillator with a resonance frequency f_o which is related approximately inversely to its equilibrium radius R_o. For air bubbles in water under one atmosphere static pressure this relationship is given by

$$R_o f_o \approx 3 \text{ Hz.m} \tag{7.47}$$

Bubbles much larger than the radius R_r which would be resonant with a sound field of frequency f pulsate in antiphase to those having radii much smaller than R_r: the phase relationship is illustrated schematically in Fig. 7.15. If the bubble is much smaller than the resonant size ($R_o \ll R_r$), or

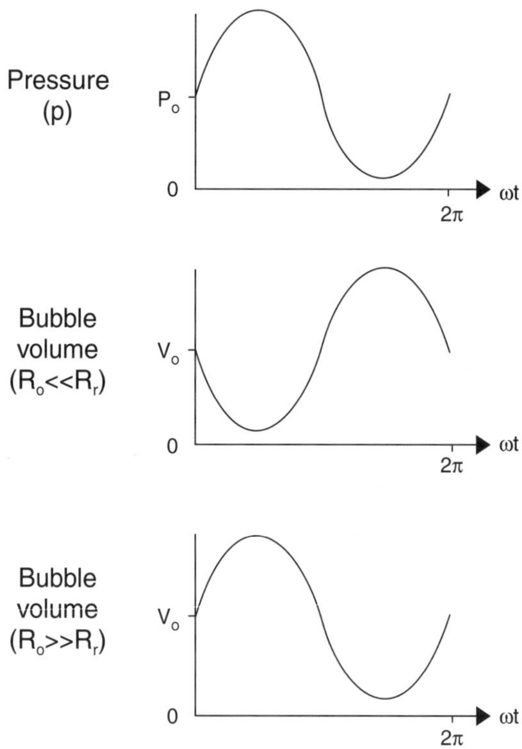

Figure 7.15 Schematic representation of the phase relationships between a driving acoustic field and the pulsations of bubbles which are much greater than, or much less than, resonance size.

if the frequency of the acoustic field is much less than the bubble resonance ($f \ll f_o$), then the bubble is compressed during the positive half-cycle of the acoustic field, and $\mathrm{d}P/\mathrm{d}V$ is negative. The conditions are quasi-static as in Question 7.6.2 (b). The volume changes undertaken by the bubble serve to reduce the bulk modulus to below that of bubble-free water, and the sound speed is reduced. If, however, the bubble is much larger than the resonant size ($R_o \gg R_r$), or if the frequency of the acoustic field is much greater than the bubble resonance ($f \gg f_o$), then the bubble is expanded during the positive half-cycle of the acoustic field, and $\mathrm{d}P/\mathrm{d}V$ is positive. The bubble presence actually contributes an expansive element under compressive forces, and the bulk modulus of bubbly water is greater than that of bubble-free water, so that the bubble presence tends to increase the sound speed. The effect is greater the larger the amplitude of bubble volume oscillation, which will be highest at resonance.

Simple expressions can be stated for the sound speed when the acoustic frequency and number density of bubbles take extreme values (Leighton, 1994, §3.8.2(b)). In the limit of low acoustic frequencies and high number densities of bubbles of the same size, the sound speed in a bubble cloud is approximately

$$c_b \approx \sqrt{\frac{\gamma p_o}{\rho_w\{VF\}}} \tag{7.48}$$

if the gas in the bubbles behaves adiabatically. In the isothermal limit,

$$c_b \approx \sqrt{\frac{p_o}{\rho_w\{VF\}}} \tag{7.49}$$

where $\{VF\}$ is the 'void fraction' (strictly the gas-volume fraction). For the example of Question 7.6.2 (b) (where the void fraction of 10^{-4} is relatively high, and which, being quasi-static, is valid for low insonation frequencies), equations (7.48) and (7.49) predict sound speeds of 1183 m/s and 1000 m/s respectively. In the opposite limit of a low number density (n_b), the sound speed in cloud of bubbles, all of radius R_o, is

$$c_b = c_w\left\{1 - \left(\frac{2\pi R_o n_b c_w^2}{\omega^2}\right)\left(\frac{(\omega_o/\omega)^2 - 1}{\left\{(\omega_o/\omega)^2 - 1\right\}^2 + d^2}\right)\right\}. \tag{7.50}$$

where d gives the bubble pulsation damping

This equation is plotted in Fig. 7.16 for $R_o = 0.1$ mm and varying n_b. At low insonation frequencies ($\omega \ll \omega_o$), equation (7.50) reduces to

$$c_b = c_w\left\{1 - \left(\frac{2\pi R_o n_b c_w^2}{\omega_o^2}\right)\right\} \tag{7.51}$$

which reduces to

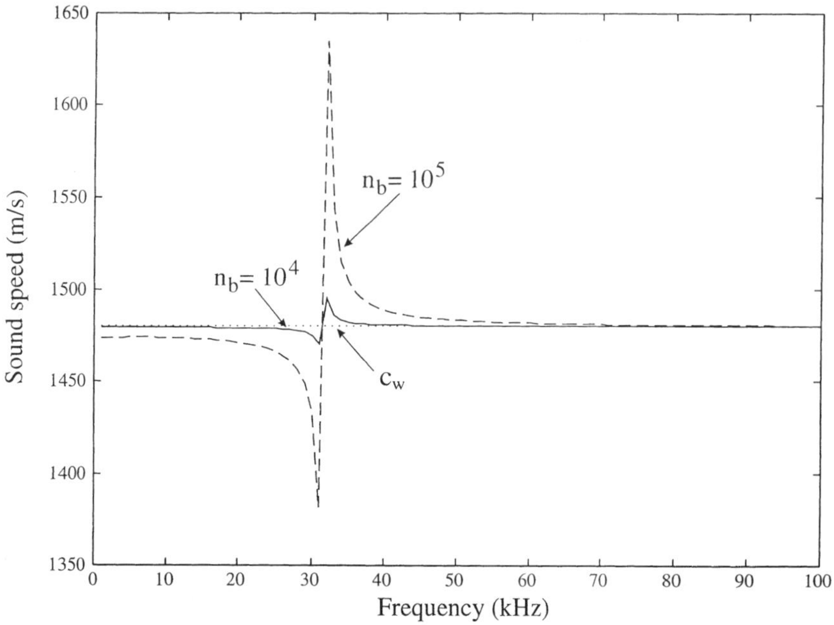

Figure 7.16 Plot of the sound speed as a function of the acoustic frequency for a population of bubbles, all having $R_o = 0.1$ mm, with the number densities shown (n_b/m^{-3}). The sound speed for bubble-free water, c_w, is shown.

$$c_b \approx c_w \left\{ 1 - \frac{1}{2} \{VF\} \frac{\rho c_w^2}{\kappa p_o} \right\} . \tag{7.52}$$

Usually there is a distribution of bubble sizes within the cloud, such that $n_b^{gr}(z, R_o)\mathrm{d}R_o$ is the number of bubbles per unit volume at depth z having radii between R_o and $R_o + \mathrm{d}R_o$, and the speed of sound is a function of both the depth and the acoustic frequency:

$$c_b(z, \omega) = c_w \left\{ 1 - \left(2\pi c_w^2\right) \int_{R_o=0}^{\infty} \frac{R_o}{\omega^2} \left(\frac{(\omega_o/\omega)^2 - 1}{\left\{(\omega_o/\omega)^2 - 1\right\}^2 + d^2} \right) n_b^{gr}(z, R_o)\,\mathrm{d}R_o \right\} . \tag{7.53}$$

(c) The oceanic bubble population

IIn section 7.2.2.2, the vertical variation in sound speed was introduced, though note was made that variations of sound speed in the horizontal plane occur as well because of changes in the properties of the seawater. Having discussed and quantified the role of bubbles in determining the sound speed, the section will close with a description of how variations in the bubble population, in both the horizontal and vertical planes, can arise. Though bubbles can be introduced into the ocean through other routes (e.g. volcanic or biological action, methane seeps, etc.), the majority of bubble entrainment occurs at the atmosphere/ocean interface through wave action. While details of the breaking and evolution of bubble population under breaking waves are not known, it is possible to gain a picture from the available evidence. Much of this comes in the form of sonar scatter from subsurface bubble clouds. Strong scatter is seen both by resonant bubbles (whose high-amplitude pulsations couple well with the sound field), and with larger bubbles (which, although they do not pulsate to a significant extent, simply represent large 'inert' scatterers because of their physical size). This effect can be seen in Fig. 7.17, where the acoustic scattering cross-section (the ratio of the power scattered spherically by a bubble to the intensity of the incident plane acoustic wave which is scattered) shows a local maximum at the resonance, but increases with the bubble cross-sectional area for the larger gas bodies.

Using such scatter, it is possible to obtain an idea of the relevant processes. Figure 7.18 shows a representation of many surface and subsurface features, relevant to a windspeed of about 10 m/s, developed by Monahan and Lu (1990). It includes the stages of development of a bubble plume, illustrating the accompanying surface features. The near-surface continuous-layer of bubbles, which occurs for windspeeds in excess of 7 m/s, is not shown.

The general pattern described by Monahan and Lu is as follows. Breaking waves form surface and subsurface features, which dissipate in time to

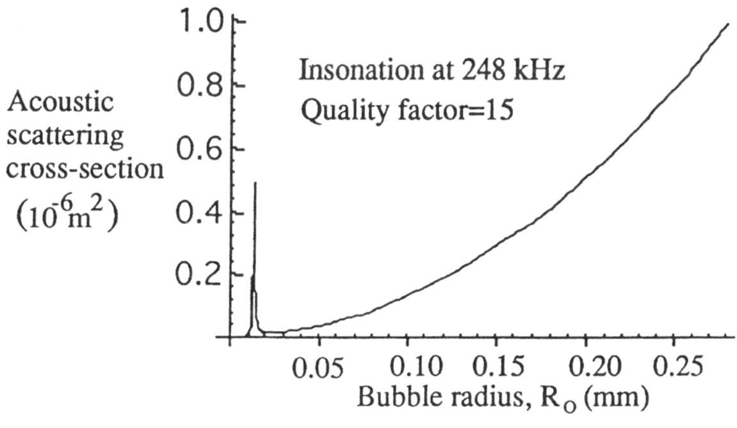

Figure 7.17 The acoustic scattering cross-section from a single bubble as a function of the radius of that bubble, for a fixed insonation frequency of 248 kHz (with a quality factor $Q = 15$ assumed). (after Leighton (1994), by permission of the publishers, Academic Press, London.)

eventually form a background bubble population; however the latter may itself have ordered structure imposed by large-scale circulations. Surface feature A is a spilling wave crest or active whitecap, the surface accompaniment to the subsurface concentrated bubble plume (type α). Type-A features are transient events, only persisting for longer than 1 s if they are regenerated by the parent spilling wave. Otherwise they immediately decay into the mature whitecap, or hazy foam patch, on the surface (type B features in Fig. 7.18), and a type β plume extending to the order of a metre below. The void fraction (i.e. % volume of bubbly water which is air as opposed to water) in type-A can be as high as 30% at the crest of the spilling wave, and from 0.01–0.02% up to 0.1–0.2% in type-β. The void fraction decreases with depth in both (Longuet-Higgins and Turner, 1974). The B and β features coexist for several seconds, dissipating together. After a β-plume is dissipated, no subsurface features exist unless it is as the underside of stabilized foam, aggregating and persisting as a result of the surface convergence of the Langmuir circulation currents.

The concept of Langmuir circulation arose from a report in 1923 of lines of floating weed observed by Langmuir as he crossed the Atlantic, and from subsequent measurements of wind-induced surface effects in Lake George (Langmuir, 1938). That this circulation can affect the distribution and motion of surface particles is evidenced by the accumulation of foam in windrows. These are demarcated by bands of buoyant material (such as foam and weed), aligned with the wind, which occur at the line of convergence of opposing flows in Langmuir circulation. Surface active agents and oils accumulate at the line of convergence and lower the surface tension, so stabilizing the foams. Such windrows can be seen after protracted steady wind. The circulatory flow results from the momentum given to the ocean or lake surface by the wind. The strongest downwellings are spaced typically at 100 m, with smaller Langmuir cells (which may contain even smaller elements) between them. Langmuir circulation can be a major mechanism in the transport of bubbles to depths (Thorpe, 1984, 1992; Thorpe and Hall, 1980, 1983). The drag between the bubble and liquid movements associated with Langmuir circulation (and turbulence, etc.) can act against the buoyant forces, and bubbles may be transported to ~ 10 m depth.

Large subsurface γ-plumes which persist for hundreds of seconds (Thorpe, 1986) are acoustically detectable after a spilling wave has vanished. As a result of the longer lifetimes, any sample of ocean like that illustrated in the figure will contain many more of these γ-type plumes than the β-type, the former being larger owing to the spread of bubbles by turbulent diffusion (Monahan and Lu, 1990). Thorpe (1982) used an echo sounder, located at a depth of 30 m and inverted so that it was upward-pointing, for below-surface observation of these clouds of small bubbles. The bubbles collected in two types of cloud distinguishable on sonograph records. *Columnar clouds* form during unstable or convective conditions when the air is cooler

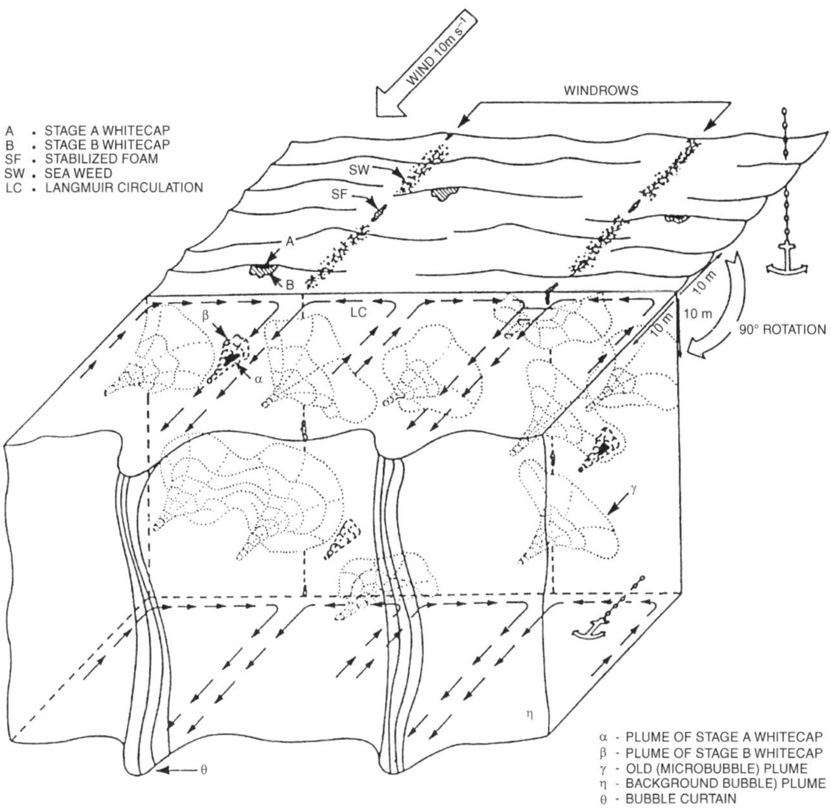

Figure 7.18 Schematic illustration of the subsurface bubble layer and associated surface features, for a windspeed of 10 m/s. For clarity, an imaginary cut is made in a volume of ocean just below the surface, and the body of water is folded forwards (the hinge at the nearest edge), the surface remaining in place. The orientation of the surface and volume features thus represented can be inferred from the anchors drawn on the figure. After Monahan and Lu (1990). © 1990 IEEE.

than the surface water temperature, presumably influenced by the vertical convection and low shear. *Billow clouds*, conversely, are tilted roll-like structures, appearing in stable conditions when the air is warmer than the surface water temperature, the heat flux through the surface stabilizing the water column. The depth of penetration of the bubble clouds increases with increasing windspeed, and for a given windspeed is greater in convective than in stable conditions. Since small bubbles have dissolved and large ones have been removed by buoyancy, the range of bubble radii present in γ-plumes is narrower than those in α or β plumes. Estimates of the void fraction suggest values of $10^{-4}\%$ to $10^{-5}\%$. The concentration decreases with depth. The γ-plumes can be billowy or columnar, though Fig. 7.18

illustrates idealised clouds, having a horizontal cross-sectional area which narrows exponentially with depth, a representation which may not be appropriate to the general case.

The positions of bubbles which have persisted for a long time become randomized, and they can be thought of as a background population, the concentration of which is uniform in the horizontal plane and falls off exponentially with depth. Thus if some given cut-off concentration is taken to mark the base of this background population, then it can be labelled as the item η on Fig. 7.18.

Persistent features, such as windrows, γ-plumes and the background bubble population are influenced by several factors. The stability of bubble constituents will be influenced by local concentrations of dissolved atmospheric gases in the water, and by any stabilizing factors. Considering the possibility of bubble stabilization by surface-active substances and solid particles (Leighton, 1994, § 2.1.2(b)) it is not surprising that bubbles produced by sea-water waves tend to persist longer than those from fresh water waves. In addition, bubbles will be moved by prevailing water motions. For example, the down-welling component of the Langmuir circulation will distort the background bubble population, generating curtains or walls, θ, where the concentration of bubbles is increased above the norm for the otherwise horizontally-uniform background.

Conclusion to section 7.2.2

This description concludes section 7.2.2, which has introduced the sources and characteristics of sound speed variation in the ocean, and the consequences for propagation. Attention was paid to the ability to detect signals at long range because of resulting effect on geometrical spreading. This spreading, along with absorption, influenced the amplitude of the signal detected remotely. However the detectability of a signal does not depend simply on the signal amplitude. It also depends on the level of noise, the sources of which are the topic of the next section.

7.2.3 Sources and applications of ambient noise

7.2.3.1 Applications of ambient noise

As described at the end of the previous section, for most of the history of underwater acoustics the significance of ambient noise has been seen largely in that it generates a masking background which hinders communications and sensing. This is important, and as such a great deal of work has been undertaken in characterizing ambient noise, which is fully described in the references given at the opening of this chapter. In this section, therefore, the range of sources of ambient noise will be briefly summarized, but details will be given only on illustrative developments from the last ten years. Three proposed ways of exploiting ambient noise as a sensor will be discussed. The first of these is 'acoustic daylight'. The other two applications continue with the theme, first discussed in section 7.2.2.5(c), of bubble entrainment at the upper ocean surface. These applications relate to investigating over-ocean rainfall, and the atmosphere/ocean gas flux.

(a) Acoustic daylight

Ambient noise in the frequency range 4–40 kHz has been exploited, using its reflection or absorption from bodies, in a passive form of sonar called 'acoustic daylight'. A truly silent body cannot be detected by conventional passive sonar, which detects the sound generated by that body itself. However 'acoustic daylight' works on the principle that a body in the ocean which makes no noise itself can scatter or absorb sound. A system was designed by Buckingham (1992) to exploit this effect to detect objects. The initial study is illustrated in Fig. 7.19. The technique was named 'acoustic daylight', since it relies on detection through the scattering by a body of the incoherent, ambient 'noise' radiation, much in the same way as the eyes 'see' using daylight.

A hydrophone was placed at the focus of a parabolic reflector, and mounted on a pier. It was able to detect the orientation of plywood targets, which could be turned either face on, or edge on, to the axis of the parabola. Compared with the edge-on condition, the face-on situation produced a rise in the level of ambient noise of about 10 dB across the frequency range of

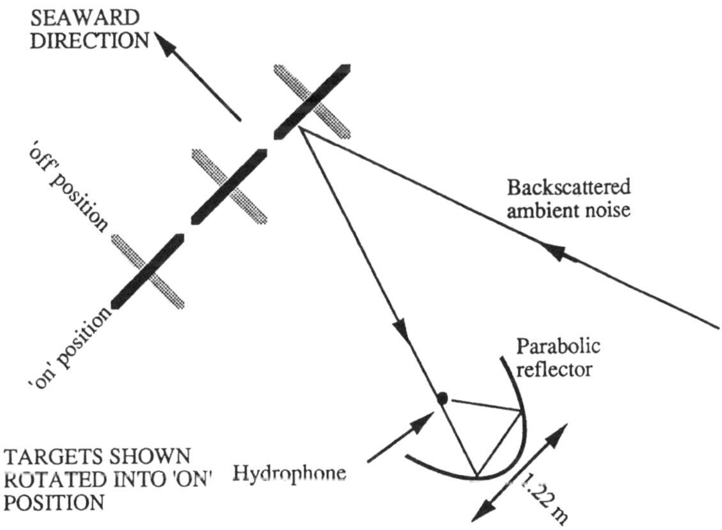

Figure 7.19 Schematic of the apparatus used in the first acoustic daylight experiment.

4–40 kHz. The detection beam was looking out to sea, so clearly the plywood was detected through backscatter of sound produced near-shore. Later versions used an array at the focus of the parabolic reflector, instead of a single hydrophone, to generate a multi-pixel image and, by forming images from the ambient noise in different frequency bands, could generate images from both back-scatter and occlusion of the emissions from different sources of ambient noise (e.g. motor boats, shellfish etc.) (Buckingham and Epifanio, 1996). The process has also been used to form images from the acoustic emissions emitted by bubbles entrained in breaking waves (Buckingham and Epifanio, 1997).

(b) Acoustic emissions from bubble entrainment

In section 7.2.2.5(b) the existence of a bubble pulsation resonance was described, and an approximate resonance relation was stated in equation (7.47). The contributory parameters to that expression are:

$$f_o \approx \frac{1}{2\pi R_o}\sqrt{\frac{3\gamma P_o}{\rho_o}} \tag{7.54}$$

where P_o is the static pressure. This equation, which assumes adiabatic gas compression and ignores surface tension and viscosity, is valid for air bubbles in seawater having radii of greater than about 10 μm. It was first introduced by Minnaert (1933), who experimentally verified the theory by comparing, with his ear, the sounds emitted by bubbles on injection into a water tank with the notes of tuning forks. Minnaert went on to refer to bubble pulsation as the possible source for 'the murmur of the brook, the roar of the cataract, or the humming of the sea'. The acoustic signal generated by each bubble is an exponentially decaying sinusoid of the natural frequency, reflecting the bubble's lightly damped oscillator character. (In fact bubbles oscillating collectively in a cloud can also generate signals at frequencies much less than their individual resonances (Carey, 1987; Prosperetti, 1988; Roy *et al.*, 1992).) A series of these, recorded from a brook, can be seen in Fig. 7.20, each event representing the entrainment of a single bubble. How such emissions can be exploited as sensors will now be discussed.

(i) Acoustic sensing of rainfall over the ocean

The underwater sound from rainfall over the oceans might be used for weather sensing. This technique is particularly useful for remote sensors, when the expense of routing a ship through an area, or the interference of the land mass on island-mounted weather gauges, make those techniques unsuitable. Laville *et al.* (1991) described the use of underwater sound to measure the rate of rainfall over the ocean as 'a very promising technique', and stated that

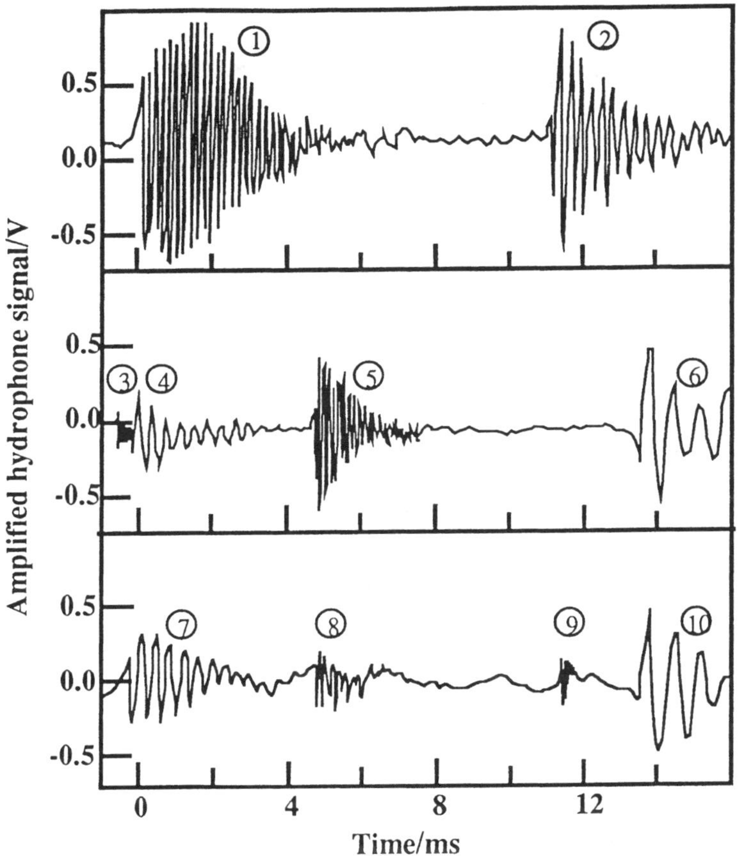

Figure 7.20 Hydrophone trace from a babbling brook. Ten bubble signatures can be identified, and the bubble size inferred from the natural frequency through equation 7.54.

> 'local measurements of rainfall rate over large lakes or oceans are needed ... to calibrate global rainfall measurements made from satellites. Above water measurements are inaccurate because of the direct effect of wave action, spray, and wind. Underwater measurements of the sound generated by rainfall are not directly affected by these surface phenomena, which makes them potentially much more accurate.'

In addition, there are applications for 'the remote sensing of rainfall from an underwater location'.

The history of measurement of the underwater sound of rain is described by Leighton (1994, §3.7.2). When a solid sphere impacts a body of water bubbles are entrained provided that the impact speed exceeds a certain threshold. In contrast, the entrainment of bubbles by liquid drop impact is far more complicated, with sometimes none, one, or many bubbles being entrained, depending on the drop size and the impact speed and angle. The first real attempt to investigate the sound produced when liquid drops impact a body of liquid under laboratory conditions was made by Franz (1959), using high-speed photography and acoustic records. He showed that the impact always generated a sharp pulse of sound as the drop initially contacts the surface. In addition, a bubble could sometimes form within the liquid as a result of the drop impact some time after the initial contact, and when this happens an exponentially decaying sinusoidal acoustic signal typical of bubble entrainment was produced. Since bubbles were not always generated by the impact, Franz was unsure of their importance. In the absence of good field measurements of rain noise on bodies of water, Franz concluded that the pulse caused by the initial impact was of more importance to the sound spectrum than the seemingly sporadic entrainment of bubbles. He thus predicted a very broad power spectrum for rain, peaking at around 3 kHz.

However, more than twenty years later, field measurements of the underwater sound of rainfall demonstrated the presence of a peak at 14 kHz. The reason for the discrepancy is that Franz studied the sound made by a shower of drops, all of which were much larger than raindrops. The source of the peak was subsequently identified as bubbles (Pumphrey *et al.*, 1989; Pumphrey and Elmore, 1990) through the following reasoning. It was found that only within a limited regime of impact velocity and drop size are bubbles entrained in a repeatable manner. In this regime of *regular entrainment*, a given impact either would or would not entrain a bubble consistently. Other combinations of drop size and speed yield other effects (Fig. 7.21). Three other classes of bubble generation could be defined, the regimes being dependent on the drop size and the impact velocity. *Irregular entrainment* involves the production of bubbles through the complex details of the splash. Pumphrey and Elmore also name this 'Franz-type' entrainment because for a given drop diameter and impact velocity in this regime one cannot predict whether or not a bubble will be entrained. *Large bubble entrainment* confines most of the volume of the crater within the bubble. This entrainment might be influenced by surface oscillations on the falling drop. What we now know to be both regular and large bubble entrainment events feature in an article by Jones (1920) who in 1915 observed drops falling into a body of water. He noticed that there were critical ranges of distances through which the drop would fall and produce a characteristic sharp 'click' (which would now be classed as regular entrainment). Occasionally a 'softer, duller sound' occurred, and a bubble was afterwards observed at the point where impact had occurred. This would now be ascribed to large bubble entrainment. Jones observed no bubble remaining behind after the 'click' events, presumably because the regularly entrained bubbles were too small or burst too rapidly for him to detect.

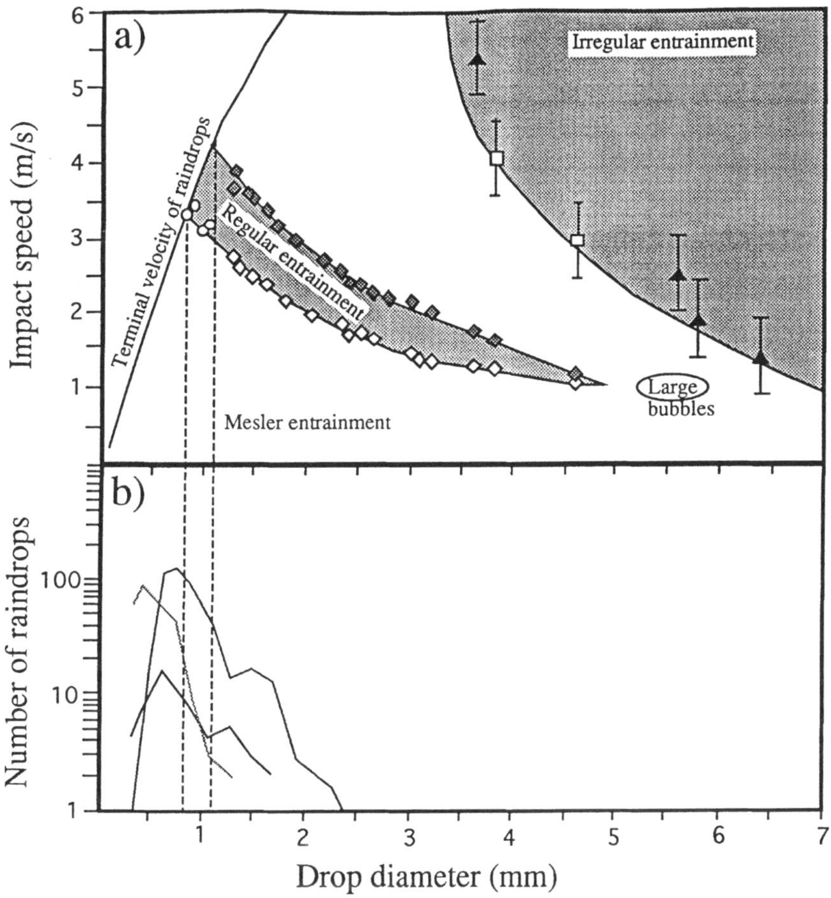

Figure 7.21 (a) Entrainment regimes (for normal incidence). (b) Typical natural raindrop spectra. Natural rainfall data for three showers, from Scrimger *et al.*, (1987). Entrainment data from Pumphrey and Elmore (1990) (Lower limit, irregular entrainment: solid triangles; lower limit, regular entrainment: open circles). All other data points from Pumphrey (1989). (after Leighton (1994), by permission of the publisher, Academic Press, London.)

Regular, irregular and large bubble entrainment all generate significant sound. *Mesler entrainment* is generally a low-velocity unpredictable phenomenon which occurs when a multitude of small bubbles ($R_0 \sim 50$ μm) is trapped in the initial stages of impact. It does not produce significant acoustic emission. The gas pockets may perhaps be trapped in the troughs of capillary waves which spread out from the point of contact on both the drop and the surface of the water body (Oğuz and Prosperetti, 1990). The peaks of the waves on the spreading drop meet those on the liquid, trapping toroidal gas pockets, which through instability break up to generate small bubbles.

The pressure field generated by regular entrainment is illustrated in Fig. 7.22. There are two distinct signals; the characteristic bubble-generated sinusoid is emitted some time ($\approx$ 20 ms) after a pressure signal which results from the initial impact of the drop. High frequencies are associated with this initial impact, and as the signal in Fig. 7.22 was filtered through an 8–50 kHz bandpass filter, there will have been some distortion, particularly to the initial impact pulse. The results should therefore not be taken to mean that the initial impact signal undergoes one or more cycles of oscillation. The initial sharp acoustic pulse, which has duration in the order of microseconds (measurements suggest 10–40 μs, though instrumentation limits the temporal resolution), is a far-field emission which decreases with distance as r^{-1}. This is followed by near-field pressure fluctuations which follows a r^{-2} dependence. The near-field pressure fluctuations may have their source in the dipole nature of the field imparted by the free surface or

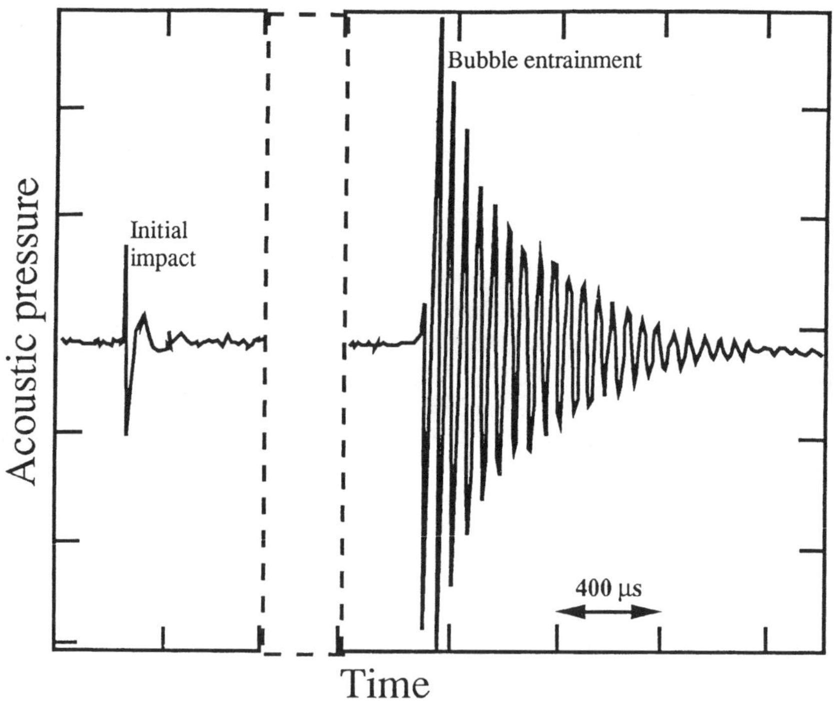

Figure 7.22 A hydrophone trace showing the underwater pressure signal emitted by a regular entrainment (after Medwin *et al.*, 1990).

in incompressible effects. While a pressure emission from the initial impact is ubiquitous, the bubble signal is of course only generated if entrainment occurs.

Fig. 7.21(b) shows, against the same drop diameter scale, the size distributions of raindrops found in natural rainfall. By comparing these populations with the terminal velocity curve plotted in Fig. 7.21(a), it is clear that a significant proportion of drops in these showers might cause regular entrainment. Since the latter is repeatable, it is then possible to determine the natural frequencies of the single bubbles generated by each such event. These are shown in Fig. 7.23 for the combinations of drop size and impact speed which cause regular entrainment at normal incidence. The terminal velocity line forms a boundary to this region, as for Fig. 7.21, with upper and lower bounds in drop size again indicated by vertical lines. Rain, falling vertically at the terminal velocity, will entrain significant numbers of bubbles having natural frequencies of 12–14 kHz. Such regular entrainment from natural rainfall is the source of the 14 kHz peak. If the impact is not at normal incidence to the liquid surface (as can happen if there is a wind), the result is similar, though only a proportion of the impacts entrain in this manner.

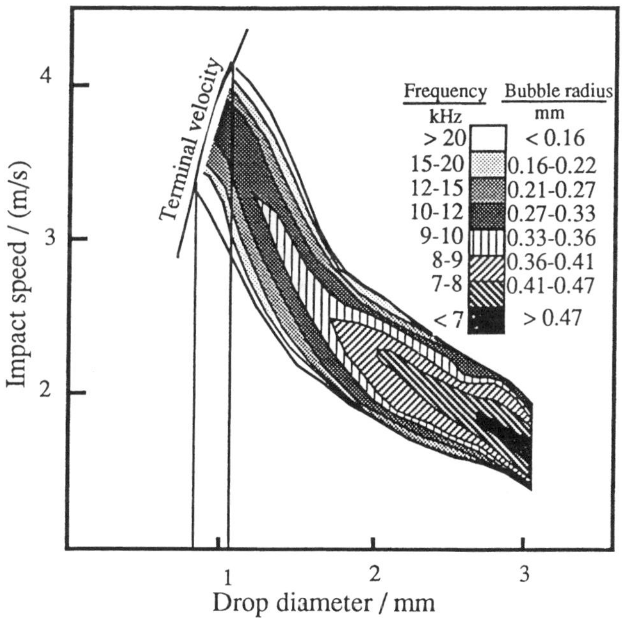

Figure 7.23 For regular entrainment at normal incidence, certain combinations of drop size and speed entrain bubbles of specific size, such that 12–14 kHz would not be unexpected from natural rain (after Pumphrey and Elmore, 1990).

(ii) Bubbles in breaking waves

The oceans act as a huge reservoir of gas into which atmospheric gases can dissolve on a global scale. By conservative estimates, 1000 million tonnes of atmospheric carbon dissolve into the seas each year. Of particular climatic significance are the atmosphere/ocean fluxes of carbon dioxide and dimethylsulphide. Bubbles clearly introduce an asymmetry into the flux. On formation through wave action, they trap atmospheric gas, and through dissolution pump it into the sea. The effect of bubbles is likely to be more important in the flux of gases which are less readily soluble in water than carbon dioxide.

Several types of breaking wave can be identified. *Surging breakers*, which develop as a wave runs up on a steep beach, are restricted to shallow waters. However two forms of breaker are not. First there are *spilling breakers*, where the broken water seems to develop more gently from an instability at the sharp crest and forms a quasi-steady whitecap on the forward slope (Fig. 7.24). Because of the presence of a significant number of air bubbles, trapped as the wave breaks gently at the crest, the whitecap is lighter than the water below. The density difference inhibits mixing with the face of the wave, and the whitecap rides on top of the sloping sea surface, retaining its identity. Longuet-Higgins and Turner (1974) proposed that a spilling breaker can be regarded as a turbulent gravity current riding down the forward slope of the wave. The component of gravity in the relevant direction draws on the distinct turbulent flow of the whitecap which, particularly at its front edge, entrains water from the laminar wave surface below and entrains air from above, so maintaining the density difference.

Second, there are *plunging breakers*, where the wave crest curls forward and plunges deeply into the slope of the wave some distance from the crest. Whether enveloped by plunging breakers (where gravity is the restoring

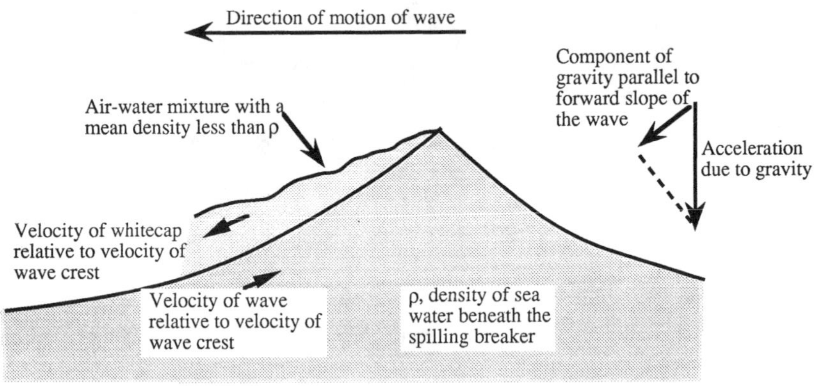

Figure 7.24 Model of a spilling breaker (after Longuet-Higgins and Turner, 1974, by permission of Cambridge University Press).

force for the wave motion), or trapped in troughs in steep capillary (where surface tension acts as the restoring force) or capillary–gravity waves, one might envisage the resulting pockets of air to resemble horizontal cylinders. Longitudinal instabilities would cause each cylinder to fragment along its length, eventually to form a number of roughly spherical bubbles. Fragmentation through shape oscillation might also occur (Longuet-Higgins, 1990; Leighton *et al.*, 1995).

The fact that a bubble, upon entrainment, 'rings' at its natural frequency indicates an exciting impulse is delivered at that time. This impulse can come from a number of sources (Leighton, 1994, §3.7.3). Two, the surface tension and hydrostatic pressures, are particularly interesting. This is because the impulse is delivered in the form of an overpressure which, in the short term, causes the newly formed bubble to be compressed towards the equilibrium radius. The overshoot of this motion sets up the oscillation and consequently generates the acoustic emission. However in the long term, it is such overpressures which drive the gas contained within the bubble into solution.

Surface tension tends to create a compressive force on a spherical bubble. This effect produces the so-called Laplace pressure (p_σ), which contributes to the degree by which the pressure of the gas in the bubble tends to exceed the pressure in the surrounding liquid. Consider the bubble of radius R. There is an internal pressure (p_{i}) within the bubble as a result of the pressure of gas (p_{g}) and the pressure of liquid vapour (p_{v}), so that

$$p_{\mathrm{i}} = p_{\mathrm{g}} + p_{\mathrm{v}} \ . \tag{7.55}$$

The pressure within a bubble at rest is greater than the pressure in the liquid immediately outside the bubble as a result of surface tension forces. If the pressure in the liquid outside the bubble, at the bubble wall, is p_{L}, then that within the bubble is

$$p_{\mathrm{i}} = p_{\mathrm{L}} + p_\sigma \ . \tag{7.56}$$

If an imaginary cut is made which divides the bubble in half, the excess pressure p_σ would tend to push the two halves of the bubble away from one another (Leighton, 1994, §2.1.1). The force which balances the excess pressure and so keeps the static bubble intact is the surface tension.

The energy associated with a surface in a liquid is given by the product of the surface tension with the surface area. Surface tension (σ) is numerically equal to the force per unit length acting perpendicular to one side of a straight line in a liquid surface. The force along the imaginary cut is therefore $2\pi R\sigma$. The force which this balances is $\pi R^2 p_\sigma$, that is the excess pressure multiplied by the effective area seen in the direction of the 'push', πR^2. Equating this to the surface tension force gives

$$p_\sigma = \frac{2\sigma}{R} \ , \tag{7.57}$$

the excess pressure inside a bubble which results from surface tension, also known as the *Laplace pressure*. This acts in conjunction with the hydrostatic pressure (p_h) to generate acoustic emission and gas dissolution. The smaller the bubble, and the shallower its depth, the more significant is the surface tension, as opposed to the hydrostatic, pressure (Leighton, 1994 §3.7.3).

Consider therefore a pocket of gas which is entrained at the ocean surface. Before the liquid meniscus is complete, the gas pressure inside is atmospheric, p_{atm} (Fig. 7.25(a),(b)). However when the bubble wall is formed, the pressure in the liquid outside the bubble is $p_{atm} + p_h$, where $p_h = \rho gh$ if h is the entrainment depth (Fig. 7.25(c)), so that the equilibrium condition of the bubble would have the internal pressure (p_i) such that

$$p_i = p_g + p_v = p_\sigma + \rho gh \ . \tag{7.58}$$

Immediately after entrainment the gas pressure equals p_{atm}, which is less than that required for equilibrium. Therefore the bubble must compress towards equilibrium, and overshoot results in oscillation. Similarly if the level of gas dissolved in the seawater is in equilibrium with the atmosphere, then it will not be in equilibrium with the gas pressure within the bubble when these oscillations have damped down; the bubble will dissolve, and a supersaturation may result.

7.2.3.2 Non-bubble sources of ambient noise

Knudsen *et al.* (1948) summarized much Second World War data by characterizing the undersea noise in the frequency range 100 Hz to 25 kHz. Wenz confirmed these so-called 'Knudsen spectra', and commented that wind-dependence between 50 Hz and 10 kHz could be caused by air bubbles. The Knudsen spectrum can conveniently be divided into three parts. There is a dominant low-frequency component below 50 Hz, a shipping component at intermediate frequencies, and above 100 Hz one which peaks at 500 Hz, then falls off at ~ 6 dB per octave. At windspeeds greater than ~ 5 m/s, whitecaps are extensively present and the Knudsen

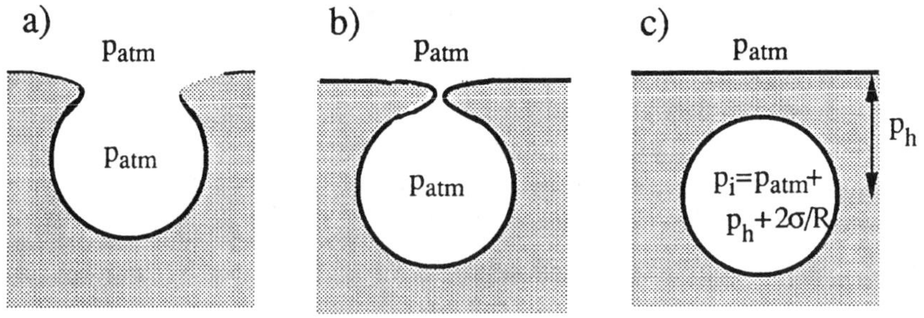

Figure 7.25 The gas pressures associated with bubble entrainment.

spectrum originates from breaking waves. However the Knudsen spectrum can still be detected at windspeeds as low as 1 m/s. The main non-bubble sources of ambient noise are detailed in the references given at the start of the chapter, and briefly outlined below. It should be noted that, in the term 'ocean ambient noise', the 'noise' need not be acoustic. It may for example arise through vibration, electrical, pyroelectric and two types of non-acoustic pressure changes; turbulent and hydrostatic

Hydrostatic pressure changes at receivers can be caused by tides (1–2 cycles/day), which can also contribute to flow noise and can occur at sensor supports. Though wave motion can change the water head above a sensor, the pressure fluctuations do not propagate to depth unless a nonlinear process associated with the surface standing waves occurs. Continuous microseisms contribute noise at around 1/7 Hz, and intermittent acoustic emissions can be generated by earthquakes, volcanoes, etc. Dynamic pressure fluctuations can arise at sensors as a result of ocean turbulence. Low frequencies (below $\sim$ 500 Hz) from ships and storms can propagate to long distances. Significant biological noise can be generated by three groups of marine animals. Some fish can emit sound; frequently cited are the croakers of Chesapeake Bay, which drum air bladder muscles. Some Crustacea (shellfish) can make noise (e.g. the claws of snapping shrimp can generate sound in the range 500–20 000 Hz). The echo-location emissions from whales, dolphins, and porpoises were mentioned earlier, and communication can also be achieved. It is now thought that the mysterious '20-cycle pulse' is generated by the Finback whale (Premus and Spiesberger, 1997). The 20 Hz tone-bursts comprises a pulse every 10 seconds, of 1 s duration, detected for several minutes (6–25 min) by bottomed hydrophones in the Atlantic and Pacific Oceans. This is followed by 2–3 mins of silence. The signal repeats, for many hours. Triangulation shows that the sources move randomly at 2–3 knots, and are high power (1–25 W), and therefore detectable 35 miles away in shallow continental shelf water.

Ambient noise in shallow water is variable in space and time. Sources include noise from wind, shipping, industrial and biological sources. In addition, deep water noise may propagate over distances into shallow waters.

7.3 NONLINEAR UNDERWATER ACOUSTICS

As sources of higher power were developed, and as our ability to analyse and process acoustic signals improved, it became clear that the description of simple linear acoustics is inadequate to explain many phenomena associated with the propagation of sound of finite amplitude. Though less amenable to simple explanation, these nonlinear phenomena have over the past few decades proved a valuable area of research. Returns have included the parametric array, which is a form of nonlinear sonar, the enhanced resolution of acoustic microscopes, and the application of the nonlinear properties of tissue for diagnostic and therapeutic purposes. Though such uses have only appeared in the last few decades, it has been the subject of consideration since the eighteenth century. Probably the first discussions were published in 1759 by Euler, who considered the nonlinear wave equation for finite-amplitude sound.

7.3.1 Nonlinear propagation of waves of finite amplitude

In Chapter 1, a derivation was given of the linearized wave equation in acoustic pressure, p:

$$\nabla^2 p = \frac{1}{c^2}\ddot{p} \ . \tag{7.59}$$

Any variable which satisfies the linearized wave equation will propagate as a waveform without loss through an ideal medium. In order for the acoustic pressure to satisfy the linearized wave equation, the above derivation requires the assumptions that fluid particle velocity fluctuations are very much less than the speed of sound in the liquid, and also that density fluctuations are negligible in comparison with the equilibrium density. Obviously density is not constant during the propagation of an acoustic wave. Particle displacements cause local variations in density which are periodic for a sinusoidal wave. Therefore the meaning of this derivation is that the smaller the amplitude of the acoustic wave, the more nearly it approximates to the case of the linear waves discussed in Chapter 1. Finite-amplitude sound waves will propagate in a fluid, but their physics is more complicated.

The same argument can be followed mathematically (Leighton, 1994, §1.2.3). The characterization of the propagation of an acoustic wave in a fluid requires three fundamental inputs enshrined in equations reflecting: first, the motion of mass in a fluid; second, the fluid dynamic properties relating such motions to the pressure gradient which causes them, as observed from a frame which moves along with an element of fluid in the flow; and third, an equation of state. This is a thermodynamic equation, the input from which is introduced into this proof from section 1.3. Stated in one-dimensional form, these three equations, are, as follows. First, the equation of continuity (see Chapter 1).

$$\frac{\partial \rho}{\partial t} + \frac{\partial (\rho u)}{\partial x} = 0 \Rightarrow \frac{1}{\rho}\frac{\partial \rho}{\partial t} + \frac{\partial u}{\partial x} + \frac{u}{\rho}\frac{\partial \rho}{\partial x} = 0 \tag{7.60}$$

where u is the particle velocity;

Second, Euler's equation for an inviscid fluid:

$$\frac{\partial P}{\partial x} + u\rho\frac{\partial u}{\partial x} + \rho\dot{u} = 0 \tag{7.61}$$

where P the total pressure.

Third, following equations (1.4), (1.5) and (1.10) from section 1.3:

$$c^2 = \partial P/\partial\rho \tag{7.62}$$

(where $\partial P/\partial\rho = \partial p/\partial\rho$ if the only fluctuating component of the total fluid pressure is the acoustic one). Combination of these three equations allows formulation of the propagation of acoustic longitudinal waves in a fluid, as now follows. The relationship

$$\frac{\partial\rho}{\partial t}c^2 = \frac{\partial P}{\partial t} \tag{7.63}$$

which is derived from equation (7.62), can be substituted into the expanded form of equation (7.60) so that the term $(\partial\rho/\partial t)/\rho$ converts into $(\partial P/\partial t)/(\rho c^2)$ to give:

$$\frac{\partial u}{\partial x} + \frac{u}{\rho}\frac{\partial\rho}{\partial x} + \frac{1}{\rho c^2}\dot{P} = 0\ . \tag{7.64}$$

If the nonlinear terms $u\rho(\partial u/\partial x)$ and $(u/\rho)(\partial\rho/\partial x)$ are negligible, equations (7.61) and (7.64) reduce, respectively, to

$$\frac{\partial p}{\partial x} + \rho_o\dot{u} = 0 \tag{7.65}$$

and

$$\frac{\partial u}{\partial x} + \frac{1}{\rho_o c^2}\dot{p} = 0 \tag{7.66}$$

where the approximations $\dot{P} = \dot{p}$ and $\partial P/\partial x = \partial p/\partial x$ are justified if the only varying component of the total pressure is the acoustic one. The introduction of ρ_0, the equilibrium fluid density, is appropriate since the second of the two assumptions made after equation (7.64) is that the proportional changes in density with position are negligible. Equations (7.65) and (7.66) combine to give the linearized wave equation by differentiating equation (7.65) with respect to x, and differentiating equation (7.66) with respect to t, and then equating the terms $\partial^2 u/(\partial t\partial x)$ and $\partial^2 u/(\partial x\partial t)$ between them to give the linearized wave equation in one dimension:

$$\frac{\partial^2 p}{\partial x^2} - \frac{1}{c^2}\frac{\partial^2 p}{\partial t^2} = 0\ , \tag{7.67}$$

which describes a pressure wave propagating at speed c. If however the nonlinear terms $u\rho(\partial u/\partial x)$ and $(u/\rho)(\partial\rho/\partial x)$, which were assumed to be negligible after equation (7.64), cannot be neglected, then equation (7.67) cannot be produced from the fundamental three equations. This occurs when proportional changes in density and $\partial u^2/\partial x$ are not small, which tends to occur as the wave amplitude increases. This is why equation (7.67) is said to refer to the propagation of waves of infinitesimal amplitude. The wave is termed 'finite amplitude' when its propagation deviates from the simple longitudinal case described in much of Chapter 1. This has two important effects when the propagation of a sinusoidal acoustic wave is considered.

First, there is a *convection* effect. In simple terms, if u/c is not negligible, then parts of the wave tend to propagate as $c+u$. The particle velocity varies throughout the wave, and so since particle velocity and acoustic pressure are in phase for such a wave (Chapter 1), the greater the local acoustic pressure, the greater the velocity of migration of that section of the wave. Therefore regions of compression tend to migrate faster than the regions of rarefaction (where u is opposite to c). The pressure peak travels with the greatest speed, the trough with the least.

Second, there is an effect which arises because, when a fluid is compressed, its bulk modulus increases. This causes an increase in sound speed, and this effect too will cause the pressure peaks to travel at greater speed than the troughs, and tend to try to catch up and encroach upon them. A continuous wave that is initially sinusoidal will therefore distort in the way shown in Fig. 7.26. The sketches illustrate the shape of a single cycle which is part of a long wavetrain of identical cycles. The plot shows the pressure as a function of distance. The cycle is initially sinusoidal, and its spectral content (shown on the right) is narrowly distributed around a single frequency (since the wavetrain is long, but not infinite). (i) As it propagates through the medium, each compressive region gains upon the preceding rarefactive half-cycle, an accumulated steepness of the waveform between the two developing. Harmonics of the frequency associated with the initial sinusoid appear in the spectrum (see section 7.3.4). (ii) After propagating a certain distance (the *discontinuity length*), the waveform includes a discontinuity; a shock wave develops, as shown in the figure by the discontinuity in the spatial pressure profile, with an associated continuum in the spectrum. (iii) With increasing time, any further compressional advance leads to dissipation and results in a reduction in the amplitude of the shock. This is because the waveform distortion has been equivalent to transferring some of the energy of the initial wave to higher frequencies, which are more strongly absorbed (section 7.2.2.4(b)). (iv) The energy transfer is not sufficient to maintain the shock, and the wave approaches a low-amplitude sinusoidal form (termed 'old age').

An additional feature, not shown in the schematic in Fig. 7.26, arises from the action of dissipation and diffraction, which cause phase shifts in the various frequency components of the wave. As a result, the distorted waveform is not symmetrical about the zero-line. In general the trough, which corresponds to negative values of the acoustic pressure, becomes rounded while the positive peak is augmented.

As outlined above, there are two contributions to the change in phase speed of the wave, the change in stiffness (or bulk modulus) and the convection. These are called the *material nonlinearity* and the *convective*

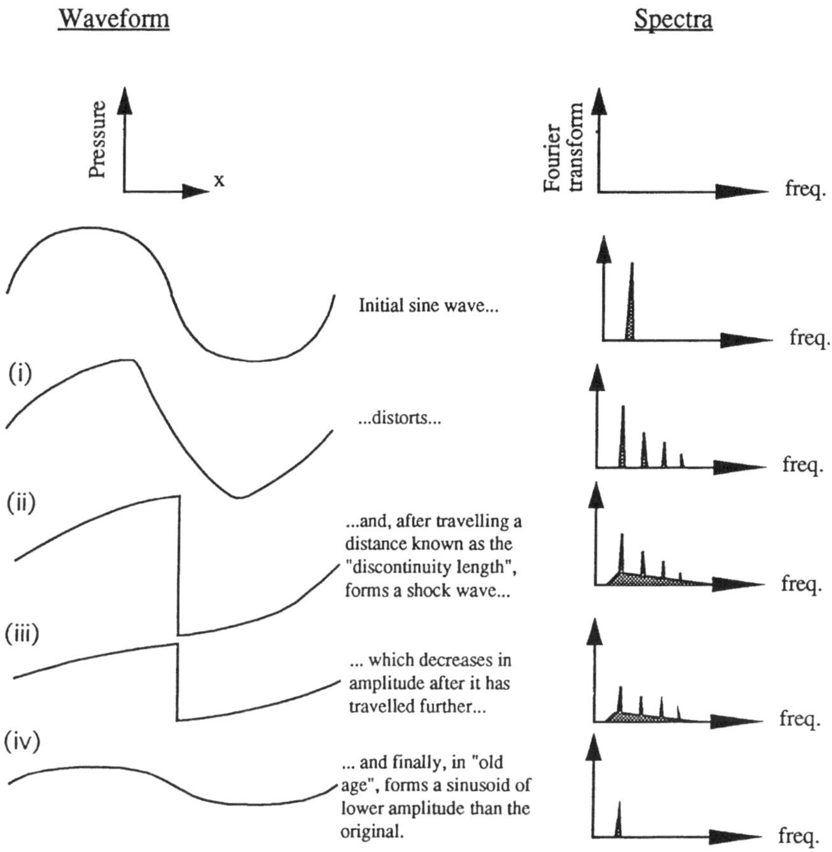

Figure 7.26 The waveforms and associated spectra relevant to this example of the non-linear propagation of finite-amplitude, initially sinusoidal, waves. The figure shows one example cycle out of an infinite wavetrain of single frequency.

nonlinearity respectively, and are formulated as follows (Leighton, 1994, §1.2.3). The phase velocity of a point in the wave having particle velocity u is

$$\text{Phase velocity (for given } u) = u + c \tag{7.68}$$

where a convection component u is added to c, the local speed of sound. This local speed has in turn been affected by the change in bulk modulus, and is related to c_0, the sound speed for waves of infinitesimal amplitude, by

$$c = c_0 + \frac{\gamma - 1}{2} u \ . \tag{7.69}$$

Though equation (7.69) applies to a perfectly isentropic gas, it may also apply to a liquid where the ratio of specific heats, γ, for the liquid is given by

$$\gamma = 1 + \frac{B}{A} \quad \text{which implies} \quad c = c_0 + \frac{B}{2A} u \tag{7.70}$$

where B/A is the second-order nonlinearity ratio of the liquid. Therefore equation (7.68) can be rewritten to include both the material and convectiv nonlinearities explicitly:

$$\begin{aligned} \text{Phase velocity (for given } u) &= u + c \\ &= c_0 + \left(1 + \frac{\gamma - 1}{2}\right) u, \quad \text{for an isentropic gas} \\ &= c_0 + \left(1 + \frac{B}{2A}\right) u, \quad \text{for a liquid.} \end{aligned} \tag{7.71}$$

The unity term in the brackets in equation (7.71) corresponds to the contribution from the convective nonlinearity, and the other term to that from the material nonlinearity. For water, the ratio $B/2A$ equals 2.5, whereas for air the isentropic gas equivalent is $(\gamma - 1)/2 = 0.2$. Therefore in water the dominant cause of the waveform distortion is nonlinearity of the equation of state (the material nonlinearity), whereas in air the distortio arises mainly through convection.

The sum of the material and convective nonlinearities is given by the parameter ζ, where from equation (7.71)

$$\begin{aligned} &\text{Phase velocity (for given } u) = c_0 + \zeta u \\ &\text{where } \zeta = \frac{\gamma + 1}{2}, \quad \text{for an isentropic gas, and} \\ &\text{where } \zeta = \left(1 + \frac{B}{2A}\right), \quad \text{for a liquid.} \end{aligned} \tag{7.72}$$

For air $\zeta = 1.2$, and for water $\zeta = 3.5$. From this one might expect the distortion to be more obvious in water than in air; however one must take into account attenuation of high frequencies. The discontinuity length, which is taken to be the distance propagated when an infinite slope first appears in the waveform, is given by

$$L_{\text{dis}} = \frac{1}{\zeta M k}, \tag{7.73}$$

where $k = \omega/c_0$ and where M is the peak acoustic Mach number of the source, the ratio of the amplitude of the particle velocity at the source to c_0. Thus the shock forms closer to the source as the coefficient ζ, which indicates the degree of nonlinearity, increases. It also forms sooner as the amplitude increases because the speed differential between the peaks and the troughs increases.

Similarly L_{dis} decreases with increasing frequency, since this signifies decreasing wavelength, and so the peaks have to travel a shorter distance to catch up with the troughs. If the point of observation is at a distance greater than L_{dis}, then as either amplitude or frequency are increased the discontinuity length decreases, the point where the shock forms moving closer to the source. This means that increasing amounts of energy can be dissipated at the shocks beyond L_{dis} before the wave arrives at the point of observation. If the energy supplied to the wave at the source is continually increased, there comes a point at which the increased dissipation between the discontinuity point and the point of observation outweighs the increase in energy supplied to the wave at the source. Beyond this critical source power, the wave ceases to be dependent upon the source amplitude. Any additional energy supplied to the wave by the source is lost at the shock front, and *acoustic saturation* is said to have occurred (Leighton, 1994, §1.2.3).

7.3.2 Radiation effects

Consider a plane wave, travelling in the $+x$ direction, approaching a wall in the yz plane of area A (Leighton, 1994, §1.1.4). The wave carries energy, such that the power flowing through any plane of area A, which does not interfere with the propagation and which is normal to the direction of propagation of the plane wave, is I. Therefore if it is assumed that the wave meets a wall of area A, aligned normal to the direction of propagation, and that the wall completely absorbs the acoustic energy of the wave, then in time Δt the energy absorbed by the wall is $IA\Delta t$. Since no wave propagates beyond the wall, the wall must have applied a force F_r in the $-x$ direction which in time Δt acted. Therefore the work done by the wall on the wave is $F_r c\Delta t$. Equating this to the energy absorbed, we obtain $F_r = (IA/c)$. From Newton's third law of motion, this must be equal and opposite to the force exerted by the wave on the wall. Therefore upon absorption the wave exerts a time-averaged *radiation pressure* in the direction of its motion of magnitude

$$p_{rad,abs} = \frac{I}{c} \tag{7.74}$$

for normal incidence of plane waves. (Demonstrations of radiation pressure are not confined to acoustics laboratories, but can frequently been seen in science fiction films, where ray guns appear to have the ability to impart a force capable of making the victim stagger backwards. Likening this motion to that observed when catching a 5 kg bag of potatoes, suggests the radiation force is 50 N, so that during firing the ray gun would need to project power W equal, from equation (7.74), to $(50c)$ N, where c is the wavespeed. Were the ray-gun to project electromagnetic radiation, such as a laser beam, the power of the gun would have to be $50 \times 3 \times 10^8 = 1.5 \times 10^{10}$ W. This is more than ten times the 1.2×10^9 W power output of a typical nuclear power station.

However since the speed of sound is lower than that of light, the equivalent power requirements would be 17 kW in air and 75 kW in water.) The force F_r exerted by the wall can also be thought of as acting upon the wave to absorb its momentum. In time Δt, the wall absorbs a length $L = c\Delta t$ of the wave, exerting an impulse $F_r \Delta t = IAL/c^2$ upon the wave, causing a change in momentum of ΔJ. Since after absorption the momentum of wave is zero, then the momentum associated with one wavelength of the wave is

$$J_\lambda = \frac{IA\lambda}{c^2} \quad . \tag{7.75}$$

If the wave is reflected, instead of being absorbed, this momentum must be not simply absorbed but reversed. The wall must exert twice as much force upon the wave, and so the time-averaged radiation pressure felt by the reflector is

$$p_{\mathrm{rad,refl}} = \frac{2I}{c} \quad , \tag{7.76}$$

for total reflection of normally-incident waves back along the line of incidence (Leighton, 1994, §1.1.4).

As the acoustic wave travels through a medium, it will be absorbed to a certain extent (section 7.2.2.4(b)). However the momentum absorbed from the acoustic field manifests as a flow of the liquid in the direction of the sound field, termed *acoustic streaming* (Leighton, 1994, §1.2.3). The above process can also be thought of as the setting up of an energy gradient in the direction of propagation, occuring because energy is absorbed from the beam during its passage through an attenuating liquid. A gradient in energy corresponds to a force, and in general such energy gradients will act upon a liquid causing a streaming flow. The force per unit volume, F/V, equals the gradient in pressure, ∇p, which results in a liquid acceleration in the direction of propagation. In the above case of a beam propagating through an absorbing liquid, the pressure gradient equals:

$$\nabla p = \frac{F}{V} = \frac{2I\alpha}{c} \quad . \tag{7.77}$$

From this equation it is clear that if both intensity and attenuation can vary spatially throughout a sound beam in a uniform medium, then so will the streaming forces and flows. An increase in either parameter will increase the streaming. As will be evident from the preceding section, the pumping of energy into higher frequencies, which are more strongly absorbed than the fundamental, means that attenuation may vary spatially in an acoustic field of finite amplitude. Finite-amplitude effects will also affect streaming through the formation of shocks. Starrit *et al.* (1989) observed the enhancement of streaming in high-amplitude diagnostic pulsed ultrasonic fields which have formed shocks in water. Streaming speeds of up to around 10 cm/s can be demonstrated from clinical ultrasonic equipment.

There is a second type of streaming. In contrast to the above examples, this is not associated with bulk flows resulting from energy gradients within the medium through which the sound propagates. Instead, it appears as circulations which occur near small obstacles placed within a sound field, or near small sound sources, or vibrating membranes or wires (Leighton, 1994). It arises from the frictional forces between a boundary and a medium carrying vibrations of circular frequency ω, and unlike the streaming described earlier, this time-independent circulation occurs only in a small region of the fluid, being generally confined to an acoustic boundary layer of thickness

$$L_{\mathrm{ms}} = \sqrt{2\eta/\rho\omega} \ , \tag{7.78}$$

where η and ρ are the shear viscosity and density respectively of the liquid. Because of the restricted scale of the circulation (often but not exclusively microscopic), it is commonly termed *microstreaming*. Microstreaming can bring about a number of important effects. The shear forces set up within the liquid may disrupt DNA (deoxyribonucleic acid), disaggregate bacteria, disrupt human erythrocytes and platelets *in vitro* and *in vivo*, and other bioeffects. Microstreaming can specifically occur as a result of the oscillations of an acoustically driven bubble in a sound field. As will be discussed in section 7.4.3.2, this can also lead to bioeffects (Leighton, 1994).

7.3.3 Self-effects

The nonlinear propagation discussed earlier may be indirectly responsible for the relative scarcity of nonlinear acoustic phenomena, compared to optical ones, as in many pure media sound is non-dispersive at frequencies for which absorption over the distance of a wavelength is small. As a result, high-pressure fields are converted into shock waves as energy is pumped into the higher harmonics, as outlined earlier. Since absorption tends to be greater at higher frequencies, the high frequency oscillations are strongly absorbed, so that strong sound waves are rapidly dissipated in the liquid. This effect can be reduced by engineering dispersion into the medium, for example by propagating the sound through a waveguide, or by introducing gas bubbles. Nevertheless there are several other nonlinear effects associated with the propagation of ultrasound, such as self-interaction and parametric phenomenon, stimulated scattering and phase conjugation (Leighton, 1994, §1.2.3). Examples of some of these are given below.

One self-interaction effect is illustrated by the *self-focusing* of acoustic beams (Fig. 7.27). Thermal self-focusing occurs because, through absorption of acoustic energy, the medium through which the sound beams passes becomes warmer. In most liquids, this causes the sound-speed to fall (or equivalently causes the acoustic refractive index to increase) so that the beam is focused in towards the axis as a result of total internal reflection at its

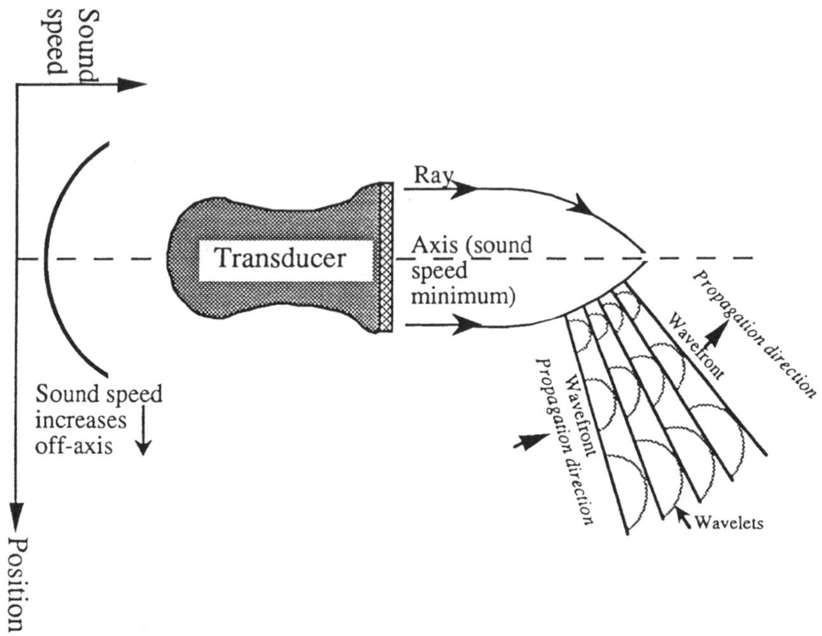

Figure 7.27 Self-focusing. At the left of the figure, the sound speed profile is shown. Compare with Figs 7.7 and 7.8.

perimeter. In water, however, the sound speed passes through a maximum at 74 °C (Fig. 7.4), so that thermal self-focusing only occurs at temperatures in excess of this. At lower temperatures, the inverse phenomenon of self-defocusing occurs (Fig. 7.28). Acoustic streaming imparts a defocusing effect by increasing the sound speed near the beam axis.

There are mechanisms other than thermal which can bring about self-focusing, including some resulting from the presence of bubbles, which can accumulate on the transducer axis through Bjerknes forces (Leighton, 1994, §4.4.1) and reduce the sound speed there (section 7.2.2.5(b)). The presence of bubbles can also bring about self-transparency, where the absorption decreases with increasing intensity, though this can arise in bubble-free media, such as glycerin, owing to the temperature-dependence of the absorption coefficient.

Self-focusing can occur not just in the bulk of a liquid, but at an interface between two media where, for example, distortion of the surface of a water sample by the beam can lead to a focusing effect simply as a result of local angling of the reflecting surface (Leighton, 1994, §1.2.3). In the extreme case, when the reflected beams are directed upwards, it can lead to *acoustic self-concentration* and the formation of a fountain.

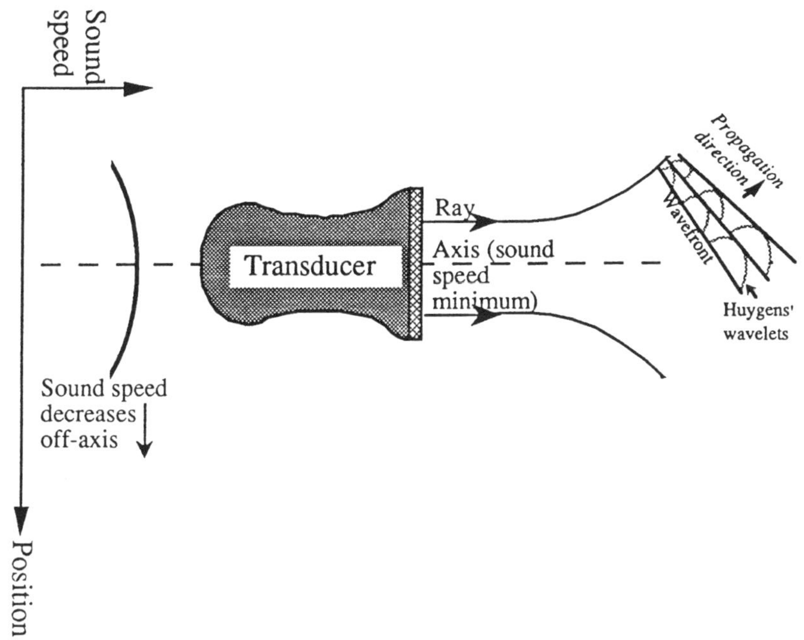

Figure 7.28 Self-defocusing. At the left of the figure, the sound speed profile is shown. Compare with Figs 7.7 and 7.8.

7.3.4 The parametric array

In section 7.3.1, the response (e.g. $\Delta\rho/\rho$) of an element of water to an acoustic wave was considered to be nonlinear at finite amplitude, tending to the linear at infinitesimal amplitudes. Even the simple power series expansion† of the force/response relationship illustrates how frequencies other than the driving frequency can arise in such situations in the steady state (a phenomenon which was discussed in that section). In that model, the general response $¥(t)$ of the oscillator is a power series of the driving force F, for example,

$$¥(t) = s_0 + s_1\mathrm{F}(t) + s_2\mathrm{F}^2(t) + s_3\mathrm{F}^3(t) + s_4\mathrm{F}^4(t) + \cdots \tag{7.79}$$

In this circumstance ¥ might stand for the response of the fluid element to pressure. Such a response might be the motion of the fluid element.

† Note: The power series expansion given in equation (7.79) is not a complete descriptor of nonlinear oscillation. For example, it will not predict subharmonics. This may be done through Fourier series expansion of the response, or by perturbation methods.

This expansion can illustrate how the increase in the second (and higher) harmonics arises when an initially sinusoidal wave propagates. If the driving force were a single-frequency, the acoustic pressure would be $p(t) = p_A \cos \omega t$. Substitution of this expression for $p(t)$ into equation (7.79) reveals how frequencies that are integer multiples of the driving frequency can arise. For example, the quadratic term F^2 will produce a harmonic at twice the driving frequency through

$$2\cos^2 \omega t = 1 + \cos 2\omega t \ . \tag{7.80}$$

The cubic term F^3 will produce a harmonic at three times the driving frequency:

$$4\cos^3 \omega t = \cos 3\omega t + 3\cos \omega t \ , \tag{7.81}$$

and so on to higher harmonics. Thus harmonics can be detected in the acoustic field after these nonlinearities have had time to develop.

The power series model (equation (7.79)) can also be used to illustrate how the use of *two* insonating frequencies can detect nonlinearities. Say the driving force F consists of the sum of two coherent forces of different frequency, i.e.

$$p(t) = p_1 \cos \omega_1 t + p_2 \cos \omega_2 t \ , \tag{7.82}$$

where $\omega_2 > \omega_1$ (the presence of phase constants would not alter the general result). After substituting equation (7.82) into equation (7.79), expansion of the quadratic component contains a term which can be expanded to generate the sum and difference frequencies:

$$2p_2 p_1 \cos \omega_1 t \cos \omega_2 t = p_2 p_1 \{\cos(\omega_1 + \omega_2)t + \cos(\omega_2 - \omega_1)t\} \ . \tag{7.83}$$

The parametric array was conceived in 1960 by Westervelt (1960) and, following examples of applications suggested by Berktay (1965), is now perhaps the most familiar application of nonlinear acoustics. If two collimated acoustic beams of finite amplitude and high frequencies, ω_1 and ω_2, are transmitted in the same direction, nonlinear interaction between them generates sound at $\omega_c = \omega_1 - \omega_2$. Thus a parametric antennae, driven at two frequencies, provides a highly directional extended source of low frequency sound. The source itself can be relatively small, and the sidelobe features greatly reduced. The low frequency, ω_c, can be detected in the medium, along with such combinations as $\omega_1 \pm \omega_c$. However these other frequencies, being higher, are more strongly attenuated and over long propagation distances the lowest frequency, ω_c, dominates. The possibility of increasing the efficiency of these systems by the addition of bubbles has been investigated.

Parametric receivers exhibit the same high directionality in the detection of sound. It can be formed from two transducers, one being the source of a collimated acoustic beam of high intensity, and the other a hydrophone positioned in this beam. The high-amplitude sound in the beam interacts nonlinearly with any acoustic signal that crosses its path, and the resulting modulation is detected by the hydrophone. The amplitudes of the sum- and difference-signals so detected are greatest when the parametric receiver is aligned with the incident signal.

7.4 THE EFFECTS OF BIOMEDICAL ULTRASOUND

Biomedical ultrasonics are employed for diagnosis and therapy. The preceding section of this chapter illustrated ways in which an acoustic beam can alter the medium through which it passes. If controlled, this can be exploited for therapeutic effect. However such effects are unwanted during diagnosis. For details, see Leighton (1994, §5.4.2).

7.4.1 The question of changes to the medium during diagnosis

Diagnostic ultrasonics can be utilized in a number of ways, including both the formation of images of regions of differing acoustic impedance using acoustic reflection, and blood flow measurements using Doppler techniques. Images may be generated using ultrasound in the approximate range 3–30 MHz, using pulse profiles as shown in Fig. 7.29(a) (the spectrum being shown in 7.29(b)). The pulses are of micro-second order, and the pulse repetition frequency (PRF) is of order 1 kHz. Such technology is used to image major organs (e.g. the heart, gall bladder), the foetus, and abnormal tissues such as tumours and cysts. By 1984 well over a quarter of a million foetuses were exposed to ultrasound annually in Britain alone.

Despite the millions of examinations over the 30 years that diagnostic ultrasound has been used clinically, there is no verified documented evidence of adverse effects in patients caused by exposure to ultrasound. However in the last decade there have been a number of public statements by those advocating caution, and those who attest to the safety, of diagnostic ultrasound. In 1984 Mr John Patten, Junior Health Minister in the United Kingdom, warned against the routine use of ultrasound in pregnancy in a letter to the Association for Improvements in Maternity Services. Quoted in the *Daily Mail* (22 October 1984) and five days later in *Lancet*, he said, ‘Given the publicity there has recently been about the possible risk of ultrasound scanning, we would not expect any health authority to be advocating screening for all mothers as a routine procedure.’ A debate followed and, for example, Ziskin (1987) commented that ‘There is nothing that I'm aware of that has a safer record than that of diagnostic ultrasound.’

Several factors need to be considered. One is the pace of development of the technology. Figure 7.30 shows the trend in output of pulse-echo and Doppler diagnostic equipment. One of the underlying reasons for the trend

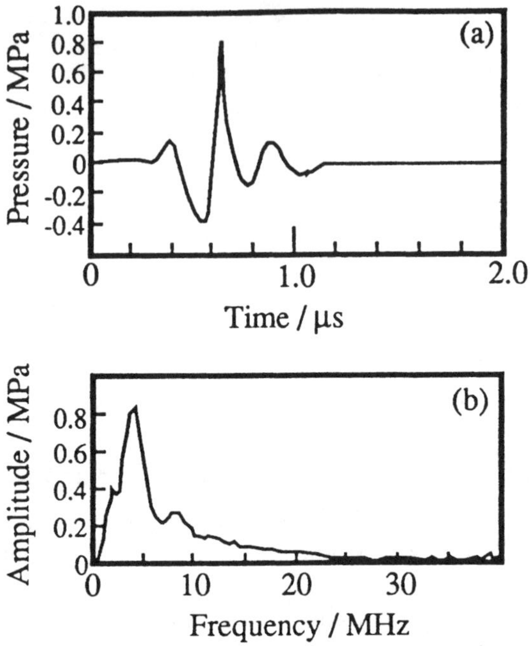

Figure 7.29 The (a) waveform and (b) spectral magnitudes measured on axis at a distance of 100 mm from transducer faceplate from diagnostic imaging apparatus (after Baker, 1991). Reprinted by permission from *Physics in Medicine and Biology*, **36**, 1457–64; copyright © 1991 IOP Publishing Ltd.

to increase intensities and acoustic pressures lies in the tendency for manufacturers to produce diagnostic systems which utilize ever-increasing frequencies, so that the shorter wavelengths will lead to an improvement in resolution. However, since these higher frequencies are more strongly attenuated (section 7.2.2.4(b)), there is an associated tendency to utilize higher acoustic pressures.

The figure also illustrates the need to define the way in which the intensity has been measured. With pulses that may be of microsecond duration but with an inter-pulse time of millisecond order, the time-average acoustic intensity may be three-orders of magnitude less than the temporal peak intensity. Similar comments apply to spatial averaging because often the fields are focused, and the wavelengths are of sub-millimetre order.

Carstensen (1987) states that

> 'temporal average intensities in some pulse Doppler devices in use today are close to levels which produce measurable heating of tissues, and the temporal peak pressures in some modern instruments approach the levels of acoustic saturation, i.e. the condition in which nonlinear losses of the

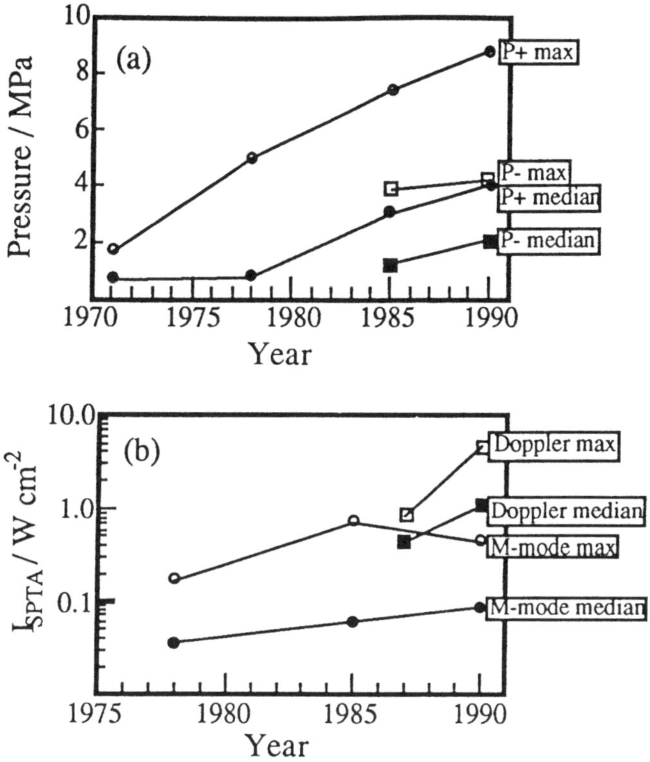

Figure 7.30 The median and overall maximum outputs from surveys of clinical devices up to 1990. (a) The change in peak negative (labelled 'P-') and peak positive (labelled 'P+') pressures measured in water for pulse-echo diagnostic equipment. (b) The change in I_{SPTA} (spatial-average, temporal-average intensity), measured in water, for M-mode operation and for pulsed Doppler modes (note that the scale is logarithmic, showing that the maximum from pulsed Doppler has increased roughly ten-fold). After Duck and Martin (1991). Reprinted by permission from *Physics in Medicine and Biology*, **36**, 1423–32; copyright © 1991 IOP Publishing Ltd.

medium set upper limits on local sound source levels regardless of the output from the source.'

Williams (1990) notes that

'the time-averaged intensities emitted by many devices in certain operating modes can exceed the time averaged intensities to which patients are being subjected during typical physiotherapeutic treatments. We are consequently in a state of confusion where the physiotherapists are convinced (albeit based on tenuous or anecdotal evidence) that their treatments change the functioning of tissues and are therefore efficacious, while the clinical users are equally convinced (based on even less evidence) that their diagnostic investigations are not changing the tissues at all. Clearly both groups of users cannot be correct in their beliefs.'

7.4.2 Therapeutic

Ultrasound is commonly employed for therapy in physiotherapy. Figure 7.31 schematically shows typical pulse profiles for physiotherapeutic applications, which usually operate at 1–3 MHz, exploiting continuous-wave insonation or pulses of around 2 ms duration with duty cycles (the ratio of the pulse duration to the inter-pulse time) of 1:2, 1:4, or 1:7. Spatial average intensities are generally less than 1 W/cm^2. There are probably over 1 million treatments per year in the English and Welsh National Health Service (ter Haar *et al.*, 1985).

In lithotripsy, high amplitude ultrasonic pulses are employed for the treatment of kidney stones and gall bladder stones which are fragmented and subsequently removed by surgery or passed through the urethra. In a typical clinical lithotripsy procedure, more than 1000 shocks with pressure amplitudes in excess of 20 MPa are deployed (Fig. 7.32). Temporal peak pressures have been reported as high as 100–300 MPa, in contrast to diagnostic ultrasonics, where the largest peaks are no more than about 10 MPa. In clinical practice, the pulse repetition frequency rarely exceeds 2 Hz, so that the time-average intensity is small.

Ultrasonic dental procedure, including scaling and filing, employ magnetostriction to generate a micrometer order displacement of the tip at 30 kHz. The mechanism of action is through direct physical contact, microstreaming and cavitation. Experiments have investigated treatments involving the application of ultrasound in surgery, where focused, high-

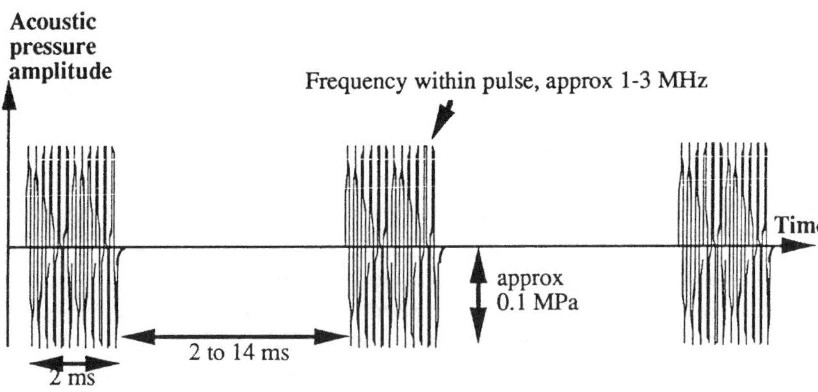

Figure 7.31 Illustration of typical physiotherapeutic output.

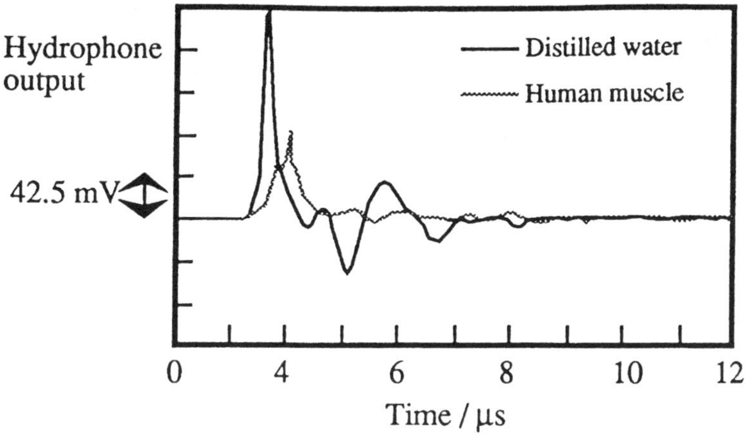

Figure 7.32 Extracorporeal lithotripter pressure waveforms obtained in (i) distilled water, and (ii) human bicep muscle, obtained by an Imotec needle hydrophone. After Finney *et al.* (1991). Reprinted by permission from *Physics in Medicine and Biology*, **36**, 1485–93; copyright © 1991 IOP Publishing Ltd.

intensity continuous wave ultrasound has been employed in ophthalmic surgery and neurosurgery (Lizzi *et al.*, 1992; Fry *et al.*, 1970). ter Haar *et al.* (1991) used high-intensity 1.7 MHz ultrasound, focused to peak intensities of 1400–3500 W cm^{-2} in 10 s exposures to arrest histologically detectable growth in rat tumours that had been exposed over the whole volume. Wound healing may be promoted by physiotherapeutic-type ultrasound. Other possible effects include a possible placebo effect and phonophoresis, where at frequencies of a few MHz and I_{SATA} of order 1 W cm^{-2} intensities, ultrasound can increase the rate of penetration of a wide range of pharmacologically active agents through the skin. See Leighton (1994, §5.4.2) for details of these various therapies.

7.4.3 The mechanisms of bioeffect

A bioeffect may be desired (e.g. in the application of an ultrasonic cell disrupter) or undesired (e.g. during a diagnostic scan). Throughout section 7.3, several mechanisms by which an ultrasonic bioeffect might be brought about have been introduced. These can be interdependent, as the following discussion illustrates. One of the major mechanisms of bioeffect is through hyperthermia, the heating of the medium as a result of the absorption of the acoustic energy (section 7.2.2.4(b)). Such absorption can be enhanced by nonlinear propagation (section 7.3.1). Nonlinear propagation can also enhance streaming by increasing the rate at which the momentum formerly carried by the acoustic wave is transferred into the liquid as the wave is absorbed (section 7.3.2). Self-focusing (section 7.3.3) can increase the wave amplitude and therefore the effects of nonlinear propagation; however hyperthermia on the transducer axis can cause either self-focusing or self-defocusing. High-amplitude ultrasound can stimulate bubble activity quite different to that discussed earlier in section 7.2.2.5(b). These mechanisms will now be discussed in terms of clinical ultrasound.

7.4.3.1 Hyperthermia

The 1988 guideline from the American Institute of Ultrasound in Medicine (AIUM, 1988) postulates that in an afebrile patient (i.e. one not having a fever) a temperature rise of 1 °C resulting from exposure to diagnostic ultrasound is 'safe'. Hyperthermia is probably the mechanism which deserves most consideration with respect to the application of diagnostic ultrasound. No hazard has been demonstrated *in vivo* with clinical ultrasound.

Experiments show that the biological consequences depend crucially on both the *magnitude* of temperature elevation, and the *duration* of it. The most sensitive tissues are those containing a large proportion of actively-dividing cells (mitosis), and abnormalities can occur in cell division and DNA replication. A wide range of abnormalities have been reported in animals, including retardation of brain changes with reduced learning performance, and low birth weight. For details, see WFUMB (1992, 1994, 1996).

Problems in testing for the effects of hyperthermia include those associated with extrapolating findings from animals to the human case. For example, species have different basal temperature (38.5 °C in rats; 39.5 °C in guinea pigs; 37 °C as in humans). It is not certain whether the absolute temperature, or the temperature increase above normal, is the relevant parameter for hyperthermic bioeffect. If the whole body is subjected to elevated temperatures, as occurs in the air incubators tests which are used in much hyperthermia research, the results may not be directly relevant to the local heating produced by, for example, *in utero* ultrasound. In the latter case, localized heating may be affected by perfusion, whereby blood flow can reduce the temperature rise caused by local heating; and by bone, ultrasonic reflection off which can enhance heating. On the other hand, whole body heating may adversely affect the maternal physiology, or it may be that protective mechanisms, stimulated in response to whole-body heating, do not occur when ultrasound heats locally.

Heating occurs not only through bulk absorption of ultrasound by tissue: losses in the transducer itself cause its temperature to rise, and this must be considered particularly during intracavitory (e.g. endovaginal) applications.

7.4.3.2 Bubble-related mechanisms

In section 7.2.3.1(b), the pulsations of bubbles, decaying in amplitude following their entrainment, was discussed. However when driven by a sound field, the pulsations, rather than decaying, can persist, the bubble pulsating in what is termed 'stable cavitation' or 'non-inertial cavitation'. Associated with such behaviour is the effect of bubbles on the sound speed, and the scattering of the sound field, discussed in sections 7.2.2.5(b) and (c) respectively. A range of other behaviours can affect the medium. These include Bjerknes forces, which are radiation forces associated with bubble oscillation, and cause translational movements of bodies suspended in the liquid (Leighton, 1994, §4.4.1). For example, platelets might be caused to accumulate around a pulsating bubble (Miller *et al.*, 1979); bubbles might group together, forming acoustic 'shields' which hinder the penetration of sound further into the liquid (Leighton, 1994, §5.3; Leighton, 1995); and bubbles may translate through the medium generating hydrodynamic shear which can cause cell lysis (Miller and Williams, 1989). Microstreaming currents generated in the liquid close to the oscillating bubble wall can cause cell lysis (Vivino *et al.*, 1985). One of the most important effects associated with stable cavitation can be *rectified diffusion* (Leighton, 1994, §4.4.3), where the equilibrium radius of the bubble grows as the pulsations draw previously dissolved gas out from the surrounding liquid into the bubble interior. This could conceivably cause a bioeffect through mechanical action or deplete the gas reservoir, but probably is most important in that it affects the size distribution of bubbles in the population, which can affect the type of cavitational behaviour they undertake. The most important such behaviour, which depends on the equilibrium bubble size, is the phenomenon of 'inertial cavitation' (formerly known as 'transient cavitation'). This will now be described.

When bubbles within a certain size range are driven strongly, they undergo a dramatic change in behaviour (Leighton, 1994, §4.2, §4.3). This transition from non-inertial to inertial cavitation (where the bubble undergoes rapid explosive growth, followed by a violent collapse, generating high gas temperatures, gas shocks, and liquid shocks) is a threshold phenomenon with respect to initial bubble radius, acoustic frequency, and acoustic pressure amplitude. Both the high gas temperatures and the gas shocks may generate free radicals (e.g. in aqueous solution, H atoms, OH radicals, and products such as H_2O_2, hydrogen peroxide). These are highly reactive and may represent a hazard (they are the sources of sonochemical reaction and of sonoluminescence). As regards the hot-spot itself, although its temperature is high, it does not contain significant energy or last for a long time. The *rebound pressure pulse* emitted into the liquid as the bubble rebounds from minimum size may cause mechanical damage to structures close to it (pressure amplitude falls off as r^{-1} since it is spherically diverging – section 7.2.2.4(a). However in a cloud of bubbles the shocks may act to greater distances through a co-operation effect between many bubbles (cloud concentration).

The threshold is such that there is a critical size range, which increases with increasing acoustic pressure amplitude and decreasing frequency, over which bubbles will undergo inertial/transient collapse if insonated by an appropriate sound field. This is because bubbles which are initially too large, though they will grow sufficiently, are then too sluggish to undergo a collapse that is sufficiently violent. Conversely very small bubbles are prevented from undergoing the initial explosive growth by the Laplace pressure (which increases rapidly with decreasing radius – section 7.2.3.1(b)). Therefore only bubbles of intermediate size will undergo transient collapse. This is illustrated in Fig. 7.33.

Erosion through violent cavitational collapse has been known for many years. One of the earliest observations was of 'pitting' in ship propellers. One cause of this is the rebound pressure pulse, which also generates noise. Submariners know that submerging will tend to reduce cavitation noise by suppressing the propeller cavitation. However, when the cavitation is strong

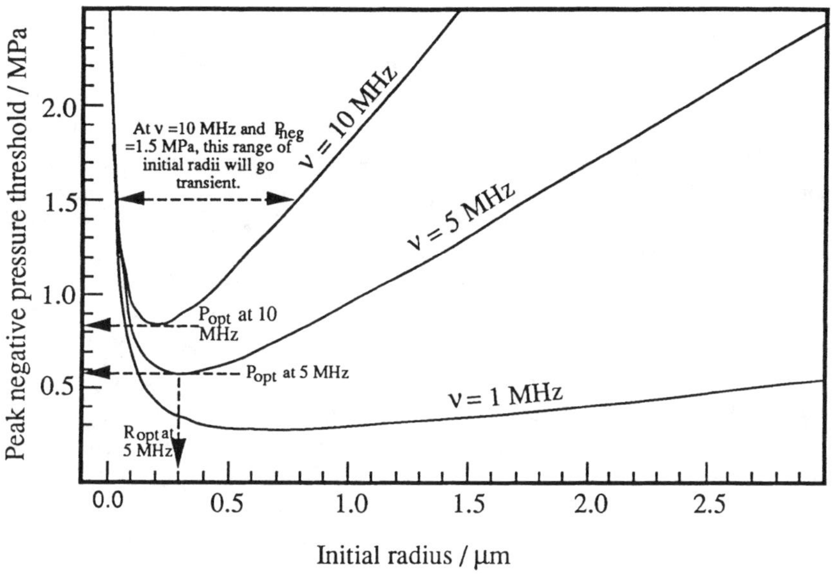

Figure 7.33 Illustration of the threshold for inertial cavitation, after the calculation of Apfel and Holland (1991), in terms of the variation in the peak negative pressure required to generate inertial cavitation from a free-floating spherical gas bubble nucleus, as a function of the initial radius of that bubble. Above the curve, inertial cavitation will occur. As the acoustic frequency increases, the threshold tends to increase for all initial bubble sizes, and the radius range which will nucleate inertial cavitation decreases. Reprinted by permission of Elsevier Science from *Ultrasound in Medicine and Biology*, **17**, 179–85; copyright © 1991 by World Federation of Ultrasound in Medicine and Biology.

and the vessel is at high speed, increasing the depth of the vessel will first cause an increase in the cavitation noise, before suppression occurs. This so-called 'anomalous depth effect' is due to the fact that an increasing static pressure tends to increases the violence of each individual collapse before, at the highest static pressures, it suppresses the growth phase. However 'pitting' can also be caused by an aspherical form of collapse, which can occur when the bubble is close to an inhomogeneity such as a solid boundary. In such circumstances, the bubble can invert on collapse such that a high-speed liquid jet can pass through the bubble and, on impacting a nearby boundary, can create damage (Leighton, 1994, §5.4.1).

7.4.3.3 Radiation forces

Direct bioeffects from radiation pressure are not often evident since the magnitude of forces tends to be smaller than, for example, the hydrodynamic stresses induced by bubble activity. A typical diagnostic scanner might generate a pulse average pressure of 1–10 kPa, with a time-average pressure of order 1–10 Pa. This can be compared with the tensile strength of the weakest tissue, which is around 50 kPa (Yamada, 1970). Modest streaming flows of 10 cm/s have been measured in water.

7.4.4 Conclusion

Which mechanism for bioeffect dominates is dependent on the parameters of a particular sound field. In general, hyperthermia tends to respond to the time-averaged acoustic intensity, since the time-scale of bulk heating is relatively slow. In contrast, bubbles respond almost instantaneously to the acoustic pressure field, so that the question of inertial cavitation will be more dependent on the temporal-peak, rather than the temporal-average, acoustic pressure. Inertial cavitation will not be an issue if one is below the threshold conditions for cavitation. *In vivo* this threshold is likely to be a nucleation threshold; however if nuclei are plentiful, as in aerated water, the relevant threshold is likely to be the one discussed in section 7.4.3.2.

This concluding discussion of the physical acoustic and fluid phenomenon associated with the potential for bioeffect from clinical ultrasound provides a useful exercise for employment of the nonlinear propagation, self-effect, and radiation force considerations given in section 7.3 (Nonlinear underwater acoustics). These in turn benefit from exploitation of the ray path, absorption, and signal-to-noise features introduced in section 7.2 (Ocean acoustics). Similarities do exist between the acoustics of millimetre-wavelength ultrasound in the human body, and sound of much greater wavelengths in the oceans.

7.5 REFERENCES

AIUM (1988) Bioeffects considerations for the safety of diagnostic ultrasound. *J. Ultrasound Med.*, **7**: S1–S38, Supplement no. 9, American Institute of Ultrasound in Medicine, Bioeffects Committee.

Albers, V. M. (ed.) (1972) *Underwater Sound*, Dowden, Hutchinson, and Ross, Inc., distributed by Academic Press.

Apfel, R. E. and Holland, C. K. (1991) Gauging the likelihood of cavitation from short-pulse, low-duty cycle diagnostic ultrasound. *Ultrasound Med. Biol.*, **17**, 179–185.

Baker, A. C. (1991) Prediction of non-linear propagation in water due to diagnostic medical ultrasound equipment. *Phys. Med. Biol.*, **36**, 1457–1464.

Berktay, H. O. (1965) Possible exploitation of non-linear acoustics in underwater transmitting applications. *J. Sound Vib.* **2**, 435–461.

Buckingham, M. (1992) Acoustic daylight: imaging the ocean with ambient noise. *Nature*, **356**, 327–329.

Buckingham, M. J. and Epifanio, C. L. (1996) Acoustic daylight imaging in the ocean: Experimental Results. In J. Papadakis (ed.) *Proceedings of the 3rd European Conference on Underwater Acoustics* (*Heraklion, Crete, 24–8 June 1996*), Crete University Press, Heraklion, 341–347.

Buckingham, M. J. and Epifanio, C. L. (1997) Bubbles and breaking waves: imaging in space, time and frequency. In T. G. Leighton (ed.) *Natural Physical Processes Associated with Sea Surface Sound*, University of Southampton, 39–44.

Burdick, W. S. (1984) *Underwater Acoustic System Analysis*, Prentice-Hall, Eglewood Cliffs, New Jersey.

Carey, W. M. (1987) Low-frequency noise and bubble plume oscillations. *J. Acoust. Soc. Am. suppl.* **82**: S62.

Carstensen, E. L. (1987) Acoustic cavitation and the safety of diagnostic ultrasound. *Ultrasound Med. Biol.*, **13**, 597–606.

Chen, C-T. and Millero, F. J. (1976) Reevaluation of Wilson's sound-speed measurements for pure water. *J. Acoust. Soc. Am.* **60**, 1270–3.

Chen, C-T., Fine, R. A. and Millero, F. J. (1977) The equation of state of pure water determined from sound speeds. *J. Chemical Physics*, **66**, 2142–4.

Clay, C. S. and Medwin, H. (1977) *Acoustical Oceanography*, John Wiley and Sons, New York.

Coates, R. F. W. (1990) *Underwater Acoustic Systems*, Macmillan, Basingstoke.

Duck, F. A. and Martin, K. (1991) Trends in diagnostic exposure. *Phys. Med. Biol.* **36**, 1423–32.

Euler, L. (1759) *Mém. Acad. Sci.* Berlin, **15**, 185–209.

Finney, R, Halliwell, M, Mishriki, S. F. and Baker, A. C. (1991) Measurement of lithotripsy pulses through biological media. *Phys. Med. Biol.* **36**, 1485–93.

Fisher, H. and Simmons, V. P. (1977) Sound absorption in sea water. *J. Acoust. Soc. Am.*, **62**, 558.

Franz, G. J. (1959) Splashes as sources of sound in liquids. *J. Acoust. Soc. Am.* **31**, 1080–1104.

Fry, F. J., Kossoff, G., Eggleton, R. C. and Dunn, F. (1970) Threshold ultrasonic dosages for structural changes in the mammalian brain. *J. Acoust. Soc. Am.* **48**, 1413–17.

Jones, A. T. (1920) The sounds of splashes, *Science*, 295–6.

Kinsler, L. E., Frey, A. R., Coppens, A. B. and Sanders, J. V. (1982) *Fundamentals of Acoustics*, John Wiley and Sons, New York, Ch. 15.

Knudsen, V. O., Alforf, R. S. and Emling, J. W. (1948) Underwater ambient noise. *J. Marine Res.*, **7**, 410–429.

Langmuir, I. (1938) Surface water motion induced by wind. *Science*, **87**, 119–123.

Laville, F., Abbott, G. D. and Miller, M. J. (1991) Underwater sound generation by rainfall. *J. Acoust. Soc. Am.* **89**, 715–721.

Leighton, T. G. (1994) *The Acoustic Bubble*, Academic Press, London.

Leighton, T. G. (1995) Bubble population phenomena in acoustic cavitation. *Ultrasonics Sonochemistry*, **2**, S123–S136.

Leighton, T. G., Schneider, M. F. and White, P. R. (1995) Study of bubble fragmentation using optical and acoustical techniques. In M. J. Buckingham and J. R. Potter (eds) *Proceedings of 3rd Meeting on the Natural Physical Processes Related to Sea Surface Sound*, World Scientific Publishing Ltd, Singapore, 414–28.

Leroy, C. C. (1969) Development of simple equations for accurate and more realistic calculation of the speed of sound in sea water. *J. Acoust. Soc. Am.* **46**, 216.

Lizzi, F. L., Driller, J., Lunzer, B., Kalisz, A. and Coleman, D. J. (1992) Computer-model of ultrasonic hyperthermia and ablation for ocular tumors using B-mode data. *Ultrasound Med. Biol.*, **18**, 59–73.

Longuet-Higgins, M. S. (1990) Bubble noise spectra. *J. Acoust. Soc. Am.* **87**, 652–61.

Longuet-Higgins, M. S. and Turner, J. S. (1974) An 'entraining plume' model of a spilling breaker. *J. Fluid Mech.* **63**, 1–20.

Lyman, J. and Fleming, R. H. (1940) *J. Mar. Res.* **3**, 134–46.

Mackenzie, K. V. (1981) Nine-term equation for sound speed in the oceans. *J. Acoust. Soc. Am.* **70**, 807.

Medwin, H. (1975) Speed of sound in water for realistic parameters. *J. Acoust. Soc. Am.*, **58**, 1318.

Medwin, H. Kurgan, A. and Nystuen, J. A. (1990) Impact and bubble sound from raindrops at normal and oblique incidence. *J. Acoust. Soc. Am.* **88**, 413–8.

Mikhalevsky, P. N. and Baggeroer, A. B. (1995) Experiment tests use of acoustics to monitor temperature and ice in Arctic Ocean. *Eos, Trans. Am. Geophys. U.*, **76**, 265.

Miller, D. L. and Williams, A. R. (1989) Bubble cycling as the explanation of the promotion of ultrasonic cavitation in a rotating tube exposure system. *Ultrasound Med. Biol.* **15**, 641–64.

Miller, D. L., Nyborg, W. L. and Whitcomb, C. C. (1979) Platelet aggregation induced by ultrasound under specialized condition in vitro. *Science*, **205**, 505.

Minnaert, M. (1933) On musical air-bubbles and sounds of running water, *Phil. Mag.*, **16**, 235–48.

Monahan, E. C. and Lu, N. Q. (1990) Acoustically relevant bubble assemblages and their dependence on meteorological parameters. *IEEE J. Oceanic. Eng.*, **15**, 340–5.

Munk, W. and Baggeroer, A. B. (1994) The Heard Island Papers: a contribution to global acoustics. *J. Acoust. Soc. Am.*, **96**, 2327.

Nystuen, J. A. (1986) Rainfall measurements using underwater ambient noise. *J. Acoust. Soc. Am.* **79**, 972–982.

Oğuz, H. N. and Prosperetti, A. (1990) Bubble entrainment by the impact of drops on liquid surfaces. *J. Fluid Mech.*, **219**, 143–179.

Premus, V. and Spiesberger, J. L. (1997) Can acoustic multipath explain finback (*B. physalus*) 20-Hz doublets in shallow water? *J. Acoust. Soc. Am.*, **101**, 1127–1138.

Prosperetti, A. (1988) Bubble-related ambient noise in the ocean. *J. Acoust. Soc. Am.* **84**, 1042–54.

Pumphrey, H. C. (1989) Sources of ambient noise in the ocean: an experimental investigation. PhD Thesis, University of Mississippi.
Pumphrey, H. C. and Elmore, P. A. (1990) The entrainment of bubbles by drop impacts. *J. Fluid Mech.*, **220**, 539–567.
Pumphrey, H. C., Crum, L. A. and Bjørnø, L. (1989) Underwater sound produced by individual drop impacts and rainfall. *J. Acoust. Soc. Am.* **85**, 1518–26.
Richards, S. D., Heathershaw, A. D. and Thorne, P. D. (1996) The effect of suspended particulate matter on sound attenuation in seawater. *J. Acoust. Soc. Am.* **100**, 1447–50.
Robinson, A. R. (ed.) (1983) *Eddies in Marine Science*.Springer-Verlag, New York.
Ross, D. (1976) *Mechanisms of Underwater Noise*. Pergamon Press, New York.
Roy, R. A., Carey, W., Nicholas, M. and Crum, L. A. (1992) Low-frequency scattering from bubble clouds. *Proceedings of 14th International Congress on Acoustics*, Beijing, China.
Scrimger, J. A., Evans, D. J., McBean, G. A., Farmer, D. M. and Kerman, B. R. (1987) Underwater noise due to rain, hail and snow. *J. Acoust. Soc. Am.* **81**, 79–86.
Starrit, H. C., Duck, F. A. and Humphrey, V. F. (1989) An experimental investigation of streaming in pulsed diagnostic ultrasound beams. *Ultrasound Med. Biol.*, **15**, 363–73.
ter Haar, G., Dyson, M. and Oakley, E. M. (1987) The use of ultrasound by physiotherapists in Britain, 1985. *Ultrasound Med. Biol.*, **13**, 659–663.
ter Haar, G., Rivens, I., Chen, L. and Riddler, S. (1991) High intensity focused ultrasound for the treatment of rat tumours. *Phys. Med. Biol.*, **36**, 1495–1501.
Thorpe, S. A. (1982) On the clouds of bubbles formed by breaking wind-waves in deep water, and their role in air-sea gas transfer. *Philos. Trans. R. Soc.* London, **A 304**, 155–210.
Thorpe, S. A. (1984) The effect of Langmuir circulation on the distribution of submerged bubbles caused by breaking wind waves. *J. Fluid. Mech.*, **142**, 151–70.
Thorpe, S. A. (1986) Measurements with an Automatically Recording Inverted Echo Sounder; ARIES and the Bubble Clouds. *J. Phys. Oceanography*, **16**, 1462–78.
Thorpe, S. A. (1992) Bubble clouds and the dynamics of the upper ocean. *Q. J. Meteorol. Soc.* **118**, 1–22.
Thorpe, S. A. and Hall AJ. (1980) The mixing layer of Loch Ness. *J. Fluid Mech.*, **101**, 687–703.
Thorpe, S. A. and Hall AJ. (1983) The characteristics of breaking waves, bubble clouds and near-surface currents observed using side-scan sonar. *Continental Shelf Res.*, **1**, 353–384.
Tolstoy, I. and Clay, C.S. (1966) *Ocean Acoustics*, McGraw-Hill, New York.
Urick, R. J. (1983) *Principles of Underwater Sound*, McGraw-Hill, New York.
Vivino, A. A., Boraker, D. K., Miller, D. and Nyborg, W. (1985) Stable cavitation at low ultrasonic intensities induces cell death and inhibits ^{3}H-TdR incorporation by con-a-stimulated murine lymphocytes *in vitro*. *Ultrasound Med. Biol.*, **11**, 751–9.
Webb, D. J., Beckles, D., de Cuevas, B. A. *et al.* (1991) An eddy-resolving model of the Southern Ocean. *Transactions of the American Geophysical Union*, **72**, 169–75.
Westervelt, P. J. (1960) Parametric end-fire array. *J. Acoust.Soc. Am.*, **32**, 934–5.
WFUMB (1992) Symposia for Safety and Standardization in Medical Ultrasound: Issues and recommendations regarding thermal mechanisms for biological effects of ultrasound. In S. B. Barnett and G. Kossof (eds), *Ultrasound Medi. Biol.*, **18**(9) (Special issue).
WFUMB (1994, 1996) *Symposia for Safety of Ultrasound in Medicine: Emphasis on non-thermal effects. World Federation for Ultrasound in Medicine and Biology*

(*draft document*), Barnett, S. (ed) (Utsunomiya, Japan, 1994; Koster Banz, Germany, 1996).

Williams, A. R. (1990) Editorial to Bioeffects special issue. *Ultrasonics*, **28**, 131.

Wunsch, C. (1981). Low frequency variability of the sea. In B. A. Warren and C. Wunsch (eds), *Evolution of Physical Oceanography*, Cambridge, MIT Press, Mass, 342–77.

Yamada, H. (1970) *Strength of Biological Materials* (F. G. Evans, ed.). Waverley Press, Inc. Baltimore.

Ziskin, M. C. (1987) The prudent use of diagnostic ultrasound. *J. Ultrasound Med.*, **6**, 415.

7.6 QUESTIONS

1 (a) A source of sound transmits into the ocean, where I_1 and c_1 are the acoustic intensity and sound speed (respectively) at a distance of 1 metre from the source; and where I_2 and c_2 are the acoustic intensity and sound speed (respectively) at a distant point X. As a result of ray divergence, there is a certain transmission loss (*TL*), defined as $TL = 10\log_{10}(I_1/I_2)$, between the source and point X. By considering two rays, which leave the source at angles of $\theta_1 + \delta\theta/2$ and $\theta_1 - \delta\theta/2$, find an expression for the transmission loss in terms of $\delta\theta$, c_1 and c_2; the length (δh) of the vertical line which extends between the two rays and intersects point X; and the horizontal distance (r) between the source and point X. Assume no absorption of the acoustic energy by the sea.

(b) Use Huygens' principle to demonstrate the formation of sound channels at *both* the sea surface *and* at about 1 km depth (the SOFAR channel), indicating the local sound speed profiles that you assume in both cases. State two physical mechanisms by which a surface channel might form, and discuss the relationship between the angle a ray in the SOFAR channel makes with the axis when they intersect, and the travel time taken for a pulse using that ray path to propagate to a long distance. Use your answer to part (a) of this question to demonstrate that spreading losses over large distances (e.g. of order 1000 km) would be less in a sound channel than for spherical spreading.

(c) Describe how this principle has been exploited in climate monitoring, including a detailed discussion of the initial feasibility study *and* brief outlines of two subsequent experiments.

2. (a) Discuss the production of underwater sound by rain, describing both the contributions from both bubble and non-bubble sources.

(b) Consider a cubic meter of bubbly water, containing $V_a = 100$ ml $= 10^{-4}$ m^3 of gas as bubbles, so that volume of water is $V_w = 1 - 10^{-4} = 0.9999$ m^3. Compare the quasi-static compression of this water with that of 1 cubic metre of bubble-free water. Hence estimate the change in sound speed which the introduction of such a bubble population would induce, were the acoustic frequency so much less than the bubble resonance as to impose quasi-static conditions.

(c) Strong acoustic backscatter is observed when 400 Hz sonar beams, angled a few degrees from the horizontal, are deployed to examine the sea surface under stormy conditions. Discuss scatter from surface waves, and from bubbles, in this context.

3. (a) Euler's equation in one spatial dimension (x) relates the total pressure (P) to the particle velocity (u) and the liquid density (ρ), as a function of time. Combine Euler's equation with the equation of continuity and the wavespeed equation (all three of which can be quoted without proof) to demonstrate, with full working, how the one-dimensional wave equation may be derived. State clearly and precisely all assumptions, and comment on whether each assumption is justified.

(b) Use a power-series approximation of the nonlinearity to demonstrate the generation of harmonics during the propagation of an initially sinusoidal acoustic wave in water. Using sketches of the waveforms and associated spectra at various positions from the transducer, discuss how the energy is distributed with respect to frequency, the mechanisms which change this distribution, and the effects on the waveform.

(c) Use the same power-series approximation of the nonlinearity associated with propagation to mathematically describe the principle of operation of parametric sonar. Describe the characteristics and advantages of parametric transmitters and receivers.

(d) Write the expression relating the phase velocity, the particle velocity (u), and the speed of sound for waves of infinitesimal amplitude (c_0), for an isentropic gas and for a liquid, introducing a constant based on the nonlinearity parameter, $B/2A$, of the liquid, and (for the gas) the equivalent measure of nonlinearity, $(\gamma - 1)/2$. Here γ represents the ratio of the specific heat of the gas at constant pressure, to that at constant volume. Since $B/2A = 2.5$ for water, whereas, for air, $(\gamma - 1)/2 = 0.2$, one might therefore expect to see the effects of nonlinear propagation in water more readily than in air. However, often the converse is true. Why is this?

4. (a) Describe how the use of diagnostic ultrasound over the last 30 years in the UK has been influenced by concern over the potential for hyperthermic bioeffect, indicating the limitations of laboratory tests of such an effect.

(b) Describe one medical application which exploits ultrasonic cavitation in humans *in vivo*, including specification of the typical acoustic parameters and (by sketching) pulse profile. Indicate any biohazard concerns associated with this procedure.

(c) Sketch a graph of the inertial cavitation threshold, using realistic values for the acoustic pressure amplitude axis and the initial bubble radius axis, for acoustic frequencies of 1 MHz, 5 MHz and 10 MHz. Describe why, above the threshold, there exists only a limited range of initial bubble sizes which can nucleate inertial cavitation. Extrapolate from this description to explain the 'anomalous depth effect' observed when submariners attempt to suppress propeller cavitation by diving the submarine to greater depths.

8

Fundamental principles of measurement and analysis techniques

R.J. PINNINGTON

8.1 INTRODUCTION

Before describing the techniques and limitations of experimental methods it is wise to reflect upon the purpose of measurement. The broad aim of scientific enquiry is the understanding of nature, and probably the first persons with sufficient leisure to pursue this extensively lived in towns in the region of the Aegean sea around 600 BC (Russell, 1991). These people made the initial steps in understanding by proposing ideals or theoretical models, simplified by assumptions, to explain their casual observations of the physical world. For a large part, these theories were not put to the test of careful experiment, for there was a widespread belief that theory rested in the perfection of heaven while to handle lowly matter was to consort with evil (Russell, 1991). This view still has its sympathizers.

It was not until the Renaissance period that scientific enquiry was supported by careful experiments leading to what is known as the scientific method. This involves an initial observation and the development of a theory which is tested by an experiment in which the chosen phenomenon is isolated from other contributions; the outcome disproves, confirms or leads to modification to the theory. In summary, theory and experiment are needed as much as a walking man requires two legs.

A theory permits a single phenomenon to be viewed in isolation from all other influences. These influences may be dismissed by the assumptions. The reality of the world, however, is that there is always interaction between everything, it and us are a whole; we are related. To be slightly disturbing, Berkeley (Russell, 1991) thought that an event may not even occur unless perceived by the observer, i.e. the event needs the observer. Implicit in the quantum mechanics view of the universe adopted in the twentieth century is the notion that the presence of the observer will always interfere with the outcome of an event. The objective then of a good experiment is to be sufficiently present for the event to occur, while interfering as little as possible.

Experimental design requires in the first instance a knowledge of the theory to be tested; the second requirement is that the *assumptions* of the theory are enforced in reality, which is only possible if the physics behind all the possible undesirable influences is understood. There are two sorts of influence on the ideal system seen in Fig. 8.1 which may be divided between the predictable and the unknown. The predictable influence is the *interaction* between the test equipment and the test piece, its effects can be mathematically subtracted or minimized by careful design. In this chapter the interaction between the mechanical and electrical measurement system with a test piece is analysed to find the conditions for minimum interference. The second sort of influence is the unknown, which is the unwanted contributions from the rest of the world to the experiment, this is called *noise*. Noise will always occur because we live in a real world; however, understanding of the possible mechanisms can assist in its reduction.

The chapter is divided into four parts: ideal models; the mechanical and electrical interaction between the measurement system and the test piece; the limitations of signal processing; and the mechanisms of noise. Some suggestions for successful dynamic testing are given in summary.

8.2 IDEAL TRANSFER FUNCTIONS

Figure 8.2 shows the essence of a mechanical system in which the components are a time-varying input force $F(t)$, response $x(t)$, and system

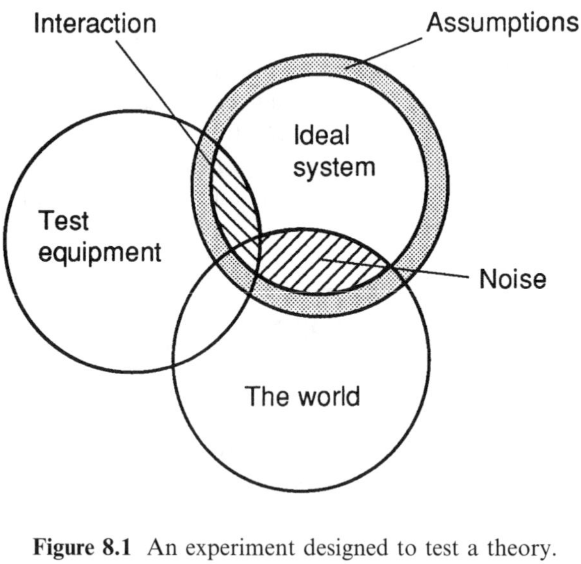

Figure 8.1 An experiment designed to test a theory.

transfer function H. For example, according to Newton's Second Law of motion, the transfer function 'mass' relates the acceleration to the force. In general, the transfer function relating the input and response may be a function of amplitude and time, but for the basic theory presented here certain assumptions are made which permit the time and amplitude dependence to be ignored. The transfer function is made amplitude invariant by assuming *linearity*, i.e. if the force $F(t)$ is multiplied by a factor n, the response $x(t)$ is also multiplied by n. This assumption is the basis of all linear analysis in which the combined response from multiple inputs is simply the addition of the responses to the individual inputs.

Linearity also implies that transfer functions can be made time independent by introducing the concept of *frequency* f (cycles/s) or ω (radians/s), where $\omega = 2\pi f$. Harmonic (single frequency) time variation may be represented by the combination of a phasor $e^{j\omega t}$ (a unit vector rotating at ω radians/s in the positive direction), and a phasor $e^{-j\omega t}$ rotating in the negative direction. The more familiar $\sin\omega t$ and $\cos\omega t$ functions may be constructed from these fundamental elements. Mathematical analysis using transfer functions is more easily conducted using phasors, although the physical sense is suspended until the final result. A phasor exists for all time, for which reason transfer functions are expressed in terms of frequency, making them *time invariant*.

The transfer function in Fig. 8.2 can now be represented as a frequency response function (FRF) by assuming linearity and applying a harmonic force, $Fe^{j\omega t}$. If the system is linear, the response $x(t) = xe^{i(\omega t+\phi)}$ occurs only at the same frequency. In fact, this represents the primary test for system linearity as the degree of non linearity can be quantified from the amplitude of the harmonics of the excitation frequency ω which appear in the response, i.e. at $2\,\omega$, $3\,\omega$, etc.The FRF H of a linear system is therefore the ratio of the response to the force at a single frequency, as seen in Fig. 8.3, i.e.

$$xe^{j(\omega t+\phi)} = HFe^{j\omega t}. \tag{8.1}$$

Linear superposition implies that the response at each frequency is independent of that at other frequencies, so any signal in the time domain $x(t)$ can be decomposed into a set of single frequency components of amplitude $X(f)$. The mathematical expression of this decomposition is the

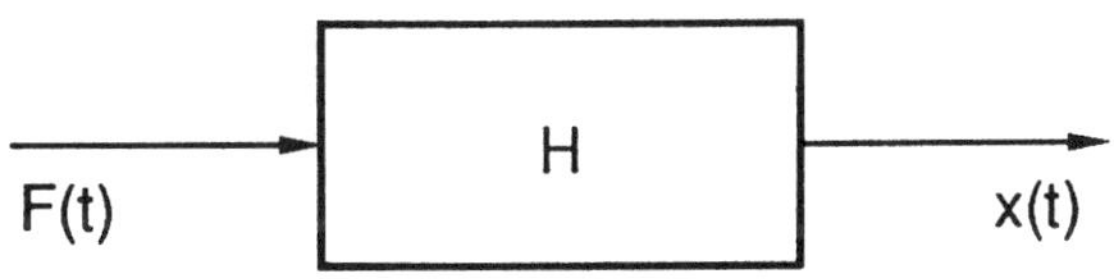

Figure 8.2 Transfer function.

Fourier integral transform, which involves an integral of the signal multiplied by the single frequency function $e^{-j2\pi ft}$ over all time (see Chapter 6), i.e.

$$X(f) = \int_{-\infty}^{\infty} x(t)e^{-j2\pi ft}\,dt. \qquad (8.2)$$

This procedure filters out the response $X(f)$ at each frequency.

If the force signal $F(t)$ is similarly processed to give the force in the frequency domain $F(f)$, the FRF appears in the same form as equation (8.1), i.e.

$$X(f) = H(f)F(f). \qquad (8.3)$$

Division at each frequency of the response by the force yields the frequency dependent transfer function

$$X(f)/F(f) = H(f). \qquad (8.4)$$

This division in the frequency domain is the underlying principle behind all frequency response testing irrespective of whether the signals are transient or continuous random time signals.

The FRF is a complex function with two components, modulus and phase, thus

$$H = |H|e^{j\phi}. \qquad (8.5)$$

The modulus describes the amplitude and the phase ϕ expresses the lead time of the response with respect to the force. Another way of representing the same transfer function uses the real (or in phase) component $\mathrm{Re}\{H\}$ and the imaginary (or quadrature) component $\mathrm{Im}\{H\}$, thus:

$$H = \mathrm{Re}\{H\} + j\,\mathrm{Im}\{H\}, \qquad (8.6)$$

where $\mathrm{Re}\{H\} = |H|\cos\phi$, $\mathrm{Im}\{H\} = |H|\sin\phi$, and $\tan\phi = \mathrm{Im}\{H\}/\mathrm{Re}\{H\}$.

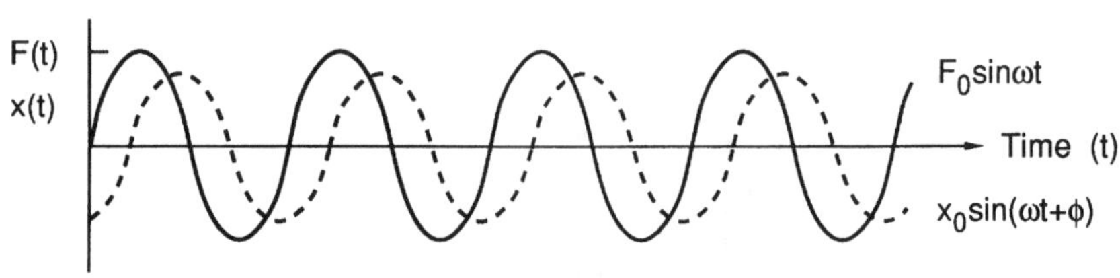

Figure 8.3 Input and output for a harmonic input to a linear system.

8.2.1 Transfer functions: the single degree-of-freedom system

The single degree-of-freedom (SDOF) system seen in Figure 8.4 is the fundamental oscillatory system consisting of a discrete mass M, a massless spring of stiffness K, and a viscous dashpot D. The associated FRF is the simplest example that embodies the essential physical principles of a vibratory system. The FRFs are displayed in three alternative ways to illustrate the controlling mechanism in each frequency region and the principles of stability and causality. This knowledge can then be used to assess the physical plausibility of any measured or predicted FRF.

The applied force $Fe^{j\omega t}$ and the resulting velocity $ue^{j\omega t}$ are related at any frequency by a particular FRF termed impedance Z where

$$Z = \frac{F}{u} = \frac{1}{j\omega}\left(K - \omega^2 M + j\omega D\right). \qquad (8.7)$$

The mobility $Y = 1/Z$ is plotted in the modulus and phase form of equation (8.5) in Fig. 8.4. The *log–log* presentation clearly describes the two frequency regimes: below the undamped natural frequency f_0, given by $f_0 = (K/M)^{\frac{1}{2}}/2\pi$, the response is controlled by the static stiffness, i.e. $|Y| \approx \omega/K$; above the natural frequency, the response is dominated by the mass, i.e. $|Y| \approx 1/\omega M$; while at resonance the stiffness and mass contributions in equation (8.7) cancel, leaving only damping to control the response, and $|Y| = 1/D$.

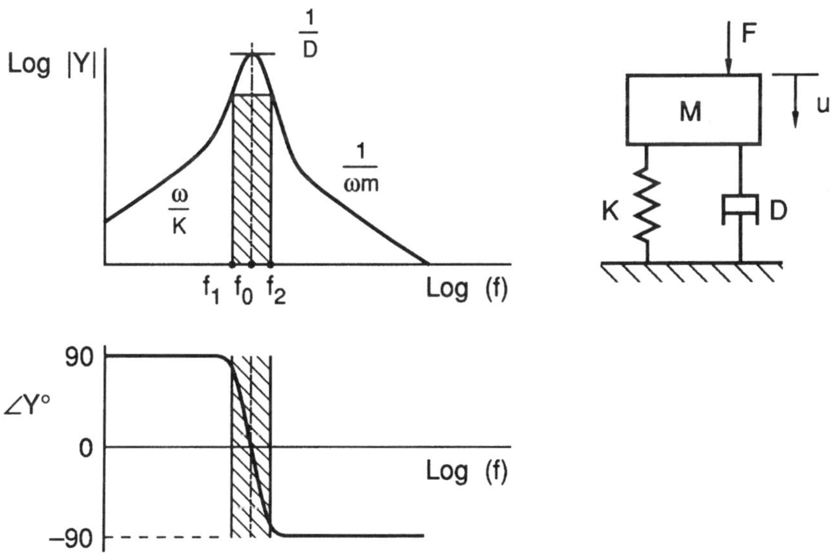

Figure 8.4 Modulus and phase of the mobility of a single degree-of-freedom system. ▨ Half-power bandwidth.

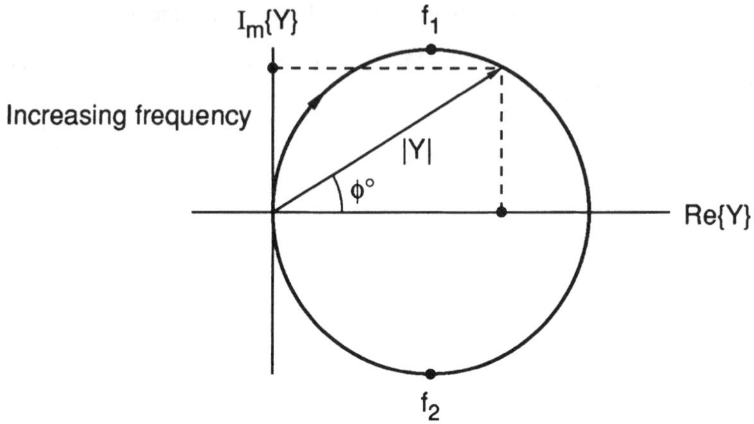

Figure 8.5 Vector diagram of a single degree-of-freedom system.

The modulus $|Y|$ and phase ϕ can also be plotted in the complex plane at each frequency to give a 'vector' or Nyquist diagram, as shown in Fig. 8.5. The vector plot of mobility gives a perfect circle with the maximum real component Re$\{Y\}$ at resonance, and equal real and imaginary components at f_1 and f_2, the *half power points*. The damping controls the response within this *half power bandwidth*, also shown in Fig. 8.4. The vector plot is a display of the resonant, damping controlled, region; it is therefore useful in the identification of modes and non linear behaviour (Ewins, 1984).

Figure 8.5 also shows the FRF as real Re$\{Y\}$ and imaginary mobilities Im$\{Y\}$, as described by equation (8.6). The mobility in this form plotted for both positive and negative frequencies from equation (8.4) is shown in Fig. 8.6.

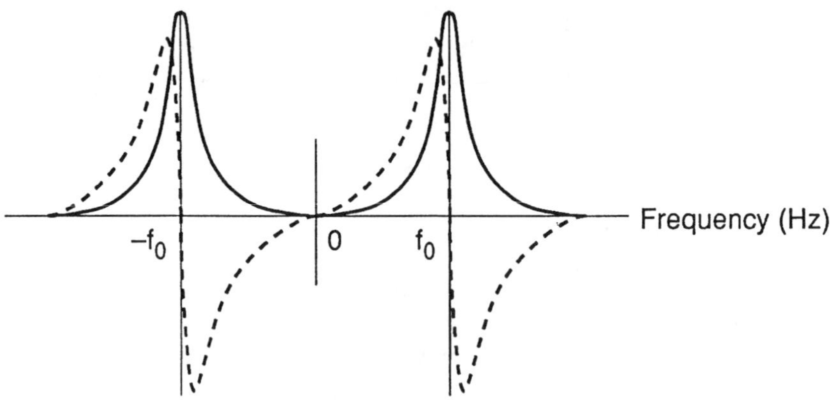

Figure 8.6 Mobility of a single degree-of-freedom system. — Re$\{Y\}$, - - - - Im$\{Y\}$.

The real component is the component of the velocity which is in phase with the force; therefore a positive value describes the power absorbed by the system. For a stable system (for which the response to an impulse decays with time, as in Fig. 8.7), the real component of mobility is therefore always positive. This is an excellent test of measured data quality because satisfaction implies uncontaminated data and a stable system. The imaginary components of mobility are associated with the stiffness below resonance and the mass above resonance; in accordance with equation (8.7), it goes through zero at the natural frequency as it changes sign.

Figure 8.6 shows the full double-sided FRF which has both positive and negative frequencies as derived directly from the Fourier transform in equation (8.2). The usual convention is to show the FRF only for positive frequencies, as for example in Fig. 8.4, as the values for negative frequencies are almost a mirror image. If, however, the intention is to transform the FRF back into the time domain to provide the *impulse response function* $h(t)$ (i.e. the response to a unit force impulse applied at $t = 0$), the double-sided FRF is required, i.e.

$$h(t) = \frac{1}{2\pi} \int_{-\infty}^{\infty} H(\omega) e^{j\omega t} \, d\omega, \tag{8.8}$$

or an equivalent single-sided integral including both $H(f)$ and $H(-f)$, i.e.

$$h(t) = \int_{0}^{\infty} \left(H(f) e^{j\omega t} d\omega + H(-f) e^{-j\omega t} \right) d\omega. \tag{8.9}$$

The relationship between $H(f)$ and $H(-f)$ can be found by observing that the impulse response function must be both real and *causal* (i.e. the response is only for $t > 0$ as it must follow the impulse at $t = 0$). Application of these two criteria to equation (8.9) gives the relationship $H^*(f) = H(-f)$ (* denotes complex conjugate) for causal systems, which is displayed in Fig. 8.6.

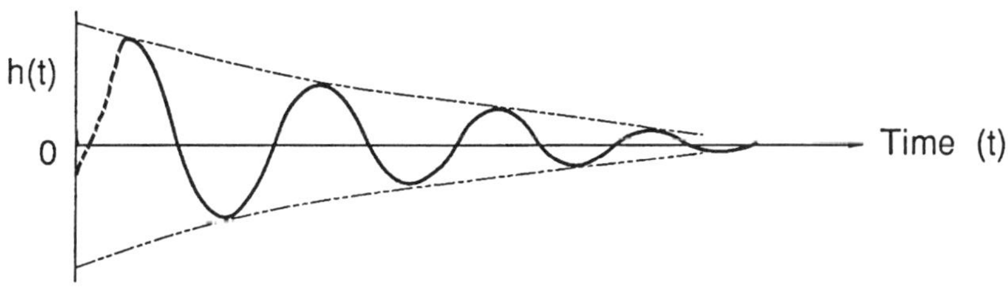

Figure 8.7 Impulse response of a single degree-of-freedom system. — viscous damping, - - - - hysteretic damping.

Some theoretical transfer functions do not obey this relationship and so will produce an invalid acausal time response. In particular, this is the case for any function that uses the common hysteretically-damped spring to represent a spring and a viscous dashpot, i.e. $K + j\eta K \equiv K + j\omega D$. Although problematic in the time domain, the hysteretic damping model is very useful for frequency domain analysis.

8.2.2 Transfer functions: finite continuous system

The second ideal element is a finite continuous system (Cremer *et al.*, 1973; Pinnington, 1988) such as a rod, beam, plate or acoustic cavity. The simplest form is the rod shown in Fig. 8.8(a) where a force F_1 at point x_1 causes a velocity u_2 at any position x_2. The response is the summation of right and left propagating travelling waves, generated at the driving point and reflected back at the boundaries. For light-damped systems, the waves in each direction have almost the same amplitude and so interfere to give standing waves at every frequency. The response u_2 due to a force F_1, and

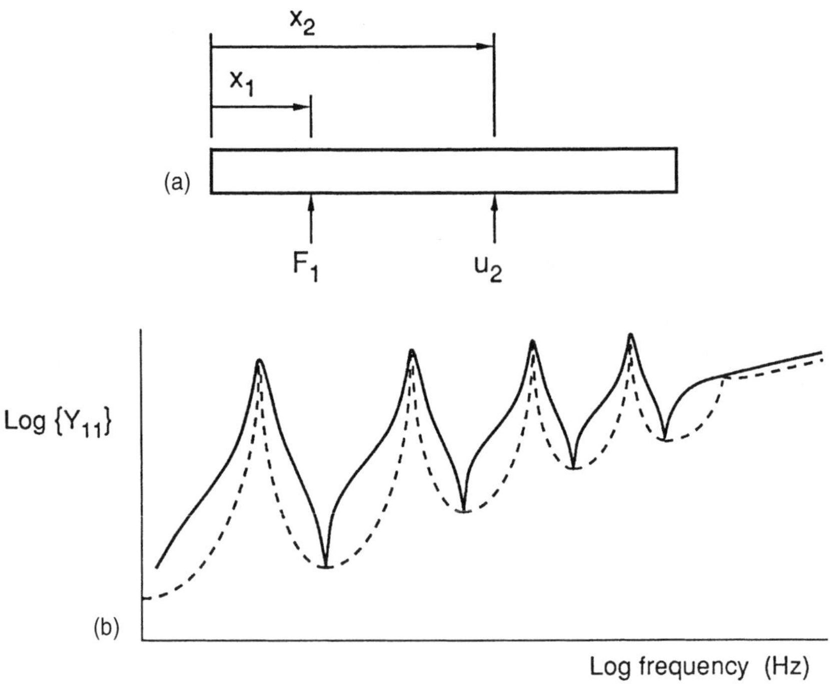

Figure 8.8 (a) Finite beam. (b) Input mobility (u_1/F_1) of a finite system. — modulus, - - - - real component.

the force F_1 are related by the *transfer mobility*, Y_{12} This can be shown, mathematically speaking, to be summed from an independent set of SDOFs, or modes, which for hysteretic damping is

$$Y_{12} = \frac{u_2}{F_1} = \sum_r \frac{j\omega q_r(x_1)q_r x_2}{K_r(1 + j\eta_r) + \omega^2 M_r}. \tag{8.10}$$

The peaks in the modulus associated with the resonance of each mode are clear in Fig. 8.8(b). For each mode r there is a modal mass M_r, loss factor η_r, stiffness K_r and mode shape (or eigenvector) $q_r(x)$. At a resonance frequency f_r, the response of one mode r becomes very large and the observed standing wave pattern corresponds closely to the associated mode shape $q_r(x)$. *Modal analysis* is a curve-fitting procedure (Ewins, 1984) to fit equation (8.10) to measured data.

When the response position and direction is the same as that of the force F_1, the *input mobility* Y_{11} is obtained. The real component of mobility of a SDOF is always positive, therefore from equation (8.10) it follows that for the finite continuous system the same is true, i.e. $\text{Re}\{Y_{11}\} > 0$. To repeat the earlier statement, this is an excellent check of data quality where measured data should take the ideal form seen in Fig. 8.8(b). In practice, there are often negative zones due to data corruption, particularly in the regions of minima where the true signal becomes very small.

At high frequencies, damping tends to attenuate the travelling waves so much that the response becomes dominated by the outgoing wave from the driving point x_1, and standing waves do not occur. As there are only outgoing waves, the system behaves as if it is of infinite extent. The input mobility for a uniform system is a smooth power law function of the general form (Goyder, 1978),

$$Y_{11} = A\omega^s e^{js\pi/2} \qquad -1 < s < 1 \tag{8.11}$$

in which A is a constant. For a stiffness $s = 1$; for a dissipative system $s = 0$; and for a mass $s = -1$. The literature gives these formula for many forms of infinite system (Pinnington, 1988).

A final observation concerning ideal transfer functions concerns the *principle of reciprocity* (ten Wolde *et al.*, 1975) which relates *transfer* FRFs in *any linear system*. Transfer FRFs define the ratio of the response at one point to the force at another point, as for Y_{12} in equation (8.10), where the force is applied at position 1 and the response measured at position 2. If the response and force positions are exchanged, giving Y_{21}, it is seen from the equation that the transfer FRF is identical, i.e. $Y_{12} = Y_{21}$. This reciprocity principle will be satisfied by good-quality data (Ewins, 1984). Reciprocity can also be used for quantifying transmission paths by removing the excitation, exciting the inactive system at the receiver position and monitoring the response at the original excitation positions (ten Wolde *et al.*, 1975).

8.2.3 Ideal sensors and actuators

Frequency response measurement requires external *excitation* while the applied force and the response must be monitored with *transducers* to transform the mechanical response quantities into an electrical voltage suitable for processing or display. A combined electro-mechanical system to perform these operations is shown in Fig. 8.9. Each unit has a system transfer function controlled by its passive mechanical properties (mass M, stiffness K and damping D) or electrical properties (inductance L, capacitance C, resistance R).

The exciter, transducer and amplifier in Fig. 8.9 have transfer functions H_e, H_t, and H_a, relating the output to the input; e.g. the applied force F is proportional to the signal voltage V_e:

$$F = H_e V_e. \tag{8.12}$$

The transducer voltage output V_t is linearly related to the vibrational quantity u by the *sensitivity* H_t, i.e.

$$V_t = uH_t. \tag{8.13}$$

An ideal transducer is not sensitive to any other quantity or motion in any other direction. However, a practical transducer has undesired sensitivities to many quantities which are termed *cross-sensitivities*. For example, a typical accelerometer for axial acceleration has a cross-sensitivity to lateral motion at about 3% of the axial sensitivity. These cross-sensitivities tend to represent the main limit to measurement accuracy as it is very difficult to manufacture transducers with lower cross-sensitivities from the example given.

The output voltage V_0 is related by a frequency independent amplitude factor from the signal conditioning H_a, i.e.

$$V_0 = uH_tH_a \tag{8.14}$$

The ideal form of transfer functions H_e, H_t and H_a, is a *constant magnitude* and *zero phase* at all frequencies, as shown in Fig. 8.10. These ideal relations, given in equation (8.14), are usually achieved in practice by careful attention to the electrical impedances as is discussed in section 8.3.7. Many commercial transducers and amplifiers also have frequency independent

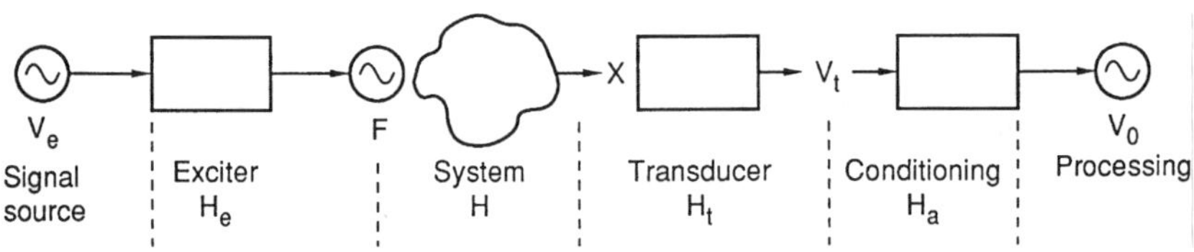

Figure 8.9 Transfer function measurement with ideal exciter and transducer.

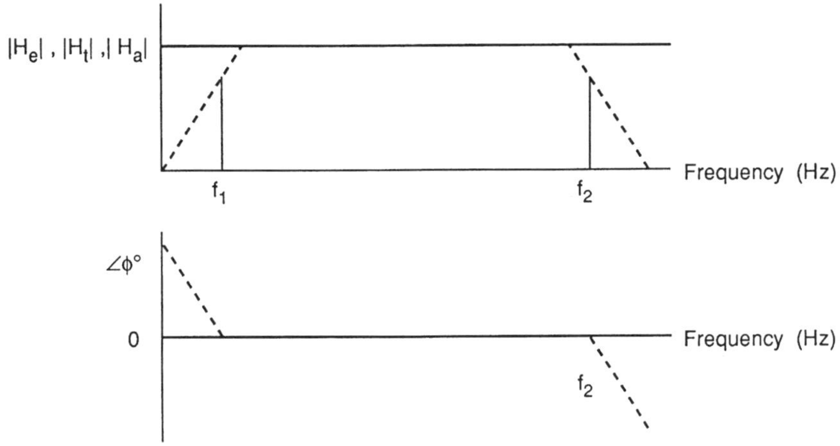

Figure 8.10 — Ideal sensor and exciter transfer functions, - - - - Real transfer functions.

FRF's, H_{a}, H_{t}, described by a single constant which permits easy calibration of the transducers.

Example of transducer calibration:

To recover the true acceleration 'a' ms^{-2} from a voltage V_0 input to an analyser, a variation of equation (8.14) is used:

$$V_0 = aH_{\mathrm{t}}H_{\mathrm{a}}.$$

If the sensitivity of an accelerometer $H_t = 2.1\,\rho\mathrm{C/ms}^{-2}$ and the signal conditioning sensitivity $H_a = 10\ \mathrm{V}/\rho\mathrm{C}$, then

$$a = \frac{V_0}{2.1 \times 10}\ \mathrm{ms}^{-2}.$$

In practice (see section 8.4), the transfer functions have a limited bandwidth $f_1 - f_2$ because of internal impedance and noise effects. It is therefore wise before conducting a test to check the bandwidth on the equipment data sheets; the values of H_{t} and H_{a} are also supplied with the equipment.

8.3 INTERACTION BETWEEN THE EXCITATION, MECHANICAL SYSTEM AND SENSORS

The previous section described ideal behaviour where each component behaves as if it were independent of all other components to which it is coupled. The transfer function of two coupled components is simply the product of the two independent transfer functions (as in equation (8.14)). In reality, there is always interaction between components, giving a more complicated frequency dependent coupled response. This coupled response can, however, be calculated from the known properties of the individual systems, permitting a correct choice of system parameters to give almost the ideal behaviour.

To investigate this possibility, the measurement systems and test piece are here represented in terms of *four pole parameters* (Snowdon, 1971) in which each element has two inputs with a force and response at each input. The response at each input is related to the forces at both inputs by the element *impedances*. Using this representation and the principles of equilibrium of forces and continuity of response, the coupled response of inter-connected elements can be calculated.

If the measurement system is thus described, it would then seem to be possible, for the mathematically confident, to subtract its contribution from the coupled result to reveal the properties of the uncoupled test piece. Unfortunately, there are always unknown inputs known as *noise*, the contribution of which is shown to increase with increasing interaction between the measurement system and the test piece, diluting the benefit of such post-test processing. The best procedure then for measuring the test piece characteristics is to set up a test with minimum interaction as this will also minimize the noise corruption.

8.3.1 Electro-mechanical analogues

Implicit in the nature of vibration excitation or transduction is the notion of electro-mechanical interaction. For example, when a current I passes through a magnetic field of flux density $B\ell$, a force F is generated where

$$F = B\ell I. \tag{8.15}$$

This is the principle of the loudspeaker or electro-dynamic exciter which are shown in section in Figs 8.11(a) and (b). The two devices have the same principle of operation: a moving element consisting of a coil of wire on a paper former oscillates in the magnetic field in the gap of the permanent magnet. The moving element is located by a suspension system. The current I in the coil generates an output force F in the case of the electrodynamic exciter, while for the loudspeaker the force moves a paper cone, thereby radiating sound. There is also a reciprocal relationship: when the conductor moves at velocity u a voltage V is generated in the conductor as described by Lenz's Law, i.e.

$$V = -B\ell u. \tag{8.16}$$

If the force and velocity are in the same direction, the generated voltage is negative in relation to that which set up the current I in the previous force generation mode, and so it is termed a *back emf* (reverse electro-motive force). The negative sign relating equations (8.15) and (8.16) describes the rather unusual 'inverse reciprocity' relationship characterizing electro-dynamic systems. Electrical or mechanical systems or piezo electric elements

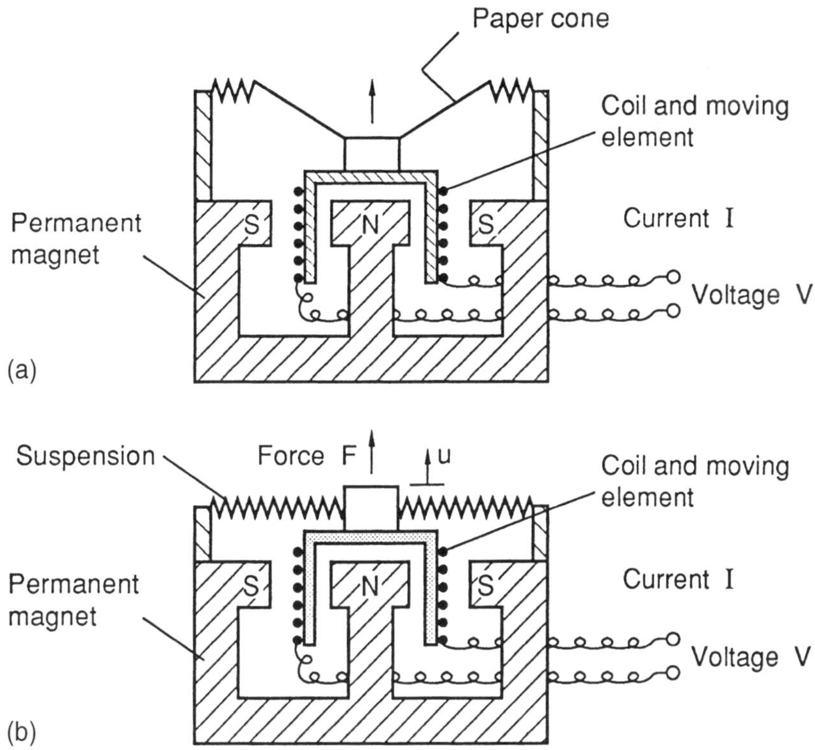

Figure 8.11 (a) Cross-section through an electrodynamic loudspeaker. (b) Cross-section through an electrodynamic vibration exciter.

satisfy the reciprocity relationship of equation (8.10), which does not involve the negative sign.

Instruments invoking a reciprocal principle can therefore be used either as transducers or as exciters; for example, the electrodynamic devices in Fig. 8.11 give an output voltage V in response to a moving element velocity u. The inherent disadvantage is that when they are in one primary mode of operation, e.g. excitation, as with equation (8.15), they are simultaneously acting as a transducer, as with equation (8.16), setting up an opposing voltage to the driving voltage. The converse situation occurs when the device is used as a transducer.

There is also mechanical interaction which derives from the mechanical impedance (which expresses the mass, stiffness and damping) of the transducer or actuator. For example, if a force source develops a maximum force F_0 when applied to an immoveable object, the same source applied to another object which accelerates at a applies a lower force F which is reduced from F_0 by the effort in accelerating its own internal mass m, i.e.

$$F = F_0 - ma. \tag{8.17}$$

A transducer of mass m will similarly load the structure with a force $-ma$, and so change the response it is supposed to measure.

The whole interaction problem involving electrical and mechanical elements can be modelled using the unifying electro-mechanical analysis. The principle is that 'force' and 'response' pairs can be used that have power, or one of its derivatives, as the product of the two quantities. The following pairs can be used: volts V, amperes I; acoustic pressure P, volume velocity Q; force F, velocity u. The electrodynamic system illustrated in Fig. 8.11 is an example, since voltage V and current I are applied at one input while force F and velocity u are applied at the other input.

8.3.2 Sources and elements

A real voltage source V_s has an internal impedance Z_s giving an *output* voltage V which is reduced by the current drawn I as shown in Fig. 8.12(a) with the usual convention. Alongside, in Fig. 8.12(b), is the equivalent convention used in this chapter for both electrical and mechanical impedances.

At the boundary between the source and receiver the voltage V_1 is equal and opposite and there is continuity of current, i.e.

$$I + I_1 = 0.$$

The voltage applied to the receiver is

$$V_1 = V_s - I_1 Z_s. \tag{8.18}$$

A similar representation can be used for force and velocity in mechanical systems.

A real force source F_s in Figs 8.12(c) and (d) has a mechanical impedance Z_{sm} giving an *output* force F reduced by the velocity u_1. Again, the forces are equal and opposite at the junction, while there is continuity of velocity, i.e. $u + u_1 = 0$.

The output force F_1 is

$$F_1 = F_s - u_1 Z_{sm}. \tag{8.19}$$

The basic electrical impedance elements (voltage/current), $j\omega L$, R, and $-j/\omega C$ are associated with inductance, resistance and capacitance. The mechanical impedance elements are similar, $j\omega M$, D, and $-jK/\omega$, expressing mass, damping and stiffness, respectively. These are shown in log–log form in Fig. 8.13(a), (b).

8.3.3 Four pole parameters

Impedance elements can be coupled so that they have two or more junctions. At each junction, there is 'force' and response; force–velocity for

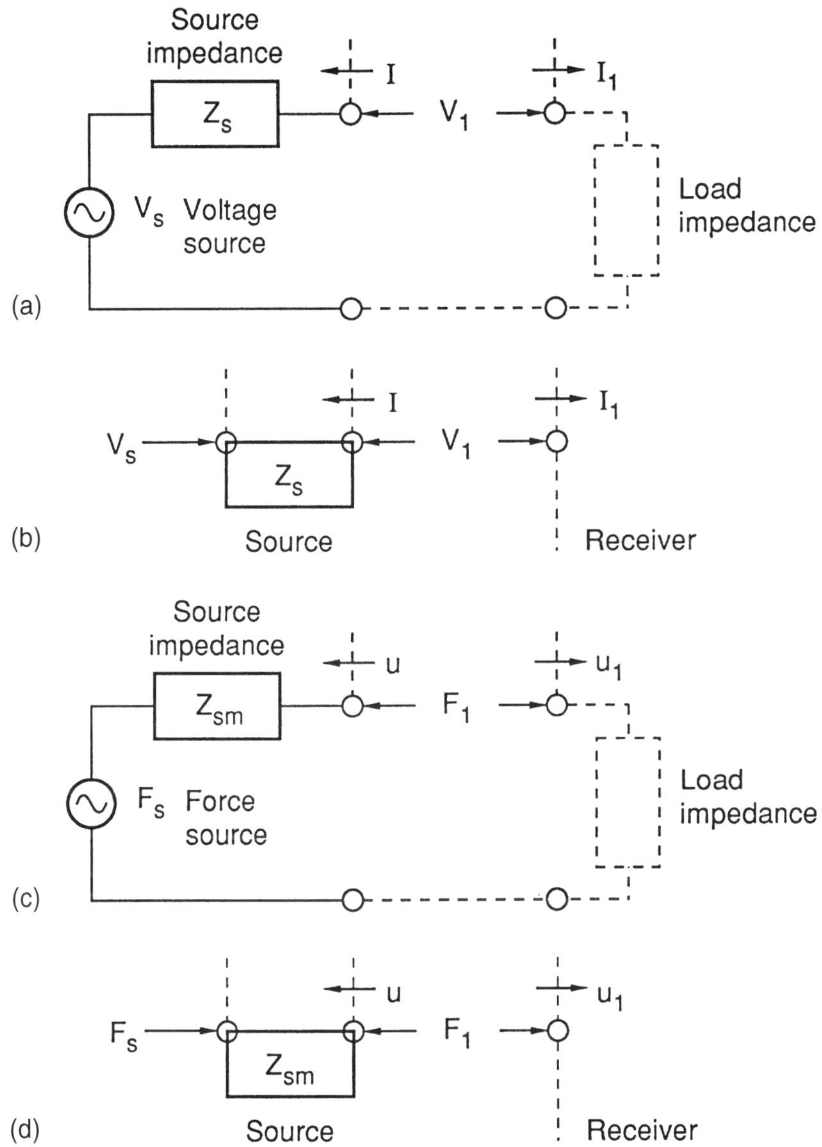

Figure 8.12 (a) Voltage source and coupling, usual convention. (b) Voltage source and coupling conditions, impedance convention. (c) Force source and coupling, electrical analogue. (d) Force source and coupling conditions, impedance convention.

mechanical circuits, voltage–current for electrical circuits. There are four possibilities shown in Fig. 8.14(a)–(d).

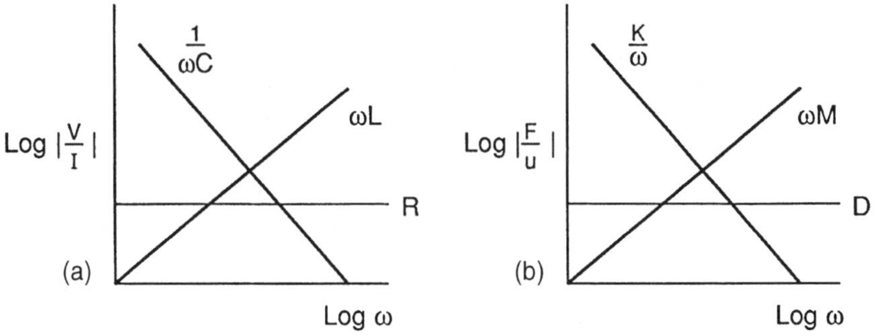

Figure 8.13 (a) Impedance elements: electrical. (b) Impedance elements: mechanical.

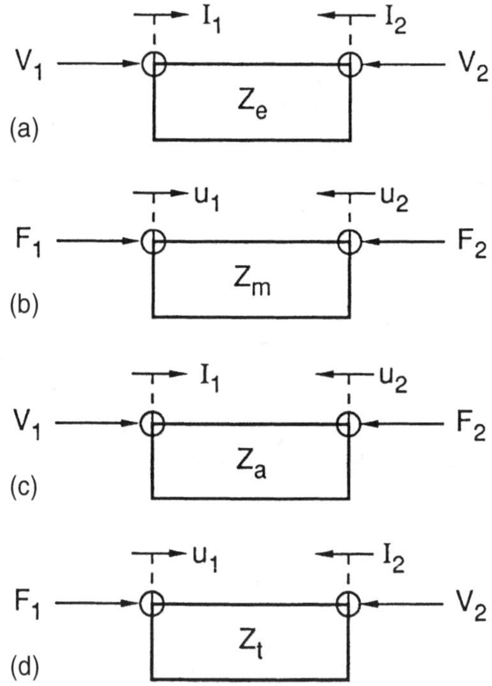

Figure 8.14 (a) Impedance circuits and matrix form: electrical. (b) Impedance circuits and matrix form: mechanical. (c) Impedance circuits and matrix form: electro-mechanical actuators. (d) Impedance circuits and matrix form: electro-mechanical transducers.

$$\begin{Bmatrix} V_1 \\ V_2 \end{Bmatrix} = [Z_e]\begin{Bmatrix} I_1 \\ I_2 \end{Bmatrix}, \tag{8.20}$$

$$\begin{Bmatrix} F_1 \\ F_2 \end{Bmatrix} = [Z_m]\begin{Bmatrix} u_1 \\ u_2 \end{Bmatrix}, \tag{8.21}$$

$$\begin{Bmatrix} V_1 \\ F_2 \end{Bmatrix} = [Z_a]\begin{Bmatrix} I_1 \\ u_2 \end{Bmatrix}, \tag{8.22}$$

and

$$\begin{Bmatrix} F_1 \\ V_2 \end{Bmatrix} = [Z_t]\begin{Bmatrix} u_1 \\ I_2 \end{Bmatrix}, \tag{8.23}$$

where $[Z_e]$, $[Z_m]$, $[Z_a]$, $[Z_t] = [Z]$:

$$[Z] = \begin{bmatrix} Z_{11} & Z_{12} \\ Z_{21} & Z_{22} \end{bmatrix}. \tag{8.24}$$

The properties of $[Z]$ are as follows: the real component of impedances on the leading diagonal (Z_{11}, Z_{22}) are positive; the transfer impedances are reciprocal, i.e. $Z_{12} = Z_{21}$. This is true for electrical systems, mechanical systems, and the piezoelectric/mechanical actuator/sensor. However, for an electrodynamic system, equations (8.15) and (8.16) indicate that $Z_{12} = -Z_{21}$. For a stable system, the determinant is greater than zero, i.e. $|Z| > 0$. In this case, the matrix can be successfully inverted to give the *mobility* $[Y]$,

$$[Y] = [Z]^{-1}, \tag{8.25}$$

which for a mechanical example using force and velocity is

$$\begin{Bmatrix} u_1 \\ u_2 \end{Bmatrix} = \begin{bmatrix} Y_{11} & Y_{12} \\ Y_{21} & Y_{22} \end{bmatrix}\begin{Bmatrix} F_1 \\ F_2 \end{Bmatrix}. \tag{8.26}$$

For 'soft' systems such as acoustic transmission lines and mechanical vibration isolators, the impedance elements can be measured directly from equation (8.24) by setting to zero (or blocking) the velocity at one input (say $u_2 = 0$), and driving the other input with the known velocity u_1. The impedance elements then become

$$Z_{11} = F_1/u_1, \quad Z_{21} = F_2/u_1, \qquad u_2 = 0. \tag{8.27}$$

A system making a measurement of the input and transfer impedance of a vibration isolator in one direction is shown in Fig. 8.15. The process is reversed for the other two terms: Z_{22}, Z_{12}.

For electrical systems, such as that shown in Fig. 8.16, an open circuit (say, at junction 2) would cause zero current ($I_2 = 0$). The impedance terms would be available from the current I_1 and voltages V_1 and V_2.

However, many mechanical systems have too great an impedance to be blocked at junctions by another object of far larger impedance. The problem

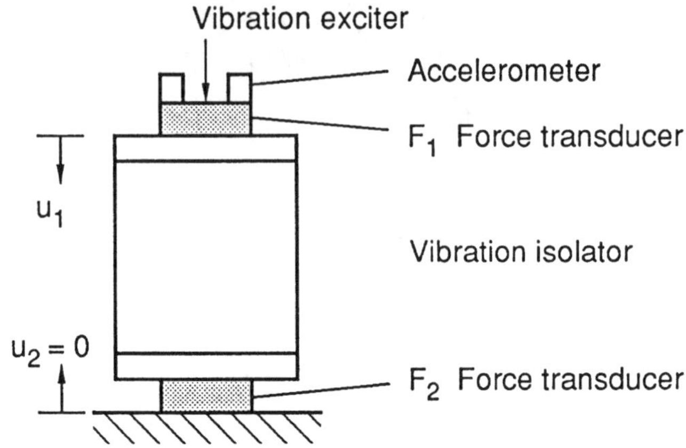

Figure 8.15 System for measurement of vibration isolator input and transfer impedances.

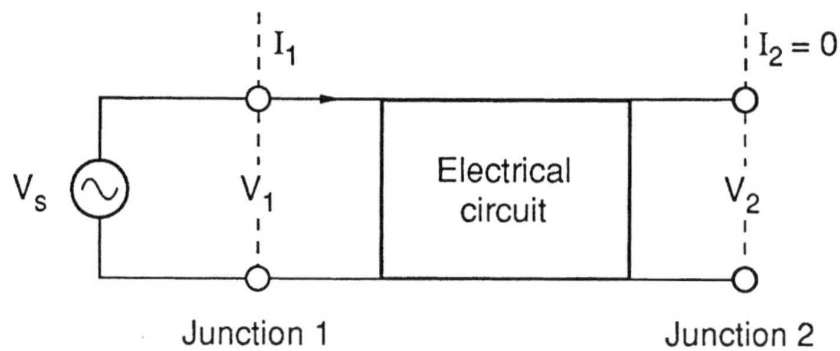

Figure 8.16 System for measuring input and transfer electrical impedances.

is resolved by measuring the *mobility* matrix of equation (8.26) which has the simple requirement of zero force at the non-active junctions. Figure 8.17 shows a system for measuring mobilities Y_{11} and Y_{12} on a test piece.

Although a matrix inversion of $[Y]$ to produce the impedance $[Z]$ is mathematically possible, the measured data is rarely of sufficient quality. The chances of a determinant $|Y| > 0$ at all frequencies giving a successful inversion improves as the damping of the structure increases the magnitude of the real components of mobility or impedance on the leading diagonal.

8.3.4 Force response relations for simple elements

There is widespread familiarity with Kirchhoff's Laws which facilitate the analysis of an electric circuit. Similar laws apply to mechanical systems. Newton's Second Law and Hooke's Law are used to define the force–

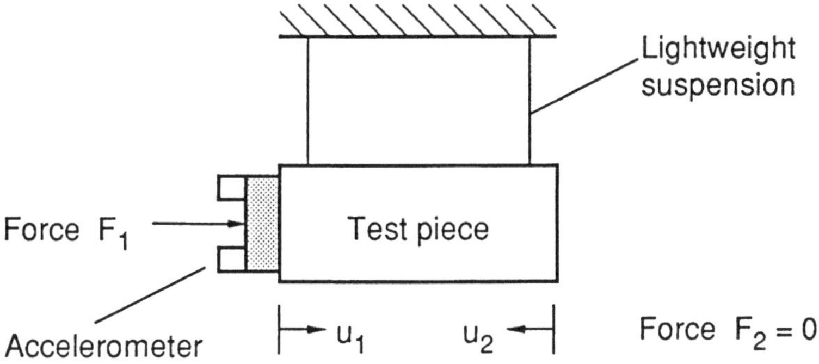

Figure 8.17 System for measuring input mobility Y_{11} and transfer mobility Y_{12}.

response relationships for the simple elements such as a mass, spring and a dashpot. These elements can then be combined using the principles of equilibrium of forces and continuity of displacement at the junction. This procedure is demonstrated here on two systems to obtain the mobility and impedance matrices. The force and response relationships for the spring, dashpot and mass are first defined using Fig. 8.18.

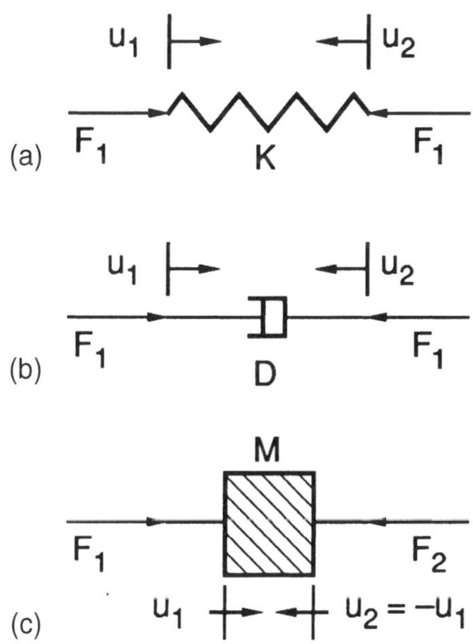

Figure 8.18 (a) Force–velocity relations for a spring. (b) Force velocity relations for a dashpot. (c) Force velocity relations for a mass.

For the spring of stiffness K, there is no mass and so an equal and opposite force F_1 is applied on each end. The Hooke's Law relation between force F_1 and relative velocity $u_1 + u_2$ is

$$F_1 = \frac{K}{j\omega}(u_1 + u_2). \tag{8.28}$$

For the massless dashpot of damping rate D shown in Fig. 8.18(b), a similar relation holds between force F_1 and relative velocity $u_1 + u_2$.
i.e.

$$F_1 = D(u_1 + u_2). \tag{8.29}$$

Newton's Second Law gives

$$F_1 - F_2 = j\omega M u_1. \tag{8.30}$$

8.3.4.1 Example 1

Now that the elements are defined, it is possible to determine the *impedance matrix* of the system shown in Fig. 8.19 to which external forces F_1 and F_3 can be applied at both ends, giving responses u_1 and u_3.

To obtain Z_{11} and Z_{31}, set the velocity $u_3 = 0$. The equation of equilibrium for the mass is given in equation (8.30), i.e.

$$F_1 - F_2 = j\omega M u_1. \tag{8.31}$$

The spring and dashpot have the same relative velocity u_1, therefore the forces from each add to give F_2 (which also equals F_3).

$$F_2 = \left(\frac{K}{j\omega} + D\right)u_1. \tag{8.32}$$

Adding equations (8.31) and (8.32) gives

$$F_1 = \left(j\omega M + \frac{K}{j\omega} + D\right)u_1. \tag{8.33}$$

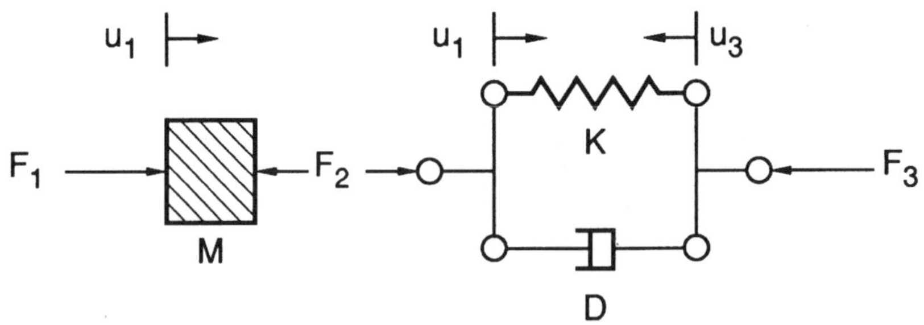

Figure 8.19 Impedance of a single degree-of-freedom system.

The bracketed term is the input impedance Z_{11}, i.e.

$$j\omega M + \frac{K}{j\omega} + D = Z_{11} = \frac{F_1}{u_1}. \tag{8.34}$$

Note that because the same velocity is applied to one side of each element, the element impedances add (in parallel). The force $F_2 = F_3$, therefore equation (8.32) yields the transfer impedance Z_{31}.

$$\frac{F_3}{u_1} = Z_{31} = D + \frac{K}{j\omega} \tag{8.35}$$

The other two impedances of the system in Fig. 8.19, Z_{13} and Z_{33}, can now be found by setting the velocity $u_1 = 0$ and applying an input u_3. In this case, the mass does not move, and therefore $F_1 = F_2$. The stiffness and dashpot impedances add in parallel to give the input impedance Z_{33} and transfer impedance Z_{13}.

$$Z_{33} = \frac{F_3}{u_3} = \frac{K}{j\omega} + D \tag{8.36}$$

and

$$Z_{13} = \frac{F_1}{u_3} = \frac{K}{j\omega} + D. \tag{8.37}$$

As a useful check it can be seen that the system is reciprocal, i.e. $Z_{13} = Z_{31}$.

8.3.4.2 Example 2

Elements can also be added in series when each is subjected to the same force. This is demonstrated in the determination of the mobility matrix of the system shown in Fig. 8.20. The external forces are F_1 and F_3 which are related to the responses u_1 and u_3 by the mobility matrix.

To obtain the mobility elements Y_{11} and Y_{13}, the force F_3 is first set to zero. The same force F_1 is applied to each element because the spring and dashpot are massless. The equations of equilibrium for each element are

$$\frac{j\omega}{K} F_1 = u_1 - u_2, \tag{8.38}$$

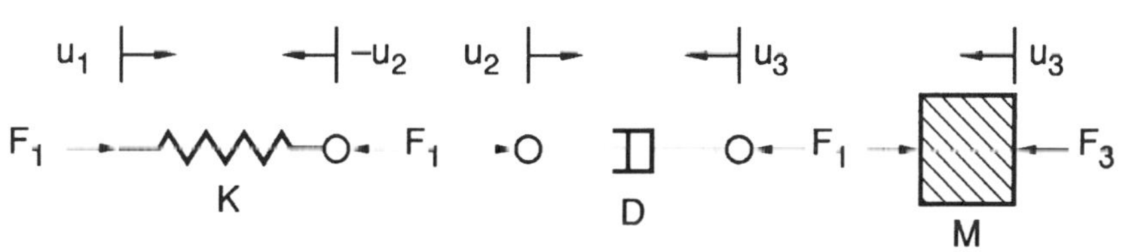

Figure 8.20 A system composed of simple elements.

$$\frac{F_1}{D} = u_2 + u_3, \tag{8.39}$$

and

$$\frac{F_1}{j\omega M} = -u_3. \tag{8.40}$$

The input mobility Y_{11} is found by adding equations (8.38)–(8.40), thus:

$$Y_{11} = \frac{u_1}{F_1} = \frac{j\omega}{K} + \frac{1}{D} + \frac{1}{j\omega M}. \tag{8.41}$$

In this case, the element mobilities (the reciprocal of the impedances) are added in series since the same force is applied to each element. The transfer mobility Y_{31} is given in equation (8.40), i.e.

$$Y_{31} = \frac{u_3}{F_1} = -\frac{1}{j\omega M}. \tag{8.42}$$

To obtain the other two mobilities, the force F_1 in Fig. 8.20 is set to zero and the force F_3 activated. The intermediate forces between elements all become zero because the elements are massless. The input mobility Y_{33} is

$$Y_{33} = \frac{u_3}{F_1} = \frac{1}{j\omega M}. \tag{8.43}$$

The transfer mobility Y_{13} is found by noting that the elements are rigid with no force across them, giving $-u_3 = u_1$, i.e.

$$Y_{31} = -\frac{1}{j\omega M}. \tag{8.44}$$

Note once more that $Y_{13} = Y_{31}$ giving a reciprocal system.

8.3.5 Interaction between the actuator and test piece

Figure 8.21(a) shows a practical arrangement to drive a mechanical test piece with an electrodynamic exciter. A voltage signal V_s is generated which has one of the ideal spectra discussed in section 8.4.1.5. This could be a sine wave, a random signal or swept sine. The signal is input to a power amplifier, the purpose of which is to give a large output current I_a, typically about 1 A peak. The current generates a force $B\ell I_a$ in the actuator which must move both the test piece and the moving actuator element. It is a good practice to include an ammeter in the circuit to monitor the current I_a, as the heating effect inside the actuator is proportional to I_a^2. The manufacturer's data indicates the maximum safe current. Typical electrodynamic actuators and loudspeakers have an impedance of about 8 Ω, and the output impedance of the power amplifier should be set to a similar value for maximum power output. The power amplifier is a linear device only below a certain voltage level; above this level the semi-conductors and inductive

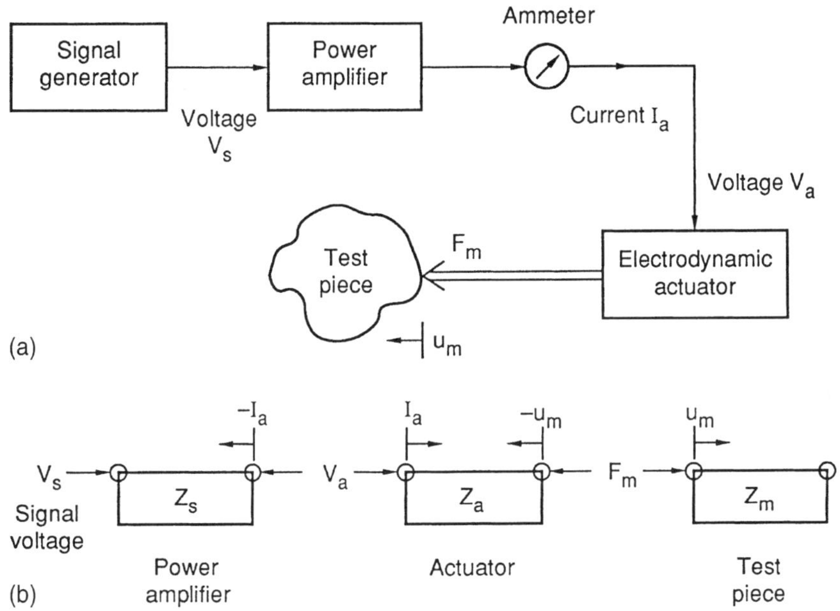

Figure 8.21 (a) Practical arrangement for exciting a test piece. (b) Impedance representation.

elements saturate and a distorted signal is produced, perceived as a buzzing sound.

In an ideal system, a force F_m is applied to the test piece which has a frequency independent relationship to the voltage V_s. However, in practice, there is electrical and mechanical ineraction between components which reduces or amplifies the output force at certain frequencies. The objective is to quantify this interaction in terms of the impedances of the amplifier, actuator and test piece. To assist in this exercise the three components are shown in Fig. 8.21(b) with the 'force' and response at each junction.

The actuation side of the measurement system will be dealt with first. The elements of a voltage source from the amplifier, the actuator and test piece are taken from Figs 8.11 and 8.14 (b,c), and linked in a chain. The coupling 'forces' are equal and opposite and there is continuity of 'velocity'. The objective is to find the force F_m on the test piece due to a constant voltage source V_s connected to the amplifier.

The voltage V_a from the amplifier output is given in equation (8.18) as

$$V_a = V_s - I_s Z_s, \tag{8.45}$$

where Z_s is the output impedance of the power amplifier.

The force F_m applied to the test piece is related to the voltage V_a by electro-mechanical impedance of the actuator in equations (8.22).

$$\begin{Bmatrix} V_a \\ F_m \end{Bmatrix} = \begin{bmatrix} Z_{a11} & Z_{a12} \\ Z_{a21} & Z_{a22} \end{bmatrix} \begin{Bmatrix} I_a \\ -u_m \end{Bmatrix}, \tag{8.46}$$

where $[Z_a]$ is the impedance of the electro-mechanical actuator.

The mechanical impedance of the test piece is Z_m, and hence

$$F_m = Z_m u_m. \tag{8.47}$$

Combining equations (8.45), (8.46), (8.47) gives the required solution to the actuation problem. For the specific example of an electrodynamic exciter, the electrical circuit in Fig. 8.22 (equation (8.45)) and the top line of equation (8.46) describes the actuator side, giving

$$V_s = (Z_{a11} + Z_s)I_a + B\ell u_m. \tag{8.48}$$

The electrical coupling of equation (8.16) is embodied in the transfer impedance $Z_{a12} = -B\ell$. The induced force due to the current is given by equation (8.15), i.e. $Z_{a21} = B\ell$ which illustrates the inverse reciprocal relationship of electrodynamic systems. Eliminating the force F_m from equations (8.46) and (8.47) yields equation (8.49),

$$B\ell I_a = (Z_{a22} + Z_m)u_m, \tag{8.49}$$

which states that the force must now be divided between the exciter $(Z_{a22}u_m)$ and the test piece $(Z_m u)$. If the exciter impedance becomes large compared to the test piece, i.e. $Z_{a22} \gg Z_m$, the force applied to the test piece will be greatly reduced. Combining equations (8.49) and (8.48) gives the transmitted force F_m as

$$F_m = \frac{F_0 Z_m}{Z_m + Z_{ae}}, \tag{8.50}$$

where $F_0 = B\ell V_s/(Z_{a11} + Z_s)$ and $Z_{ae} = Z_{a22} + (B\ell)^2/(Z_{a_{11}} + Z_s)$. Re-arranging equation (8.50) using the substitution of equation (8.47) gives

$$F_m = F_0 - Z_{ae}u_m, \tag{8.51}$$

which is the same form as the real force source of equation (8.19). This formulation would embody all the electrical and interaction effects and could be applied in conjunction with equation (8.47) for a description of the

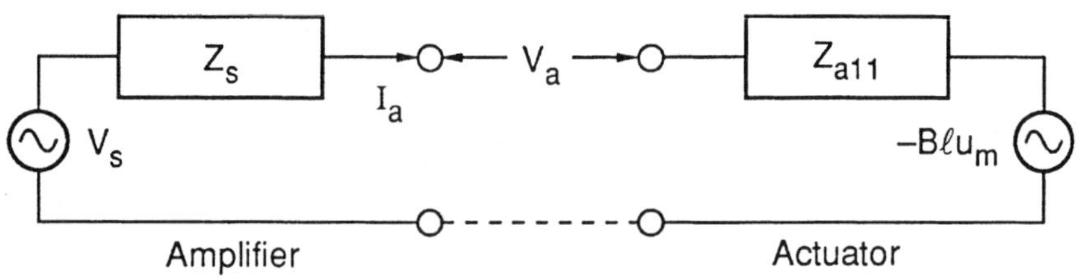

Figure 8.22 Electrical circuit of the coupled amplifier and actuator.

actuation system and test piece. The quantities F_0 and Z_{ae} could be measured directly from the output of the actuator. F_0 is the blocked force, obtained by setting $u_m = 0$. Z_{ae} is the effective output mechanical impedance of the actuator, the $(B\ell)^2$ term is real, indicating a damping effect. The resonant, or free vibration, response of the coupled actuator and test piece can be controlled by this component. If free vibration measurements are to be made on the test piece, the electrical circuit should be opened to prevent this form of energy dissipation.

The coupled system becomes very efficient at resonances when $\text{Im}\{Z_m + Z_{ae}\} = 0$, giving a large force output F_m, seen at point b in Fig. 8.23. This is the basis for transmitting signals from sonar systems using a stack of piezoelectric crystals as an actuator.

There are, however, problems when a uniform force spectrum over a broad band of frequencies is required. At the resonance frequency ω_r of the uncoupled test piece, the impedance Z_m in equation (8.50) takes a minimum value $\text{Re}\{Z_m\} = D_r$. The resulting minimum value at point a in Fig. 8.23(a) is

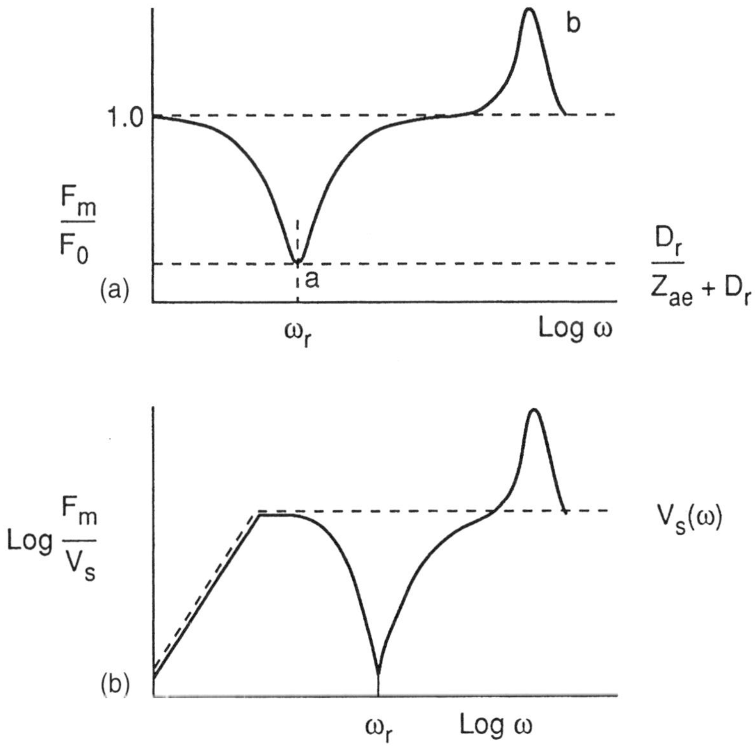

Figure 8.23 (a) Force change due to coupling. (b) Transmitted force/voltage spectrum. — ideal signal, - - - - force output.

$$\frac{F_\mathrm{m}}{F_0} = \frac{D_\mathrm{r}}{D_\mathrm{r} + Z_\mathrm{ae}}. \tag{8.52}$$

The minimum therefore deepens with decreasing system damping D_r and increasing actuator impedance Z_ae. The problem here is that at resonance the test piece becomes very 'lightweight', and most of the force available from the actuator is spent moving itself, rather like trying to throw a feather, when the velocity developed is limited by the mass of the hand rather than by that of the feather. This effect has consequences on signal reproduction. A broadband output from a signal generator may look like $V_\mathrm{s}(\omega)$ in Fig. 8.23(b).

The signal is clearly not well represented at the test piece resonance. To compensate for the dip, the voltage must be controlled by a feedback system to create a *constant current source* I_s proportional to the original signal, as the output force follows the current, i.e. $F_\mathrm{m} = B\ell I_\mathrm{s}$.

8.3.6 Frequency response testing in the presence of noise

A frequency response test on a system would require a force transducer (Fig. 8.25) to measure F_m and a response transducer to measure u_m. In principle, the ratio of the quantities is always described by equation (8.47) and should not be influenced by the interaction problem. However, this is not true if there is noise present (see section 8.5), as it can corrupt data if the signal levels become low (McConnell, 1990; White and Pinnington, 1982).

Figure 8.24(a) shows a transfer function measurement between the force F_m and the response in which noise of spectrum $G(\omega)$ is added to the force transducer output and noise of spectrum $N(\omega)$ is added to the velocity transducer. The measured mobility Y_m is therefore

$$Y_\mathrm{m} = \frac{u_\mathrm{m} + N}{F_\mathrm{m} + G}. \tag{8.53}$$

This may be re-expressed in terms of the force source F_0 using equation (8.50) and (8.47), i.e.

$$Y_\mathrm{m} = \frac{1 + \frac{N}{F_0}(Z_\mathrm{m} + Z_\mathrm{ae})}{Z_\mathrm{m} + \frac{G}{F_0}(Z_\mathrm{m} + Z_\mathrm{ae})}, \tag{8.54}$$

as plotted in Fig. 8.24(b).

At the minima areas, the impedance Z_m of the test piece is large and so noise in the *response* becomes dominant if

$$\left|\frac{N}{F_0}\right| \geq \mathrm{Re}\{Y_\mathrm{m}\} + j\mathrm{Im}\{Y_\mathrm{m}\}. \tag{8.55}$$

The real component of mobility $\mathrm{Re}\{Y_\mathrm{m}\}$ (seen in Fig. 8.8) is the smallest part of the transfer function and so is the part most influenced by the noise

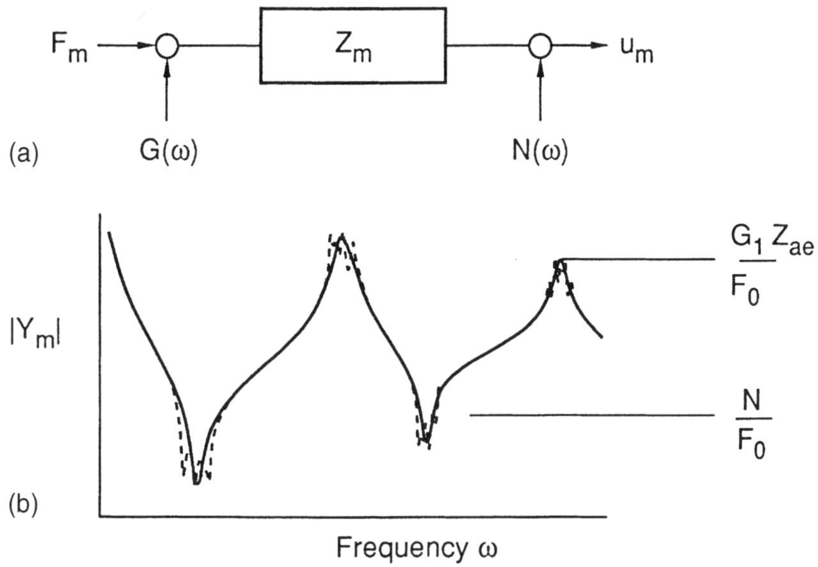

Figure 8.24 (a) Transfer function measurement in the presence of noise. (b) Measured transfer function with noise corruptions.

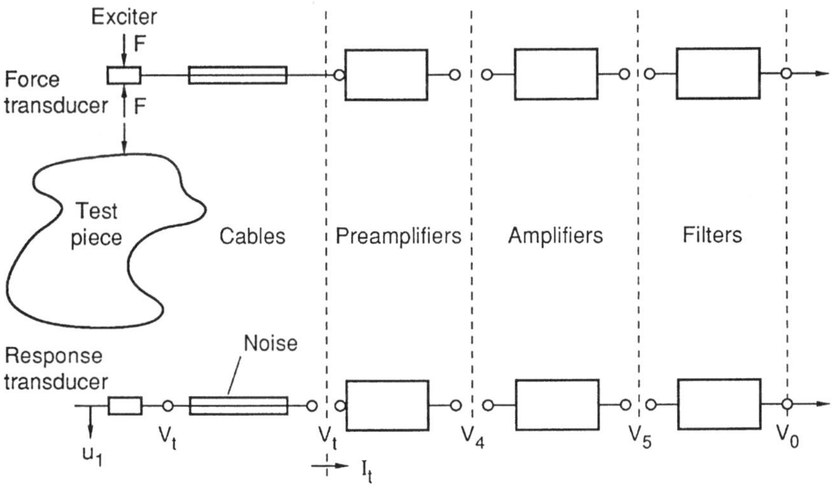

Figure 8.25 Instrumentation required for a frequency response measurement.

in the response. To reduce the noise corruption, it is necessary to use larger forces, or more sensitive transducers, or to apply some signal averaging. At the resonance peaks, the test piece impedance becomes a minimum controlled only by the damping D_r; $Z_m \approx D_r$. The denominator of equation (8.54) is therefore significantly corrupted if

$$\left|\frac{G}{F_0}\right| > \frac{D_r}{|Z_{ae}|}. \tag{8.56}$$

Therefore, problems can be expected if the actuator becomes large, particularly for lightly damped structures. The general principle then, in vibration testing, is to minimize the mechanical interference by keeping actuators as lightweight as possible and minimizing any restraint in translation or rotation.

8.3.7 Interaction between the test piece and the sensing system

There are two types of interaction which will be considered independently. These are the mechanical interaction between the test piece and the whole transducer system, and the electrical interaction between the transducer and the remainder of the conditioning system.

8.3.7.1 Electro-mechanical interaction between the test piece and the transducer

It is assumed that the test piece is acted upon by a force source F_s, but the transducer system is represented on a single electro-mechanical element as shown in Fig. 8.26.

An ideal transducer would measure the response u_1 without applying a seismic force F which makes its own contribution to u_1. The objective here is to see how the output voltage V_t is influenced by this largely mechanical interaction. The best example of this effect is the seismic accelerometer which is attached to the structure, thereby adding mass and altering the velocity of the test piece. There are non-contacting transducers available that use light, capacitance, inductance, or ultrasound, which may not significantly load the structure, but these tend to be expensive and lack

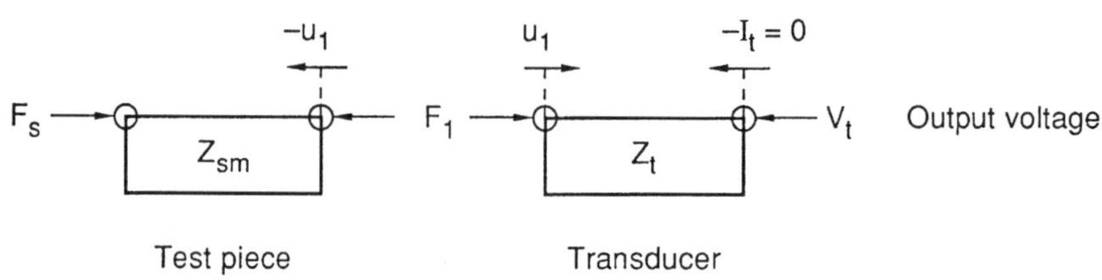

Figure 8.26 Mechanical interaction between the test piece and transducer.

sensitivity at high frequencies as they usually monitor displacement rather than acceleration. Electrical interaction can also occur, for example, by electro-magnetic effects, but these are already included in the transducer – at system Z_t.

The interaction force F_1 is given in equation (8.19) as

$$F_1 = F_s - u_1 Z_{ms}, \tag{8.57}$$

while the electrical system is a version of equation (8.22), i.e.

$$\begin{Bmatrix} V_t \\ F_1 \end{Bmatrix} = \begin{bmatrix} Z_{t11} & Z_{t12} \\ Z_{t21} & Z_{t22} \end{bmatrix} \begin{Bmatrix} -I_t \\ u_1 \end{Bmatrix}. \tag{8.58}$$

The output voltage V_t is developed across the very high electrical impedance of the conditioning circuit causing negligibly small current, i.e. $I_t \approx 0$. The force F_1 and current I_t can be eliminated from equations (8.57) and (8.58) giving a transducer velocity u_1 of

$$u_1 = F_s/(Z_{t22} + Z_{ms}). \tag{8.59}$$

The presence of the transducer of mechanical impedance Z_{a22} perturbs the response from the ideal value of

$$u_1 = F_s/Z_{ms}. \tag{8.60}$$

The most usual type of loading from a transducer is seismic, e.g. for an accelerometer at mass $m_a, Z_{a22} = i\omega m_a$, influencing the coupled response in two ways: at the low frequency resonances, and at very high frequencies. At very high frequencies, the ideal response seen in Fig. 8.10 becomes limited by the transducer mass, as seen in Fig. 8.27.

At lower frequencies, the resonance peaks are shifted downwards by a factor determined by the total mass of all the transducers on the test piece $\sum m_a$ and the mass of the test piece, M (Ewins, 1984; McConnell, 1995). The

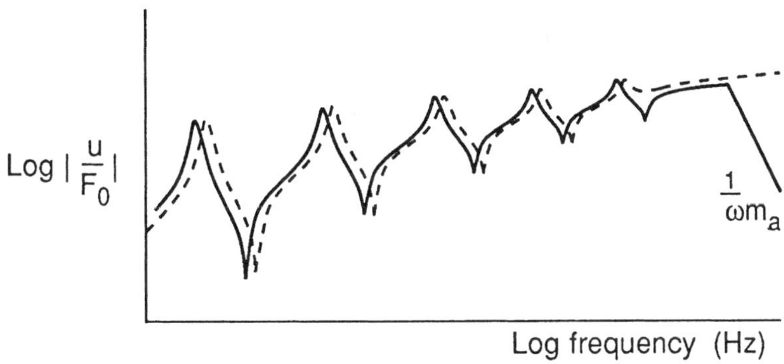

Figure 8.27 Mechanical interaction between test piece and transducer.

natural frequency of a single mode of an ideal system is given by equation (8.10), as

$$f_r = \frac{1}{2\pi}\sqrt{\frac{K_r}{M_r}}. \tag{8.61}$$

Increasing the mass decreases the resonance frequencies to f_m, as seen in Fig. 8.27. For a single transducer at position x,

$$f_m = \frac{1}{2\pi}\sqrt{\frac{K_r}{M_r + m_a q_r^2(x)}}, \tag{8.62}$$

where $q_1(x) = \sin\frac{r\pi x}{\ell}$ for a simply-supported beam of length ℓ. The modal mass M_r describes a discrete ideal system with the same kinetic energy as a distributed system of mass M, i.e. for a one-dimensional system the modal mass is half the total mass:

$$M_r = M\int_0^{\ell} q_r^2(x)\mathrm{d}x \approx \frac{M}{2}. \tag{8.63}$$

The influence of the transducer also varies with the position, as seen in equation (8.62).

Moving a single transducer to measure mode shapes will therefore give unreliable results because of the corresponding change in natural frequency.

8.3.8 Electrical interaction between the transducer and the signal conditioning

The electrical part of the transducer in Fig. 8.28 can now be coupled to the conditioning amplifier, the purpose of which is to provide a frequency independent transfer function relating the output voltage to the velocity u. This is the case for ideal transducers, described earlier in equations (8.12), (8.13), (8.14). It will be seen here that this ideal relationship only occurs under certain impedance conditions.

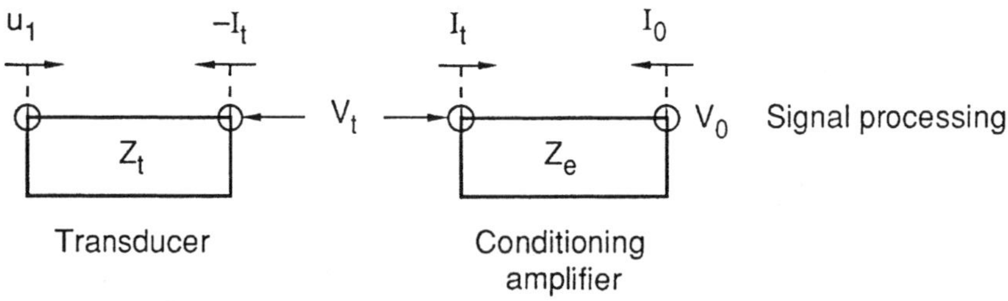

Figure 8.28 Electrical interaction between the transducer and conditioning circuits.

The transducer electrical output V_t is given in equation (8.57), i.e.

$$V_t = -I_t Z_{t11} + u_1 Z_{t12}, \tag{8.64}$$

where Z_{t12} is the transducer *sensitivity* relating the input signal u_1 to the output voltage V_t. The other term involving the current I_t describes the interaction with the electrical circuit. The requirement on the connected electrical circuit $[Z_e]$ is as follows: (i) to amplify the voltage V_2 to the larger value V_0 required by signal processing, this task is performed by an *amplifier*; (ii) to set the interacting current to a negligible value–this is performed by a *preamplifier*; (iii) to provide filtering to limit the bandwidth of the signals (described in section 4). These three stages are shown in Fig. 8.25 for a frequency response measurement requiring two channels. They may be separate units combined in a single *conditioning amplifier*, as depicted in Fig. 8.28. Some transducers incorporate these other stages using microchip technology. Nevertheless the same functions must be performed.

The conditioning is described by the electrical transfer function $[Z_e]$.

$$\begin{Bmatrix} V_t \\ V_0 \end{Bmatrix} = \begin{bmatrix} Z_{e11} & Z_{e12} \\ Z_{e21} & Z_{e22} \end{bmatrix} \begin{Bmatrix} I_t \\ I_0 \end{Bmatrix}. \tag{8.65}$$

It can be assumed that I_0 is almost zero as the output of an electrical impedance of any voltage measurement device is extremely high. Therefore, the output voltage V_0 is given from equations (8.64) and (8.65) as

$$V_0 = u_1 \cdot H_t \cdot H_a \cdot \frac{1}{1 + Z_{t11}/Z_{e11}}, \tag{8.66}$$

where $H_t = Z_{t11}$ is the transducer sensitivity, $H_a = Z_{e21}/Z_{e11}$ is the amplification factor of the conditioning which should be a constant.

This summarizes the essential requirement for effective coupling between components. This is that the *output impedance of one stage* (for the transducer Z_{t11}) *is much less than the input impedance of the next stage* (for the conditioning, this is Z_{e11}). Under these conditions, the interacting current is small and so all the voltage generated by the transducer is developed across the terminals of the next stage, and equation (8.66) takes the form of the ideal measurement systems of equation (8.14).

$$V_0 = u_1 H_t \cdot H_a \tag{8.67}$$

The response u_1 can then be calculated from the output voltage by simple division by the product of the sensitivity and amplification factor.

8.3.8.1 Transducer cables

Figure 8.25 shows a connecting cable between the transducer and the preamplifier. If this has an impedance of Z_c, a voltage drop $V_t - V_t' - I_t Z_c$ occurs over the length. Increasing the cable length will increase this problem. However, provided the preamplifier input impedance $Z_{e11} \gg Z_c$, this effect can be ignored, as is the case when *charge amplifiers* are used in conjunction with piezoelectric transducers. When a preamplifier is incorporated into the transducer as with a *condenser microphone*, the problem does not exist. As a rule, though, it is good practice to keep cables as short as possible to reduce the influence of electro-magnetic pick-up (discussed in section 8.5).

8.4 FREQUENCY LIMITATIONS DUE TO MECHANICAL AND SIGNAL PROCESSING

The linear transfer function relationship $H(f)$ in equation (8.4) appears at first sight to be available simply by the division of response $X(f)$ by the force $F(f)$. In practice, however, the presence of *noise* limits the bandwidth of reliable data, as illustrated in Fig. 8.29.

Extraneous signals called *noise*, occurring for many reasons, represent the unwanted contribution of the real world on an ideal experiment. The level of this noise background or *noise floor* can be measured by conducting the experiment without the desired response signal, and measuring the residual output. Some confidence can then be placed on test signals larger than this noise floor. It is unfortunately not possible to increase signal levels indefinitely to float clear of the noise as all transducers and instrumentation have a maximum level of operation usually limited by saturation of some sort, which causes *non linear* behaviour. The ratio of the maximum to minimum signal, shown in Fig. 8.29, is called the *dynamic range*. Also seen in the figure is the *bandwidth* which limits the *frequency range* of all instruments. This bandwidth is controlled by (i) the physical properties of the instruments, and (ii) the signal processing limits.

8.4.1 Bandwidth limitations in transducers and exciters

The primary requirement of an exciter or transducer is *sensitivity*, i.e. output per given input, for example, Bℓ for electrodynamic devices in equations (8.15) and (8.16). The sensitivity can therefore be increased with size ℓ.

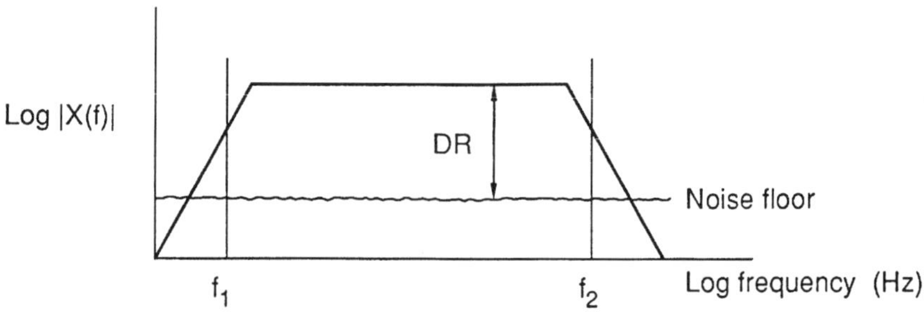

Figure 8.29 Bandwidth and noise limitations for transducers.

Inevitably, increasing size imposes other limitations. It was seen to increase the measurement-test piece interaction in the previous section, it also decreases the *frequency bandwidth* as discussed here.

8.4.1.1 Force and pressure transducers

The measurement of dynamic force and dynamic pressure is related to the *strain* or relative displacement within a transducer element.

Consider the force transducer in Fig. 8.30. The output voltage is proportional to the force provided the transducer behaves as a simple spring. Under these conditions of ideal operation, the sensitivity is a constant within the bandwidth $f_1 - f_2$ in Fig. 8.29. However, at high frequencies, a resonance will occur within the transducer, for example, when the length ℓ of a spring is equal to half a wavelength, i.e. at frequency

$$f_0 = \frac{c}{2\ell}, \tag{8.68}$$

where c is the phase speed of waves travelling along the spring axis.

Figure 8.31 shows the useful range of operation between 0 and f_2 where $f_2 < 0.3 f_0$, in which the frequency response is flat and the phase is zero. Equation (8.68) suggests that the frequency range increases with decreasing

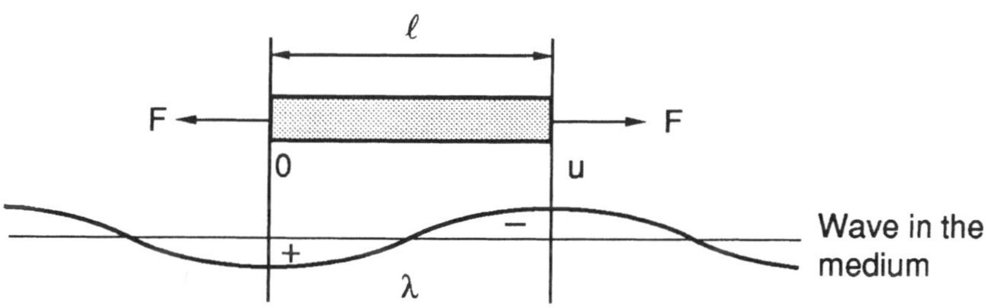

Figure 8.30 Force or pressure transducer in a medium.

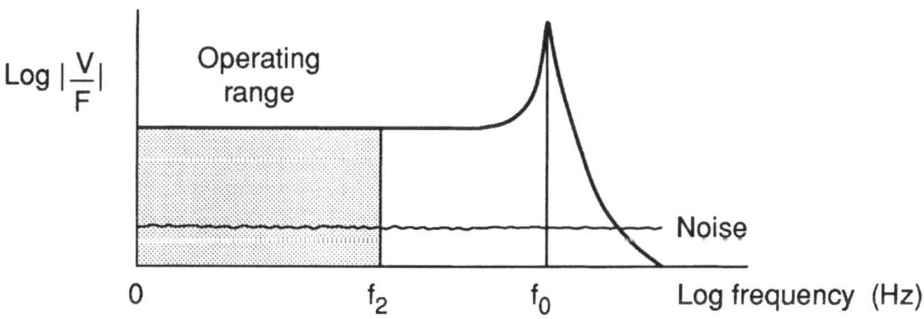

Figure 8.31 Transducer bandwidth limited by internal resonance.

length ℓ and increasing wavespeed inside the transducer. Indeed, the ideal transducer is massless, and occupies only a point in space.

The upper frequency is also limited when half a wavelength within the medium is equal to the transducer length ℓ, as shown in Figs 8.30 and 8.32.

Below this frequency the wavelength is much greater than the transducer causing a positive strain over the whole length, resulting in a constant sensitivity. However, when more than half a wavelength occupies the length cancellation of the strain reduces the output.

8.4.1.2 Exciters

The same remarks can be made for vibration or sound sources as these are often reciprocal devices. The output is increased with size, but the upper frequency limit is reduced by increasing size.

8.4.1.3 Response transducers

Response transducers also have the same limitations, however there are a few other considerations determined by the quantity measured. The kinematic quantities are acceleration $a(t)$, velocity $u(t)$, and displacement $\xi(t)$.

Acceleration is normally monitored using seismic devices called accelerometers, the principle of which is shown in Fig. 8.33; a force is developed across a strain sensing element of stiffness K by the acceleration a_1 acting upon a seismic mass M. The charge output q_1 is thus proportional to acceleration a_1, and has the sensitivity curve of Fig. 8.31. Displacement must in contrast be measured relative to a fixed datum position and usually non-contacting transducers employing principles such as ultrasound, photoelectricity, capacitance, electromagnetism or laser interferometry are used. Velocity, likewise, is usually measured with fixed reference devices using electrodynamics or laser doppler shift. The three kinematic quantities are related by the time differential. For a single frequency acceleration $a(t)$,

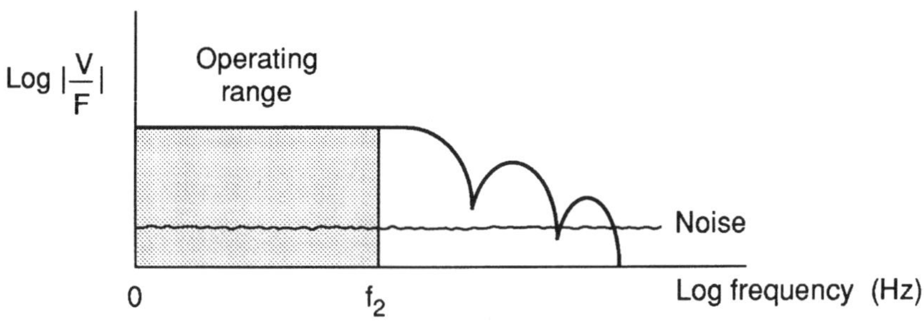

Figure 8.32 Transducer bandwidth limited by external wavelength.

$$a(t) = a_0 \sin \omega t,$$

$$u(t) = \frac{a_0}{\omega} \cos \omega t, \tag{8.69}$$

and

$$\xi(t) = -\frac{a_0}{\omega^2} \sin \omega t.$$

In the general frequency (Fourier) domain,

$$u(f) = \frac{a}{j\omega}(f)$$

and

$$\xi(f) = -\frac{a}{\omega^2}(f). \tag{8.70}$$

The relations are plotted in log–log form in Fig. 8.34.

Transfer functions between force and the kinematic quantities of displacement, velocity and acceleration are therefore related in the same way as equation (8.70). The terms of these relationships are given in Table 8.1. Also given are the inverse relationships where a force is the response to a given kinematic input, e.g. impedance describes force/velocity.

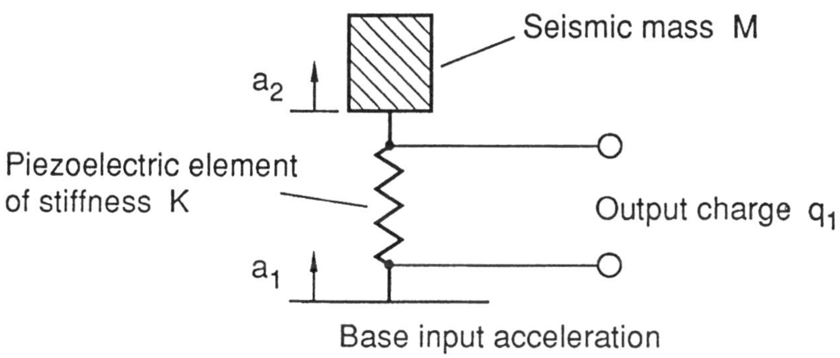

Figure 8.33 Seismic accelerometer with piezoelectric strain element.

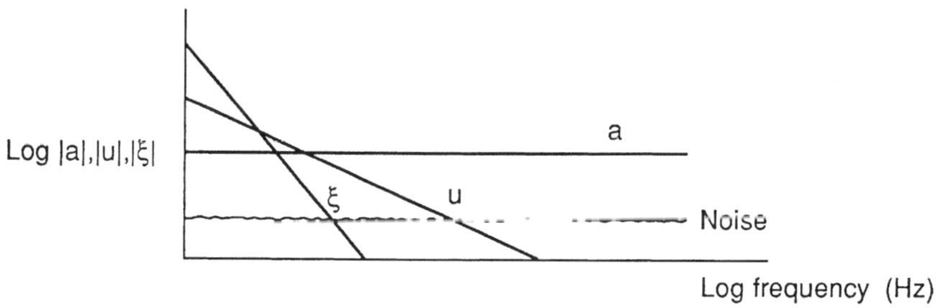

Figure 8.34 The output of acceleration, velocity, and displacement transducers.

Table 8.1 Relationship between different forms of transfer function

$\frac{\text{Displacement}}{\text{Force}}$	Receptance	$Y/j\omega$
$\frac{\text{Velocity}}{\text{Force}}$	Mobility	Y
$\frac{\text{Acceleration}}{\text{Force}}$	Accelerance	$j\omega Y$
$\frac{\text{Force}}{\text{Displacement}}$	Dynamic stiffness	$j\omega Z$
$\frac{\text{Force}}{\text{Velocity}}$	Impedance	Z
$\frac{\text{Force}}{\text{Acceleration}}$	Apparent mass	$\frac{Z}{j\omega}$

If the acceleration spectrum is constant (as is often nearly the case in practice), the velocity decreases in proportion to frequency while displacement decreases with the square of frequency. At low frequencies, the displacement is large (hence the blurred visual image of vibrating objects), and the acceleration is small. At high frequencies, the situation is reversed and the acceleration is dominant. The velocity signals and displacement ultimately become buried in noise at high frequencies. For this reason and for the convenience of the seismic device, *accelerometers* are the commonest form of vibration sensor. Conversion to velocity or displacement if required is best done in the frequency domain using equation (8.70).

8.4.1.4 Presentation of frequency data

Sound and vibration instrumentation has a dynamic range covering many orders of magnitude. To display this data a vertical logarithmic scale is employed. A horizontal logarithmic scale is also used in conjunction to give '*Bode diagrams*', as, for example, in Fig. 8.34. Slopes of any power of frequency appear as straight lines, the gradient describes the power law.

The mechanisms controlling the behaviour, e.g. mass, damping, or stiffness, can be identified from the slope as in Fig. 8.13. Another way of presenting logarithmic data is the *decibel*. The decibel is a *ratio* of a power-like quantity W to a reference quantity W_{ref}, i.e.

$$\text{Power (dB)} = 10 \log \frac{W}{W_{\text{ref}}} = 10 \log W - 10 \log W_{\text{ref}}. \qquad (8.71)$$

Power is proportional to the square of linear quantities such as voltage or pressure p, therefore

$$\text{Pressure (dB)} = 20 \log p - 20 \log p_{\text{ref}}. \tag{8.72}$$

The chosen reference should be stated if this convention is used, as this will define the units. For frequency response data there is no particular reason for using decibels, and the logarithmic scale is less confusing. Data should be presented using the standard units.

8.4.1.5 Force signals and spectra

Vibration and acoustic exciters and transducers have the limited bandwidth shown in Fig. 8.29. The bandwidth of accurate data is also limited by the signal processing parameters discussed in the next section. It is shown that the system should only be excited within this bandwidth because energy at other frequencies interferes with the measurement as *noise*. The aim, therefore, is to provide the maximum force possible without undue non-linearity in any component, while maintaining a uniform spectrum within the required bandwidth. This bandwidth can be marshalled with high-pass and low-pass filters to give the ideal spectrum. Four types of signal that have this spectrum are summarized in Fig. 8.35.

The sine wave (a) is stepped in frequency to produce a flat spectrum (Broch, 1980). The chirp (b) or rapid frequency is a short duration transient, and is similar to the sound used by a bat for navigation (White and Pinnington, 1982). The impulse (c) is generated by a hammer, explosion, starting piston, or spark source; its duration T_0 determines the useful bandwidth, i.e. $\approx 1/T_0$ Hz. The random signal (d) is slightly different from (b) and (c) in that in a single time period T an imperfect spectrum is generated; only after several averages is the 'true' spectrum built up (Broch, 1980; Newland, 1975). Signals (a), (b) and (d) are usually generated using a voltage source, amplifier and exciter, as discussed in the interaction problem of section 8.3.

8.4.2 Signal processing aspects of frequency response testing

Most testing utilizes digital signal processing in which an analogue signal $x(t)$ shown in Fig. 8.36 is represented as a sequence of N points sampled at $1/\Delta$ samples/s over a time duration T, i.e. $N\Delta = T$. The sampling rate $1/\Delta$ limits the 'upper frequency' f_2 of accurate data, while the time duration T controls the *frequency resolution* Δf and hence the lower frequency limit f_1 of accurate data. As the number of points N is a constant, there is always a trade-off between the upper and lower frequencies for accurate data. The purpose here is to show how these two parameters influence the accuracy of transfer functions measured using either single frequency techniques or Fourier transform techniques (Newland, 1975; Otnes and Enockson, 1978).

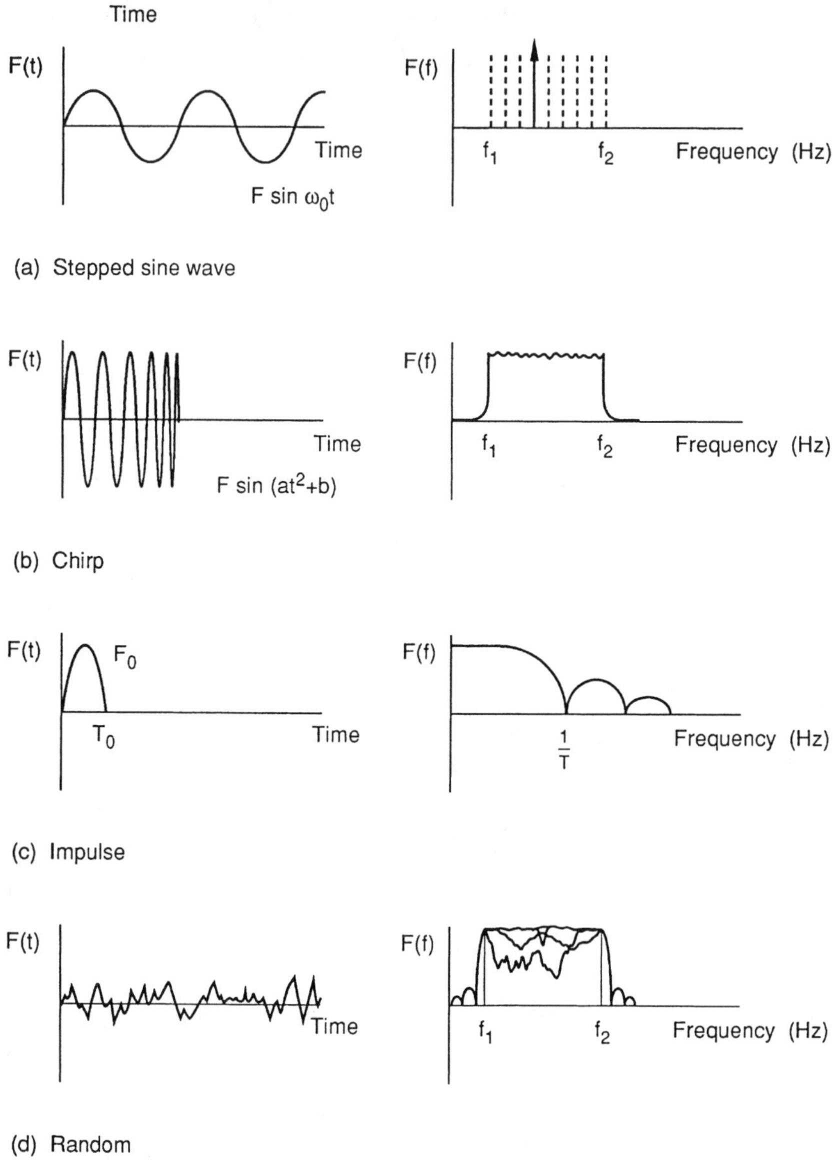

Figure 8.35 (a) Force signals: stepped sine wave. (b) Force signals: chirp. (c) Force signals: impulse. (d) Force signals: random.

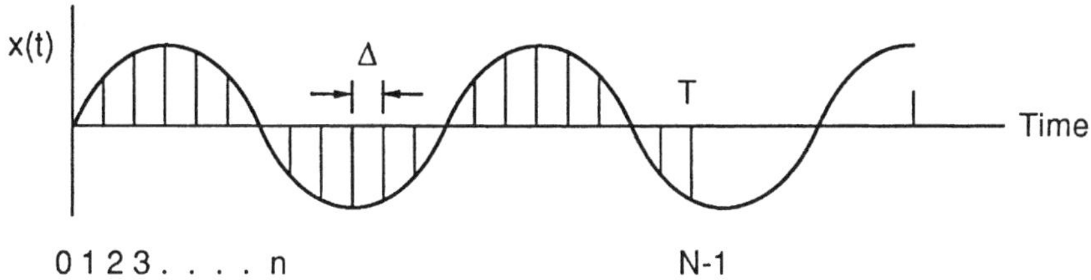

Figure 8.36 Analogue and digital signal representations.

8.4.2.1 Time domain measurement of frequency response

The most direct way of measuring the transfer function H in equation (8.1) is to excite the structure with a harmonic signal $f(t)$ at ω rad/s and measure the response $x(t)$ as seen in Fig. 8.37. The real force can be decomposed into positive and negative frequency components.

$$f(t) = \frac{F}{2}e^{j\omega t} + \frac{F^*}{2}e^{-j\omega t} \tag{8.73}$$

where * denotes complex conjugate. For $f(t) = F\cos\omega t$, F is purely real. The response $x(t)$ is likewise

$$x(t) = \frac{F}{2}He^{j\omega t} + \frac{F^*}{2}H^*e^{-j\omega t}. \tag{8.74}$$

If both of these components are multiplied by the unit phasor $e^{-j\omega t}$ and integrated for a finite number of periods T, only the positive frequency component is selected, since the $e^{-2j\omega t}$ term averages to zero, thus:

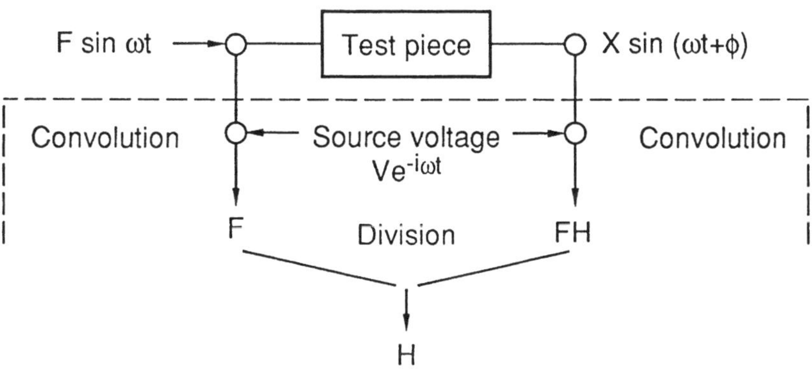

Figure 8.37 Single frequency testing with an analyser.

$$\frac{F}{2} = \frac{1}{T}\int_0^T f(t)\mathrm{e}^{-j\omega t}\mathrm{d}t \tag{8.75}$$

and

$$\frac{FH}{2} = \frac{1}{T}\int_0^T x(t)\mathrm{e}^{-j\omega t}\mathrm{d}t. \tag{8.76}$$

Division of equation (8.76) by (8.75) yields the complex transfer function H.

The multiplication with the signal $\mathrm{e}^{-j\omega t}$ acts as a narrow band filter which removes 'noise' due to signals at other frequencies. All the excitation energy is also focused at the one frequency providing a large signal level. This method is therefore effective for measuring transfer functions in very noisy environments, for example, on a running engine (although engine speed is not truly constant). Another form of noise is 'non linearity' which may occur due to friction or other mechanisms. The output of this process would be the linearized version of the transfer function.

The process shown in equations (8.75) and (8.76) is the same as for the Fourier transform in equation (8.2), only the operation is at a single frequency and for a limited time duration T. For non-resonant behaviour, steady-state conditions are established almost instantaneously after the first return of waves to the driving point from the system boundaries. However, at resonance, sufficient time T must be allowed to build to the resonant level at which point the rate of energy supply is matched by the rate of energy dissipation. The time T will be inversely proportional to the resonance bandwidth as described in equation (8.81). After a single frequency has been processed, the driving frequency is incremented by Δf and the process is repeated. The frequency interval Δf should be less than any resonance half power bandwidth, i.e. $\Delta f < 2\zeta f_n$ where ζ is the viscous damping factor, f_n is the resonance frequency. This avoids frequency resolution problems described in section 8.4.2.3.

8.4.2.2 Frequency domain measurement of frequency response

The signal processing procedure described above is rather time-consuming; therefore another method is often used in which all the required frequencies are excited simultaneously over the interval T and the response is measured over the same interval T. Conversion to the frequency domain of both force and response signals is achieved using the *discrete Fourier transform* DFT which is almost the same as the integral for single frequencies in equations (8.75) and (8.76). All signal processing methods, whether transient or 'continuous random', have therefore the same limitations due to finite sample rate $1/\Delta$ and time duration T, as described below.

The ideal frequency decomposition of any time signal is the Fourier integral transform of equation (8.2). The finite interval version of equation (8.2) is

$$X(f) = \int_0^T x(t)\mathrm{e}^{-j2\pi ft}\mathrm{d}t \tag{8.77}$$

which is almost the same as the orthogonality relationship used in determining the components of the Fourier series. The signal within period T is therefore assumed to be repeated at intervals of T, i.e. $x(t + rT) = x(t)$, with integer r, as seen in Fig. 8.38 for the decaying transient.

This periodic signal contains only frequency components (harmonics) at intervals of $1/T$, as for the Fourier series. The frequency f at the hth harmonic is k/T, as seen in Fig. 8.39.

If substitution is made for time $t = n\Delta, \mathrm{d}t = \Delta, f = k/T$, equation (8.77) becomes a summation of harmonics spaced at $1/T$, i.e.

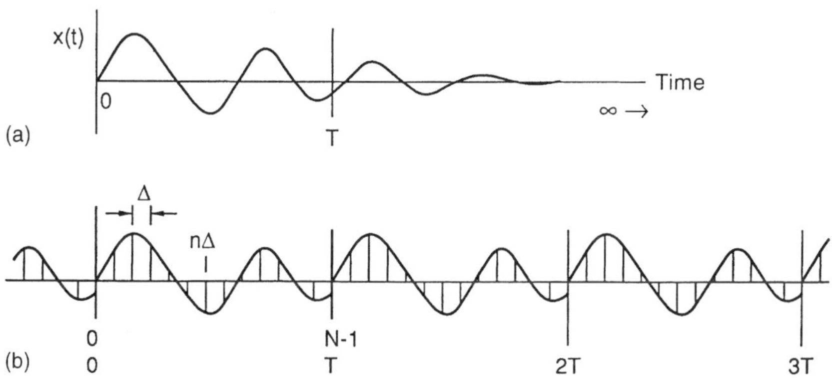

Figure 8.38 (a) True time signal. (b) Signal seen by the DFT.

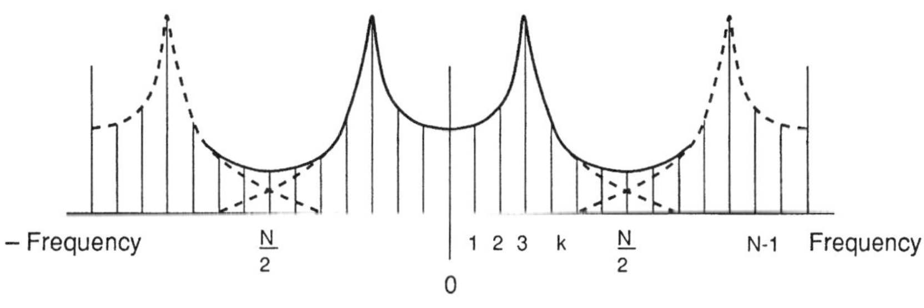

Figure 8.39 Frequency response obtained from the division of two DFTs.

$$X\left(\frac{K}{N\Delta}\right) = \sum_{0}^{N-1} x(n\Delta)e^{-j2\pi\frac{kn}{N}} \tag{8.78}$$

The force signal $f(t)$ is processed in an identical manner to give $F\left(\frac{k}{N\Delta}\right)$. The division gives the frequency response function of Fig. 8.39, i.e.

$$H\left(\frac{k}{N\Delta}\right) = \frac{X\left(\frac{k}{N\Delta}\right)}{F\left(\frac{k}{N\Delta}\right)}, \tag{8.79}$$

which is now discussed.

8.4.2.3 Frequency resolution

The frequency points are spaced at intervals of $1/T$ which is termed the *frequency resolution*. Greatest resolution is required where there is most detail, which is in the region of the resonances. This area may be examined using the vector plot in Fig. 8.40, obtained from discretized data as compared to the ideal circle from infinite resolution.

If the ideal circle can be fitted to the discrete points the damping of the mode can be calculated using the properties of a damped single degree of freedom system, i.e.

$$\frac{f_2 - f_1}{2f_0} = \zeta \tag{8.80}$$

or

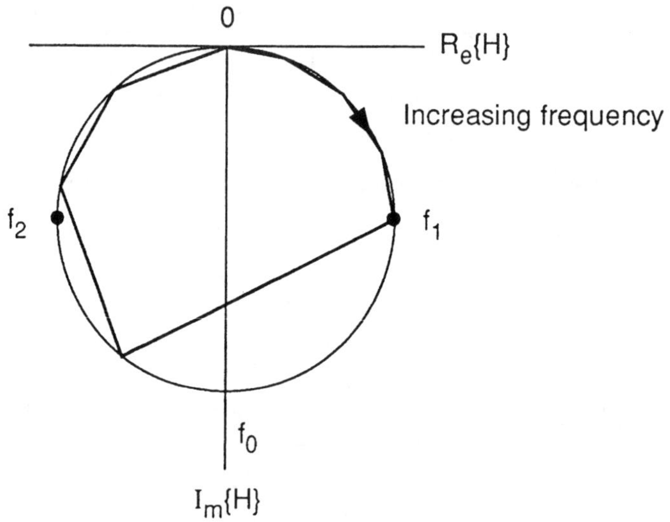

Figure 8.40 Vector plot of a single resonance using discrete data.

$$\frac{f_2 - f_1}{f_0} = \eta.$$

ζ is the viscous damping ratio and η is the hysteretic damping (loss) factor. For a circle to be fitted, the frequency resolution $1/T$ must be finer than the half power bandwidth $f_2 - f_1$, which for a single degree of freedom system could be calculated from equation (8.7) to be given by

$$\frac{1}{T} < 2\zeta f_0 \quad \text{or} \quad \frac{1}{T} < \eta f_0. \tag{8.81}$$

For a constant loss factor η describing the material damping, the required time interval T *decreases* with increasing frequency. Figure 8.41 shows measured accelerance data from 0.68 s of data (a) and 2.05 s of data (b). For frequencies less than 200 Hz, the data with the longer time interval gives the better description. Equation (8.81) shows that the lowest resonance frequency that can be measured accurately depends on the damping and the time interval T.

8.4.2.4 Data truncation in transient testing

If a transient force such as the chirp or impulse in Fig. 8.35 were applied to a system the vibration of each mode $x_0(t)$ would be a decaying harmonic signal, represented by

$$x_0(t) = X_0 e^{j\omega_0 t} \mathrm{e}^{-\zeta\omega_0 t}, \tag{8.82}$$

where ω_0 is the natural frequency and ζ the viscous damping ratio.

The ideal frequency representation of this signal requires the Fourier transform of an infinite time interval $T = \infty$, as in equation (8.2). However, a finite time T is used in practice, *truncating* the data as in Fig. 8.38. The frequency response at ω_0 is found from the multiplication with a signal $\mathrm{e}^{-j\omega_0 t}$, as

$$X(f_0) = \int_0^T X_0 \mathrm{e}^{-\zeta\omega_0 t} \mathrm{d}t \tag{8.83}$$

or

$$X(f_0) = \frac{X_0}{\zeta\omega_0}\left(1 - \mathrm{e}^{-\zeta\omega_0 T}\right). \tag{8.84}$$

The truncation of the tail beyond T reduces the resonance peak in the frequency response to $H(f_0)$, thereby simulating the addition of damping to the system. For only a 10% error in resonance peak height the time T and bandwidth $\zeta\omega_0$ are related by equation (8.81), as was also the case for the frequency resolution.

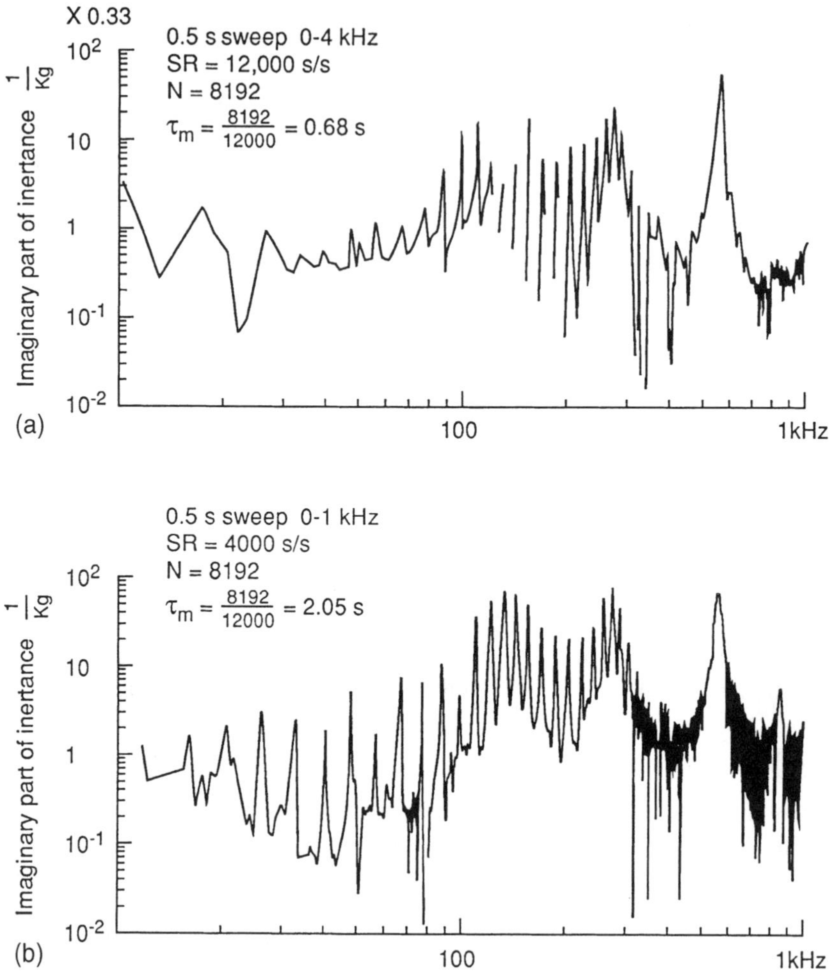

Figure 8.41 Measured transfer function showing low frequency sample length τ_m effect and high frequency sample rate (SR) effect.

If data truncation occurs, there will be a non-zero value at the end of the time window T of Fig. 8.38. The DFT transforms this into the sin $\omega_0 T/\omega_0 T$ modulation superimposed onto the ideal response seen in Fig. 8.42. This feature is also seen in the measured data of Fig. 8.43 in which the truncation was associated with the ringing of the resonance at 20 Hz.

Data truncation problems can be avoided by exciting the system with a force $f(t)$ repeated exactly every time interval T. In this case, the DFT becomes a true Fourier series as the data in each interval T is identical. The chirp in Fig. 8.35(b) is particularly amenable for this purpose.

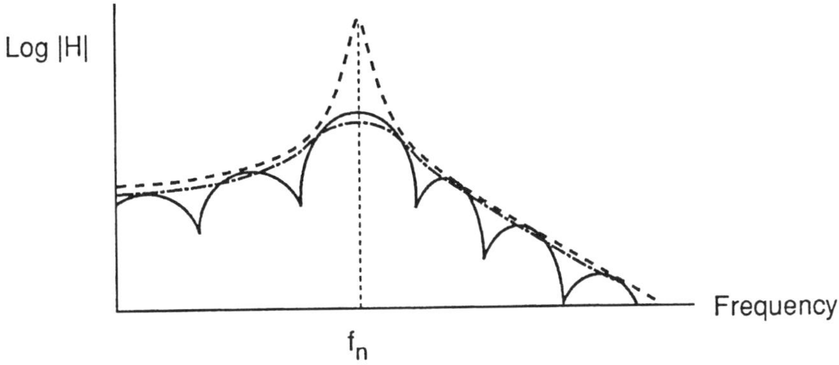

Figure 8.42 Data truncation effects on measured data. - - - - ideal, — truncated with boxcar window, -·-·- truncated with Hanning window.

8.4.2.5 Data truncation using random signals

The most common form of excitation is band- limited white noise which is a random time signal with equal energy in each equal frequency band. The processing is very similar to that for the transient using time windows of finite length T. The applied force and response are displayed in Fig. 8.44 with three sequential time periods T.

An estimate of the transfer function is formed from the DFTs of the simultaneous force and response over a single time period 0–T using equation (8.79). To improve the quality, the process is repeated and estimates are taken from as many subsequent time periods as required; the final result is the average of these estimates. The same frequency resolution and truncation criteria hold as before, if violated these also give a reductio in resonant peak heights and the sin $\omega_0 T/\omega_0 T$ truncation effects seen in Fig. 8.42 occur. It is, however, common practice to apply a shaped time window seen in Fig. 8.44 to all the data, which drives the signal at the beginning and end at each period T to zero. This 'smoothes' the frequency data as in Fig. 8.42 where the dips are removed but the true resonant level is not attained because the window has artificially added damping.

8.4.2.6 Aliasing

The time signal of Fig. 8.36 is sampled at intervals of Δ. For the waveform of a single frequency f_2 to be represented, at least two samples per wavelength are required, i.e. $f_2 < 1/2\Delta$; f_2 is called the Nyquist frequency and is the folding line at point $N/2$ in Fig. 8.39. The data at frequencies higher than f_2 in Fig. 8.39 and equation (8.78) is the mirror complex conjugate of that below, i.e.

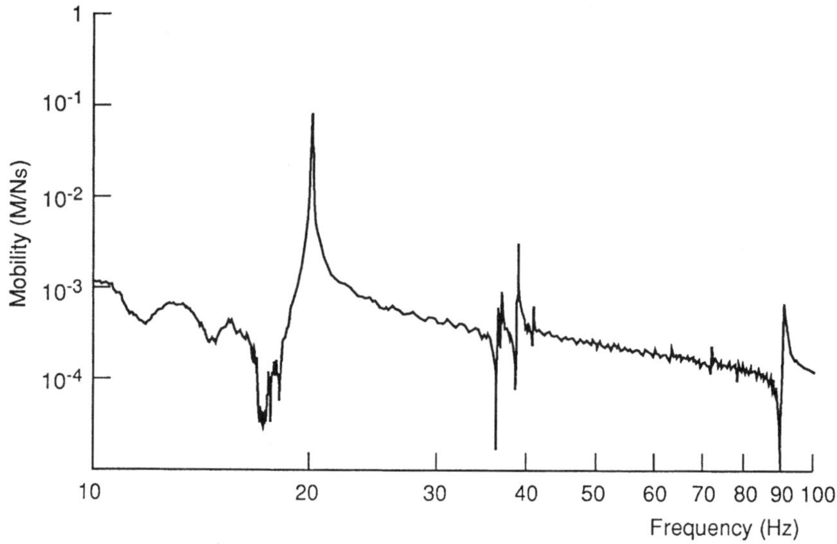

Figure 8.43 Truncation effect beating of side bands.

$$H(N-k) = H(-k) = H^*(k). \tag{8.85}$$

The data above and below the Nyquist frequency (or with positive and negative frequencies) combine to form real causal time signals as in equation (8.9).

As frequencies only 'exist' between 0 and f_2, the contribution from any higher frequencies must be folded back into this range and become a form of *noise* called *aliasing*. It is therefore good practice only to excite frequencies in the desired range $0{-}f_2$ and also to use anti-aliasing filters set to frequency f_2. All force and response channels should have the same filter setting, otherwise phase shifts occur between channels.

Although aliasing should be avoided by this means, the Nyquist criterion is not adequate for good quality data because only two samples/wavelength gives a poor representation of amplitude and so three samples/wavelength, therefore

$$f_2 < 1/3\Delta \tag{8.86}$$

is the criterion often employed in commercial analysers. Figure 8.41(a) shows good data at 1 kHz for a sample rate, $1/\Delta = 12$ kHz, while Fig. 8.41(b) indicates that even a sample rate of 4 kHz is not giving good data at the minima.

In summary, equation (8.81) should be used for the lower frequency f_1 of believable data and equation (8.86) for the upper frequency f_2. The total

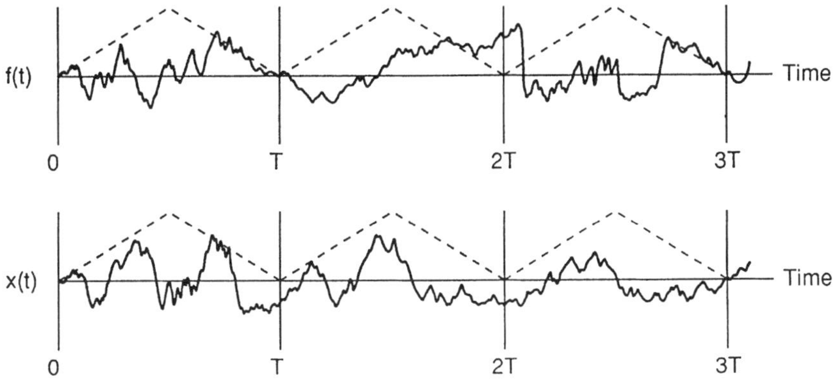

Figure 8.44 Force, $f(t)$, and response, $x(t)$, for random signals. - - - - time window.

range depends upon the number of points N in the transform which from equations (8.81) and (8.86) must for accurate data be

$$N > \frac{3f_2}{2\zeta f_1}. \tag{8.87}$$

If the damping ratio ζ is of the order of 0.01, as is typical for many structures, a 1000 point transform will only give accuracy over one decade of frequency as seen, for example, in Fig. 8.41.

8.5 PRACTICAL DETAILS

The objective of a good experiment must be to try to measure a single independent phenomenon. This phenomenon must be isolated from the rest of the physical world, the encroachment of which onto the experiment is called *noise*, rather like weeds in the garden. If the physical principles behind the noise are understood, these unwanted interactions can be identified and reduced. To this end, some of the mechanisms are listed below, although an open mind should be maintained towards any possibility (White and Pinnington, 1982).

8.5.1 Noise sources

1. *Electronic noise*. The presence of electrical currents causes agitation at an atomic level within semi-conductors. This is seen as a broadband white noise on an analogue oscilloscope. This activity is at millivolt level and so the desired measured signal should be kept large around 1 volt. This

electronic noise is truly random and so may be reduced by averaging repeated signals.

2. *Temperature fluctuation*. Piezoelectric elements are sensitive to temperature; small random temperature fluctuations cause noise which limits the lowest useable frequency of piezoelectric accelerometers, typically to about 20 Hz.
3. *Triboelectric noise*. Gross movement of the cable coupling the transducer to the preamplifier generates large fluctuating signals. These cables should be secured with tape.
4. *Hum*. Electromagnetic waves, usually associated with the mains electricity supply, generate harmonics of the supply frequency. These may be reduced with a complete screen, earthed at a *single point*. The most sensitive area is the sensor and cable before the pre-amplifier where voltages are low. Multiple earth points in the circuit can result in *ground loops*, which are hum induced in the screen. A break in the screen, for example, at a cable connection point will cause hum.
5. *Cross talk*. This is electromagnetic radiation from other signal channels. A large current in the excitation circuit can induce corresponding voltages in the sensor circuit. This is perplexing as the input and output are *correlated* giving good *coherency*. In fact, suspicions should be aroused if the coherency is too good, for example in signals from piezoelectric accelerometers at very low frequencies.
6. *Cross-sensitivity*. An ideal transducer should only be sensitive to one quantity in a single direction; in reality they are to a lesser degree sensitive to motions in all six degrees of freedom, pressure and base strain. These cross-sensitivities tend to be of the order of 3% and should be supplied by the manufacturer. The most commonly observed problems are force transducers acting as accelerometers and accelerometers sensitive to lateral motion or base strain. As with the cross talk, this is a correlated noise source and cannot be defeated using coherency.
7. *Digitization noise*. The digitization process causes noise of two sorts: that associated with digitization in the time domain (discussed in section 4.2) and that occurring because the instantaneous signal level is represented in binary form as a discrete number of bits. If signals are allowed to become too low they will approach the bit level and will be represented inaccurately, the error in relation to the true level is 'noise'.

8.5.2 Calibration

The discussions in the previous section should lead to the conclusion that there is no reason for naively trusting that the output from a measurement system represents the desired quantity. The only way to attain this confidence is to *calibrate* the complete measurement system by comparing the output with a known ideal output. This is difficult as a reliable reference is required, preferably at all the frequencies of interest. In vibration testing, this can be done by exciting a compact mass and monitoring the force and acceleration. The transfer function should appear as Fig. 8.10, the level $F(f)/a(f) =$ mass kg and the phase $= 0°$. In acoustics, the same thing can be done using two microphones within a very small air-filled, rigid-walled chamber of volume V. One microphone is used as a volume source $Q(f)$ and the other as a pressure sensor $p(f)$. Then $p(f)/Q(f) = B/V$ where B is the fluid bulk modulus.

Some commercial devices are available that test the system with a known reference at a single frequency; an example is the B&K pistonphone for testing condenser microphones.

8.5.3 Test procedure

By way of a summary of the previous sections some steps are suggested in setting up a practical test [9, 12].

1. Form a theoretical expectation of the experiment in terms of frequency range, levels and interactions with the measurement system.
2. Minimize the interactions as much as possible by using lightweight and compact sensors if they are supposed to be massless; massive and rigid supports are required if no motion is assumed in the boundaries; a lightly suspended structure should be employed if boundaries are assumed to be free.
3. The exciter should be as small and lightweight as possible while delivering enough signal, but not so much as to cause non linearity. The exciter should only be driven over the required range, using a constant spectrum signal.
4. Ensure all channels are identical to reduce phase shift.
5. Monitor signals with an analogue oscilloscope to see all types of noise, and to check that the signal levels are sufficient for the input of the processing device, e.g. about 1 V.
6. Check the system for extraneous noise on the oscilloscope.
7. Calibrate the complete system, using a seismic mass for mechanical systems.

8.6 REFERENCES

Broch, J. T. (1973). *Acoustic Noise Measurements*, Brüel & Kjaer .

Broch, J. T. (1980). *Mechanical Vibration and Shock Measurements*, Brüel & Kjaer, Naerum, Denmark.

Cremer, L., Heckl, M. A. and Ungar, E. E. (1998). *Structure-borne Sound*, 2nd edn, Springer-Verlag, Heidelberg, Berlin, London.

Ewins, D. J. (1984). *Modal Testing: Theory and Practice*, John Wiley & Sons Inc, New York, Chichester.

Goyder, H. G. D. (1978). PhD Thesis, University of Southampton.

McConnell, K. G. (1995). *Vibration Testing, Theory and Practice*, J. Wiley & Sons Inc, New York, Chichester.

Newland, D. E. (1975). *Random Vibration and Spectral Acoustics*, Longman, New York, London.

Otnes, R. K. and Enockson, L. (1978). *Applied Time Series Analysis, Vol. 1: Basic Techniques*, J. Wiley & Sons.

Pinnington, R. J. (1988). Approximate mobilities of build-up structures. *ISVR Technical Report No. 162*, University of Southampton.

Russell, B. (1991). *History of Western Philosophy*, Routledge, London, Ch. 1, 3, 15, 16.

Snowdon, J. C. (1971). Mechanical four-pole parameters and their application. *Journal of Sound and Vibration*, **15**, 307–23.

ten Wolde, T., Verheij, J. W. and Steenhoek, H. F. (1975). Reciprocity method for the measurement of mechano-acoustical transfigurations. *Journal of Sound and Vibration*, **42**, 49–55.

White, R. G. and Pinnington, R. J. (1982). Practical application of the rapid frequency sweep technique for structural frequency response measurement. *Aeronautical Journal*, **88**, 179–97.

8.7 QUESTIONS

1. Plot out a Bode diagram (log–log) of a single-degree-of-freedom system giving the value in the stiffness controlled region, the damping controlled region and mass controlled region for mobility, receptance, impedance and apparent mass (use equation (8.7)).
2. Plot out the point mobility in Bode diagram form, modulus and phase of the infinite system described by equation (8.11), i.e. a plate ($s = 0$), a beam in bending ($s = -\frac{1}{2}$), a membrane ($s = 1$).
3. A piezoelectric hydrophone is connected in series with a charge amplifier and an amplifier. If the hydrophone sensitivity $H_t = 3.2\,\rho c/\text{Pa}$, the charge amplifier sensitivity $H_a = 10\,\text{mv}/\rho\text{c}$, and the amplification factor is 10, calculate the output voltage V_0 in terms of the acoustic pressure p acting on the hydrophone (equation (8.14)). If the pressure $p = 0.1$ Pa, calculate the voltage V_0 in volts. Express the pressure in terms of dB re 10^{-6} Pa (equation (8.72)).
4. A single-degree-of-freedom system with impedance Z_m (equation (8.7)) has a mass $M = 1$ kg, stiffness $K = 10^5$ N/m and hysteretic loss factor $\eta = 0.01$ (equation (8.9)). Calculate the natural frequency and minimum response value. This system is driven by an exciter represented as a single degree of freedom system of mass $m = 0.1$ kg, stiffness $k = 2 \times 10^4$ N/m and hysteretic loss factor $\eta = 0$. The exciter and the test piece are rigidly coupled by a force transducer as shown in Figure Q8.1. Neglecting the electrical effects, sketch out the transmitted force F_m for the system in terms of the impedances and the ideal blocked force F_0 (using equation (8.50)). What are the frequencies of maximum force and minimum force?

5. Write out the procedure for performing a frequency response test on a steel beam weighing about 3 kg. The test is between 10 Hz and 1 kHz. The first natural frequency is 30 Hz, the loss factor η is 0.01. An electrodynamic exciter is to be used in conjunction with piezoelectric transducers. Comment on both noise effects, signal processing parameters and calibration.

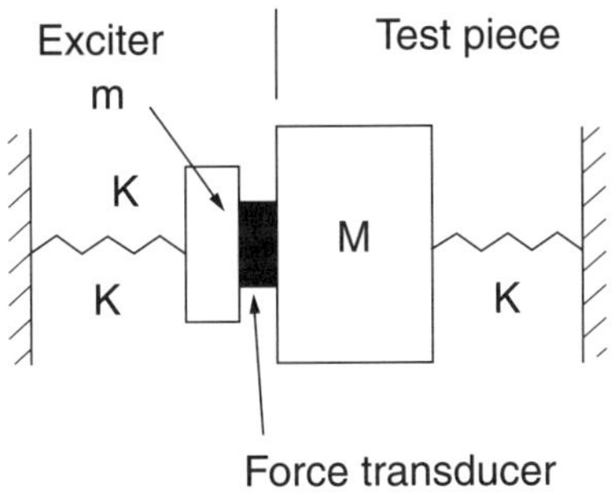

Figure Q8.1 Exciter test piece system.

Appendix: Lists of symbols

Chapter 1

A	complex amplitude
a	radius of sphere
A_{I}	imaginary part of complex amplitude
A_{R}	real part of complex amplitude
c_0	sound speed
$\boldsymbol{D}$	vector dipole strength
$\boldsymbol{d}$	vector distance
E_{k}	acoustic kinetic energy
E_{p}	acoustic potential energy
e	total acoustic energy density
e_{k}	acoustic kinetic energy density
e_{p}	acoustic potential energy density
$\boldsymbol{f}$	fluctuating force vector
f	frequency (Hz)
G	Green function
γ	ratio of specific heats
I	time-averaged acoustic intensity
$\boldsymbol{I}$	time-averaged acoustic intensity vector
$\boldsymbol{i}, \boldsymbol{j}, \boldsymbol{k}$	unit vectors
j	$\sqrt{-1}$
k	wavenumber
L	length
λ	wavelength
L_{p}	sound pressure level
L_{w}	sound power level
n	integer
p	acoustic pressure
p_0	ambient pressure

p_{ref}	reference pressure
p_{tot}	ambient pressure plus acoustic pressure
Q	monopole strength
θ	angle
q	volume velocity (source strength density)
R	gas constant
ρ	fluid density
r	radial coordinate
ρ_0	ambient density
ρ_{rms}	rms pressure
ρ_{tot}	ambient density plus acoustic density
S	area of cross-section
T	period
t	time
T_a	absolute temperature
U	complex velocity amplitude
u	acoustic particle velocity
$\boldsymbol{u}$	acoustic particle velocity vector
V	volume
W	sound power
ω	angular frequency (rad/s)
x	linear coordinate
$\boldsymbol{x}, \boldsymbol{y}$	coordinate position vectors
z	specific acoustic impedance
∇	del operator

Chapter 2

a	magnitude of step force
a_1, a_2	displacement amplitude of lumped masses
B	displacement amplitude of vibration
c	viscous damping coefficient
C	viscous damping matrix
c_h	hysteretic damping coefficient
D	dissipation function
e	experimental constant
f	frequency (Hz) or component of force
f_n	undamped natural frequency (Hz)
F	force
F_0	amplitude of harmonic force
$\hat{F}$	area under impulsive force time history
$g(t)$	impulse response function
h	length of unstretched spring
I_G	moment of inertia about centroidal axis
I_P	moment of inertia about axis through arbitrary point P

ℓ	distance between supporting spring and centroid
ℓ'	distance between supporting spring and arbitrary point P
j	$\sqrt{-1}$
k	spring stiffness
K_d	dynamic stiffness matrix
K	stiffness matrix
m	point mass
M	mass matrix
q	displacement vector
Q	force vector
t	time
t_p	duration of transient force
T	periodic time or kinetic energy
T_{60}	reverberation time
U	strain energy
v	velocity amplitude of vibration
x	displacement of mass from datum position
x_0	harmonic displacement amplitude
x_{st}	static displacement
X	displacement vector
y	displacement of base
z	relative displacement between base and mass
α, β	quadrature and in-phase components of displacement
δ	logarithmic decrement
δ_{st}	static deflection due to gravity
ε	duration of impulsive force
ϕ	phase angle
η	hysteretic loss factor
μ	mass ratio
θ	angular displacement
θ_o	harmonic angular displacement amplitude
ω	angular frequency (rad/s)
ω_a	natural frequency of absorber mass–spring system
ω_n	undamped natural frequency
ω_d	damped natural frequency
ω_m	frequency for maximum response (resonant frequency)
ψ	impulse response time
ζ	viscous damping ratio
Ω_n	natural frequency of main mass–spring system

Chapter 5

a	absorption of room (m^2)
A	complex wave amplitude
B	complex wave amplitude, plate bending stiffness

c_0	speed of sound
c_{ph}	phase speed
c_g	group speed
c_b	bending wave speed
c_l, $c_{l'}$, $c_{l''}$	longitudinal wave speeds
c_t	trace speed
C	damping coefficient, complex wave amplitude
C_{crit}	critical damping coefficient
d	cavity depth, diameter
e	energy density
E	Young's modulus, isolator effectiveness
$E' = E(1+j\eta)$	*complex Young's modulus*
f	frequency (Hz)
f_c	critical frequency (lowest coincidence frequency)
F	force
$F(j\omega)$	Fourier coefficients (also A_n, B_n)
f_0	mass-air-mass resonance frequency
h	porosity, plate thickness
$h(t)$	impulse response function
i	integer
I	intensity (active)
$\boldsymbol{I}$	active intensity vector
I_i	incident intensity
I_r	reflected intensity
I_{ref}	reference intensity
j	$\sqrt{-1}$
k	wavenumber
k_b	free bending wavenumber
$\boldsymbol{k}$	wave vector
K	spring stiffness
l	length, layer thickness, spacing, airway width
L	length
L_p	sound pressure level
L_I	sound intensity level
L_w	sound power level
L_A	A-weighted L_p
m	mass per unit area
M	mass, Mach number
n	integer, modal density subscript (normal)
$\boldsymbol{n}$	unit normal vector (outward or inward)
N	rotation speed (rev/s or rev/min), number of samples
0	subscript – mean fluid quantity (equilibrium)
p	integer, pressure, sound pressure
p_{ref}	reference sound pressure

p_r	reverberant sound pressure
p_{rms}	root-mean-square sound pressure
Q	volume velocity of source (m^3/s)
R	sound reduction index
$R(0)$	normal incidence value of R
R_d	diffuse incidence value of R
R_f	field incidence value of R
$R(\phi)$	value of R for wave incidence angle ϕ
r	distance, flow resistivity, ratio of characteristic specific acoustic impedances
r_n	real part of normal specific acoustic impedance (resistance)
$\boldsymbol{r}$	position vector
s	structure factor
S	surface area
t	time, trace (subscript)
T	period, reverberation time, absolute temperature, force transmissibility
TL	transmission loss
u	x-directed fluid particle velocity
u'	volume flow per unit cross sectional area
$\boldsymbol{u}$	fluid particle velocity vector
U	flow speed
v	normal velocity of a plate/shell surface
V	volume, wave particle velocity amplitude
ω	z-directed fluid particle velocity, angular frequency (red/s)
W	sound power, power
W_{rad}	radiated sound power
W_{diss}	dissipated power
W_{ref}	reference sound power
x	Cartesian co-ordinate
x_n	imaginary part of normal specific acoustic impedance (reactance)
$\boldsymbol{x}$	position vector
y	Cartesian co-ordinate
Y	mobility
z	Cartesian co-ordinate, complex specific acoustic impedance
z_c	characteristic specific acoustic impedance
α	acoustic absorption coefficient, attenuation constant
$\alpha(\phi)$	acoustic absorption coefficient for plane wave incident at angle ϕ
α_d	acoustic absorption coefficient for diffuse incidence
β	phase constant
γ	complex propagation constant
κ_0	bulk modulus

ζ viscous damping ratio (fraction of critical damping)
η hysteretic loss factor
η_{rad} radiation loss factor
η_{d} dissipation loss factor
η_{mg} mechano-acostic efficiency
θ angle (rad)
λ wavelength
λ_{t} trace wavelength
ρ fluid density
ρ_{m} solid material density
ρ_{0} mean (equilibrium) fluid density
σ radiation efficiency (ratio)
τ time delay, sound power transmission coefficient
τ_{d} diffuse incidence value of τ
ϕ phase angle, angle of plane wave incidence (to normal to surface)
ω angular frequency (rad/s)
ω_{o} natural frequency of oscillator (undamped)
ω_{d} natural frequency of oscillator (damped)
ω_{c} angular critical frequency, band centre frequency
$\dot{x}, \ddot{x}$ etc time derivatives of x
x', x'' etc space derivatives of x
$\overline{x}$ mean value of x (time-average)
$\overline{x^2}$ mean square value of x (time)
$< x >$ space average of x
$< x >_y$ average of x over domain y
$< \overline{x^2} >$ space-averaged mean square value of x
x^* complex conjugate of x
$F\{x\}$ Fourier transform of x
$\mathrm{Re}\{x\}$ real part of x
$\mathrm{I_m}\{x\}$ imaginary part of x
x_{rms} root-mean-square value of x
$|x|$ magnitude of x
X modulus of complex amplitude X
$\angle x$ phase of complex amplitude X
bold vector
$\int_s$ integral over domain s
lg $\log_{10}$
ln natural log
$x.y$ scalar product
$\overline{xy}$ mean value of product of x and y

Chapter 6

$A_x(t)$ instantaneous amplitude

ADC	analogue-to-digital converter
C_n	complex Fourier series coefficient
C_{xy}	coincident spectral density
$E[.]$	expected value operator
f	frequency (Hz)
$F(x)$	probability distribution function
$G(f)$	spectral density function (single-sided)
$G(\omega)$	spectral density (single-sided)
$h(t)$	impulse response function
m.s.e.	mean square error
$p(x)$	probability density
$p(x,y)$	joint probability density function
$Q_{xy}(f)$	quadrature spectral density
$R(\tau)$	autocorrelation function
$R_{xy}(\tau)$	cross-correlation function
$S(f)$	spectral density
$S(\omega)$	spectral density
$\hat{S}_{xx}(f)$	raw power spectral density estimate
$\tilde{S}_{xx}(f)$	smoothed power spectral density estimate
$\hat{S}_{xy}(f)$	raw cross-spectral density estimate
$\tilde{S}_{xy}(f)$	smoothed cross-spectral density estimate
t	time
$\mathrm{Var}[x]$	variance of x
$\hat{x}(t)$	Hilbert transform of $x(t)$
$X(e^{j2\pi f\Delta})$	Fourier transform of a sequence $x(n\Delta)$
$X(f)$	Fourier integral (transform) of $x(t)$
$x(n)$	data sequence
$\delta(t)$	Dirac delta function
γ_1	coefficient of skewness
γ_2	coefficient of kurtosis
$\gamma_{xy}^2(f)$	ordinary coherence function
$\phi_x(t)$	instantaneous phase
μ_x	mean value
$\rho_{xy}(t)$	cross-correlation coefficient
σ^2	variance
χ_n^2	chi-squared random variable with n degrees of freedom
ω	angular frequency (rad/s)

Chapter 7

A	arca
A_1	a specific area
A_2	a specific area
B/A	the nonlinearity parameter of a liquid
c	sound speed

c_0	sound speed in the linear limit (i.e. for waves of infinitesimal amplitude)
c_1	speed of sound at depth marker d_1
c_2	speed of sound at depth marker d_2
c_a	speed of sound in air
c_b	speed of sound at depth h_b or speed of sound in bubbly water
c_i	speed of sound at depth marker d_i (when the wave is travelling horizontally)
c_w	speed of sound in (bubble-free) water
d	width of the region where cylindrical spreading occurs , or damping constant of a bubble
d_1	value of depth when speed of sound is c_1
d_2	value of depth when speed of sound is c_2
d_i	value of depth when speed of sound is c_i (when the wave is travelling horizonatally)
F	a general forcing term for an oscillator
F	force
f	frequency of an insonifying field (Hz)
f_o	natural frequency of a bubble of radius R_o
g	acceleration due to gravity
h	depth
h_b	a particular depth
I	intensity
I_1	a specific value of intensity
I_2	a specific value of intensity
I_{SATA}	spatial average temporal average intensity
J_λ	the momentum associated with one wavelength of a wave
K	bulk modulus
k	wavenumber
K_b	bulk modulus of bubbly water
K_w	bulk modulus of bubble-free water
l_a	a specific change in depth
l_b	a particular change in depth
L_{dis}	the discontinuity length
L_{ms}	acoustic boundary layer thickness
M	the peak acoustic Mach number of a source
n_b	number density of bubbles
n_b^{gr} (z, R_o)	number of bubbles at depth z per unit volume having radii between R_o and $R_o + dR_o$.
p	acoustic pressure
P	total fluid pressure
p_σ	Laplace pressure
p_0	acoustic pressure amplitude

P_0	static pressure in a fluid
p_g	gas pressure within a bubble
P_h	hydrostatic pressure
p_i	internal pressure within a bubble
p_L	pressure in the liquid at the bubble wall
$p_{rad,\ abs}$	the radiation pressure for absorption of plane waves at normal incidence
$p_{rad,\ refl}$	the radiation pressure for reflection of plane waves at normal incidence
p_v	vapour pressure within a bubble
r	a distance (radius of curvature of a ray, or distance coordinate in spherical coordinates)
R	bubble radius
R_a	radius of curvature of a specific ray
R_b	radius of curvature of a specific ray
R_o	equilibrium radius of a bubble
R_r	equilibrium radius of a bubble which would be resonant with the insonifying field
S	salinity
s_0	coefficient in a power series expansion of an oscillator's response
s_1	coefficient in a power series expansion of an oscillator's response
s_2	coefficient in a power series expansion of an oscillator's response
s_3	coefficient in a power series expansion of an oscillator's response
s_4	coefficient in a power series expansion of an oscillator's response
T	temperature (°C)
t	time
t_0	datum time
t_1	a time value
t_2	a time value
TL	transmission loss
u	particle velocity
V	volume
VF	volume fraction
W	power
x	distance coordinate with repect to plane wave propagation
X	distance to observation point
x_a	a particular horizontal distance
ζ	a parameter denoting the effect of both material and convective non-linearities on the phase speed

α	absorption coefficient
θ	an angle
ω	circular frequency of insonifying field
ρ	density
κ	polytropic index of a gas
γ	ratio of the heat capacity of a gas at constant pressure to the heat capacity of the gas at constant volume
η	shear viscosity
σ	surface tension
λ	wavelength
ω_0	angular natural frequency of a bubble of equilibrium radius R_o
ρ_0	equilibrium density
ω_1	a circular frequency associated with parametric sonar
ω_2	a circular frequency associated with parametric sonar
θ_a	the angle of a particular ray to the horizontal when it crosses the SOFAR axis
α_{ab}	logarithmic absorption coefficient
ρ_b	density of bubbly water
ω_c	a circular frequency a ssociated with parametric sonar
ρ_w	density of bubble-free water
Ɏ	the gradient of sound speed with respect to depth, or response of a general oscillator

Chapter 8

ω	angular frequency (rad/s)
ϕ	phase (radians)
η	hysteretic loss factor
ζ	viscous damping ratio
ω	displacement
Æ	time interval
λ	wavelength
a	acceleration
B	flux density
C	capacitance
c_0	speed of sound
D	damping constant
f	frequency (Hz)
j	$\sqrt{-1}$
k	integer counter
K	stiffness
$h(t)$	impulse response function
$H(f)$	frequency response function
I	current

F	force
G	noise in force
ℓ	length
L	inductance
N	noise in response, integer
M, m	mass
p	pressure
q	generalized coordinate
Q	volume velocity
r	mode counter
R	resistance
s	slope number
u	velocity
x	kinematic quantity in time, distance
X	kinematic quantity in frequency
t	time
T	time
Y	mobility
Z	impedance
V	voltage
W	power

Index

Entries in **bold** refer to figures; those in *italic* refer to tables.